1950

W9-DIW-259

AMINE OXIDASES
AND
METHODS
FOR
THEIR STUDY

QP601
K25

AMINE OXIDASES
AND
METHODS
FOR
THEIR STUDY

R. KAPELLER-ADLER, DR.PHIL., D.S.C.

HONORARY LECTURER IN THE DEPARTMENT OF PHARMACOLOGY OF THE
UNIVERSITY OF EDINBURGH, SCOTLAND

WILEY-INTERSCIENCE

A DIVISION OF JOHN WILEY & SONS

NEW YORK · LONDON · SYDNEY · TORONTO

NO LONGER THE PROPERTY
OF THE
UNIVERSITY OF R. I. LIBRARY

Copyright © 1970, by John Wiley & Sons, Inc.

All rights reserved. No part of this book may be reproduced by any means, nor transmitted, nor translated into a machine language without the written permission of the publisher.

Library of Congress Catalogue Card Number: 77-101967

SBN 471 45676 4

Printed in the United States of America

10 9 8 7 6 5 4 3 2 1

UNIVERSITY OF R.I. LIBRARY

To My Husband

Preface

An attempt has been made by summarizing current knowledge on amine oxidases and the many and diverse techniques applied by different workers in their studies of this subject to present a general picture of this most fascinating group of enzymes.

Current theories on enzyme action and regulation of enzyme activity are presented in the general introduction to Chapter II on amine oxidases.

It is a pleasure to express my special thanks to Dr. C. P. Stewart for his critical reading of the manuscript.

This monograph was written during my tenure of an award from the Distillers' Company, Ltd., Grants Committee, which is gratefully acknowledged. I wish to thank Professor R. W. Kellar for his interest in this work.

I am also indebted to Mrs. L. M. Ellis for her very careful preparation of the typescript.

Finally I wish to record my gratitude to Dr. G. A. Hamilton and Dr. F. Buffoni for their very valuable information in regard to the model systems that they devised for the oxidation mechanisms of amino acids and amine oxidases, respectively.

R. KAPELLER-ADLER

Edinburgh, Scotland
October 1969

Contents

AMINE OXIDASES

AND

METHODS

FOR

THEIR STUDY

CHAPTER I

Introduction

The amine oxidases are enzymes of great biological interest, since most of their natural substrates belong to the group of biogenic amines. As the latter term indicates, these compounds are formed *in vivo* in various metabolic processes. They occur in microorganisms as well as in higher forms of life. It is significant that many of these biogenic amines seem to be involved in regulatory enzymic mechanisms, vital to the higher mammals.

On reviewing the vast literature that has been published on amine oxidases since a monoamine oxidase ("tyramine oxidase") was first described by Hare in 1928 (1) and histaminase was discovered by Best in 1929 (2), the author has realized that, unfortunately, despite intensive and very competent *in vivo* and *in vitro* studies carried out in many different laboratories, no clear, well-defined picture of the fundamental character of these enzymes can be obtained from the enormous volume of work reported. No unequivocal data are to be found in the literature with regard to the classification of the many types of amine oxidases, their substrate specificities, the kinetics of their activities, or their physiological substrates. Opinions are also divided with respect to their prosthetic groups, the significance of their biological occurrence, and their *in vivo* reactions to various inhibitors. The classical concept of a monoamine oxidase or diamine oxidase type has gone out of date, since it has become obvious that some monoamine oxidases easily attack various diamines, and diamine oxidases act well on some monoamines.

It is therefore concluded that unless we approach the problem of the amine oxidases from the angle of present-day enzymology and try to obtain and to express kinetic data in a standard fashion, thereby making results consistent and comparable with each other, and further-

1

more try to find a common denominator for the diverse activities of these enzyme systems by viewing them in the modern light of multiplicity of enzymes, the prevailing confusion will persist.

It is the aim of this presentation to describe in detail well-recognized methods and to suggest new parameters of investigation that may help to elucidate and clarify some problems of amine oxidases. However, since analytical methods cannot be considered by themselves, but only within the general framework of all other aspects involved, a rather concise discussion of various amine oxidase systems precedes the description of the relevant analytical procedures.

CHAPTER II

Amine Oxidases

1. GENERAL INTRODUCTION: CURRENT THEORIES OF ENZYME ACTION

A. X-Ray Crystallography Studies of Enzymes

The announcement in 1926 by Sumner of the first isolation of an enzyme (urease) as a crystalline protein (3) met with considerable scepticism. This work was, however, soon followed by the classical isolation of crystalline proteolytic enzymes by Northrop and his colleagues (4, 5). Since that time several hundred enzymes have been obtained in a highly purified form and more than 100 of these in the crystalline state; they all have proved to be proteins (6).

In 1934 Bernal and Crowfoot (7) were the first to take X-ray diffraction photographs of protein crystals. They predicted that the technique of X-ray crystallography would allow more detailed conclusions to be drawn about protein structure than previous physical or chemical methods had permitted. But, in spite of the great ingenuity shown by many crystallographers, no real progress in the determination of protein structure was made for about 20 years since there was no means of solving a fundamental problem of protein crystallography—namely, the phase problem. In 1954, however, Green, Ingram, and Perutz (8), following a technique suggested by Bokhoven et al. (9), reported a solution of the phase problem for the centric data in hemoglobin. This was achieved by the addition of heavy-metal atoms to specific loci in the protein crystals. This introduction of the method of isomorphous replacement (8) provided the necessary breakthrough to the determination of protein structure. A complete solution of the phase problem was accomplished

3

by Blow (10), who extended the technique of Green et al. (8) to include acentric data. This opened the way to the determination of protein structures at the atomic level.

A brief theoretical outline of protein crystallography is now presented.

In crystals, there is a regular three-dimensional lattice of unit cells; each unit cell has the same relationship to its neighbors, and each cell has the same content. The unit cell may consist of one or a few atoms or molecules. For large and complex molecules, such as proteins, exact dimensions and geometrical relationships of unit cells are of critical importance. The only method at present available to obtain this information is the technique of X-ray analysis. X-Rays have very short wavelengths of the order of interatomic distance. Thus, the X-ray beam of 1.542 Å, produced by electron bombardment of copper, is used for the work on protein structure. When X-rays strike an atom, they are diffracted in proportion to the number of extranuclear electrons in the atom. Accordingly, heavier atoms produce stronger diffraction than lighter ones. The crystal may be regarded as a three-dimensional pattern of electron density, which shows high values near the centers of atoms and low or zero values in between. When a protein crystal is placed in an X-ray beam, the X-rays are diffracted by the crystal and may be recorded on a photographic film. The most striking feature of such a photograph is that the diffracted beams lie on a regular grid. This regularity of the grid indicates clearly the underlying order in the crystal. The other important feature of the photograph is the problem of the intensity of the individual beams. There is an overall decline in the intensities from the center of the photograph to its edges, but within this limitation the individual intensities appear to be randomly distributed, many beams being too weak to register on the film at all. The diffraction pattern is completed by imagining that other films with an identical grid pattern (although not with an identical intensity pattern) are placed at regular intervals above and below the photographic film described, their radii gradually decreasing as the distance of the layers both above and below the said film increases so that the boundary of the entire set of layers forms a sphere.

Such X-ray diffraction pictures permit measurement of the intensities (amplitudes) but not the phases of the diffracted X-rays . This lack of information is known as the phase problem. The relative importance of phases and amplitudes has been examined by Srinivasan (11), who found that the phases are of dominating importance and that little progress can be made without them.

There are several methods for the determination of phases. The most important technique in the study of protein crystals is that of isomorphous replacement. This requires the production of protein crystals that are identical except for the introduction of a heavy atom in the structure—for example, by replacing a hydrogen by a mercury atom.

If a crystal structure is known in detail, its diffraction pattern may easily be calculated, including the phase information. Alternatively, the contribution of any part of the structure—for example, of a single atom—to the total diffraction can be calculated (12). It is noteworthy that in crystals of small molecules, the constituent atoms are usually close together so that the molecules in the interior of the crystal are entirely isolated from the external environment of the crystal. In protein crystals, however, the large globular protein molecules are unable to pack together without having holes in the structure that are rather large in comparison to the water or alcohol molecules from which they are crystallized. In these circumstances the holes are filled with mother liquor of crystallization, which forms an integral part of the crystal structure. If the liquid is allowed to evaporate, the crystals shrink and the crystalline order disappears to a large extent. In a typical protein crystal this liquid comprises 30 to 50% of the volume of the unit cell. Whereas part of the liquor forms an ordered layer, hydrogen bonded to the protein surface, the rest seems to possess the properties of a normal liquid. All the liquid regions are conjoined and are continuous with the general body of the mother liquor. This aspect of protein crystals confers on them two properties important to the formation of protein-heavy metal complexes. First, each molecule in a protein crystal is available to reaction by the diffusion of reactants through the liquid region, just as they would be in dilute solution. Second, when a protein–heavy metal complex is formed, the atoms may occupy a region previously filled with liquid and may be therefore accommodated without otherwise disturbing the structure of the crystal (12).

When a heavy-metal atom is attached to each molecule in a protein crystal, the unit-cell dimensions will remain unaltered, since the basic structure of the crystal is unchanged. Accordingly, the spacing of the diffraction lattice will be the same as it was before the addition of the heavy atom. The electron density of the cell, however, will be radically altered at the site occupied by the heavy atom. If the metal is mercury, the change will be an additional peak of electron density containing 80 electrons, an order of magnitude greater than the normal atomic constituents. If, however, only half the protein molecules in the crystal have mercury attached to them, the average electron density change will be 40 electrons. This large, localized change in electron density is manifested in an alteration of the intensities of the diffracted beams in the diffracted lattice. The only data available at this stage are obtained by measuring the intensities of the native protein as well as of the heavy-atom derivative. A complete solution can be obtained by using two different protein–heavy metal complexes (12). The question of how heavy the additional atom should be, to be usefully applied in phase determinations has been discussed by Crick and Magdoff (13). Phillips (14) has found that the addition of one mercury atom to a protein of a molecular weight of 30,000 will bring about an average fractional change in intensity of 26%. The usefulness of such changes will greatly depend on the precision of the method of determining the intensities. The standard error in measuring intensities is about 10% so that for the protein just quoted the average fractional intensity change would be only about twice the likely error. Nevertheless even such small changes have facilitated satisfactory phase determinations (12).

A heavy-atom derivative that is perfectly isomorphous may be defined as one in which the only change in electron density between it and the native protein crystal is a peak at the site of the heavy-atom substitution (12) .

Any heavy atoms introduced into the protein structure may act as anomalous scatterers and will usually be the only important anomalous scatterers in the structure. Anomalous scattering differences in proteins are very small and will generally be close to the limit of experimental error (15) . In practice, nonisomorphism may be caused by either a distortion of the protein molecules themselves or by a disturbance of their normal mode of packing in the unit cell. Thus, the binding of a heavy-atom compound may displace one or more side chains or perhaps disturb the secondary structure of the protein in the vicinity of its binding site. When substitution takes place at the active center of an enzyme, this may result in the molecule taking up an "active" conformation, which would differ from the normal conformation (16–18).

Heavy-atom substitution may cause translation or rotation of the protein molecules in the unit cell, or both. In other instances it may cause a decrease in the degree of order in the crystal; this effect, however, is usually detected, since it causes a diminution of the average intensity of the outer part of the diffraction pattern. Also readily detected are changes in the dimensions of the unit cell that may frequently be brought about by translation or rotation of the protein molecules (12). Crick and Magdoff (13) have suggested more reliable methods for the detection of nonisomorphism. These authors have demonstrated that all types of nonisomorphism will interfere more seriously with phase determination as resolution is increased. Thus, the difficulty of determining whether derivatives are sufficiently isomorphous seems to be a serious one, particularly for high-resolution work, since the subsequent use of such derivatives in phase determinations may produce errors in the phases of sufficient magnitude to render the interpretation of the electron-density map impracticable (12).

The development of X-ray crystallography has undoubtedly revolutionized the approach to studies of the structure of biological molecules. The elucidation of the three-dimensional structure of a protein has had a number of important effects. Thus, it has become possible to correlate and interpret manifold chemical and physicochemical studies of the protein and to relate the structure of the protein molecule to its biological function. Although proteins exhibit a diversity of form and function, it appears that there are common factors that guide their folding, interactions with other molecules, and catalytic processes.

Lysozyme was the first enzyme to be revealed in complete three-dimensional detail as result of the crystallographic studies of Blake and his colleagues (17, 18) and Phillips (19, 20). Resolution of other enzymes—such as ribonuclease, chymotrypsin, carboxypeptidase, and papain—soon followed. The identification of the active site in lysozyme was achieved largely by crystallography at 2-Å resolution without the aid of previous extensive modification studies. This provides an example of what can be accomplished by crystallographic interpretations of active sites with hardly any use of conventional protein chemistry.

Lysozyme is a β-1,4-glycosidase and shows a specificity for N-acetylamino sugar groups. The enzyme has a molecular weight of 15,000 and is composed of 129 amino acid residues joined together in a single polypeptide chain (21, 22). It contains no prosthetic groups and does not require metal ions or cofactors for its action. Hence, the entire catalysis and specificity appears to be effected through the amino acid residues. Particularly striking in the lysozyme structure are the large regions without any ordered secondary structure. That the structure of lysozyme is determined by interactions of amino acid side chains seems to follow from the fact that the molecule can be unfolded and refolded (23). These interactions lead to a reproducible three-dimensional structure which does not have a highly ordered secondary structure.

The most characteristic feature of the lysozyme molecule revealed by X-ray diffraction analysis is the deep cleft through the middle of the molecule, which Phillips and his colleagues (17–20) were able to identify with the active site. Competitive inhibitors and presumably the substrate fit into this cleft, as revealed by isomorphous crystals of inhibitor-enzyme complexes. Furthermore, from X-ray analysis it has been possible to ascertain the nature of the binding sites of inhibitors to the enzyme and to postulate a reasonable mechanism of action of the enzyme (17, 19, 20).

The three-dimensional structure of ribonuclease A, obtained at a 2-Å resolution, has been published by Kartha, Bello, and Harker (24); and that of ribonuclease S, determined at a 3.5-Å resolution, by Richards, Wykoff, and their co-workers (25, 26). Unlike the case of lysozyme, there had been extensive protein modification studies on ribonuclease before its three-dimensional structure was known, so that the predictions of physical and chemical techniques could be checked against the structure. It is noteworthy that the protein-modification techniques were vindicated to a great extent. The three-dimensional models published for ribonuclease S by Wyckoff et al. (25, 26) reveal a cleft in the molecule similar to that of lysozyme, and across this cleft there are two histidine residues, histidine 12 and histidine 119, which had been implicated in the catalytic activity of ribonuclease by the modfication studies of Crestfield, Stein, and Moore (27). The distance between these two histidines in the crystallographic model may be close to the distance predicted by the modification experiments. Moreover, the crystallography revealed that these two histidines were near the binding site of uridylic acid so that their identification with catalysis as deduced from the modification studies has become highly plausible. Furthermore, deductions about buried and exposed residues were similarly supported; for example, the finding that the N–3 atom of histidine 12 is alkylated in preference to N–1 is explained by the observation that the N–1 atom is buried and N–3 exposed.

The three-dimensional models deduced by physical methods and deductive reasoning turned out, however, to be largely incorrect (28, 29).

The general assumption that the three-dimensional structure of a molecule depends on the covalent links holding the chains together and also on noncovalent bonding between chains was supported by studies of Wyckoff and his associates (26). As in the case of lysozyme, the position of the binding of some substrates and inhibitors was identified by crystallographic means in a way that would have been impossible by protein modification techniques (24).

Wyckoff, Richards, and Tsernoglou (30) have developed a flow-cell technique that may greatly facilitate the search for heavy-atom derivatives. The crystals are fixed in position in thin-walled polyethylene capillaries by cotton wadding, and solutions of heavy-metal complexes may then be flowed past the crystal while it is being examined with an X-ray diffractometer. In this way it is possible to follow the extent of substitution and to observe when saturation is reached. This method could have a widespread applicability, particularly in the case of proteins containing unstable inhibitors, which could be replenished as they were used up. The method also permits very exact measurements of the intensity changes that occur on heavy-metal substitution, since the same crystal in the same setting may be used for measurements of both the unsubstituted and the heavy-metal-substituted protein, thus minimizing the differences that might result from the use of different crystals.

A recent important achievement in crystallography has been the elucidation of the structure of chymotrypsin. The structure of bovine α-chymotrypsin, a protease containing 241 amino acid residues, has been determined at 2-Å resolution by Matthews, Sigler, Henderson, and Blow (31), and that of π-, δ-, and γ-chymotrypsin at 5-Å resolution by Kraut and his associates (32). The structure of the enzymically inactive precursor chymotrypsinogen had been determined by Kraut and his co-workers (33) in 1964; however, it was not then fully understood. Now, chymotrypsinogen can be related to the active chymotrypsins and thus offers an insight into the activation process. The crucial step in the activation of chymotrypsinogen is the cleavage of an arginyl–isoleucine peptide bond by trypsin to form enzymically active π-chymotrypsin. Subsequent cleavages by chymotrypsin give rise to α-chymotrypsin, which consists of three polypeptide chains, known as the A, B, and C chains, connected by two interchain disulfide bridges (34, 35); γ- and α-chymotrypsin are interconvertible crystal forms of the same covalent structure (36).

The crystal form of α-chymotrypsin used by Matthews and his colleagues (31) was first described in 1938 by Bernal, Fankuchen, and Perutz (37). According to the electron-density map of Matthews et al. (31), the crystallographic asymmetric unit consists of two chymotrypsin molecules, each of a molecular weight of 25,000, which are nearly identical in their tertiary structure. The parent protein in the latter authors' work was tosyl-α-chymotrypsin and not the native enzyme, since these workers found it easier to obtain heavy-atom derivatives of the tosylated enzyme. One of the isomorphous derivatives they used was pipsyl-α-chymotrypsin (31). Moreover, the tosyl group at serine 195 also proved useful as an unambiguous marker of the active site, since chemical studies had clearly established the importance of this residue in catalysis. The interpretation of the Fourier synthesis was further facilitated by selective diiodination of tyrosine 171 in the crystal. Four of the five cystine bridges were readily recognized. The amino acid sequence determined by Hartley and Kauffman (38) and by Keil, Sorm, and Prusik (39) was found to be indispensable to the interpretation of the electron-density map. The clear appearance of many aromatics at appropriate positions in the chain confirmed the correctness of the interpretation.

A rather accurate description of the technique for X-ray diffraction measurements leading to the determination of the three-dimensional structure of α-chymotrypsin by the method of isomorphous replacement is given by Matthews et al. (31) to which the reader is referred.

The three-dimensional structure of α-chymotrypsin again reveals the details of a protein on which a vast amount of protein modification has been performed. The structure has not yet been reported in as great detail as has ribonuclease. What has been revealed, however, again agrees well with the modification literature. Serine and histidine are near each other as modification literature has indicated (40–42).

The molecule is compact, the overall dimensions being $45 \times 35 \times 38$-Å. Much of the interior is composed of regions in which the hydrophobic residues of the A, B, and C chains interact. The folding of the molecule is complex. Only eight residues at the carboxyl terminal end of the C chain are in an α-helical conformation. All charged groups, except for two important residues, point toward the solvent. The chains tend to be fully extended and often run parallel to each other, separated by about 5-Å.

A finding of particular interest that agreed with previous studies performed with modification techniques was the identification of an electrostatic bond between the aspartic acid and the free α-amino group of isoleucine 16. The activation of chymotrypsinogen as revealed by Neurath (43) showed that the splitting of the lysine 15–isoleucine 16 bond in chymotrypsinogen converts the precursor to an active chymotrypsin. Oppenheimer, Labouesse, and Hess (44) had obtained evidence for a conformational change identified with the protonation of the isoleucine 16 which was similar to the conformational changes occurring in the conversion of chymotrypsinogen to chymotrypsin (45). The crystallographic finding that aspartate 194 forms an electrostatic bond with isoleucine 16 throws some light on many of these facts. Thus a conformational change would be required to form such a bond. Moreover, the structure places isoleucine 16 near the active site where its protonation would be likely to affect the enzymic activity. The interaction of aspartate 194 with the positively charged α-amino-group of isoleucine 16, mentioned above, is striking. This ion pair is located in the interior of the molecule. They are the only two polar groups in chymotrypsin that are not located on the surface of the molecule. Both residues are hydrogen bonded to nearby side-chain and main-chain groups. Unlike some other proteins, chymotrypsin does not have a deep cleft at its active site.

The involvement of serine 195, histidine 57, and isoleucine 16 in the active center was predicted by chemical studies. Serine 195 was identified by peptide mapping as the site of phosphorylation (46). The participation of histidine 57 was directly demonstrated by affinity labeling (47, 48) as was the proximity of methionine 192 (49).

As mentioned above, the structures of π-, δ-, and γ-chymotrypsin, and of chymotrypsinogen have been determined at 5-Å by Kraut and his co-workers (33). At this resolution, however, the polypeptide chain cannot be followed unequivocally. It is hoped that in the near future higher resolution studies and improved phases for chymotrypsinogen and the chymotrypsins will produce electron-density maps that will make it possible to deduce detailed mechanisms of activation, binding, and catalytic processes (50).

The elucidation of the structure of carboxypeptidase A at 2-Å resolution by Lipscomb and his associates (51–53) is a most interesting recent achievement. In addition, these workers have determined the structure of an enzyme–substrate complex at 2.8-Å resolution (53). These studies by Lipscomb and his colleagues have provided a striking example of conformational changes taking place on catalysis and have led to the proposal of a catalytic mechanism (53).

The compact molecule of the enzyme consists of several helical regions, a large central twisted sheet of extended polypeptide chains and a tortuously folded coil. The α-helical segments, which comprise about 30% of the molecule, are located on the outside. There are four pairs of parallel and three pairs of antiparallel extended chains. About 20% of the residues are in an extended-chain conformation.

The active site was identified by difference Fourier studies of complexes of the enzyme with substrate or inhibitor. There is a zinc atom at the active site; it appears to have three protein ligands, one of which is an imidazole nitrogen of a histidine side chain. Lipscomb and co-workers (53) found little change in the structure of this enzyme when zinc was removed, and no change in the structure when mercury replaced zinc. Since both of these changes produced a protein inactive toward peptide substrates, and since zinc is apparently not essential for the structure of the protein, these workers concluded that zinc must play a catalytic role.

Various interactions involved in the binding of glycyl-L-tyrosine, a substrate that is only slowly hydrolyzed by the enzyme, have become evident from the difference Fourier studies (53). Thus, the terminal carboxylate group of this substrate interacts with the guanidinium group of an arginine side chain. The aromatic ring of the substrate is located in a rather large pocket, and the carbonyl group of the peptide bond of the substrate is probably bonded to the zinc atom. Moreover, a tyrosine is located in such a way that it could interact with the side chain of the penultimate amino acid of a substrate.

Most significant was the observation of Lipscomb and his colleagues, mentioned above, that when a substrate, (e.g., glycyl-L-tyrosine) binds to the active site of the enzyme, a widespread conformation change is induced. The phenolic hydroxy group of a tyrosine residue moves about 14 Å so that the hydroxy group is close to the peptide bond of the substrate. This motion involves a rotation of the side chain about the $C\alpha$-$C\beta$ bond as well as a motion of the peptide backbone. Furthermore, the guanidine group of an arginine side chain moves about 2 Å on binding the carboxylate group of the substrate. These tyrosine and arginine residues of the protein are located in the tortuous-coil sections of the molecule. The involvement of a tyrosyl residue in the active center was previously shown by selective chemical modification (54) and by iodination in the presence and absence of substrate (55). Moreover, the electron density near this tyrosine side chain agrees with the sequence of the active-site peptide isolated in the iodination study (55).

The mechanism of catalysis proposed by Lipscomb and his co-workers (53) involves the zinc atom, the tyrosine side chain, mentioned above, and the carboxylate group of an adjacent aspartate or glutamate residue. These workers suggest that the carboxylate group acts as a nucleophile to form a transient anhydride with the carbonyl group of the substrate. The tyrosine hydroxy group is in a position to donate its hydrogen atom to the peptide amide nitrogen. In this scheme, the zinc serves to orient the substrate and to polarise the carbonyl group thereby facilitating the formation of an anhydride and its subsequent hydrolysis.

Koshland and Neet (56) investigated the relations between crystallography and protein-modification findings and arrived at the conclusion that the identification of active sites is not only more informative but more certain by the crystallographic method. The position of certain amino acid side chains is revealed in a manner that is impossible to achieve by modification studies. Some ambiguities such as the function of zinc in carboxypeptidase A may be clarified only by X-ray analysis, since protein-modification methods do not readily distinguish between a group essential to catalysis and a group essential to the maintenance of the three-dimensional structure of the protein. Taking into consideration the possibility of conformational changes induced by substrate, the identification of active sites by protein modification-protection studies is probably more ambiguous than identification by direct observation of bound substrate. It should be stressed that crystallography has verified the major conclusions obtained in protein-modification work. Thus, the identification of residues at the active site, for example the histidines at the active site of ribonuclease, is confirmed by the crystallographic studies. Furthermore, the conclusions about the general features of protein structure such as groups remote from the active site, and inside and outside groups, are also confirmed.

Not all protein modification and physicochemical studies, however, are corroborated by crystallography. Koshland and Neet (56) point out that the crystallographic technique has shortcomings just as the protein-modification methods have and will require complementary kinetic and modification studies to solve, for example, the question concerning the role of the individual amino acid residues in enzyme activation. It has become obvious that much of the protein-modification work, which was very useful in the past, is now superseded by crystallography. Active sites can now be identified more rapidly by crystallography when supported by critical chemical experiments than by the previous modification techniques. With the development of automated crystallographic techniques and the chemical problems of isomorphous replacement solved, it will probably be quicker to determine a three-dimensional structure than to perform laborious modification studies. Although knowledge of amino acid sequences has been essential for the resolution of the structure reported so far, it appears to be quite possible that in future such structure will be solved only with X-ray data and the knowledge of the composition of tryptic peptides. Koshland and Neet (56) conclude that "ironically a time may come when three-dimensional structure can be calculated from known folding principles and a known sequence, at which point sequencing may be the shortest way to a protein structure."

The excellent reviews on the subject of X-ray crystallography of proteins by Phillips (14), Holmes and Blow (15), Stryer (50), Crick and Kendrew (57), Dickerson (58), and Kraut (59) are recommended.

B. Regulation of Enzyme Activity

The crystallography breakthrough that has led to the delineation of the three-dimensional structure of a number of enzymes has substantiated to a significant extent many past predictions with regard to protein structure and function. Moreover, recent studies on enzyme pathways, enzyme flexibility, and enzyme conformation, have greatly contributed to the understanding of various fundamental metabolic processes.

Modulation of enzyme behavior by varying the concentration of specific metabolites was first recognized in studies made on the regulation of biosynthetic pathways. Thus, mutant strains in which a late stage of a biosynthetic sequence had been blocked have been found to accumulate an intermediate prior to the blocked step. In many cases, however, this accumulation could be prevented when the normal end product of the sequence was supplied at a level allowing rapid growth (60). The possibility that such feedback control might be general and might be operative at a molecular level was previously discussed by many workers in this field (61–63). As early as in 1950 Gots (64) suggested that a biosynthetic block may be caused by an inhibitor rather than by mutation. The use of isotopic techniques by Abelson and his colleagues (65) has led to the discovery that incorporation of endogenous carbon into an amino acid was specifically depressed when the amino acid was available in the medium.

Using ^{14}C Abelson (66) showed that in the intact *Escherichia coli* cell threonine is converted to isoleucine and that exogenous isoleucine abolishes this conversion. Furthermore, when working with a mutant strain unable to synthesize threonine, Umbarger (67) observed that the amounts of growth obtained by limiting the levels of threonine increased on the addition of isoleucine. This finding led Umbarger (68) to the discovery that threonine dehydrase is strongly and specifically inhibited by L-isoleucine. On comparing such effects to technological control devices, Umbarger (68) suggested that these effects may be of general importance in biosynthetic regulation. Almost at the same time as Umbarger (68), Yates and Pardee (69) observed that aspartate transcarbamylase was inhibited by cytosine derivatives. These authors interpreted this result as an example of a feedback-control mechanism. These independent findings by Umbarger and by Yates and Pardee led to the rapid discovery of many similar cases illustrating the feedback-control mechanism.

In a subsequent paper Umbarger and Brown (70) reported in greater detail on the *E. coli* threonine dehydrase, discussed further the possible general significance of this kind of regulatory control, and compared its effects with those of mass action and of control of enzyme synthesis. In 1961 three groups of workers reported independently that the enzymes aspartate transcarbamylase (71), threonine dehydrase (72), and the enzyme catalyzing the reaction of ATP with phosphoribosyl pyrophosphate in histidine biosynthesis (73) lost under various conditions sensitivity to their inhibitors but retained catalytic activity. On account of these findings Monod and Jacob (74) as well as Gerhart and Pardee (75) independently proposed the existence of separate but interacting sites for substrate and inhibitor. Monod and Jacob (74) stressed the fact that the inhibitor need not be a steric analog of the substrate and referred to this type of interaction as allosteric inhibition. The word "allosteric" has, however, since been employed in various contexts, often with no evident sign of its original definition. In 1962 Gerhart and Pardee (75) suggested that binding of the inhibitor leads to conformational change in the enzyme, with a concomitant alteration of affinity for substrate at the active site. Changeux (72, 76) proposed a model involving two types of sites, each able to bind either substrate or inhibitor, but with the various permutations producing different effects. Monod, Changeux, and Jacob (77) studied a selected group of regulatory enzymes and concluded that the evidence, although not fully conclusive, favored entirely separate, rather than overlapping, substrate and inhibitor sites. Stressing that specific modulation of a catalyst could be expected to arise only by mutation and selection, these authors considered such behavior to be a fundamental attribute of living systems and suggested a basic similarity in the mechanisms responsible for enzyme modulation, repression of protein synthesis, and hormone action (77).

As to the prevailing nomenclature Koshland and his co-workers (78) define the term "catalytic power" as the ability of the enzyme to accelerate a chemical reaction. Although specificity and catalytic power are both characteristics of enzyme catalysis, Koshland et al. (78) find it convenient to separate these properties by a specific term for the acceleration of the bond-forming and bond-breaking process. The term "catalytic factor" is used to define the quantitative ratio of the enzymic and nonenzymic rates under the same conditions. The name "catalytic residues" is used to specify the amino acid residues directly involved in the covalent bond changes during enzyme action, for example, the serine 195 and histidine 57 residues of chymotrypsin. The term "specific residues" designates those residues that are involved both in the binding of substrate and in subsequent processes of the chemical reaction that are not directly involved in covalent changes, for example, the methionine 192 residue of chymotrypsin. The term "contact residues" is defined by Koshland (79) as referring to those residues that contain at least one atom within a bond distance (about 2 Å) of the ligand under discussion and may include both catalytic and specificity residues. Hence, a contact residue for the substrate in lysozyme would be alanine 107, which makes a hydrogen bond with the N-acetyl side chain of the glucosidic substrate.

The term "kinetic specificity" refers to the specificity in steps following the binding step, and "binding specificity" is related to specificity in the binding process (80). Whereas the term "active site" is defined as that general region in the neighborhood of the catalytic residues that are identified with the binding and reaction of substrates, "allosteric site" designates a topologically separate site which lacks catalytic activity but which can exert regulatory control over the active site (77). Since most enzymes are composed of subunits and exhibit cooperative effects, these phenomena play an important part in the action of effectors (56). A small molecule that interacts with an enzyme, thereby changing the catalytic behavior of the latter, has been called effector, modifier, or modulator, its abbreviation for use in symbols being M (60). The effector which is either a part of the active site or is adjacent to the active site is called autosteric effector to distinguish it from an allosteric effector, which operates at a distant site (56). The terms "effector" and "modifier" are used interchangeably for any ligand that affects the activity of an enzyme but is not itself changed during the enzyme action (60, 81). A positive effector increases the activity or binding of a second ligand, a negative effector decreases it (81). The word "ligand" is used to describe a small molecule bound to the protein by noncovalent forces; it includes activators, substrates, and inhibitors (56). The term "subsite" is used to signify a portion of an active or allosteric site or a portion of a site identified with binding a particular moiety of one of the substrates— for example, the nicotinamide binding subsite of the NAD binding site (56).

The term "regulatory site" is defined by Koshland and Neet as a site which binds effectors that regulate enzyme activity but which differs from the active site (56). The term "regulatory site" is frequently used interchangeably with "allosteric site," but the extension of the term "allosteric" to include cooperative effects between distant identical sites would require the definition of a new term that would specifically indicate a different type of site. Thus, under these definitions both hemoglobin and aspartyl transcarbamylase would be allosteric proteins but of the two compounds only aspartyl transcarbamylase possesses a regulatory site (56). "Conformational change" is used in the classical organic sense to indicate a change in the average positions of the atomic nuclei, but covalent bond changes are not included. Thus, the unfolding or rotation about bonds of a protein is a conformational change, whereas the dissociation of a proton is not, even if the latter is concomitant with a conformational change. Furthermore polarizations of electrons without change in the positions of the atomic nuclei are not conformational changes (56).

Although, as mentioned above, the crystallographic determination of the three-dimensional structure of an enzyme, the extensive modification of amino acid residues in enzymes, the isolation of enzyme mutants, and detailed physicochemical studies have greatly elucidated the behavior of enzymes, the phenomenon of their enormous catalytic power is not yet clearly understood. Many hypotheses have been proposed to explain the extraordinary catalytic power of enzymes. One of the earliest concepts to explain enzyme efficiency, known as the proximity effect, was the idea that a ternary complex between enzyme and two substrates improved collision probabilities between reactants. When a method was designed for the quantitative evaluation of this concept (82, 83), the results were rather disappointing. Whereas the theoretical acceleration achieved by the juxtaposition of two substrates was found to be substantial in some cases, a decrease in velocity rate was actually obtained in many other instances (83). The proximity effect by itself seems to be insufficient to explain enzyme action. The sigmoid and bell-shaped curves observed when V_m is plotted against pH have led to the proposal of an "acid base catalysis" (40, 82, 84, 85). Impressive circumstantial evidence in support of such a concept has been obtained in some enzymes, and the bell-shaped curve has been explained by suggesting that a nonprotonated group is operating as a general base and a protonated group is operating as a general acid.

Jencks (86) has pointed out, however, that there are a number of reactions in which general acid and base catalysis could not participate. Although the general acid base catalysis may contribute to enzyme action, this theory cannot explain the enormous catalytic powers of enzymes (56). It should be recalled here that catalytic factors of 10^9 to 10^{12} were calculated for enzymic catalysis (82).

Swain and Brown (87) advanced a variant of the acid-base catalysis, the "push-pull hypothesis," to explain enzyme efficiency. This theory suggests that an acid and a base acting together may be more efficient than either acting alone. However interesting this hypothesis may be for enzyme action, the evidence for it in nonenzymic systems has been shown to be rather feeble (88). Various other hypotheses—such as a "covalent intermediate catalysis," the "rack-and-strain theories," the "microscopic environment effect," the "ion-pair effects," the "orientation effects"— remain fascinating suggestions for enzyme action but cannot per se explain the enormous catalytic powers of enzymes. The reader may be referred to the discussion of the latter theories in a recent masterful review by Koshland and Neet (56) on the catalytic properties of enzymes.

CONFORMATIONAL FLEXIBILITY OF PROTEINS

The "lock and key," or "template," hypothesis advanced by Emil Fischer (89) to explain enzyme specificity postulated that the substrate must be adsorbed on the surface of a rigid protein, where the catalytic groups would act on the bonds to be formed or broken in the substrates. The existence of such an enzyme–substrate complex was supported by the kinetics of the reaction and the extreme specificity of enzymes. Although this template theory was able to elucidate most of the kinetic phenomena, certain important observations remained unexplained. To clarify some anomalous findings the "induced-fit theory" was proposed by Koshland (90–92). According to the induced-fit theory, small molecules may induce conformational changes, when binding to the enzymes. Furthermore, a precise orientation of catalytic groups is required to bring about reaction; finally, whereas substrates are capable of inducing the proper alignment of residues, nonsubstrates are not.

It should be mentioned here that the term "induced-fit" was originally applied to suggest both a complementary fit between protein and ligand in the enzyme-ligand complex and the induction by ligand of a new protein conformation, which would not be present in significant amounts in the absence of ligand. This term was proposed to contrast with the template theory in which the enzyme was postulated to be in the active conformation in the absence of substrate (56).

The new and the old theories share certain similarities; but there are also certain differences between them. Thus, whereas both theories postulate a fit between substrate and enzymes, that is, a specific enzyme substrate (ES) complex and, furthermore, both hypotheses assume that a single conformation of the substrate is selected in the ES complex, they differ in many other respects. Representative of these differences is, for example, the explanation for the missing reactivity of a substance which is a "potential substrate"; that is, a compound that possesses the reactive group of a substrate but is unable to react, such as an ester that fails to react with an esterase enzyme. This lack of reactivity can be explained on the template hypothesis by the assumption (a) that there is a steric hindrance, the molecule being unable to fit into the active site because it is too large, or (b) that the molecule lacks appropriate groups by which it would be attracted to the enzyme surface or may even contain groups that repel the enzyme protein, or (c) that the molecule is bound to the surface in a nonproductive orientation which would not facilitate a juxtaposition of the bond to be broken and the catalytic groups on the enzyme surface. The induced-fit theory suggests that some molecules may not react even with flexible proteins for the same reason as given above for the template theory, but in addition it assumes that substrates must induce a favorable change in the conformation of the enzyme protein. Binding per se is not sufficient, since the proper alignment of catalytic groups must first be induced. Moreover, whereas the older template theory assumes that the average position of the atoms in the protein remains the same before and after ligand binding, the induced-fit theory suggests that there is a meaningful change in the average positions of atoms and that this change is capable of affecting enzyme activity. On desorption of the ligands from the "flexible" enzyme the protein returns to its initial conformation.

Koshland's theory of a flexible enzyme (90–92) was explored and verified in subsequent research work. Direct evidence was provided for protein conformation changes under the influence of bound ligand, and a rationale was found for sequential binding of substrates. Furthermore, a role was assigned for activators that did not take part directly in the enzyme action and an explanation provided for noncompetitive inhibition reactions as well as for the sharp breaks in temperature-activity curves, to mention only a few findings (78, 92–94).

The theory of a flexible enzyme has led to a number of consequences, and various important conclusions in regard to this theory have been drawn in recent studies. Lately, O'Sullivan and Cohn (95) correlated proton relaxation spectra and the reactivity of SH groups with the activity of creatine kinase toward substrates. In their very extensive investigations these authors observed five separate conformational states of the enzyme which paralleled the velocity of the substrate reaction, thus indicating that good substrates gave the larger conformational change. This finding seems to be in accord with the suggestion made in regard to enzyme flexibility that the final conformation state of the protein is a sum of complementary interactions between a ligand and a protein (56). Similar observations were made on substrates, activators, and inhibitors of enzymes, such as phosphoglucomutase (96), threonine deaminase (97), and citrate synthetase (98); for example, in the case of phosphoglucomutase, it was shown that substances similar to the substrates, such as inorganic phosphate and glycerol phosphate, brought about a conformational change in the protein, which, however, was much smaller than that produced by the substrates glucose-1-phosphate and glucose-6-phosphate (96).

Potassium and other monovalent ions seem to be essential in conformational changes in the case of pyruvate kinase (99, 100). When tetramethylammonium chloride was used as a control, no change in the properties of the protein was observed, which strongly indicates that specific sites for sodium and potassium binding were responsible for the conformational change.

The size of the conformation change does not seem to correlate with the size of ligand. A rather considerable conformation change induced by a very small ligand has been observed, for example, on the binding of oxygen to cytochrome c (101, 102). Although the extent of conformational change necessary to affect partially or completely enzyme activity is unknown, studies on thiol subtilisin, however, have suggested that the alignment of key residues may be extremely exact. The most plausible explanation for the lack of reactivity of the thiol subtilisin may be found possibly in the change in geometry caused by the change from an oxygen to a sulfur atom. Should this suggestion prove to be correct, only a fraction of an Angstrom unit change in the orientation of a catalytic group would be capable of producing very large changes in reaction velocity (56).

The first unequivocal demonstration of induced fit indicating that the act of binding of substrate to enzymes induces a necessary alignment of catalytic groups on the enzyme surface has been provided by Lipscomb and his co-workers in their recent work on carboxypeptidase (51). As mentioned above, these workers have shown that when a substrate, glycyltyrosine, binds to the active site of carboxypeptidase, a widespread conformational change is induced, involving the movement of a tyrosine residue through 14 Å (51). It has not yet been established, however, that all enzymes exhibit flexibility. Koshland and Neet (56) suggest a gradation from enzymes whose specificity pattern or control characteristics require considerable flexibility to small compact enzymes, which have no need of such flexibility. The latter category may include such enzymes as lysozyme, ribonuclease, and chymotrypsin, which are all small molecules, contain disulfide cross linkages, and are already bathed in one of their substrates, water. Whereas evidence for conformational changes of groups involved in catalysis of lysozyme and ribonuclease will have still to be established, in the case of chymotrypsin there appears to be good evidence obtained by Oppenheimer and his co-workers (44) for a conformational change identified with protonation of a group of pK 8.5, possibly the isoleucine which is N-terminal for the B-chain.

The problem of the necessity for all enzymes of conformational changes for specificity and control during enzyme action will have still to be solved. A very important consequence of a flexible enzyme was the observation that a compound, not itself directly involved in the enzyme action, could induce a conformational change leading to reaction. Thus, Inagami and Murachi (103) were able to demonstrate such an effect for the trypsin-catalyzed reaction. Whereas acetyllysine ethyl ester is a good substrate for trypsin, acetylglycine ethyl ester reacts extremely sluggishly with this enzyme. Adding ethylammonium ions to the trypsin solution activated the acetylglycine ethyl ester hydrolysis by a factor of nine. Furthermore, they were able to show that the ethylammonium ion did not change the binding constant of glycine ethyl ester but rather changed its reactivity after it was bound to the enzyme surface. This finding the authors explained on the basis of the induced-fit theory by suggesting that the glycine compound lacked the components necessary for induction of a conformational change by the lysine side chain. Ethylammonium would thus be an example of an autosteric effector (56).

The low reactivity of water in enzymic reactions in which it might be expected to compete with the acceptor molecule could be explained by its lack of sufficient structure to induce a conformational change. Chance, Theorell, and their co-workers (104–106) found in such enzymes as alcohol dehydrogenase, which require conformational changes for their action, that the inhibition of these conformational changes slowed down the reaction. Similar effects have been observed for ribonuclease (107) and carboxypeptidase (108). The reaction of all these enzymes was found to be slower in the crystal form than in solution, which seems to indicate that the packing of the molecules in the crystal hinders a conformational change.

The binding of oxygen to hemoglobin is also known to be slower in the crystal than in solution (106). The results obtained agree well with the observation that crystals prepared in the absence of substrate frequently crack up on binding the substrate. Alternatively, the cracking phenomenon could be explained by suggesting that substrate binds in the interstices of the crystal in such a way that the crystal lattice is deformed. Considering, however, the observed conformational changes of proteins in solution and the fact that so many crystals have been found to undergo significant changes in form, or actual cracking on binding of substrate, the more general explanation seems to be a change in conformation leading to disintegration of the crystal form. In this context we should mention work by Haurowitz (109), published as long ago as 1938, in which the author reported that hemoglobin underwent a change in crystal form on binding oxygen. Muirhead and Perutz (110) established unequivocally that there is an actual shift in the position of the peptide chains with the binding of oxygen.

The concept of a flexible enzyme has been found to be most useful in regard to the discussion of regulation of enzyme activity (56).

Hemoglobin, although not an enzyme, displays many properties that are relevant to enzyme action. Because of this behavior, hemoglobin has been called an honorary enzyme. The "cooperative effect" (111) that is characteristic of many enzymes composed of subunits was initially discovered in hemoglobin. The term "subunit" is usually referred to any subdivision of polypeptide chains within a polymeric protein (56). The term "protomer" is defined as the identical repeating unit in a protein that contains a finite number of identical subunits (81); for example, in aspartyl transcarbamylase the presence is assumed of four protomers, each containing a catalytic and regulatory polypeptide chain. Whereas the term "protomer" is very convenient for proteins of the aspartyl transcarbamylase type, it is prone to ambiguity in a protein, such as hemoglobin, which is composed of two different types of peptide chains displaying similar functions. In such a case "subunit" instead of "protomer" may be a better term. The terms "homotropic" and "heterotropic interactions" (81) specify interactions between like and unlike ligands, respectively.

When a plot of fractional saturation of oxygen binding sites in hemoglobin versus oxygen concentration is made, a sigmoid curve is obtained in contrast to the usual hyperbolic curve of Michaelis-Menten kinetics. Because such a curve would be obtained, if the first molecule of ligand facilitated the binding of the second molecule, the phenomenon was termed cooperative. Today it is more precisely called a positive homotropic effect. This type of a phenomenon was previously explained by Pauling on the basis of electrostatic interactions (112). However, the finding that in the hemoglobin molecule the hemes are far apart in the three-dimensional structure, and because of the widespread explanation of similar phenomena in enzymes by conformational effects, it appears probable that the conformational explanation is the more likely one (56). Furthermore, oxygen binding undoubtedly induces changes in quarternary structure of the protein; that is, the rearrangement of the subunits in regard to each other, as shown by X-ray crystallography (110). The reactivity of side chains such as the SH groups with alkylating reagents and the susceptibility of peptide bonds to proteolytic digestion (113–115) changes under the influence of oxygen. Although these effects could be possibly explained by rearrangement of subunits, it seems most probable that further crystallography will reveal conformational changes within the subunits (56).

Two general models relating in detail the subunit structures of the protein to its allosteric and cooperative effects have been proposed, and numerous theories to elucidate various features of these effects have been published. The reader is referred to a very clear presentation of the relevant literature by Koshland and Neet (56).

Monod, Wyman, and Changeux (81) have elaborated a model for allosteric and cooperative effects based on the suggestion of a "symmetry principle" in proteins. On consideration of the potential form of binding between proteins containing identical subunits, Monod and his co-workers arrived at the following conclusions: (a) the protomers will be arranged in a symmetrical fashion, and (b) this symmetry will be conserved during changes in protein conformation; that is, the change in conformation of one subunit is accompanied by equal changes in all identical subunits. Furthermore, these workers assumed that the conformational change occurs through an isomerization between two states, the R- and T-states of the protein. If one of these states, for example, the R-state, binds substrate preferentially and is present in small amounts, addition of substrate will tend to pull the equilibrium in the direction of the R-state, and the mathematical evaluation of this assay will show that the saturation curve will be sigmoid. The action of inhibitors and activators can be easily explained on this model by assuming that inhibitors preferentially bind to the T-state and activators to the R-state. Although this model is not entirely limited to two states, its postulates that the protein is designed to conserve symmetry, and that the ligands tend to pull the equilibrium toward preexisting states involve the assumption that two or a few states exist. The model of Monod and his co-workers (81) predicts positive and negative heterotropic effects, but only positive homotropic ones.

Koshland and his colleagues (116–118) have approached the problem of allosteric and cooperative properties of proteins on the basis of the induced-fit theory suggesting that there is a flexible interaction between ligand and protein, which may induce a new conformation of the subunit. This interaction may in turn affect the shape and stability of the neighboring subunits. The extent of the change would depend on the nature of the interaction between the subunits and the amount of distortion induced by the ligand. If the conformation change in one unit is very loosely coupled to the conformation of the neighboring subunits, the change may not affect its neighbors and Michaelis-Menten kinetics will hold true. If there are closely coupled subunit interactions, a change induced in one subunit will cause partial or equal changes in the neighboring subunits and in general the changes will take place sequentially with ligand addition. Thus, hybrid conformational states will be observed. This model predicts both positive and negative homotropic and heterotropic effects (119, 120) and it suggests many final conformational as well as intermediate states, since the structure stabilized involves a complementary interaction of ligand and protein rather than the stabilization of a predetermined structure. The model proposed by Koshland et al. (116–118) has been proved to be general in application and to lead to simultaneous changes in subunit conformations as a limiting case (118). At present, it is rather difficult to draw conclusions in regard to the correctness of either of these theories, the symmetry model or the sequential model (56). The behavior of some enzymes, such as aspartyl transcarbamylase or yeast glyceraldehyde phosphate dehydrogenase, can be explained on the basis of the symmetry model, whereas other enzymes, such as rabbit muscle glyceraldehyde phosphate dehydrogenase and isocitric dehydrogenase, require a sequential model (56). The observation that a particular protein under particular conditions obeys one particular model seems to Koshland and Neet (56) "to be a reflection of a special case of a far more general phenomenon, i.e., Nature's utilization of varying subunit interactions to tailor the kinetic properties of a protein to the need of the organism."

The dissociation of a protein into subunits undoubtedly imposes individual problems on the kinetics of the reaction and may provide a very good tool for studying the extent of interactions of this kind. The conformational flexibility of proteins may be considered to be a universal property of the latter molecules which seems to enable them to adapt themselves to their widespread functions in the living system (56).

In this context it should be recalled that it is known that the amino acid side chains of proteins do not by themselves catalyze reactions at velocities approaching those of enzymes. Enzymologists are thus confronted by the paradox that a highly specific structure is apparently maintained by an assembly of nonspecific and variable interactions (56, 121). Modification studies and X-ray crystallography have revealed, however, that the amino acid residues are present on the enzyme surface in their expected form and do not react with each other to form new structures, such as oxazolines. Some combined action of these ordinary side chains may be responsible, therefore, for the catalytic power of many enzymes as well as for the potentiating effect on the catalytic power of those enzymes that possess prosthetic groups. Although some characteristics of the catalytic power can be identified in the form of the proximity effect, covalent catalysis, general acid-base catalysis, and ion-pair effects, the combined action of all these factors only accounts for part of the velocity of enzymic reactions. An unknown factor still required to explain the high turnover number of enzymes will be possibly found in such concepts as those dealing with a precise orientation of catalytic groups and substrates, or with microheterogeneity of the protein surface, which can enhance the polarizing effect of charged or dipolar groups, or with strain induced by the interactions of protein and substrate. Each of these factors may play a more or less important role in different enzymes. Together these factors may produce a large catalytic power (56).

It is to be hoped that more profound work will be able to solve one of the most fascinating biochemical problems—that is, that of the catalytic power of enzymes.

In regard to the question of enzyme evolution Koshland and Neet (56) have put forward a rather interesting theory. According to these authors, the key feature of enzymes might have been initially of template-type specificity. Conformational changes, if they at first existed at all, would have been accidental and would not have displayed any catalytic or specificity characteristics of proteins. Enzymes of the latter type might have shown inherent specificity inefficiency; for example, they would have been unable to discriminate between hydroxylic nucleophiles and water, between acetic acid and propionic acid, between valine and isoleucine. On the other hand, a flexible enzyme by offering an opportunity for added specificity in such reactions might have given its organism a competitive advantage. As for reasons of specificity the number of flexible enzymes increased, an enzyme, such as trypsin, might have been activated for reaction with a "deficient substrate," such as glycine ethyl ester, by a small molecule metabolite, such as the ethylammonium ion. The added facility for control of such a system might then have given this type of effector system a selective advantage. It is possible that from this form of process feedback control and kinetic control might have originated within a simple polypeptide chain.

The tendency of subunits to associate would have resulted in an association of unlike proteins. The conformational changes induced by the substrates at the active site of one enzyme of such a complex could have then influenced the conformation of the active site of the second enzyme, whereby the activity of the latter would have been either increased or decreased. Either of these effects might have had a selective advantage. If enzymes catalyzing two consecutive steps in a common pathway enhance the activity of each other, a considerable gain in activity might be obtained. An inhibition, however, would also be advantageous, if one of the enzymes in the complex catalyzed a final reaction in which feedback control could operate. The induced conformational changes might be mutual, in that either enzyme 1 or enzyme 2 of the complex could be active at any time but not both together. In such a case the advantages of regulatory control might be offset by impairment of one of the metabolic pathways. Gene duplication of the polypeptide chain of enzyme 1 might then produce more effective systems. Thus, a polypeptide chain 1 might arise that had retained catalytic properties but was no longer attracted to the polypeptide chain of enzyme 2 and therefore would escape adverse regulatory effects from that protein. On the other hand, polypeptide chain 1 might mutate to lose its catalytic function and become more and more optimally designed to interact with enzyme 2 and to act as its regulatory unit. One would then expect regulatory sites to be in most cases mutated active sites, and homologies will be observed between regulatory subunits and other enzymes, just as homologies are observed between trypsin and chymotrypsin. It might also be feasible that the mutual inhibition of enzyme 1 and enzyme 2 is of advantage and that some so-called regulatory subunits are indeed enzymes for which the second substrate is yet to be determined. Furthermore, the association of enzyme subunits with each other would be advantageous to the organism, since it would bring about cooperative effects which sensitize the system to regulatory control by producing enzymes more responsive to changes in metabolite concentrations. The association of enzyme subunits may, however, also produce proteins with antagonistic subunit effects, thus making them less sensitive to changes of metabolites. Either of these effects may be of advantage in different situations (56).

The identification and realization of this and other similar hypotheses will be one of the exciting tasks of workers in this field in the near future.

2. CLASSIFICATION OF AMINE OXIDASES

Many competent reviews have been published on amine oxidases (122–131).

The group of amine oxidases comprises enzymes that oxidatively deaminate amines (mono-, di-, and polyamines) with the stoichiometric formation of one molecule each of aldehyde, ammonia, and hydrogen peroxide.

$$RCH_2NH_2 + O_2 + H_2O \rightarrow RCHO + NH_3 + H_2O_2.$$

In this reaction two electrons are probably transferred from the amine to the molecular oxygen, which is reduced to hydrogen peroxide (132). The NH_2 group is changed thereby into an imino group which by the ensuing hydrolysis is split off in the form of ammonia.

$$RCH_2NH_2 + O_2 \rightarrow RCH = NH + H_2O_2,$$
$$RCH = NH + H_2O \rightarrow RCHO + NH_3.$$

The mechanism of the electron transfer is not yet fully understood. It is known, however, that oxygen cannot be replaced by any other hydrogen acceptor (126).

Evidence has been obtained of the formation of an aldehyde during the action of purified hog-kidney histaminase on histamine (133).

In 1940 Zeller (134, 135) was the first to attempt a classification of amine oxidases into two groups, the monoamine oxidases and the diamine oxidases. This classification was based on supposed substrate specificities of those enzymes.

The first doubts about the significance of this type of classification arose when it became clear that long-chain aliphatic diamines were substrates of monoamine oxidase (MAO) and not of diamine oxidase (DAO) (123, 136).

Further, in 1947 Steensholt (137) reported the occurrence of an amine oxidase in rabbit liver that apparently oxidized not only mescaline, a typical monoamine, but also histamine and cadaverine; moreover, it was inhibited by typical DAO inhibitors. This behavior of the rabbit-liver mescaline oxidase was found to be contrary to that of the mescaline-oxidizing mitochondrial enzyme of hog-kidney cortex as well as of hog liver. The latter enzymes, in their reaction pattern toward inhibitors, resemble a monoamine oxidase rather than a diamine oxidase (138, 139).

In 1953 Hirsch (140) encountered in beef and sheep plasma an amine oxidase showing great affinity for spermine and spermidine. This enzyme was also found to be inhibited by such typical DAO inhibitors as cyanide, semicarbazide, and hydroxylamine.

Tabor and his co-workers (141) purified a beef-plasma amine oxidase 150 to 200 times and found it most active on spermine, spermidine, benzylamine, as well as on various aliphatic monoamines. Whereas tyramine and mescaline were rather poor substrates, tryptamine, serotonin, epinephrine, and norepinephrine were not attacked at all. This plasma enzyme did not show any activity on histamine, and of the diamines only the decamethylenediamine was oxidized. Moreover, the purified beef-plasma enzyme differed from the particulate classical liver MAO (1, 122, 123) not only in substrate specificity but also in its behavior toward inhibitors, for Zeller et al. (142) had reported that liver and brain MAOs were both inhibited by iproniazid, but were resistant to isoniazid, whereas Tabor's plasma MAO was inhibited by both drugs.

Yamada and Yasunobu (143) succeeded in obtaining from beef plasma a crystalline MAO preparation that showed the same substrate and inhibitor specificities as Tabor's plasma spermine oxidase (141).

Another amine oxidase occurring in hog serum was described by Kolb (144). This enzyme strongly degraded histamine, but its effect on cadaverine was very small. Blaschko et al. (145) confirmed the observation of Kolb. In 1964 Buffoni and Blaschko (146) obtained the hog-plasma amine oxidase in crystalline form and found that its best substrate was benzylamine, with mescaline and histamine serving also as good targets. The enzyme was inhibited by carbonyl reagents.

Hill and Mann (147) purified extensively an amine oxidase occurring in pea seedlings. The enzyme degraded aliphatic diamines as well as aliphatic monoamines, phenylalkylamines, histamine, spermidine, and agmatine. It was inhibited by carbonyl reagents.

From consideration of those manifold and rather confusing observations Blaschko and his co-workers (145) put forward in 1959 another scheme of classification of amine oxidases, based on differences in inhibitor specificities instead of substrate specificities. Two main groups of amine oxidases were thereby distinguished, those resistant to carbonyl reagents and those inhibited by carbonyl reagents.

Group I: Amine oxidases resistant to carbonyl reagents.

1. Classical MAO, an intracellular, mainly mitochondrial, insoluble enzyme, present in many vertebrate and invertebrate tissues (1, 123). Amine oxidases of this group act on primary, secondary, and tertiary amines, and on long-chain aliphatic diamines.

2. Mouse-liver histaminase (148).

Group II: Amine oxidases inhibited by carbonyl reagents.

1. Classical histaminase (2, 149), DAO (122, 150), preponderantly an intracellular, probably mitochondrial, soluble enzyme (151).

2. Plasma enzymes, extracellular enzymes.

(a) Ruminant-plasma amine oxidase, spermine oxidase (140, 141, 152, 153).

(b) Nonruminant-plasma amine oxidase, benzylamine oxidase (144, 146, 153, 154).

3. Rabbit-liver amine oxidase, mescaline oxidase (137, 138, 139, 144, 155).

4. Plant amine oxidase, pea-seedling amine oxidase (147).

5. Bacterial amine oxidase, polyamine oxidase (156).

Amine oxidases belonging to Group II oxidize primary amines but do not act on secondary amines.

A similar classification of amine oxidases, also based on differences in inhibitor specificities, was proposed by Zeller et al., also in 1959 (157,

158). According to Zeller, the enzymes attacking aliphatic compounds with terminal amino groups can be separated by means of semicarbazide into two groups:

Group 1. Semicarbazide-resistant MAOs with only one kind of receptor in their active center.

Group 2. Semicarbazide-sensitive DAOs with two types of receptors in their active center.

Group 1 of the semicarbazide-resistant amine oxidases includes the classical MAO (1, 123). The rather heterogeneous group 2 of semicarbazide-sensitive amine oxidases includes the classical histaminase (2, 149), DAO (149, 150), the beef- and sheep-plasma spermine oxidase (138, 140, 141, 152, 153), plasma benzylamine oxidase (144, 146, 153, 154), and mescaline oxidase (137–139, 144, 155). Zeller (157, 158) points out that in his classification scheme semicarbazide must not be replaced by the term carbonyl reagent, since many carbonyl reagents, such as monosubstituted alkyl and arylhydrazines, act on both enzyme groups efficiently, whereas only monosubstituted acylhydrazines and hydrazine itself solely inhibit enzymes of group 2.

According to Zeller (158), both groups of amine oxidases represent "whole families of homologous enzymes." These enzymes share many properties (159), have some substrates in common—for example, histamine, 1,4-methylhistamine, 1,5-methylhistamine, tyramine, and mescaline (126, 128)—and, as mentioned above, the enzymes of both groups are inhibited by monosubstituted alkyl- and arylhydrazines (157, 158).

Zeller and his co-workers (160) consider hydrazine and its derivatives to be pseudoamines.

CHAPTER III

Amine Oxidases Not Sensitive to Carbonyl Reagents and Not Inhibited by Semicarbazide

1. CLASSICAL MONOAMINE OXIDASE*

A. Occurrence

Monoamine oxidase (MAO) has been found in all classes of vertebrates so far examined: mammals, birds, reptiles, amphibians and teleosts (161). The enzyme occurs in many different tissues, particularly in glands, plain muscle, and the nervous system (162). In man, the parotid and submaxillary glands seem to be the richest source of MAO (163). A high MAO activity is always shown by the liver (161), but the kidney, intestine, aorta, stomach, and pancreas may also contain rather high concentrations of the enzyme (161, 164). MAO occurs in the human heart (165, 166) and lung (167). Low MAO activity was encountered in skeletal muscle (162), but this observation should be confirmed. In the human and animal brain MAO activity is to be found in the central gray matter and the hypothalamic region (168–172). It is also present in the nerve cells and capillary walls, but not in glial cells or in nerve fibers (169, 173, 174). The enzyme has been encountered in the male sex organs (164, 175). In the uterus MAO activity is high (164), but this seems to be related to the functional state of the organ. The human

---

* E.C.1.4.3.4. Monoamine: O_2 oxidoreductase, deaminating.

placenta (176) as well as that of other mammals (177) has been found to contain MAO. More recent observations suggest that erythrocytes (178), blood plasma (179), and blood platelets (180) also show some MAO activity. Robinson (181) reports on a rather high MAO activity in the nictitating membrane and iris of cats and rabbits. Whereas in newborn human babies the mean MAO activity of the kidney cortex and medulla was found to be about one-half the value for adults (182, 183), the MAO content of the liver and the intestinal mucosa of newborn babies was fully equal to that of the respective adult tissues (162). An increase in MAO activity in aging tissues was reported by Birkhäuser (168).

In invertebrates Blaschko and his co-workers (184–186) have demonstrated a wide occurrence of MAO, a very high enzymic activity having been encountered in the hepatopancreas of two species of cephalopods, *Sepia officinalis* and *Octopus vulgaris* (184). This finding of MAO in mollusks is of particular neurophysiological interest, since it has been suggested by Welsh (187) that serotonin, one of the main substrates of MAO, acts as a neurotransmitter in those lower animals. Hence, it would be important to study the role played by MAO in that neurohumoral process.

The enzyme also occurs in plants. Werle and Roewer (188,189) succeeded in preparing active MAO extracts that showed a great affinity for butylamine, from the leaves, stalks, and roots of *Salvia uliginosa*. Moreover, Korzenovsky et al. (190) reported on a bacterial MAO that acted on tyramine as substrate, from cell-free extracts of *Sarcina lutea*.

B. Properties

It is now well established that the classical intracellular MAO is characterized by its location in mitochondria, its oxidative action on monoamine substrates, its inability to deaminate histamine and short-chain diamines, its sensitivity to octanol and certain heavy metals, and its resistance to carbonyl reagents, to hydrazine and semicarbazide, as well as to carbon monoxide. Furthermore, a very characteristic property of this enzyme is its action on primary as well as secondary amines. The intracellular distribution of MAO in rat liver was extensively studied by two groups of workers, Cotzias and Dole (191) and Hawkins (192). Results obtained by differential centrifugation suggested that the greater part of the rat-liver MAO was located in the mitochondria. More recent investigations (193, 194) of the intracellular distribution of MAO in rat liver have confirmed these results, since it was found that 70% (193) to 76.7% (194) of the total MAO content of the rat-liver homogenate was of mitochondrial origin. The activity of the mitochondrial MAO seems to depend to a certain extent on the structure of the mitochondria;

it was found, for example, that swelling of the rat-liver mitochondria stimulated the oxidation of tyramine (195). Unsaturated fatty acids have been shown to be involved in the regulation of mitochondrial swelling (196). Gorkin et al. (197, 198) report that sodium oleate in low concentrations (10^{-4} M) inhibits the enzymic effect of rat-liver mitochondrial MAO on tyramine, whereas the deamination of benzylamine is not affected at all. Purification experiments on rat-liver mitochondria (199, 200) and mitochondria of rat brain (201) have demonstrated that MAO is tightly bound to the insoluble structure of mitochondrial membranes. All attemps to separate MAO from the insoluble particulate matter have been so far unsuccessful (197).

Partially purified MAO was, however, prepared in various laboratories by treatment of the mitochondrial membranes with detergents, or with ultrasonic waves, or by disruption of the mitochondria by homogenization and subsequent treatment with Triton X–100. For details of some purification procedures of MAO see Chapter VI.2.A.

Unlike other mitochondrial enzymes, MAO is very stable. Heating for 10 minutes at 50°, however, destroys 50% of its activity. Its pH optimum depends on the purity and the origin of the enzyme. With tyramine as substrate, the optimum pH of guinea-pig-liver MAO has been found to be 7.3 (130).

C. Prosthetic Groups

Little is known about the chemical constitution of the mitochondrial MAO. All indications of possible prosthetic groups have until recently been indirect. Since mitochondrial MAO does not discriminate in its action between primary amines and their N-methylated dervatives as substrates, and, furthermore, is not inhibited by carbonyl reagents, this enzyme probably does not require any pyridoxal as a prosthetic group (130, 202). On the other hand, the inhibition of MAO by quinacrine, a known inhibitor of flavoenzymes (203), observed by Lagnado and Sourkes (204) as well as studies made on riboflavin-deficient rats (205, 206), suggest that MAO may have a flavin prosthetic group. If the ability of an amine oxidase to act on a secondary amino group is considered to be one of the criteria for its classification as an intracellular MAO (202), then the spermidine dehydrogenase of *Serratia marcescens* should be mentioned here. This interesting microbial enzyme attacks spermidine at a secondary amino group with the formation of Δ^1-pyrroline and 1,3-diaminopropane (207). For its reaction with oxygen, the spermidine dehydrogenase requires the presence of flavin adenine dinucleotide (FAD) and an additional electron carrier (208).

Lately, Nara et al. (209) and Erwin and Hellerman (210, 211) have independently reported the occurrence of a flavin in purified preparations of beef-liver and of beef-kidney mitochondrial MAO, but both groups of workers failed to identify the flavin fully. A direct indication of the absence of pyridoxal from intracellular MAO, and of the presence of FAD in purified preparations of hog-brain mitochondrial MAO has been very recently provided by the work of Tipton (212–214). A purified enzyme preparation of hog-brain MAO (214; see Chapter VI.2.B.g) has been investigated for the presence of pyridoxal and pyridoxal-5-phosphate by subjecting it to pronase digestion, followed by acid hydrolysis (215), and by determining the ability of the acid hydrolysate to reactivate *Streptococcus faecalis* L-tyrosine apodecarboxylase in the presence of ATP (216). No detectable amounts of pyridoxal phosphate were encountered in purified preparations of hog brain. This finding supports earlier reports indicating that pyridoxal phosphate is not involved in MAO activity (130, 202). It should be pointed out, however, that those observations were based on indirect evidence such as lack of sensitivity of MAO to carbonyl reagents or the absence of effect of a pyridoxine-deficient diet on the MAO activity in rat liver (130, 202).

Tipton (214) has been able to isolate from purified preparations of hog-brain mitochondrial MAO a fluorescent material that proved to be FAD.

Fluorescent material could be liberated from the purified MAO either by heating an enzyme solution at 100° for 12 min or by treatment of such a solution with trichloroacetic acid (214). In this context it should be pointed out that the purification procedure used for hog-brain mitochondrial MAO (see Chapter VI.2.B.g) involves alcohol fractionation as a final step. Attempts to extract fluorescent material by trichloroacetic acid treatment of a less pure MAO preparation that had been obtained without alcohol fractionation were largely unsuccessful since only negligible amounts of flavin could be extracted. This finding is of interest since other workers (209, 211) who tried to liberate flavin from their MAO preparations were unsuccessful.

The flavin component from supernatants of boiled MAO extracts was identified as FAD in two ways, either by ascending chromatography and comparison of the fluorescent spot obtained with that produced by authentic FAD, or by titration with the FAD-specific D-amino acid apooxidase. From reactivation experiments with samples of the purified supernatant from boiled MAO, a molecular weight of 120,000 per molecule of FAD was calculated. Furthermore, the fluorescence emission spectrum of the material extracted from hog-brain MAO by trichloroacetic acid treatment was found to be identical with that of a sample of authentic FAD. On calculation of the FAD content of the enzyme from the fluorescence of this trichloroacetic acid supernatant, a value of 1 mole FAD per 114,000 g of enzyme protein was obtained (214). Incidentally, the average molecular weight value of 116,000 for hog-brain MAO, calculated from results obtained by different estimation methods (214) seems to be in agreement with the value of 102,000 obtained from gel-filtration data (213). This average molecular weight value also compares well with the value of 100,000 reported by Erwin and Hellerman (211) for beef-kidney MAO, although these authors calculated a considerably higher molecular weight (290,000) for the latter enzyme from ultracentrifuge studies.

The relative ease with which flavin can be removed from the hog-brain enzyme is not in agreement with results obtained with beef-liver and beef-kidney MAO. This finding led Nara et al. (209) to the assumption that the flavin may be covalently bound to the enzyme. However, the properties of the MAO flavin group seem to depend on the degree of purification. Thus the FAD could not be extracted from an only partially purified preparation of hog-brain MAO by techniques which quantitatively extracted it from the purified MAO (214).

The difficulty of extracting FAD from preparations of beef liver and kidney (209) may have been due to impurities, or to species or organ differences between the enzymes (214). As to the effect of substrates on the fluorescence of the FAD component of MAO, Tipton (214) found that addition of n-amylamine, which is a good substrate of MAO, caused a significant decrease in the flavin fluorescence. Histamine and spermidine, neither of which is a substrate of MAO, did not significantly reduce the flavin fluorescence.

By applying a technique that Swoboda (217) devised for the removal of FAD from glucose oxidase without irreversibly inactivating the enzyme, Tipton (214) has been able to split MAO into apomonoamine oxidase and FAD. Preparations of the apooxidase thus obtained had not detectable enzyme activity and showed negligible fluorescence. The inactive apoenzyme could be partly reactivated by incubation with FAD, but not with FMN. The phenothiazine derivative, chlorpromazine, was found to inhibit competitively the reactivation of monomine apooxidase by FAD. This observation agrees well with that of Gabay and Harris (218). As a result of their thorough study of the competitive inhibition of the reactivation of D-amino acid apooxidase by a number of phenothiazines, the latter authors suggested that phenothiazines may act by inhibiting flavoenzymes (218). Tipton (214) regards the production of the hog-brain MAO apoenzyme and its specific reactivation by authentic FAD as a decisive proof of the involvement of this cofactor in the activity of MAO.

Further support for the assumption that mitochondrial MAO is a flavin dinucleotide enzyme has been recently provided by Yasunobu et al. (219) and Nara, Yasunobu, et al. (209, 220, 221). These workers found that purified beef-liver mitochondrial MAO preparations contained FAD covalently attached to the enzyme and suggested that the mitochondrial MAO holds 3 moles of FAD per mole of enzyme, based on the assumed molecular weight of MAO of 300,000. The finding that the beef-liver mitochondrial MAO is a flavoenzyme seems to be in agreement with observations of Hawkins (206), who reported a decreased MAO level in the liver of riboflavin-deficient rats. The results obtained by Yasunobu et al. (209, 219) also seem to explain in part the finding of Gorkin et al. (222) that proflavin is an inhibitor of the rat-liver MAO.

Extensive investigations of beef-liver mitochondrial MAO preparations by Yasunobu and his colleagues (209, 219–221) have demonstrated that this enzyme contains copper, mainly in the cupric state, which is essential for activity.

In their electron-paramagnetic-resonance (EPR) studies on beef-liver mitochondrial MAO Nara, Gomes, and Yasunobu (220) were able to demonstrate that the EPR spectrum of the latter enzyme differed considerably from that of hog-kidney diamine oxidase (DAO) (223) and of plasma amine oxidase (224; also see Chapter IV.1.D). Whereas in the very similar EPR spectra of the DAO and beef-plasma amine oxidase the copper bonds show considerable ionic character, the EPR spectrum of beef-liver mitochondrial MAO is quite different. It resembles more that reported for cytochrome oxidase (225), in which the copper-protein bond is covalent.

Further differences in the copper-protein bonds of the beef-liver and beef-plasma amine oxidases are indicated by the fact that cuprizone is a competitive inhibitor of the plasma amine oxidase, but a mixed inhibitor (i.e., partially competitive and partially noncompetitive) of the beef-liver enzyme (220).

McEwen (226) has reported that cuprizone is also a mixed inhibitor of human plasma amine oxidase.

Yasunobu et al. (209, 219–221) make the interesting observation that, in contrast to the DAO type of amine oxidases, which are claimed to be copper–pyridoxal phosphate dependent, the beef-liver mitochondrial MAO seems to be so far the only known example of an enzyme requiring for its activity both copper and flavin as prosthetic groups.

Gabay and Valcourt (227), who obtained from rabbit liver a highly purified mitochondrial MAO preparation, studied the effects of metal chelating agents on this enzyme and suggested that Cu^{2+} or other metal ions may be involved in the enzyme function. However, these authors thought that is was not necessary to accept with certainty that a divalent metal ion is a functional component of the catalytically active center, since the metal ion may be involved in the maintenance of an effective structure of the enzyme, as is the case with many oxidative enzymes (228, 229).

In contrast, to the above findings Erwin and Hellerman (210, 211) and Lagnado and Sourkes (204) failed to detect the presence of significant concentrations of copper in mitochondrial preparations of bovine-kidney and rat-liver MAO, respectively; this observation may indicate that copper is-not required for MAO activity. Similarly, Tipton (214) obtained very low copper values for purified hog-brain MAO preparations, which seemed to suggest that there were considerably fewer copper atoms in these preparations than there were molecules of enzyme. Moreover, Cu^{2+} was found to be unable to reactivate MAO. The latter finding is in contrast to work with rat-liver mitochondria by Coq and Baron (230) in which the inhibition of MAO by prolonged dialysis was shown to be reversed by the addition of Cu^{2+}. More fundamental work is required to elucidate the function of copper in the activity of amine oxidases.

Friedenwald and Herrmann (231) were the first to suggest that MAO possessed a sulfhydryl group, essential for its activity, the SH group being possibly involved in an electron transfer. The enzyme was inhibited by p-chloromercuribenzoate, but a complete reversal of inhibition was obtained by adding glutathione or cysteine in the presence of cyanide. Later work (232) showed that cyanide had reacted with the p-chloromercuribenzoate used in the experiment, and this apparently accounted for the complete reactivation. Since glutathione reactivated the enzyme also in the absence of cyanide and since MAO was found to be inhibited also by organoarsenicals as well as by iodoacetate, it was concluded that a sulfhydryl group was essential for its activity (233). The inhibition of MAO activity by mercaptide-forming reagents was confirmed in experiments with partially purified enzyme preparations (234–237).

Nara and his colleagues (220, 221) have found that their purified beef-liver mitochondrial MAO preparations contained sulfhydryl groups that are not vicinal. Gabay and Valcourt (227) have recently reported that their highly purified rabbit-liver mitochondrial MAO preparations were very sensitive to the action of p-chloromercuribenzoate. They obtained an almost 100% inhibition of the enzyme by 10 μM concentrations of this compound. This result is in marked contrast to the findings of Nara et al. (220), who observed an inhibition of only 37.5% by 5 mM p-chloromercuribenzoate. According to Gabay and Valcourt (227), this difference may be due to the fact that their rabbit-liver preparation was subjected to repeated column chromatography. Gorkin (237) has suggested that repeated application of column chromatography causes a distinct increase in the sensitivity of highly purified MAO preparations.

MAO has been found to be strongly inhibited by the chelating agent cysteamine (238). This reversible inhibition can be avoided by a pre-incubation of MAO with bivalent metal ions, (e.g., Co^{2+}). Moreover, when tyramine is used as the substrate, the irreversible inhibition of MAO by iproniazid can be prevented by pretreatment of the enzyme with cysteamine (238). It has been suggested that in the action of cysteamine on MAO an interaction with a metal ion on the active site of MAO may take place (239).

That metal ions may indeed be involved in the structure of mitochondrial MAO has been frequently suggested (220, 234, 239–241).

D. Substrate Specificity and Active Center

Primary and secondary amines are readily oxidized by MAO provided that the substituent in the secondary amine is a methyl group (128). The ability to act on methylated amines can be used as a criterion for the presence of a "true" MAO (128). The rate of oxidation of secondary (i.e., methylamine) derivatives is usually very high, sometimes even higher than that of primary amines. The tertiary amines are oxidized more slowly (242), the oxidation rate being higher in some species than in others (243).

The MAO substrates comprise aliphatic amines and the more important naturally occurring amines, carrying cyclic substituents (128). In the homologous series of normal primary aliphatic amines, $CH_3(CH_2)_nNH_2$, maximal rates are achieved with amylamine and hexylamine. The lower oxidation rates of amines with longer chains is probably due to their low solubility and their tendency to form micelles. MAO does not act on methylamine but in some species ethylamine is slowly oxidized.

Whereas the typical substrates of DAO, the tetra- and pentamethylene diamines (putrescine and cadaverine) are not attacked by MAO, long-chain diamines from heptamethylene diamine onward are oxidized by this enzyme. The diamine most rapidly attacked by MAO is the compound with 13 methylene groups (123). Competition experiments have shown that not only does the rate of oxidation of the diamines increase with the number of methylene groups but also their affinity for MAO. The reason for this behavior may be that in the short-chain members of this series the second amino group probably interferes with the attachment of the amine to the enzyme. With the increasing intramolecular distance between the two amino groups the disturbing action of the second amino group on the enzyme–substrate reaction becomes less effective (128, 136, 162).

A series of amines that are substrates of MAO is arranged according to the decreasing rate of the enzymic reaction (244) as follows: dopamine, tyramine, 3-methoxytyramine, tryptamine, 5-hydroxytryptamine, phenylethylamine, normetanephrine, metanephrine, epinephrine, norepinephrine, 1,4-methylhistamine, kynuramine. Aromatic amines, such as aniline, with the amino group directly attached to the benzene ring are not attacked by MAO (123). The reaction rate increases with the number of methylene groups in the side chain (163).

On introduction of a methyl group in the α-position to the amino group of an aliphatic or aromatic amine, the ability to serve as a substrate of MAO is lost (164). With an increasing number of methoxy substituents in the benzene ring the affinity of such compounds for MAO decreases. Thus, mescaline [β-(3,4,5-trimethoxyphenyl) ethylamine] is not a substrate of classical MAO (137, 139, 155, 164).

The active center of an enzyme is generally defined as that region of the enzyme surface in which the binding, interaction, and chemical alteration of reactants takes place (245). It may consist of several specific sites—such as substrate site, enzyme site, and activator site. An active site may be very small and simple or it may comprise a large region of the enzyme, itself containing many interacting groups and multiple binding points. Three main types of active sites are known: (a) a site consisting only of polypeptide chains, (b) a site consisting of a complex of a metal ion with protein, and, (c) a site consisting of a nonpeptide prosthetic group firmly bound to the protein. Enzymic catalysis is considered to be most intimately related to the polypeptide composition of the active center (245).

The active region is usually only a small part of the total enzyme protein; it has been reported that the remainder of the protein is without importance in enzymic catalysis (56; see also Chapter II.1.B). On the basis of irradiation experiments on trypsin Augenstine (246) concluded that the active surface area was 2.5% of the total trypsin area. However, in living cells it may be that few enzymes are free in solution and much of the protein may be involved in the organization of the enzymes into functionally efficient structures. The noncatalytic part of the enzyme could then be important in the operation of more complex metabolic sequences. Moreover, those regions of the enzyme that surround or underlie the active center, although not directly involved in the reaction, may be able to influence the activity of the center through steric or electrostatic effects. Nonetheless, the total protein is not always essential for activity, since active fragments have been obtained from such enzymes as aldolase, papain, ribonuclease, trypsin, and pepsin. Likewise, the intact structure of the total protein does not seem to be of essential importance since cross-linkages between polypeptide chains, such as pepsin S—S bridges, have been broken without loss of activity (245). The most extensive degradation of an enzyme without loss of activity was reported by Hill and Smith (247), who removed two-thirds of the 180 amino acid residues from papain by means of leucine aminopeptidase.

The presence of certain amino acids at an active site can be demonstrated by various techniques, such as the reaction of amino acid side chains with group-specific reagents, whereby a reduction or loss of activity would indicate, but not definitely prove, that the particular group is associated with the active site; the isolation of amino acid or peptide fragments containing a stable inhibitor after subjecting the enzyme-inhibitor complex to proteolysis; and the determination of the ionization constants of the active site groups from the variation of enzyme activity with pH (6, 245).

It is probable that histidine is an important constituent of the active site of many enzymes, since it may function efficiently in acid-base catalysis, for its pK_a is near neutrality (245). Indirect evidence for the participation of histidine in enzyme catalysis comes from studies on the photo-oxidation of the imidazole ring in enzymes with visible light in the presence of methylene blue. This treatment selectively destroys the histidine residues and causes a progressive fall in enzyme activity (248, 249).

Some groups involved in catalysis may be present at the active center in especially reactive states. Thus, the SH group at the substrate site of papain appears to be not an ordinary SH group but one that is thermodynamically at a higher energy level and kinetically more reactive (250).

At this point it should be emphasized that the active site includes the water molecules that are associated with the protein groups and the atmosphere of ions that surrounds the charged groups on the site. During the reaction of a substrate or an inhibitor with the active site, water as well as ions must often be displaced, and thus contribute to the final energy of interaction and the enzyme kinetics. If water and ions were not displaced, they would cause a modification of the configuration of the active site and the electrical fields of the protein groups (245).

In his studies on the active center of MAO Zeller (251–253) pointed out the existence of a striking structural analogy between the molecules of the optimal substrates of MAO, for example, phenylethylamine (I) and of its most active inhibitors [e.g., trans-2-phenylcyclopropylamine (II)]. Accordingly, the most suitable substrate or inhibitor of MAO is characterized by a two-carbon chain, containing an α-hydrogen atom and an α-amino group, and substituted in the β-position by an aromatic ring.

Whereas the MAO substrate (I) contains a second α-hydrogen atom in its side chain, in the MAO inhibitor molecule (II) the second α-hydrogen atom is substituted by an alkyl group.

Zeller (253) suggests that one α-hydrogen is essential for establishing the enzyme–substrate complex or the enzyme–inhibitor complex, whereas the second α-hydrogen of the substrate participates in the process of dehydrogenation, according to the scheme:

$$\text{R–CH–NH}_2 + -\overset{..}{\text{X}}=\text{Y}- \rightarrow \overset{\text{HN–|H|}}{\underset{|}{\overset{|}{\underset{\text{H}}{\text{RC–|H|}}}}} \quad \overset{..}{\text{X}}- \overset{..}{\text{Y}},$$

where $-$X$-$Y is part of the active center, in which X serves as an acceptor for the α-carbon and Y for a proton derived from an α-hydrogen atom (251). The two hydrogen atoms |H| are being received by a hydrogen acceptor.

Many properties of the active center of amine oxidases have been studied by Zeller and his colleagues (159) with the help of specifically selected substrates and inhibitors. The active center of amine oxidases seems to extend over several Ångström units (159), and its structure may be asymmetrical, possibly because it consists partly of L-amino acids (254–257).

Since some amine oxidases such as beef-plasma spermine oxidase (143), hog-plasma benzylamine oxidase (146), and hog-kidney DAO (258) have now become available in crystalline form, it is hoped that future research work will be directed toward the complete elucidation of their three-dimensional structures by means of X-ray crystallography. At this point it should be recalled that, for example, the full identification of the active site in lysozyme was achieved mainly by crystallography without the aid of previous extensive modification studies (17–20; see also Chapter II.1.A).

E. Biogenic Monoamines

The identification of various metabolites of the catecholamines and the discovery of drugs that alter the synthesis, storage, release, metabolism, transport, and physiological activity of the catecholamines greatly advanced our knowledge of the biochemistry, physiology, and pharmacology of the adrenergic nervous system (259–266). Of the monoamines that are substrates of MAO (see Chapter III.1.D) the most interesting ones from the physiological point of view seem to be the biogenic amines that occur in brain such as the catecholamines, norepinephrine, and dopamine, and the indoleamine serotonin (267). In 1954 Vogt (268) was the first to report on the uneven regional distribution of norepinephrine and the striking effect on its concentrations in mammalian brain by certain drugs that affect behavior. These early findings stimulated great interest in the biochemistry and pharmacology of central catecholamines. Vogt measured catecholamines in cat and dog brain by bioassay and first demonstrated their comparatively high concentrations in specific anatomical areas. She found the highest concentrations of norepinephrine in the hypothalamus. This regional localization supports the now widely accepted concept that catecholamines are located in central neurons and may play a part in their functions. In contrast to norepinephrine, epinephrine is encountered in the brain only in a very low concentration. Epinephrine occurs mainly peripherally in the adrenal medulla (267). Montagu (269) and Weil-Malherbe and Bone (270) were the first to present evidence that L-dopamine is a normal constituent of the mammalian brain. Whereas large amounts of dopamine are found in the basal ganglia, it is present in only low concentrations in most other brain areas (267). The presence of serotonin in brain extracts was discovered by Twarog and Page in 1953 (271). Shortly afterward Amin, Crawford, and Gaddum (272) reported that the distribution of serotonin in the mammalian brain was very similar to that of norepinephrine (268). Serotonin has been found also to occur in high concentrations in various peripheral tissues (268, 273).

By means of the recently developed histochemical fluorescence technique (274–276), applied to studies on the regional distribution of biogenic amines in the brain, evidence has been obtained that the biogenic monoamines are stored in neuronal elements and not in perivascular nerves (277, 278).

Although in earlier work in the field of naturally occurring sympathomimetic amines it was suggested that norepinephrine may function as a chemical transmitter substance at the terminals of the peripheral sympathetic nervous system (273, 279), the role of this and other biogenic amines in the central nervous system is not yet clearly understood. It is believed that at the synapses the chemical transmitter brings about, on its release from the presynaptic nerve endings, changes in the post-synaptic neuronal membrane potential, thus generating a nerve impulse. Although it has been suggested that norepinephrine, dopamine, and serotonin may each function directly as a transmitter substance in the central nervous system (280), none of them has been identified so far with certainty as a chemical neurotransmitter. More work is required for the elucidation of the mechanism by which these and biochemically related compounds affect neuronal functions (281, 282).

An essential contribution to the understanding of the pharmacology of sympathomimetic amines was, however, provided by the theory of two different receptor types of adrenergic nerve endings proposed by Ahlquist (283–285). According to this theory, the adrengeric receptor may be characterized as the part of certain effector cells that enables those cells to detect epinephrine and related compounds and to respond to them (286). It is understood that the receptor can be described only in terms of effector response to the administration of drugs (286).

The idea of an adrenergic receptor was originally expressed by Dale (287), who observed that ergot alkaloids blocked the effects of epinephrine only partially. Further studies (288) revealed the existence of two separate sets of responses to catecholamines. In 1948 the now widely used nomenclature for two different types of adrenergic receptors at smooth-muscle receptor sites was first proposed (283); for the sake of convenience these were termed adrenergic α-receptors and adrenergic β-receptors. Whereas the specific agonist for α-receptors is phenylephrine, l-m-hydroxy-α-[(methylamino)methyl]benzylalcohol (**III**), for β-receptors it is isoproterenol, 3,4-dihydroxy-α-(isopropylaminomethyl) benzylalcohol (**IV**) (286, 289).

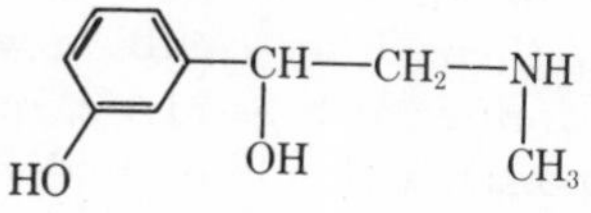

III

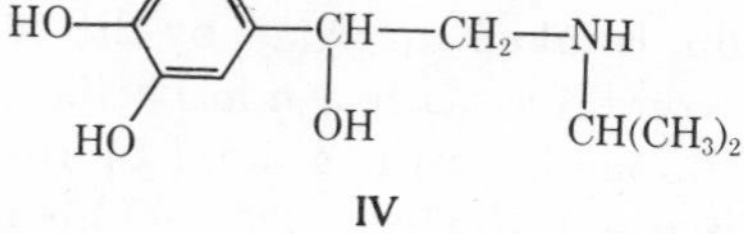

IV

Moran and Perkins (290, 291) were the first to use the terms "adrenergic α-receptor blocking agent" and "adrenergic β-receptor blocking agent." Substances that specifically block responses to phenylephrine, partially block responses to epinephrine, and have no effect on response to isoproterenol are termed adrenergic α-receptor blocking agents, whereas the term "adrenergic β-receptor blocking agents" is used for compounds that specifically block the responses to isoproterenol, impede in part responses to epinephrine, and have no effect on responses to phenylephrine (286).

In this context it should be mentioned that β-receptor blocking agents do not block adrenergic responses associated with α-receptors, nor do they block responses to such substances as digitalis, calcium ions, acetylcholine, serotonin, or histamine (286).

Recent observations in studies on intracellular localization of catecholamines and related metabolites in nerve tissue as well as on the functional correlation of catecholamines with the activity of nerve-end organs has greatly contributed toward our understanding of these biogenic amines (259, 292, 293). On the basis of currently available data (259, 261–266) Bloom and Giarman (281) have lately tried to elucidate the life cycle of a biogenic amine in brain in six steps, as follows: 1. The precursor amino acid (e.g., tyrosine or tryptophan) is transported across the bloodbrain barrier and presumably also across the neuronal cell membrane. This step is enzymically catalyzed by appropriate permeases.

2. Synthesis of the amine takes place in the neuron by a process involving hydroxylation and decarboxylation.

3. The amine formed is subsequently taken up by a membrane-bound granule and is stored in association with a material that serves as the binding substance. The amine in this form is inactive but seems to exist in dynamic equilibrium with other not yet identified pools of the amine in the neuron.

4. Physiologic release of the amine into an "active" form might then occur. The amine may then either combine with a protein that has a special affinity for the amine, such as MAO, or diffuse out of the cell.

5. A physiologic response is then initiated by the interaction of the amine with a specific receptor.

6. Dissociation of the amine from the receptor site takes place with termination of the physiologic response. The amine either succumbs to enzymic destruction, is taken up again by the cell, diffuses away from the receptor site, or combines with a nonspecific protein.

The formation of dopamine from L-dopa by the enzyme L-dopadecarboxylase has been discussed by Holtz (294). This reaction appears to be very efficient and proceeds rapidly. Although no strict correlation exists between the regional distribution of the L-dopadecarboxylase and the catecholamines in the brain, it is remarkable that the highest enzyme activity is to be found in the caudate nucleus, the hypothalamus, and the mesencephalon (170, 295, 296).

Norepinephrine may be synthesized from tyrosine through the intermediates dopa and dopamine (297) and may be stored within the nerve in the intraneuronal granules. These granules are observed by electron microscopy to occur at presynaptic endings and may, in response to nerve impulses, release their content into the synaptic cleft.

In this context studies on the rate of synthesis of dopamine and norepinephrine from the precursor amino acids both *in vitro* and *in vivo* should be mentioned. On incubation in a medium containing ^{14}C-L-tyrosine, large amounts of dopamine were formed by slices of caudate nucleus, but not in other areas of the cat brain (298, 299). Contrary to this observation, little norepinephrine (299) or none at all (298) was formed in the caudate nucleus from ^{14}C-L-tyrosine *in vitro*. Glowinski and his colleagues (300) obtained the same result with *in vivo* experiments. These authors were able to demonstrate that ^{3}H-dopamine introduced into the lateral ventricle of the rat was readily taken up by the caudate nucleus and other brain areas. Whereas 30 to 50% of ^{3}H-dopamine was converted to ^{3}H-norepinephrine in most of the brain areas, only small amounts of ^{3}H-norepinephrine were found in the caudate nucleus. These studies seem to indicate that in the striatum the metabolism of norepinephrine shows quite different biochemical characteristics from those of dopamine.

It has been suggested by Hornykiewicz (297) that in the striatum dopamine and norepinephrine may be localized in separate structures. At this point it should be recalled that in 1957 Blaschko (301) suggested that, besides its role as the immediate precursor of norepinephrine in the body, dopamine might have a physiological function of its own. At the same time Carlsson and his associates (302) envisaged the possibility of central actions of dopamine. Further investigations revealed that dopamine may be involved in the proper functioning of certain extrapyramidal centers of the brain (297). Thus, in patients with Parkinson's disease and those with postencephalitic parkinsonism a dopamine deficiency was demonstrated in extrapyramidal centers (303, 304). Patients with Parkinson's disease were also found to excrete less dopamine in the urine than normal subjects (305). Furthermore, a significant observation was made in a patient with hemiparkinsonism. The dopamine depletion was found to be more pronounced in the striatum contralateral to the side of symptoms (306). All these findings seem to support the idea of a functional role for dopamine (see Section 1.F).

Epinephrine is synthesized from norepinephrine in the adrenal medulla by enzymic methylation of the amine nitrogen. The activity of the enzyme involved in this synthesis, the phenylethanolamine-N-methyltransferase, seems to be potentiated by adrenocortical steroids (307). The blood-brain barrier effectively prevents the entry of epinephrine into the brain, except possibly in the region of the hypothalamus (308).

Serotonin is synthesized from the precursor amino acid 5-hydroxytryptophan by decarboxylation and may exist, like norepinephrine, in the neuron in a bound as well as in a free form. Serotonin is metabolized by MAO with the formation of 5-hydroxyindoleacetic acid (267).

Exogenous catecholamines accumulate in catecholamine-containing neurons by a mechanism commonly called uptake (263). In the first stage of this process the catecholamines are transferred across cell membranes into the intraneuronal space by active transport (309, 310). In the second stage intraneuronal redistribution and retention of amines in specific intracellular storage sites takes place (263). The accumulation of exogenous norepinephrine, administered into the peripheral circulation, occurs only to a very small extent in the brain and only in parts with a minimal blood-brain barrier for catecholamines, such as the area postrema and parts of the hypothalamus (311, 312). The problem of the blood-brain barrier has been solved by the direct introduction of radioactive catecholamines into the brain (263). In many studies it has been shown that radioactive catecholamines introduced into the cerebrospinal fluid accumulate in the brain (313–315). The degree of concentration of ^{3}H-norepinephrine or ^{3}H-dopamine paralleled the concentration of endogenous catecholamines in various areas of the brain, the greatest accumulation being found in the hypothalamus and the smallest in the cerebral cortex and in the cerebellum (316). Since high concentrations of radioactive norepinephrine were encountered in the corpus striatum, which contains large amounts of dopamine but very little norepinephrine, it seems that exogenous norepinephrine may accumulate in dopamine-containing neurons as well as in norepinephrine neurons (316). On application of ^{3}H-dopamine, the greatest accumulation of this amine occurred in the corpus striatum where dopamine neurons predominate. In addition there was a marked concentration of endogenous radioactive norepinephrine in various parts of the brain, especially in the hypothalamus and medulla oblongata (316).

Iversen and Simmonds (282) have recently suggested that the storage level of norepinephrine in adrenergic neurons may be maintained at a relatively constant value by a well-regulated balance between the rates of utilization of this amine and its biosynthesis. Glowinski and Axelrod (317) introduced a convenient procedure for the study of rates of norepinephrine turnover in rat brain. When small quantities of $(\pm)$-^{3}H-norepinephrine of high specific activity (2–10 C/millimole) are injected into cerebrospinal fluid by the intraventricular or intracisternal routes, 10–20% of the injected dose is selectively accumulated and retained in central catecholamine-containing neurons (317). When measurements are taken 2–8 hr after such injections, tritiated norephinephrine in the brain has been found to decline in a simple exponential manner (318). Since the level of endogenous norepinephrine in the brain remains constant during this period, the disappearance of labeled amine reflects the rate of amine utilization and replacement by newly synthesized, non-labeled norepinephrine. Determinations of the rate of decline of specific activity of tritiated norepinephrine can therefore be used for the assessment of the rate of turnover of norepinephrine in the brain or in discrete areas of the brain.

In 1963 Feldberg and Myers (319) put forward a new theory about the mechanisms involved in the regulation of body temperature by the hypothalamus. On the basis of results obtained in simple pharmacological experiments on cats these authors suggested that specific systems of cathecholamine-containing neurons in the hypothalamus are involved in temperature regulation, changes in body temperature being brought about by the release in the hypothalamus of the amines norepinephrine, epinephrine, and serotonin (319–321). In recent studies (282, 322) Iversen and Simmonds investigated the turnover rate of norepinephrine in the hypothalamus of rats that had been exposed to various environmental temperatures. Preliminary results obtained showed that exposure of animals to an environmental temperature of 32° during the period after administration of ^{3}H-norepinephrine caused a rise of 3.2° in rectal temperature and a threefold increase in the rate of norepinephrine turnover in the hypothalamus, whereas the rest of the brain was not significantly affected. On the other hand, exposure to 9° did not produce any significant change in the rate of norepinephrine turnover in the hypothalamus or other brain areas, although the rectal temperature fell by 1°. These findings appear to be consistent with Feldberg's theory (319–321) of temperature regulation, mentioned above. The results of Iversen and his co-workers (282, 322) also agree with those obtained by Feldberg and Lotti (320), which indicate that hypothalamic adrenergic systems of the rat brain are chiefly involved in mechanisms of heat loss. Further experiments are in progress in Iversen's laboratory aiming at the localization and identification of the adrenergic thermoregulatory system within the hypothalamus (282).

In vitro and *in vivo* studies appear to have revealed that norepinephrine is actively accumulated and retained in brain tissue. Likewise, an active accumulation of dopamine seems to occur in brain (263). The uptake process may be an important central mechanism for the removal of physiologically released extraneuronal norepinephrine, as is the case in the peripheral sympathetic system. Although it seems to be difficult to study this reuptake process directly in the central nervous system, it has been shown that drugs that inhibit the uptake of exogenous norepinephrine decrease the accumulation of ^{3}H-norepinephrine formed in the brain from ^{3}H-dopamine (323). If the labeled norepinephrine is formed intraneuronally and is physiologically released, its decreased concentration reflects a blockage of the reuptake mechanism (263). Studies on the disappearance of radioactive norepinephrine seem to suggest that this amine is stored in brain mainly in a form with a rapid turnover and to a much lesser extent in a more tightly bound form with a slower turnover (263).

The distribution of the serotonin nerve terminals in the central nervous system, especially in the telencephalon and the diencephalon, is not known in as great detail as that of the catecholamine nerve terminals (324). No specific distribution pattern for serotonin nerve terminals in the hypothalamus could be obtained, since the histofluorescence method (274–276) is less sensitive for serotonin than for the catecholamines, the fluorescence yield for this amine being lower than that for catecholamines. Application of serotonin derivatives showing better fluorescence may help to overcome this problem.

Information is still sparse about the physiological regulation of norepinephrine synthesis in the central nervous system. The conversion of tyrosine to dopa is probably the rate-limiting step in the synthesis of norepinephrine in the brain as it is in the peripheral sympathetic system (325, 326). Since norepinephrine can inhibit *in vitro* tyrosine hydroxylase, the enzyme essential for the conversion of tyrosine to dopa (327), and since norepinephrine turnover is decreased in the presence of an MAO inhibitor, it has been suggested that norepinephrine may act as a modulator of its own synthesis by exerting a negative feedback effect at the enzymic step (328). Furthermore, neuronal activity in the brain may also influence the synthesis of norepinephrine. Thus, such activity induced in the brain by released norepinephrine or by drugs may exert a positive feedback effect on norepinephrine synthesis (263).

Various physiological and pharmacological aspects of the extensive literature on biogenic monoamine metabolism has been recently discussed in a very competent presentation by Bloom and Giarman (281), to which the reader is referred.

During recent years a significant contribution to elucidation of the possible relationship between the biogenic monoamines and affective state in man has come from studies of the change in metabolism of brain monoamines brought about by clinically active psychotropic drugs (267).

In 1956 the important discovery that reserpine (**V**), a rauwolfia alkaloid, depletes the brain monoamines was made by Holzbauer and Vogt (329) for norepinephrine and by Pletscher, Shore, and Brodie (330) for serotonin. Soon afterwards Carlsson and his colleagues (331) reported that reserpine also depletes the brain of its dopamine content and causes a deficiency of this amine in the striatum, substantia nigra, and pallidum. The release of the monoamines by reserpine is not confined to brain tissue, but occurs at all storage sites throughout the body. The effects of reserpine have been extensively studied in *in vivo* as well as in *in vitro* systems (329–333). By a not yet elucidated mechanism reserpine seems to interfere with the intraneuronal binding of the catecholamines and serotonin, permitting them to diffuse freely through the cytoplasm onto mitochondrial MAO (332). This may result in inactivation of these amines by oxidative deamination and, thus, in depletion of tissue amine stores (266). The present view of the mechanism of reserpine action is that it impairs the storage capacity of the tissues for the catecholamines and serotonin, but not their synthesis. Pirch, Reech, and Moore (334) have recently reported that in rats, depleted of their brain monoamines by reserpine, the restoration of serotonin proceeded more rapidly than that of norepinephrine.

V

Reserpine (**V**) induces sedation in animals, a state that some authors consider to be comparable to that of depression in man (335). Reserpine-induced sedation in animals is associated with decreased brain levels of norepinephrine, dopamine, and serotonin (336, 337).

A new theory with regard to the reserpine-induced release of monoamines from tissue stores has been lately advanced by Shore (338), according to whom reserpine competes with norepinephrine for a carrier mechanism, responsible for the uptake of the amine by the storage organelles. Under the action of MAO inhibitors the endogenous norepinephrine, no longer metabolised by MAO, displaces reserpine from this carrier mechanism and this process diminishes the monoamine depletion. This interesting hypothesis awaits confirmation.

F. Inhibitors

In his extremely competent work *Enzyme and Metabolic Inhibitors* Leyden Webb (245) points out on page one that "the application of enzyme inhibitors to cells or their aggregates is one of the ways in which the scientist attempts to penetrate into the nature of living material and its transformations." This approach implies the introduction of an interference into a complex system for the sake of a better understanding of the normal or original state.

Leyden Webb maintains that a normal state of a biological system should be regarded as a perfect balance of complex chemical and functional activities, the possibilities of life depending on a precise coordination of the rates of these various processes. The incorporation of an inhibitor into a biological system leads to an alteration of the rates of some of these processes and thus to a disturbance of the perfect balance.

Usually only the results of the imbalance induced by the inhibitor can be determined. Thus, if the inhibitor upsets the balance to the detriment of the organism, it is generally termed a poison. On the other hand, when the change of balance favors the survival of the total system, the inhibitor may be regarded as a drug.

The actions of inhibitors on living matter may be indeed manifold. Thus, the inhibitor may interfere with the entrance of the substrates into the cell by altering permeability or blocking transport mechanisms, so that the total potential energy available to the cell may be decreased. The inhibitor may also exert an effect on any of the metabolic pathways —such as the oxidative degradation of substrates, the formation of high-energy phosphate, the synthesis of protoplasmic components or the utilization of high-energy substances in synthetic processes (245). In regard to the functional activity of the cell, the inhibitor may act either on the reactions involved in the utilization of energy or it may affect directly the functional systems themselves. It should be noted that, since these intracellular processes are closely related, a primary inhibition of the oxidation of substrate, for example, may reduce to a varying extent the activity of reactions such as the formation of high-energy substances, synthesis, function, and appearance of metabolites. It should be remembered here that the actions of many inhibitors are not restricted to a single pathway, since inhibitors are seldom completely specific.

A useful approach to the problem of enzyme inhibition appears to have been made by the direct distortion of pathways of enzymic reactions brought about by the introduction of a chemical substance, blocking a pathway or diverting a sequence into a new channel, interfering with coupled reactions, or obstructing the accumulation of energy. This interesting technique may often induce a rather specific attack on some particular reaction or phase of metabolism; it may then often be possible to isolate from the cell the specific enzyme or enzyme system acted on by an inhibitor, and to study in detail the mechanism of the interference, and thus to gain some insight into the effects exerted by the inhibition within the cell. Leyden Webb (245) emphasizes that it is the relative specificity of enzyme inhibitors that turns them into one of the most powerful tools in many and diverse fields of biological research. With the discovery of new inhibitors undoubtedly a still greater specificity will be attained. The more our knowledge in regard to the exact nature of the effects of inhibitors will advance, and the more selective these effects become, the greater is the likelihood of achieving a clear understanding of the energetics of the cell.

In some animal species the MAO inhibitors produce both behavioral excitation and elevated levels of norepinephrine and serotonin. This elevation of monoamine levels in the brain is probably caused by impairment by the enzyme inhibitor of normal metabolic deamination (267). There has been some controversy over attempts to relate the behavioral excitation to increased concentrations of a specific amine (339). Spector and his colleagues (340, 341), when studying the effects of various inhibitors, doses, and animal species were able to demonstrate that in those species in which MAO inhibitors increased the levels of norepinephrine and serotonin, the observed behavioral excitation was temporally correlated with the increase in levels of norepinephrine, whereas in species in which there was an increase in serotonin without an increase in norepinephrine, no behavioral excitation was observed.

Since Zeller and his colleagues (142, 342) reported in 1952 on the strong *in vitro* and *in vivo* inhibition of MAO activity by the antituberculosis drug iproniazid, the list of compounds synthesized and tested as potential MAO inhibitors has grown enormously. These compounds have provided a stimulus for an extraordinarily large variety of studies in the fields of biochemistry, pharmacology, psychopharmacology, and clinical research and therapy.

Aspects of both the *in vitro* and the *in vivo* inhibition of MAO activity have been well documented in many recent reviews and symposia to which the reader may refer (125, 130, 244, 266, 267, 281, 343–348).

Werle (130) distinguishes two main classes of MAO inhibitors:

1. The group of competitive inhibitors, which comprises hydrazine derivatives, members of the β-phenylisopropylamine series, the harmala alkaloids, and choline-p-tolyl ether.

2. The group of noncompetitive inhibitors including the amidines.

It should be pointed out that the *in vitro* effect of some of those MAO inhibitors often does not parallel their *in vivo* action (349, 350).

a. COMPETITIVE MAO INHIBITORS

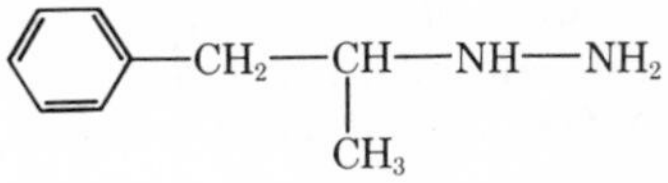

VI

Pheniprazine, β-phenylisopropylhydrazine **(VI)** inhibits MAO *in vitro* about 50 times as strongly as iproniazid. It is also a potent *in vivo* MAO inhibitor and exerts an extremely prolonged effect (351, 352). In the intact animal this drug shows a higher selectivity for the brain than for the liver (352). This may be due to the presence of the benzene ring in its structure, since isopropylhydrazine does not display this selectivity for the brain MAO. Phenylisopropylhydrazine shows stereospecificity, the D-derivative having a stronger inhibitory effect on MAO than its antipode.

Iproniazid, 1-isonicotinyl-2-isopropylhydrazine **(VII)** is the best studied

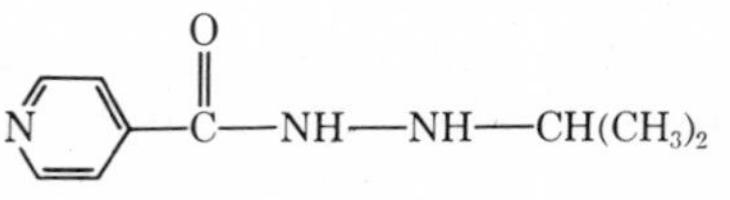

VII

inhibitor of MAO. It belongs to the group of hydrazine derivatives, which strongly inhibit MAO *in vitro* as well as *in vivo*. A 10^{-4} M concentration of iproniazid produces a 90 to 100% inhibition of mitochondrial MAO activity in mouse, rabbit, cat, dog, beef, and human kidney, liver, and brain (160). It seems likely that *in vivo* iproniazid is first hydrolyzed to yield isonicotinic acid and isopropylhydrazine, the latter compound being a stronger inhibitor of MAO than iproniazid (128, 353). On incubation with erythrocytes, iproniazid loses its ability to inactivate MAO. This may possibly be due to the presence in the erythrocyte membrane of an inactivator of iproniazid (354).

Iproniazid is a therapeutic agent with a rather broad spectrum of activity. It has been applied in the treatment of depression, hypertension, and angina pectoris. It is noteworthy that only hydrazine derivatives that exert a marked MAO inhibition *in vivo* have a psychostimulant, hypotensive, and antianginal effect (346). Zeller and his collegues have observed that iproniazid, when administered to schizophrenics, caused simultaneous changes in the behavior of those patients as well as in their serotonin metabolism (355, 356).

Iproniazid has also been instrumental in the discovery that octopamine, first found by Erspamer (357) in the posterior salivary glands of *Octopus vulgaris,* is a normal constituent of mammalian urine and is also present in various mammalian tissues, including the heart and the brain (358, 359). When MAO was blocked with iproniazid or pheniprazine, octopamine was found to be excreted in the urine of rats, rabbits, and man instead of *p*-hydroxymandelic acid, which normally occurs in mammalian urine (358). Apparently, octopamine is so rapidly oxidized by MAO that it has escaped detection until the enzyme could be successfully blocked. In this connection it should be recalled that Sjoerdsma (360) in his study of human hypertensive subjects has found a consistent relationship between MAO inhibition and orthostatic hypotension. The orthostatic hypotensive effects of MAO inhibitors in man have long been considered paradoxical, since on administration of iproniazid an increase in the levels of pressor amines has been demonstrated in the brain (361). Several theories have been advanced for the explanation of the orthostatic hypotensive effect of MAO inhibitors. One of them, proposed by Sjoerdsma (360), suggests that a weak pressor substance, for example, dopamine, might accumulate at nerve endings, and completely block the pressor action of norepinephrine, which is released neurogenically, when the patient stands up. In their study on the mechanism of sympathetic blockage by MAO inhibitors Kopin and his colleagues (362) have indeed obtained significant results, according to which the weak pressor amine appears to be octopamine, the formation of which by β-hydroxylation of tyramine is said to be enhanced by MAO inhibition. Kopin et al. (362) suggest that the "partial sympathetic blockade is produced as a consequence of the replacement of norepinephrine by octopamine which acts as a false inactive neurotransmitter." This suggestion, however, should be confirmed (363).

Furthermore, iproniazid has helped to throw more light on the mechanism of reserpine, a rauwolfia alkaloid, known to deplete the brain and other tissue stores of serotonin and catecholamines (332; see Section 1.E.). It was shown in many studies that iproniazid not only counteracted the reserpine-induced sedation and elicited marked cerebral stimulation, but also prevented the usual rapid decrease of biogenic amines in various tissue stores (244).

Iproniazid has also proved to be useful in investigations on the mechanism of chlorpromazine, a drug with an apparently second-order action on amines, due to its effects on the membrane permeability of cells and storage granules.

Chlorpromazine, 2-chloro-10(3-dimethylaminopropyl)phenothiazine **(VIII),** produces a striking syndrome of tranquilization without altering

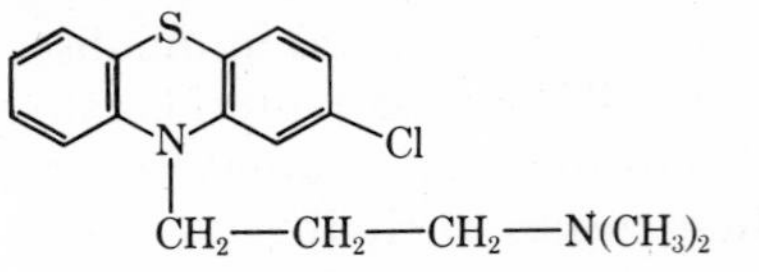

$$CH_2—CH_2—CH_2—N(CH_3)_2$$

VIII

the concentration of brain monoamines (364). This observation seems to suggest that the release of a central neurotransmitter may be inhibited by this drug or that interference with postsynaptic receptors may occur (263). The pharmacological and biochemical effects of chlorpromazine appear to be rather complex (365, 366). The effects produced by chlorpromazine concern increases of catecholamine metabolites in the brain (365–368), inhibition of the amine depletion caused by reserpine (364, 369–371), and the increase of monoamines brought about by MAO inhibitors (364, 372). Most of these effects seem to be at least partly related to the hypothermia induced by chlorpromazine (365, 366, 371, 372).

Extensive studies by many workers reveal that chlorpromazine decreases membrane permeability to amines at various cellular and subcellular sites (364–366). Furthermore, it inhibits the accumulation *in vivo* of exogenous norepinephrine in peripheral tissues (373), and of exogeneous dopa and serotonin in the brain (366). *In vitro,* chlorpromazine prevents the accumulation of exogenous norepinephrine in brain slices (374), and in adrenal granules (375, 376).

It should be recalled here that many biogenic amines—unlike their precursors, the corresponding amino acids—do not pass easily the lipoidal layers of the cell membrane, probably because of their distinctly ionic nature (344).

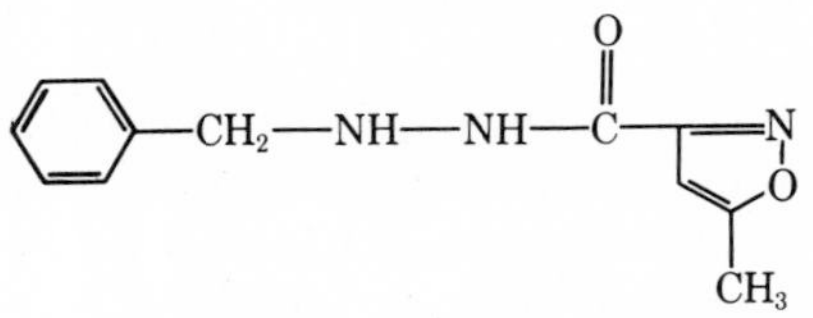

$$—CH_2—NH—NH—C—$$

IX

Isocarboxazid, 1-benzyl-2-(5-methyl-3-isoxazolylcarbonyl)-hydrazine (IX), is *in vitro* as well as *in vivo* a much stronger inhibitor than iproniazid and is less toxic than the latter (244, 361). This compound is known to be rapidly degraded. Thus, 70% of the drug is excreted in the urine in the form of hippuric acid eight hours after its administration (244).

This drug has been therapeutically used in Parkinson's disease. Since it has been established that the latter condition is characterized by a subnormal excretion of dopamine in the urine (305) as well as by a decreased content of dopamine in the caudate nucleus and putamen (303; see Section 1.E), attempts have been made to counteract this deficiency (377). When parkinsonism patients were given large doses of L-dopa, akinesis disappeared, partly or completely, for a few hours. This period of improvement could be prolonged by a simultaneous administration of L-dopa and the MAO inhibitor isocarboxazide (377).

Reports in the literature on the value of L-dopa as a therapeutic agent in the management of parkinsonism have been rather varying. This may have been due to the fact that many authors have used DL-dopa instead of the L-isomer, and some others used L-dopa in inadequate dosages. Two recent independent reports, however, have fully vindicated the role of L-dopa as an important therapeutic agent in the treatment of patients suffering from Parkinson's disease. Cotzias and his colleagues (378) have recently demonstrated that many patients with parkinsonism may be substantially improved by the administration of L-dopa. The maximum therapeutic effect and the maximum tolerated dose varied from individual to individual, but normally fell within the range of 3–16 g daily.

Calne and his associates (379) have lately reported on the very successful treatment of patients with postencephalitic parkinsonism with oral L-dopa administered over a period of six weeks in maximum tolerated doses. They found that akinesia may be dramatically relieved by the use of L-dopa. There was also general improvement in posture, gait, and facial expression. On the other hand, such features of parkinsonism as tremor were little improved by L-dopa.

$$\text{C}_6\text{H}_5\text{—CH}_2\text{—}\underset{\underset{\text{CH}_3}{|}}{\text{CH}}\text{—NH}_2$$

X

$$\text{C}_6\text{H}_5\text{—}\underset{\underset{\text{OH}}{|}}{\text{CH}}\text{—}\underset{\underset{\text{CH}_3}{|}}{\text{CH}}\text{—NH—CH}_3$$

XI

The α-alkylated arylalkylamines, amphetamine, β-phenylisopropylamine (**X**), and ephedrine, L-α(1-methylaminoethyl)benzylalcohol (**XI**), are both competitive inhibitors of MAO, but their action is readily reversible. Both substances are used therapeutically (128). Of the two stereoisomers of amphetamine, the D-form is twice as effective in inhibiting MAO activity in the human brain, and the parotid and submaxillary glands as the L-isomer (380). Ephedrine is a weaker MAO inhibitor than amphetamine because of the β-hydroxy group in its side chain (123).

Amphetamine is a short-acting sympathomimetic psychic stimulant that has been used for many years with indifferent results in the treatment of depression (381). There is evidence that amphetamine may both release physiologically active norepinephrine from nerve cells, and also block the inactivation of norepinephrine by cellular reuptake (382, 383). In their studies on the metabolism of ^{3}H-norepinephrine in the brain, Glowinski and Axelrod (382) have found that amphetamine increases the concentrations of ^{3}H-normetanephrine and decreases the content of deaminated, tritiated catechols (see Section 1.I). Many of the behavioral effects of amphetamine are potentiated by imipramine (**XIX**) (384).

Large doses of amphetamine significantly lower the concentration of brain norepinephrine in animals (385–387), but they exert only small and variable effects on dopamine and serotonin concentrations (386), 387). Amphetamine also accentuates the decrease in brain norepinephrine brought about by stress (388). Whereas many findings seem to suggest that amphetamine may release norepinephrine and thus potentiate the effects of this catecholamine at receptor sites, there is evidence that amphetamine may exert also a direct action at receptors (389, 390).

Pargyline (N-methyl-N-2-propynylbenzylamine, (**XII**), phenelzine (β-phenylethylhydrazine, (**XIII**), and tranylcypromine (2-phenylcyclopropyl-amine, (**XIV**), are also potent MAO inhibitors.

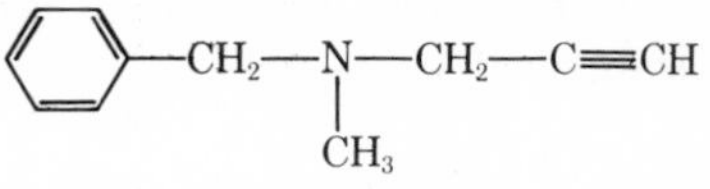

XII

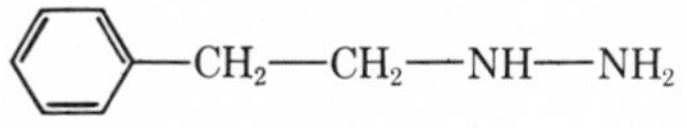

XIII

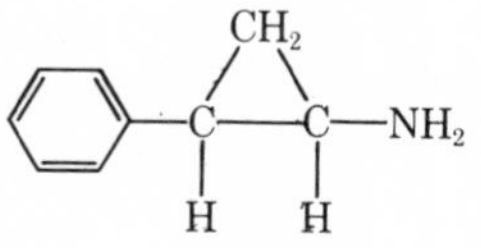

XIV

Tranylcypromine inhibits *in vitro* and *in vivo* much more strongly than does iproniazid. The *trans* form of 2-phenylcyclopropylamine is a more potent inhibitor of MAO than the *cis* form. Both stereoisomers inhibit the beef-liver mitochondrial MAO to a much larger extent than they do that of rabbit liver (251).

Choline-*p*-tolyl ether (**XV**) is a potent MAO inhibitor. At concentrations of 10^{-3} M this substance completely inhibits MAO (391).

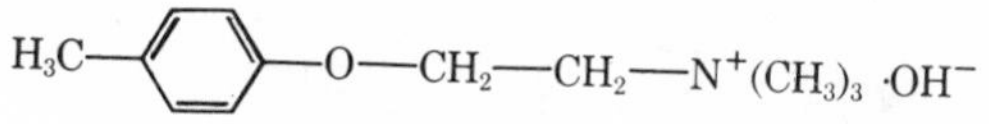
H3C—
—O—CH2—CH2—N+(CH3)3 ·OH−

XV

Another group of very strong MAO inhibitors is that of the harmala alkaloids; harmaline (**XVI**) is the most potent inhibitor of this series (392).

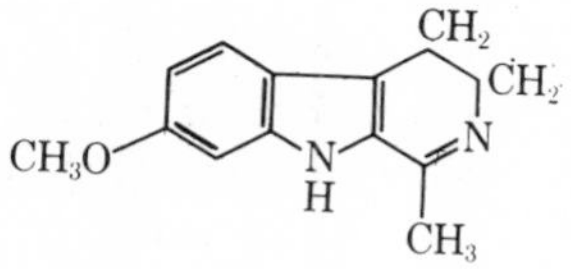
CH2
CH2
N
CH3O
N
H
CH3

XVI

Its *in vitro* effect is reversible, and *in vivo* it is a fast-acting drug (244). Applied *in vivo,* harmaline protects an animal against the actions of some of the long-acting MAO inhibitors (393).

b. NONCOMPETITIVE MAO INHIBITORS

This group comprises amidines, monoisothiourea derivatives, and diisothiourea derivatives.

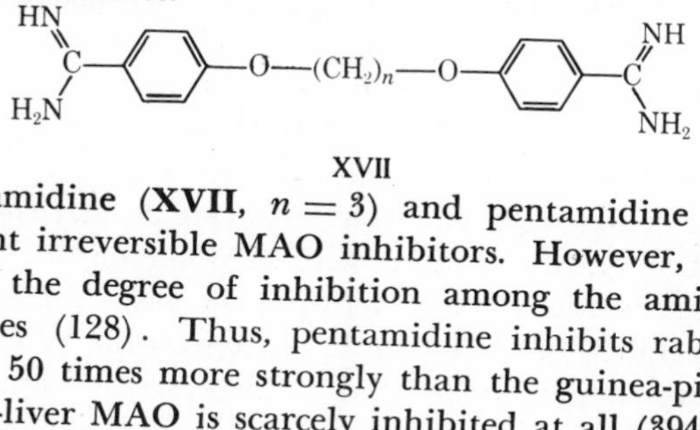
HN
C
H2N
—O—(CH2)n—O—
NH
C
NH2

XVII

Both propamidine (**XVII**, $n = 3$) and pentamidine (**XVII**, $n = 5$) are very potent irreversible MAO inhibitors. However, there are great differences in the degree of inhibition among the amine oxidases of different species (128). Thus, pentamidine inhibits rabbit-liver amine oxidase about 50 times more strongly than the guinea-pig-liver enzyme, whereas squid-liver MAO is scarcely inhibited at all (394). In a specific homologous series the length of the polymethylene chain seems to determine the magnitude of the inhibitory effect. The optimal chain length, however, differs from species to species.

In the homologous series of S-n-alkylisothioureas ($CH_3(CH_2)_nSC(=NH)NH_2$) the inhibitory effect of the short-chain members is feeble and easily reversed, but with increasing chain length the inhibitory action increases and gradually becomes irreversible (395). Again, in the same homologous series of inhibitors different degrees of reversibility may be encountered.

Methylene blue (**XVIII**) is a fast-acting, reversible MAO inhibitor (396). Octyl alcohol also actively inhibits MAO (123).

$(CH_3)_2N$ —— S^+ —— $N(CH_3)_2$ $\cdot Cl^-$

XVIII

G. Mechanism of MAO Inhibition

In his study on liver-mitochondrial MAO Davison (397) observed that the inhibition of this enzyme by iproniazid is progressive and first order, and requires the presence of oxygen. He therefore suggested that the first step in this inhibition reaction may be a dehydrogenation of iproniazid to 1-isopropylidene-2-isonicotinyl hydrazine in a way similar to that postulated for the oxidation of the substrate. As already mentioned, in all substrates of MAO the α-carbon atom must be unsubstituted. Amines in which one of the hydrogen atoms of the α-carbon atom is substituted are inhibitors of amine oxidase (256, 398, 399).

a. IN VITRO MAO INHIBITION

Zeller (353) has outlined a few factors of importance for the interpretation of results obtained in *in vitro* inhibition experiments with MAO. Thus, substrates protect MAO against the action of iproniazid, or of cis- and trans-phenylcyclopropylamine (160, 397) only when they are added simultaneously with the inhibitor. The higher the substrate concentration, the less effective is the blockage of MAO. These observations seem to suggest that iproniazid and other hydrazine derivatives are bound to that part of the active center with which the substrate usually combines. When the inhibitor is brought into contact with MAO before the addition of the substrates, the inhibition effect of the drug increases about tenfold. This reaction is independent of the substrate concentration. When the inhibitor reaches the active site, it apparently remains irreversibly attached to it. Accordingly, prolonged dialysis of the iproniazid-MAO complex restores only a small fraction of the original MAO activity (160, 251). The longer MAO remains exposed to the action of the inhibitory drug before the substrate is added, the greater (up to a point) is the effect of inhibition, maximal blocking being reached after a preincubation period of 8 to 13 min (160, 397). Since the substrates almost instantaneously form a complex with the active site of MAO, they are able to protect those enzyme molecules that have not been inactivated by iproniazid. In experiments on the competition between substrate and inhibitor the affinity of the substrate for MAO is the decisive factor. Thus, tryptamine is known to be more strongly bound by MAO than tyramine (400), and, accordingly, MAO was found to be four to six times more resistant to an inhibition by 2-phenylcyclopropylamine in the presence of tryptamine as substrate than in that of tyramine.

Similar observations were made by other workers (401) when α-methyltryptamine was used as an inhibitor. Determination of the inhibitory efficiency revealed that it increased tenfold when tryptamine was replaced by serotonin as substrate.

Green (239) has pointed out the existence of many striking similarities between the inhibition of MAO by hydrazine derivatives and the decomposition of the latter, catalyzed by cupric ions. Both processes require oxygen and both can be retarded or prevented by compounds capable of chelating with or being oxidized by cupric ions. Davison (402) was the first to notice a decrease in inhibition when MAO and one of its inhibitors, iproniazid or isopropylhydrazine, were incubated under nitrogen instead of oxygen. Furthermore, Gorkin has recently reported (238) that the chelating agent cysteamine prevents the irreversible inhibition of MAO by iproniazid when added before the inhibitor. On the basis of these and many similar observations Green (239) advanced the hypothesis that the inhibition of MAO by hydrazine derivatives may result from a copper-catalyzed liberation of free radicals near the active center of the enzyme with the copper constituting part of the enzyme. That metal ions may be involved in the structure of particles containing mitochondrial MAO has been suggested by Gorkin and his colleagues (200). More fundamental work is needed for the elucidation of the problem of MAO inhibition.

b. IN VIVO MAO INHIBITION

In animals and man all types of MAO inhibitors, regardless of their chemical constitution, cause some typical *in vivo* effects—such as increase of endogenous monoamines as well as of monoamines from exogenous sources, antagonism toward monoamine releasers, or change in the excretion pattern of monoamines and their metabolites (346). As mentioned above, MAO inhibitors exert a number of known pharmacological effects, such as psychostimulation, lowering of blood pressure, and antianginal action. The mechanisms by which these effects are induced require further elucidation.

An increase in the endogenous monoamine level of the brain and other tissues occurs only after an extensive inactivation of MAO (of at least 85%). Intensity, onset, and duration of the inhibitory effect will depend on the species, the tissue, the chemistry of the drug, the mode of administration, and the monoamine (346).

Imipramine and other related tricyclic derivatives have been found to be clinically the most effective of the antidepressant drugs (403, 404).

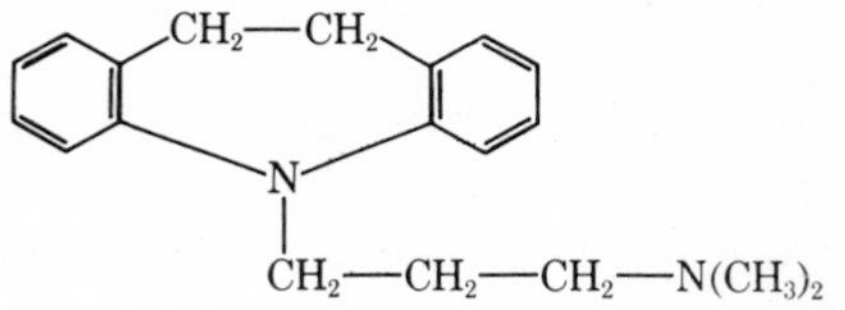

CH₂—CH₂ structure

XIX

Imipramine, 5-[3-(dimethylamino)propyl]-10,11-dihydro-dibenz[*b,f*]-aze-pine, (**XIX**) is an antidepressant. Its mode of action is not clearly understood. Thus, although it is not a stimulant like amphetamine nor does it inhibit MAO or catechol-O-methyltransferase, yet, in various experimental systems it has been found to potentiate both the response to sympathetic nerve stimulation and many of the peripheral effects of exogenous norepinephrine in animals and in man (403, 405, 406). Likewise, Sigg and his co-workers (405) have recently observed a potentiation by imipramine of the effects of serotonin in animals.

Hertting, Axelrod, and Whitby (373) have found that in animals imipramine interferes with the uptake into peripheral tissues of infused norepinephrine, and have suggested that imipramine may decrease the permeability of cell membrane or storage-granule membrane to norepinephrine. Thus, potentiation by imipramine of the effects of norepinephrine may result in part from impairment of the inactivation of free norepinephrine at the synapse by cellular reuptake. Such a mechanism would provide a basis for the theory of sensitization of central adrenergic synapses, advanced by Sigg (407) to account for the antidepressant action of imipramine. Recently, Glowinski and Axelrod (408) provided experimental evidence that imipramine inhibits norepinephrine uptake in the brain as it does in the peripheral tissues. They also found that the chemically related tricyclic antidepressants such as desmethylimipramine and amitriptyline also inhibited the uptake of norepinephrine in the brain (408).

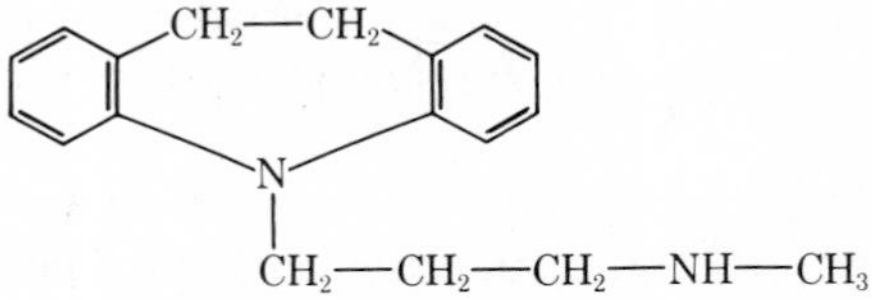

CH₂—CH₂ structure

XX

Desipramine, 5-(3-methylaminopropyl)-10,11-dihydro-dibenz[*b,f*]azepine (**XX**) is a demethylation product of imipramine and has been considered to be the pharmacologically active form of imipramine. However, this is not certain and desipramine is rather less active clinically than imipramine.

Amitriptyline hydrochloride, 5-(3-dimethylamino propylidene)-dibenzo [*a,d*][1,4]cycloheptadiene, (**XXI**) is widely used clinically.

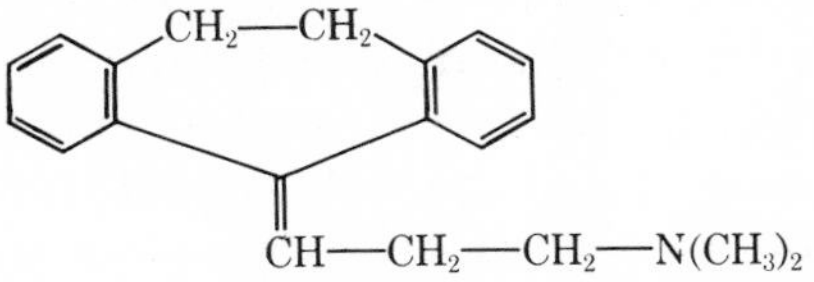

XXI

In this context it is interesting to note that Schanberg et al. (409), on intracisternal administration of ^{3}H-norepinephrine into the rat brain, found that imipramine and its derivative desmethylimipramine increased the brain concentrations of the O-methylated metabolite ^{3}H-normeta-nephrine, whereas chlorpromazine, a tranquillizer, did not. It has been suggested by various authors that physiologically active norepinephrine, which has interacted with receptors, may be inactivated by O-methyl-ation, but evidence supporting this interesting hypothesis is sparse (259, 410). Eisenfeld et al. (411) have observed that drugs which block adren-ergic receptors decrease the formation of normetanephrine in peripheral tissues.

At this point it should be noted that prior treatment with imipramine can prevent reserpine-induced sedation in animals. Sulser and his co-workers (412) found this phenomenon to depend on the availability and rate of release of catecholamines, since, when animals are only partially depleted of norepinephrine stores, prior treatment with imipramine does not prevent reserpine-induced sedation. This finding seems to be com-patible with the theory that imipramine may exert its antidepressant action through potentiation of the effects of catecholamines at adrenergic-receptor sites in the brain (335, 407, 412, 413).

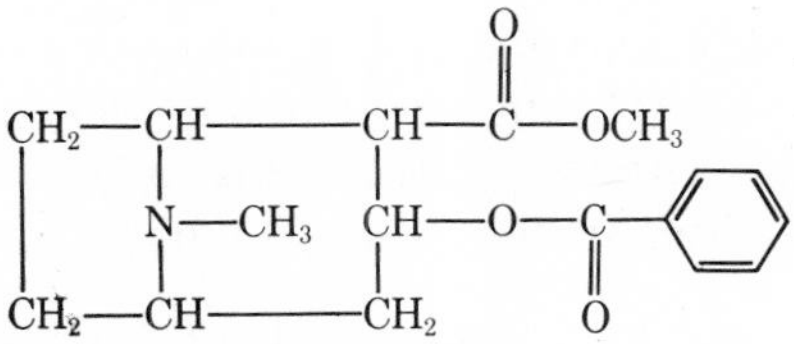

XXII

The alkaloid cocaine (**XXII**) is obtained from the leaves of a South American plant, *Erythroxylon coca*. Coca leaves have been used by the natives of Peru as a cerebral stimulant from time immemorial and were first introduced into Europe for this purpose.

Cocaine potentiates the action of epinephrine. It produces a marked vasoconstriction by a potentiation of norepinephrine released at sympathetic nerve endings. One explanation of the potentiation is that it is brought about by an inhibition of MAO; another explanation is that cocaine occupies the tissue binding sites for norepinephrine and in this way increases its concentration at the receptor sites. Cocaine has a powerful stimulant action on the brain (373, 414–416).

The MAO inhibitors affect mainly those amines that are good substrates of MAO. In this process the intraneuronal as well as the extraneuronal monoamines appear to be increased in such tissues as the heart and brain (346).

According to Ganrot and his colleagues (417), the psychostimulating effect of MAO inhibition may be indeed due to an increase of the monoamine content in the brain.

There has been recently (418) increasing evidence of many alarming reactions—for example, acute hypertension or hypotensive collapse in depressive patients treated with potent MAO inhibitors—on ingestion by those patients of ripened cheese, yeast extracts, or broad beans. These sympathomimetic effects, produced by cheese and other foods, are probably due to their high content of tyramine as well as to the almost complete blockage of oxidative deamination, induced by the potent MAO inhibitor used. The mechanism of this rather paradoxical reaction awaits clarification.

In this context it should be mentioned that in a very recent study (1968) Rand and Trinker (419) have provided evidence for their concept that MAO inhibitors enhance the pressor responses of the indirectly acting sympathomimetic amines, tyramine and amphetamine, not by interfering with the metabolism of endogenous norepinephrine, as suggested by Goldberg (420), Sjöqvist (421), and Smith (422), but by retarding the metabolism of tyramine and amphetamine in the liver. As is well known, the liver not only is a rich source of mitochondrial MAO (123, 128), but it also contains in its microsomes an enzyme system, quite distinct from MAO, that oxidatively deaminates amphetamine (423, 424). The latter enzyme system seems to be the principal route of amphetamine metabolism in the rabbit, monkey, and man (423, 424).

H. Metabolic Fate of MAO Inhibitors

It is obvious that on *in vivo* administration MAO inhibitors become involved in many reactions that influence their ultimate inhibitory action. Horita (425) investigated the influence of the route of administration of several hydrazine derivatives on rat brain and liver enzymes, and found distinctly different patterns of inhibition with different modes of administration. Whereas subcutaneous administration of those inhibitors resulted in a greater inhibition of brain MAO than of liver MAO, after oral administration the inhibition of the liver MAO was predominant. Free inhibitor molecules seem to be rapidly eliminated by oxidative processes (426), and this may explain why such reversible, fast-acting inhibitors as harmaline (244, 392, 393) and methylene blue (396) protect the organism against irreversible MAO inhibitors. Similar oxidative reactions may be responsible for the transformation of a weak *in vitro* inhibitor of MAO (e.g., 1-benzyl-2-isopropyl hydrazine) to a very potent *in vivo* agent (427). Removal of either the benzyl or the isopropyl residue apparently results in a powerful MAO inhibition.

In recent studies on the oxidative deamination of various biogenic monoamines by mitochondrial MAO, Gorkin (428) has been able to demonstrate that injections into rats of certain tricyclic dyes result in a selective *in vivo* inhibition of the deamination of various monoamines in tissues such as brain, liver, and heart. This author also found striking species differences in the sensitivity of mitochondrial MAO toward the inhibitory effects of harmine and its derivatives. Some of the potent MAO inhibitors used in clinical medicine, for example, propargyline and phenelzine, also cause a somewhat selective inhibition of deamination of various biogenic monoamines.

I. Biological Function of MAO

Oxidative deamination and O-methylation are the two enzymic pathways of catecholamine metabolism in the peripheral sympathetic system (429) and also occur in the central nervous system (162, 313, 318, 430, 431; see Section 1.A). MAO is found widely and rather evenly distributed in the central nervous system, with the hypothalamus showing somewhat greater MAO activity than other parts of the brain (170). Catechol-O-methyltransferase (COMT) is also widely distributed in brain tissue (432); its cofactor, S-adenosylmethionine, is formed in the brain (432, 433). MAO is located in brain mitochondria (170, 311, 434), mainly at synaptic nerve endings (435). The precise localization at synapses of COMT, which is a soluble enzyme, has so far not been determined (432). Some COMT activity has been encountered in the synaptosomal fraction, the enzyme having been solubilized by osmotic shock (436).

Monoamine oxidase plays an important intracellular role in the regulation of the uptake and storage of biogenic monoamines as well as in the action of amine-depleting drugs (437, 438). This role seems to be well demonstrated by the rapid increase in norepinephrine levels, that is induced by MAO inhibitors, in the brain (380, 439–442), sympathetic ganglia (443), and heart (444). Since MAO is localized in mitochondria, and, since MAO inhibitors bring about in central neurons a marked rise in norepinephrine concentrations (380, 439–442) and in fluorescence (445, 446), it is likely that MAO plays an essential part in intraneuronal metabolism of norepinephrine in the brain. Whereas evidence has been provided for the peripheral sympathetic system that COMT is an important factor in the extraneuronal metabolism of norepinephrine (447), the evidence for the role of COMT in the central nervous system is only indirect. Recent findings (448) indicate, however, that COMT is very effective in metabolizing extraneuronal norepinephrine. The cellular localization of these two processes and the importance of each in the initial metabolism of norepinephrine is not yet clearly understood (263).

The metabolic function of MAO in the animal body is to break down potent biogenic amines by oxidative deamination. This reaction is followed by the further oxidation of the aldehyde formed to the corresponding carboxylic acid. The identification and quantitation of such acidic urinary constituents has helped considerably in the elucidation of the biological function of MAO in man and other mammals (128, 449–451).

In 1964 Neff, Tozer, and Brodie (452) described a specific transport system for the removal from the brain of the rat of 5-hydroxyindoleacetic acid (5-HIAA), an acid metabolite of serotonin. The existence of such a system in the rat was deduced from the observation that on administration of probenecid to rats there was an increase in the concentration of 5-HIAA in the brain.

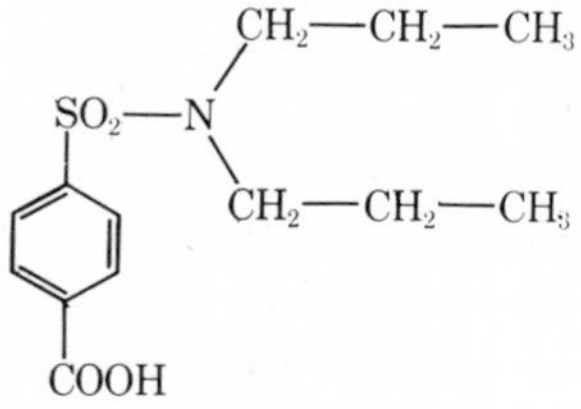

$$\begin{array}{c} CH_2-CH_2-CH_3 \\ / \\ SO_2-N \\ \backslash \\ CH_2-CH_2-CH_3 \end{array}$$

COOH

XXIII

Probenecid, p-(dipropylsulphamoyl) benzoic acid (**XXIII**), is a uricosuric agent that is rapidly absorbed when given by mouth.

Probenecid is a competitive inhibitor of acid transport in the renal tubules and is thought to act by interfering with the formation of an intermediate complex involved in the transport of acids through the renal tubular cell.

An active transport system for the removal of acid metabolites from the brain is also present in the mouse (453), but not in the rabbit (454, 455). This transport system is concerned with the removal from the brain of 4-hydroxy-3-methoxyphenylacetic acid (homovanillic acid, HVA), an acid metabolite of dopamine (453, 455). The presence of such an active transport system in the rat and the mouse and the possibility of its inhibition by probenecid has facilitated studies on the rates at which acid metabolites of monoamines are formed in the brain. Neff, Tozer, and Brodie (456) have obtained evidence that the accumulation of 5-HIAA in the brain, after the application of probenecid to the rat or mouse, can be taken as a measure of the turnover rate of serotonin. In the rat, the accumulation of 5-HIAA after a maximally effective dose of probenecid was linear for at least 90 min, and in the mouse for at least 60 min.

Whereas the application of similar methods to the study of turnover of norepinephrine in the brain has proved to be more difficult, the effect of probenecid on the acid metabolites of dopamine has been found to be useful in studies on drug-induced changes in dopamine metabolism in the striatum of the mouse (453, 457). It is hoped that the use of probenecid to block the egress of metabolites of monoamines from the rodent brain will prove to be a very useful means for the study of the metabolism of monoamines in the brain. However, it should be stressed that the effect of probenecid can be used only as an index of the rate of synthesis of an acid metabolite at a site where the active transport system can function (458).

At this point it should be mentioned that active transport mechanisms may be also involved in the removal of organic acids from the cerebrospinal fluid to the blood. Pappenheimer, Heisey, and Jordan (459) and Davson, Kleeman, and Levin (460) demonstrated that certain organic acids, when introduced in low concentrations into a ventriculocisternal perfusate, were removed by an active transport system. In the cat (460) and the rabbit (461) significant active absorption probably occurs from the choroid plexus of the lateral ventricles. Lately, Guldberg, Ashcroft, and Crawford (462) have demonstrated the presence of homovanillic acid and 5-hydroxyindoleacetic acid in the cerebrospinal fluid of the dog and have produced some evidence that these endogenous organic acids are removed from cerebrospinal fluid to blood by a transport mechanism acting in the caudal part of the intraventricular system (463). Using a method of repeated sampling of ventricular cerebrospinal fluid, developed by Ashcroft et al. (464), Guldberg (465) carried out a careful study on the metabolism of catecholamines and serotonin in the cerebrospinal fluid of dogs. His evidence indicates that the phenolic acids HVA and 5-HIAA in the cerebrospinal fluid of the dog are derived from brain tissue. There exists a good relationship between the acid concentrations in the lateral ventricular cerebrospinal fluid of the dog and those of the caudate nucleus. The concentration of an acid in a defined region of the cerebrospinal fluid system seems to be largely determined by the contribution from adjacent brain structures and the efficiency of its removal mechanisms from cerebrospinal fluid to blood. Drug-induced changes in concentrations of cerebrospinal fluid acids may reflect the metabolism of the parent cerebral amines. Guldberg suggests that estimates of concentrations of amine metabolites in the cerebrospinal fluid may be conveniently used as an index of the turnover of cerebral mono-amines (465).

At this point studies by Glowinski and Axelrod (382) on the metabolism of ^{3}H-norepinephrine (see Section 1.F.a) should be recalled. When the endogenous stores of rat-brain norepinephrine were labeled by these authors with ^{3}H-norepinephrine, and amphetamine was administered to the animals, the pattern of radioactive metabolites that was obtained (382) was found to be consistent with the theory that amphetamine may release brain norepinephrine in a physiologically active form. More normetanephrine and a smaller amount of deaminated tritiated catechols were found; this suggests that norepinephrine had been released without exposure to intraneuronal MAO, and was metabolized primarily by extra-neuronal COMT. The decrease of the deaminated catecholamine metab-olites might have been caused partly by inhibition of MAO by am-phetamine (256, 466). Inhibition of the oxidative deamination of ^{3}H-norepinephrine by low concentrations of amphetamine *in vitro* has been recently reported (323). However, amphetamine does not decrease con-centrations of O-methylated deaminated metabolites in the brain (323). Thus, the striking syndrome of rapid central excitation brought about by amphetamine does not appear to be fully explained by its limited ability to inhibit MAO.

All these data are consistent with the hypothesis that amphetamine may increase the release of norepinephrine either directly or as an indirect effect of increased neuronal activity. The decrease of brain norepinephrine caused by amphetamine may be due, at least partly, to a prevention of reuptake of physiologically released norepinephrine. Thus, in peripheral sympathetically innervated tissue the accumulation of exogenous nor-epinephrine is known to be impaired by amphetamine (373, 414, 467).

In the central nervous system there is a similar impairment of the accumulation of exogenous norepinephrine in brain slices (374, 468, 469), *in vivo* in the whole brain (382), and to various degrees in different parts of the brain (323, 468). Since in these studies the uptake process could not be measured separately, the decreased accumulation of norepinephrine could have been due to a blockage of uptake, or to an active release of norepinephrine by amphetamine, or to both.

The administration of precursors has provided further indirect evidence that neuronal-membrane uptake mechanisms may be inhibited by amphetamine. Catecholamine stores were replenished to a lesser extent by dopa administered after amphetamine to animals treated with reserpine and an MAO inhibitor, as measured by histochemical and biochemical assays (468). Furthermore, when radioactive dopamine was introduced into the intact brain after amphetamine treatment, less endogenously formed radioactive norepinephrine was found than in normal animals. In this experiment, the accumulation of norepinephrine may have been due to a specific inhibition by amphetamine of the norepinephrine reuptake process (323), but enhanced release of norepinephrine may also have been a contributing factor. The decreased amount of labeled norepinephrine may have been due to an inhibition by amphetamine of dopamine β-hydroxylase. Goldstein and Contrera (470) found that *in vitro* dopamine β-hydroxylase was inhibited by amphetamine, but only at very high concentrations.

It should be mentioned here that usually increases of brain norepinephrine are paralleled by increased accumulations of normetanephrine (302, 471). Thus, in rats that had been treated with pheniprazine, endogenously formed radioactive normetanephrine, normally encountered only in the soluble supernatant, was detected in part in a synaptosomal brain fraction (472, 473).

MAO inhibitors do not seem to affect the distribution of norepinephrine between particulate and soluble subcellular fractions (474, 475). In the cat brain after iproniazid treatment, no appreciable changes were found in the metabolism of labeled norepinephrine, injected into the lateral ventricle (313). The significant increase in normetanephrine observed in some species on MAO inhibition may suggest that the extraneuronal pathway predominates. Baldessarini (476) was able to demonstrate that brain accumulations of S-adenosylmethionine, the methyl donor utilized by COMT, were decreased by several MAO inhibitors. When large doses of methionine were administered, no decrease in brain concentrations of S-adenosylmethionine was observed. These findings seem to support the concept that MAO inhibition stimulates the utilization of the methylation pathway. It is not known, however, whether the accumulation of normetanephrine in the brain, partly located in synaptosomes, is a contributing factor to the central effects of MAO inhibitors (263).

Although there are many relatively nontoxic compounds that exert specific and lasting inhibitory effects on MAO, very few such compounds are known to be involved in the specific inhibition of COMT. Although many compounds displaying various chemical structures are known to inhibit COMT *in vitro,* only a small number of them have an effect *in vivo* (477–480). Several compounds exert short-lasting effects on brain O-methylation, but also produce many side effects by interfering in several ways with the metabolism of norepinephrine. Whereas the most effective inhibitors of brain COMT activity *in vivo* have proved to be pyrogallol and several dopacetamide derivatives, in peripheral tissues tropolones have found to be more active (259, 481).

Pyrogallol is without effect on the accumulation of catecholamines in the brain (440, 474) except when it is administered directly into the lateral ventricle of the rabbit brain (482). Whereas dopacetamide derivatives decreased the concentration of normetanephrine and of 3-methoxytyramine in the mouse brain (479), a tropoloneacetamide did not affect the concentration of exogenous norepinephrine in the rat brain; it markedly reduced, however, the formation of norepinephrine (382). After the administration of catecholamine precursors, the dopacetamide compounds increased the accumulation of dopamine and norepinephrine in the brain and potentiated the central excitation these compounds induce (259). After the administration of dopa, pyrogallol potentiated the accumulation of brain dopamine, but was without effect on the concentration of norepinephrine (474). A significant decrease in brain concentration of S-adenosylmethionine was observed on the administration of pyrogallol. This inhibitory effect may result from a competitive utilization of the methyl donor, whereby pyrogallol itself is methylated (476).

For further elucidation of the catecholamine metabolism the development of more effective, longer lasting, and less toxic COMT inhibitors is urgently required.

At this point it may be recalled that Axelrod (483) was the first author who recognized that the deaminating oxidase reaction is not necessarily the first step in the breakdown and biological inactivation of some monoamines. According to the scheme proposed by Axelrod and shown below, norepinephrine, discharged from neuronal endings in a physiologically active form as a result of either nerve impulses or the action of sympathomimetic drugs, is inactivated mainly by cellular reuptake or by enzymic conversion by COMT to form normetanephrine. In a secondary deamination reaction by MAO, normetanephrine is transformed into 3-methoxy-4-hydroxymandelic acid (vanillylmandelic acid). In contrast, norepinephrine, released intracellularly, either spontaneously or by reserpine-like drugs, seems to be inactivated mainly by mitochondrial MAO, with the formation of the deaminated catechol metabolite 3,4-

dihydroxymandelic acid. MAO thus appears to be a regulator of tissue levels of norepinephrine (259, 266, 484, 485). On secondary O-methylation by COMT the latter compound is changed to vanillylmandelic acid. In this connection it may be pointed out that vanillylmandelic acid is the major urinary metabolite of norepinephrine in man.

Scheme of enzymic pathways of norepinephrine metabolism.

It should be mentioned here that Spector and his co-workers (486) presented evidence that MAO is the principal enzyme in the inactivation process of norepinephrine and serotonin in the brain. These authors also suggested different physiological roles for MAO and COMT. Whereas MAO appears to be responsible for the metabolism of the biogenic monoamines in tissues, COMT is involved particularly in the inactivation of exogenous, extraneuronal, circulating catecholamines (486, 487).

The idea of a "protective" function of MAO for the *in vivo* enzymic deamination processes of highly active biogenic monoamines has been borne out, as mentioned above, by recent clinical and experimental findings on the effects of tyramine-rich foods and sympathomimetic drugs such as amphetamine or ephedrine, in conditions of almost complete blockage of MAO activity induced by one of its potent inhibitors, such as tranylcypromine, pargyline, or phenelzine (418, 488–490).

Reserpine alters the catabolism of brain norepinephrine. As in peripheral tissues (447), reserpine causes a decrease in the levels of normetanephrine and an increase in deaminated metabolites in the whole brain (382, 491) as well as in various areas of the brain (323). The altered pattern of metabolites seems to suggest that after reserpine application, more norepinephrine is exposed directly to intraneuronal MAO, and that less endogenous norepinephrine is released and methylated extraneuronally to form normetanephrine. Inhibition of MAO partially protects brain amines from the depleting effects of reserpine (474, 486, 492–494). These results seem to indicate that reserpine interferes with a form of intraneuronal storage that normally protects norepinephrine from the action of MAO. It follows that in intact neurons norepinephrine may exist in at least two states, and that some of the norepinephrine encountered in the supernatant subcellular fraction is rapidly depleted by reserpine. Initially, reserpine might expose to destruction by MAO a fraction of the intraneuronal amines that are normally either held loosely by vesicles, or in a protected form in the cytoplasm (263).

At this point it should be mentioned that low concentrations of reserpine prevent the accumulation of labeled norepinephrine in cerebral slices (309). Likewise, reserpine prevents the accumulation throughout the brain of exogenous norepinephrine administered intraventricularly (323, 382). In peripheral tissues the decreased concentration of exogenous epinephrine after reserpine treatment appears to be due to a failure of retention, rather than to a block of the initial uptake across the neuronal membrane (333, 445, 495, 496). Inhibition of MAO causes *in vivo* an accumulation of exogenous norepinephrine in the brain (448), and in nerve terminals MAO inhibition allows exogenous norepinephrine to accumulate *in vitro* in the presence of reserpine (497). Equally, α-methylated amines that are not attacked by MAO accumulate in central neurons *in vitro* (445) and *in vivo* in the presence of reserpine (498). This accumulation of amines seems to indicate that, as in the peripheral sympathetic system, reserpine does not act by preventing their uptake at central neuronal membranes (263).

It is noteworthy that reserpine does not seem to interfere directly with the norepinephrine-synthesizing enzymes. Thus, loading amounts of dopa reverse both the depletion of brain catecholamines (493) and the behavioral syndrome (302) brought about by reserpine. This reversing effect becomes even more pronounced in the presence of MAO inhibitors

(439, 474) . Furthermore, Glowinski and his associates (448) have recently reported that radioactive tyrosine and radioactive dopamine can both be converted to norepinephrine in the reserpine-treated brain. As the same amount of radioactive dopa is synthesized from ^{14}C-tyrosine in the brains of reserpine-treated and of normal rats (448), it is obvious that reserpine fails to inhibit this rate-limiting step of norepinephrine synthesis. However, since reserpine removes the substrate dopamine from the site of norepinephrine synthesis, it may indirectly limit norepinephrine formation (263).

Recently, Glowinski and Baldessarini (263) have drawn certain conclusions in regard to the importance of various pharmacological agents for the metabolism of brain catecholamines. Various drugs induce striking changes in the concentration of brain catecholamines, and various mechanisms may lead to either accumulation or depletion of brain norepinephrine. Furthermore, significant metabolic changes may occur even without appreciable alterations in norepinephrine concentration. Various metabolic alterations may also occur in the rate of catecholamine turnover, and the net effect observed may be the result of complementary or antagonistic actions of the drug applied. It appears that compounds which decrease synthesis, alter storage sites or compete for them, or block, release, may decrease central nervous system neural activity, probably by limiting the availability of norepinephrine at the synapse. In contrast, the administration of precursors of norepinephrine, or of drugs that activate release, inhibit reuptake, or prevent enzymic inactivation, is followed by a syndrome of central excitation, presumably due to increased availability of norepinephrine at the synapse. The authors point out that these interpretations are complicated by the uncertain role of norepinephrine as an excitatory or inhibitory transmitter in the brain, since norepinephrine has been found to exert both excitatory and inhibitory effects in various areas of the central nervous system (260).

The biological role of MAO does not seem to be confined solely to the inactivation of biogenic monoamines (398) . Thus, it has been shown by Barondes (499) that many different MAO substrates and their aldehyde metabolites stimulate *in vitro* glucose-1-^{14}C oxidation to ^{14}CO$_2$ by beef anterior pituitary slices. The effect of the aldehyde metabolites is not blocked by MAO inhibitors. Likewise, Pastan and Field (500) have reported on an *in vitro* stimulation of glucose oxidation in thyroid by serotonin. The physiological significance of these findings is not known. The as yet unexplained effects of MAO inhibitors on carbohydrate metabolism *in vivo* (489) may be due to interference by the inhibitors with some yet unknown function of MAO (237). More studies on the nature of mitochondrial MAO are needed.

2. MOUSE-LIVER HISTAMINASE

Kobayashi (148) prepared from a homogenate of albino-mouse liver an enzyme that metabolizes histamine and cadaverine. The properties of this enzyme are similar to those attributed to MAO rather than to DAO. It is found in the mitochondrial insoluble fraction instead of in the soluble one, is insensitive to the specific DAO inhibitors, semicarbazide and aminoguanidine, but is sensitive to iproniazid, which inhibits both MAO and DAO. Furthermore, the isolated mouse-liver enzyme metabolizes tyramine, histamine, and cadaverine, thus displaying both MAO and DAO activities. Incubation of the mouse-liver homogenate with radioactive histamine gave imidazole-4-acetic acid as the major end product with no other histamine metabolite present in reasonable amounts. The significance of Kobayashi's findings are discussed in Chapter V.

CHAPTER IV

Amine Oxidases Sensitive to Carbonyl Reagents and to Semicarbazide

1. CLASSICAL HISTAMINASE, DIAMINE OXIDASE*

A. Introduction

The slow disappearance of injected histamine had been observed by a number of workers (501–503) before the enzyme system involved in the catabolism of histamine was first described by Best in 1929 (2) and further elucidated, and termed histaminase by Best and McHenry in 1930 (504). The latter workers considered the enzyme to be specific for histamine. In 1938, however, Zeller (505) reported that hog-kidney preparations and various other animal-tissue extracts were capable of destroying by oxidative deamination not only histamine but also di- amines, such as putrescine, cadaverine, and agmatine. He substituted the term "diamine oxidase (DAO)" for "histaminase." Subsequently, not only have the terms "histaminase" and "DAO" been used interchange- ably, but also the substrates histamine, cadaverine, and putrescine have been used indiscriminately. Unfortunately, these practices have created much confusion in the interpretation of many enzymic assays. Further- more, recent work on the relation between pH and rate of enzymic oxidation has led to the suggestion that at least two kinds of histaminase may exist (506, 507), distinguishable by the molecular species of hista- mine upon which they act. One kind acts on histamine in the dicationic

---

* E.C. 1, 4, 3, 6. Diamine: O_2 oxidoreductase, deaminating.

form, in which not only the terminal NH_2 group but also the nitrogen in the imidazole ring carries a positive charge. The other acts on the monocationic form, in which the nitrogen in the ring carries no charge. Both types of histaminase are present in the hog, the dicationic-substrate form in the kidney, the monocationic-substrate form in the blood (146, 508).

Zeller's opinion (505) that histaminase and DAO always go together has been contested by the present author (509). The question of homogeneity of the histaminase occurring in various tissues is discussed in Chapter V.

In view of the great difficulties of any attempt to classify a given amine oxidase within the group of diamine oxidases, Zeller himself has suggested (126) that until a rational system of classification of diamine oxidases is proposed, it is best to characterize a given diamine oxidase by its origin and by one of its best substrates. The author fully agrees with this suggestion, and this is followed in the present chapter. Histamine is regarded as the best substrate of classical histaminase, and cadaverine and putrescine as the best substrates of classical DAO.

B. Occurrence

Unlike MAO, most diamine oxidases are recovered from tissue homogenates in the supernatant on removal of particulate matter after centrifugation (126).

Whereas in bacteria and higher plants the diamine oxidases seem to be distributed more widely than the monoamine oxidases, the reverse seems to hold true for the animal kingdom. The microorganism *Pseudomonas aeruginosa* is capable of adaptive production of DAO, when cadaverine is offered as the substrate (510, 511). When grown in a nutrient containing either histamine or putrescine as the only nitrogen source, *Achromobacter* produced diamine oxidases showing a marked preference for either of these diamines (512). Bachrach (513) found that *Serratia marcescens* was able to form a spermidine oxidase that did not attack spermine. Roulet and Zeller (514) reported that the DAO activity did not remain constant in the course of the growth of cultures of *Mycobacterium smegmatis* but decreased semilogarithmically during the first 10 days.

With regard to higher plants, DAO occurs in the leaves and other tissues of *Atropa belladonna* (515) and in rather large quantities in the seedlings of Leguminosae species, with only small amounts of the enzyme being present in the roots and stems (516). In embryonic peas a DAO activity can be detected within 5 hours after the onset of germination. The DAO content of the germinal layers is highest on the fifth day of germination and is about 10 times as large as the DAO content of the most active animal-tissue extracts (517). Werle and his co-workers (517) suggest that the DAO formation is induced by the embryo and that a much higher DAO activity is found in pea seedlings germinating in the dark than in those exposed to light.

In vertebrates DAO activities vary from species to species (122). DAO occurs in the kidney (504, 518), intestinal mucosa (519, 520), liver (138, 521), and lung (504, 522). This enzyme was further found in skin (122); in the prostate and mammary glands (122, 176); in some body fluids, such as blood (122, 145); in the intestinal lymph (523, 524); and in sperm plasma (122, 176). In man and other mammals the adult kidney contains more DAO than the fetal kidney (130). In the kidney of rodents hardly any DAO activity was found (525, 526). But the brain of rodents apparently does not display any DAO activity either, judging from recent results obtained on incubation of ^{14}C-labeled putrescine with brain homogenates from four rodent species (527).

These findings seem to confirm previous reports (528, 529) on the presence in the brain and other mammalian tissues of another enzyme system, differing from DAO, that will degrade histamine in its mono-cationic form, but will not attack short-chain diamines, such as 1,4-di-aminobutane (putrescine) or 1,5-diaminopentane (cadaverine). This and various other histamine-degrading systems are discussed in Chapter V.

It is noteworthy that fish brain seems to contain an amine oxidase showing all the properties of a true DAO. In the brain of the carp *(Cyprinus carpio)* a distinct distribution of DAO was observed (527), with the highest enzymic activity having been found in the diencephalon.

In 1937 Danforth and Gorham (530) were the first to report on the occurrence of large amounts of histaminase in the human placenta, and in the following year Marcou et al. (531) demonstrated the presence of histaminase in the blood of pregnant women. A considerable number of publications on the subject of histaminase activity in human pregnancy appeared subsequently (532–545). On the basis of an extensive examination of various parts of the human placenta and fetal membranes Swanberg (542) suggested that the placental histaminase most likely originates from the maternal part of the placenta, the decidua, and that the fetal elements play no part in the formation of the placental enzyme. This view of Swanberg found considerable support in results of his investigations of histaminase activity in placentae of some animal species, such as hog, horse, rabbit, guinea pig, and rat. Thus, whereas the placentae of hog and horse were found to be completely devoid of hista-minase activity, the placentae of rabbit and guinea pig did contain varying, if small, amounts of the enzyme. The histaminase activity in the rat placenta, however, turned out to exceed by about tenfold that of the human placenta. It should be recalled here that the placentae of hog and horse are free of maternal elements. It is now generally recognized, that the 100- to 1000-fold increase in histaminase content of the blood plasma of pregnant women as compared with that of nonpregnant individuals originates from the decidual part of the placenta and that this enzyme is not to be found in the fetal blood (531, 533–545). Such a dramatic rise in the plasma level of histaminase has not been observed in pregnant animals. Ahlmark (534) reported on a small rise in the plasma-histaminase level of pregnant guinea pigs and rats as compared with nonpregnant animals, but this increase was attributed by Anrep and his co-workers (539) to bacterial contamination. In 1964, however, Kobayashi (546), using a very sensitive radioassay procedure for the estimation of DAO (547), and ^{14}C-putrescine as substrate, was able to demonstrate a 50- to 100-fold rise in DAO activity in the blood plasma of pregnant rats as compared with nonpregnant ones. The difference between Ahlmark's and Kobayashi's findings may be due to the fact that Ahlmark used histamine and Kobayashi putrescine as substrates for the estimation of the deaminating enzyme in the blood plasma.

It may be worthwhile recalling here that it has been repeatedly shown by the present author that in human pregnancy cadaverine and putrescine were far better substrates of the placental and blood-plasma amine oxidase than histamine (535–538). This also applies to the human amniotic fluid, which has been found to exert a very large effect on cadaverine and putrescine but a much smaller one on histamine (548). The DAO content of the amniotic fluid by far exceeded that of the simultaneously examined pregnancy serum, the specific activity of the liquor amnii enzyme having been found to be about a 100 times higher than that of the maternal serum (548). These findings compare well with those of Kobayashi and his colleagues (549), who also showed that the DAO activity in the amniotic fluid was higher than that in corresponding maternal plasma. Swanberg (542) was the first to demonstrate the presence of a histaminase activity in the human amniotic fluid but found that the enzyme content of the latter was lower than that of the maternal plasma. Anrep and his co-workers (539) as long ago as 1947 tested the amniotic fluid for histaminase activity, but failed to demonstrate its presence.

The DAO content of the cells and body fluids does not remain constant throughout the life of an organism. Hence, different activities are observed during the embryonic, postembryonic, and adult periods and, as shown above, during pregnancy (122, 129, 550).

C. Properties

On fractionation of hog-kidney cortex in sucrose solution DAO (histaminase) is found predominantly in the cytoplasmic fraction, and is not precipitated by centrifugation at 140,000 g for several hours. In fresh kidneys, however, 10% of enzyme activity may be recovered in the mitochondrial fraction (135). In the rabbit liver DAO is found only in the mitochondrial fraction (151), from which it is liberated with the help of certain detergents (135). Purification procedures for the hog-kidney and the placental enzyme carried out in various laboratories are discussed in Chapter VI, Sections 3.A and 3.B.

A considerable purification of the hog-kidney enzyme was obtained by Tabor (551), and even more extensively purified preparations have been described by Goryachenkova et al. (552), by Kapeller-Adler and MacFarlane (553), and by Mondovi et al. (554–556). Recently, Yamada and his co-workers (258) have described a method by means of which they have obtained from hog kidney a crystalline preparation of DAO (see Chapter VI.3.A.d).

The purified hog-kidney enzyme, when subjected to electrophoresis on cellulose acetate strips at various pH values, ranging from 4.6 to 8.6, traveled as a distinct homogeneous band, to be located in the region between the α_2- and β-globulin fractions of human serum (553). Its isoelectric point was between 5.0 and 5.15, and the optimum pH for activity with histamine as substrate was 6.8. The enzyme is stable between −20 and +62° (553). Blaschko et al. (145) found that the optimum pH for the reaction of the hog-kidney enzyme with histamine as substrate was at pH 6.3, whereas the oxidation of cadaverine at this pH was very slow. Crude hog-kidney DAO is readily inactivated by pepsin in the presence of HCl, by trypsin, and by papain (2, 130, 150).

Attempts at purification of human placental histaminase, diamine oxidase, have been so far greatly impeded by the difficulty in obtaining a complete separation of the enzyme from hemoglobin and its derivatives. By adaptation to extracts of human placentae of a modern technique, developed by Henessey and his co-workers (557) and based on the use of DEAE-cellulose at various hydrogen ion concentrations, not only a complete separation of hemoglobin from histaminase, diamine oxidase, but also a considerable purification of the latter enzyme was achieved (538; see Chapter VI.3.B).*

Smith (558) has recently extracted histaminase from exsanguinated human placentae and purified the enzyme by salt fractionation, ion-

---

* The author also has unpublished data (1964).

exchange chromatography, and gel filtration. His purest enzyme preparation was, however, still contaminated with haptoglobin-methemoglobin.

The human placental histaminase, DAO, differs significantly from the hog-kidney enzyme by not being able to withstand heating at 60° (538).

D. Prosthetic Groups

Zeller (505) was the first to suggest that the formation of hydrogen peroxide in the action of DAO on its substrate pointed to the presence of FAD as a prosthetic group. It was also previously suggested that DAO was a pyridoxal enzyme (402, 559). These suggestions were based on the fact that many substances known to inhibit pyridoxal enzymes are inhibitors of DAO (402), and on the observation that the DAO activity of lung and intestinal mucous membrane of rats decreases in vitamin B_6 and B_2 deficiencies. The effect is reversed by the treatment of the animals with these vitamins (560). These observations recently received further support when spectrofluorometric evidence was obtained, suggesting the presence of both prosthetic groups, FAD and pyridoxal phosphate, in the purified hog-kidney enzyme (553).

Kapeller-Adler and MacFarlane (553) have suggested two reaction mechanisms either of which may be operative when histaminase acts on histamine.

In scheme 1, which is similar to that suggested by Braunstein (561), a two-step reaction is proposed, involving in its first step a transamination between the substrate and the pyridoxal form of histaminase with the formation of a Schiff base, the enzyme-bound pyridoxylidene azomethine. Electronic displacements result in dissociation of the α-hydrogen atom and shift of double bonds, leading to the tautomeric azomethine. This is hydrolyzed to imidazoleacetaldehyde and the pyridoxamine phosphate enzyme. In the second step of this mechanism the pyridoxamine phosphate form of the enzyme undergoes dehydrogenation by the second prosthetic group of histaminase, FAD, to the enzyme-bound pyridoxylimino phosphate compound, and the reaction mechanism is completed by reoxidation of the leucoflavin by oxygen to FAD, and hydrolysis of the enzyme-bound pyridoxylimino phosphate to the pyridoxal phosphate form of histaminase, with the formation of ammonia and hydrogen peroxide.

With purified DAO preparations from hog kidney and clover seedlings Goryachenkova (560) failed to achieve net transamination between diamines and added pyridoxal phosphate or free pyridoxal.

In scheme 2, a reaction mechanism is shown, similar to that proposed by Werle and von Pechmann (516) for the action of plant DAO on its

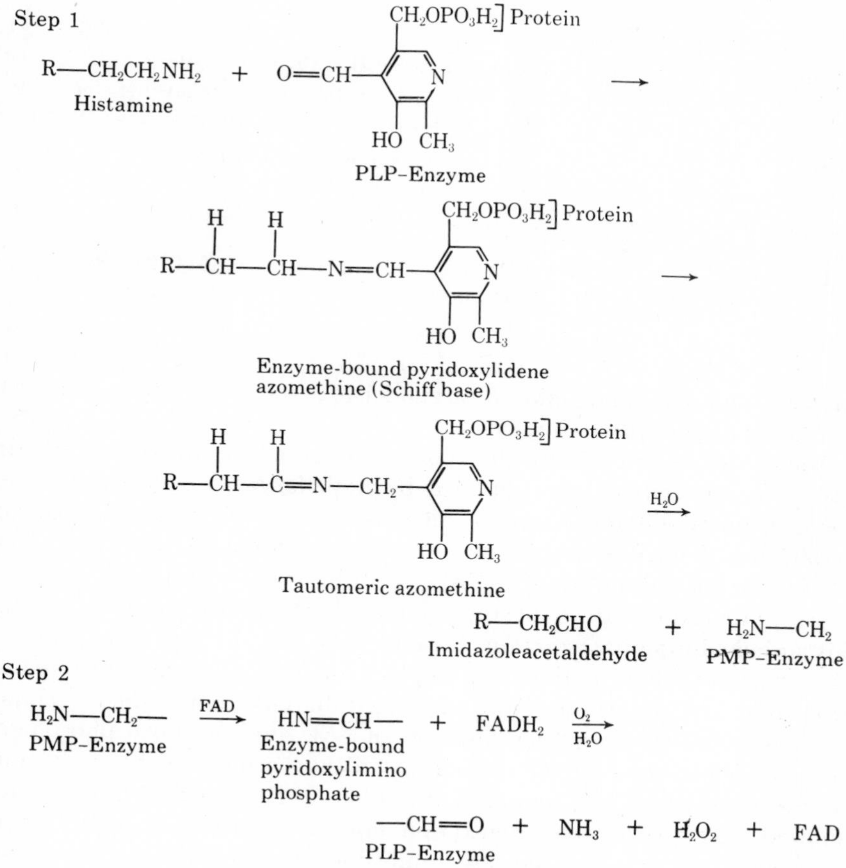

Step 1

$$R—CH_2CH_2NH_2 \quad + \quad O{=}CH—\overset{\displaystyle CH_2OPO_3H_2\big]\ Protein}{\underset{\displaystyle HO \quad CH_3}{\underset{\displaystyle N}{\bigcirc}}} \quad \longrightarrow$$

Histamine

PLP–Enzyme

$$R—\overset{H}{\underset{|}{C}}H—\overset{H}{\underset{|}{C}}H—N{=}CH—\overset{\displaystyle CH_2OPO_3H_2\big]\ Protein}{\underset{\displaystyle HO \quad CH_3}{\bigcirc}} \quad \longrightarrow$$

Enzyme-bound pyridoxylidene
azomethine (Schiff base)

$$R—\overset{H}{\underset{|}{C}}H—\overset{H}{\underset{|}{C}}{=}N—CH_2\text{-}\overset{\displaystyle CH_2OPO_3H_2\big]\ Protein}{\underset{\displaystyle HO \quad CH_3}{\bigcirc}} \quad \xrightarrow{H_2O}$$

Tautomeric azomethine

$$R—CH_2CHO \qquad + \qquad H_2N—CH_2$$

Imidazoleacetaldehyde   PMP–Enzyme

Step 2

$$\underset{\text{PMP–Enzyme}}{H_2N—CH_2—} \xrightarrow{FAD} \underset{\substack{\text{Enzyme-bound} \\ \text{pyridoxylimino} \\ \text{phosphate}}}{HN{=}CH—} + FADH_2 \xrightarrow[H_2O]{O_2}$$

$$—CH{=}O \quad + \quad NH_3 \quad + \quad H_2O_2 \quad + \quad FAD$$

PLP–Enzyme

R = Imidazole
PLP = Pyridoxal-5-phosphoric ester
PMP = Pyridoxamine-5-phosphoric ester
FAD = Flavin adenine dinucleotide

substrates. According to scheme 2, the enzymic deamination of histamine proceeds first by condensation of histamine with the PLP form of histaminase, with the formation of a Schiff base which, however, does not undergo tautomeric rearrangement, but is directly dehydrogenated by the flavin prosthetic group, FAD, to the α, β-unsaturated azomethine complex with the formation of the leucoflavin, FAD·H$_2$. The reaction cycle is completed by reoxidation of the leucoflavin by oxygen with the formation of hydrogen peroxide, and stepwise hydrolysis of the α,β-unsaturated azomethine complex to the pyridoxal phosphate-form of hista-

$$R\text{---}CH_2CH_2NH_2 \; + \; O\text{==}CH\text{---} \longrightarrow$$

Histamine   PLP-Enzyme

$$\underset{\text{Azomethine (Schiff base)}}{\overset{\overset{\displaystyle H \quad\quad H}{\displaystyle |\quad\quad |}}{R\text{---}CH\text{---}CH\text{---}N\text{==}CH\text{---}}} \xrightarrow{\text{FAD}} \longrightarrow$$

$$\underset{\alpha,\beta\text{-unsaturated azomethine}}{R\text{---}CH\text{==}CH\text{---}N\text{==}CH\text{---}} \; + \; FADH_2$$

$$FADH_2 \; + \; O_2 \; \longrightarrow \; FAD \; + \; H_2O_2$$

$$R\text{---}CH\text{==}CH\text{---}N\text{==}CH\text{---} \; + \; H_2O \longrightarrow$$

$$\underset{\text{Free enzyme}}{O\text{==}CH\text{---}} \; + \; \underset{\alpha,\beta\text{-Dehydroamine}}{R\text{---}CH\text{==}CH\text{---}NH_2} \; \xrightarrow{H_2O} \; \underset{\text{Imidazoleacetaldehyde}}{R\text{---}CH_2CHO \; + \; NH_3}$$

$$R = \text{Imidazole}$$

minase, ammonia, and imidazoleacetaldehyde, the latter formed via the imidazole-α,β-dehydroamine.

Some additional evidence that both FAD and PLP are the prosthetic groups of histaminase was obtained by Kapeller-Adler and MacFarlane (553) in many different assays leading to a successful, reversible, partial resolution of purified hog-kidney histaminase, followed by reactivation of the enzyme on addition of FAD or PLP.

An important feature of the pyridoxal phosphate catalysis of the histaminase-histamine reaction became apparent during the study of the effects of added pyridoxal phosphate on pure histaminase preparations that had been subjected to prolonged dialysis (553). When attempting to determine the optimal amounts of FAD and pyridoxal phosphate, to be added to the dialyzed enzyme in order to achieve best reactivation of the enzyme, it was found, that, unlike FAD, pyridoxal phosphate showed a very distinct concentration optimum, which, if greatly overstepped, could cause even an inhibition of histaminase action on histamine (553). In this context it should be pointed out that a very sharp substrate optimum was also essential in order to achieve best histaminase action on histamine, since in the presence of superoptimal concentrations histamine strongly inhibited its own enzymic degradation (553). Thus, it appears that both the substrate of histaminase, histamine, and one of the prosthetic groups of this enzyme, pyridoxal phosphate, may under not fully controlled experimental conditions act as inhibitors. These findings (553), as well as those indicating the presence in purified hog-kidney histaminase of pyridoxal phosphate as a prosthetic group (553, 562), have been recently confirmed by Mondovi and his colleagues (556, 563, 564).

In purified hog-kidney histaminase preparations Mondovi et al. (555, 556), Kobayashi (565), and Goryachenkova and Ershova (566, 567) could not establish the existence of FAD as a prosthetic group (553). It is possible, however, that FAD is not closely bound to the apoenzyme and hence might have been lost during the various purification procedures carried out by the workers, mentioned above, as happens on purification of D-amino acid oxidase (568).

Thus, evidence has so far remained sparse about the hydrogen transfer from the substrate to oxygen, except that oxygen cannot be replaced by other hydrogen acceptors. Recently, copper has been considered as a participant in the activation of oxygen (147, 224, 555, 556, 564, 569–571).*

In 1963 Yamada and his co-workers (224), using electron-paramagnetic-resonance (EPR) spectroscopy found that crystalline beef-plasma amine oxidase displayed an EPR signal typical of complexes of cupric copper. The observed signal accounted for 2.7 g-atoms of cupric copper per mole of enzyme compared with 3.7 g-atoms per mole determined chemically. When reducing substances, such as dithionite, were added under anaerobic conditions to the enzyme, the cupric signal disappeared, which indicates that the copper was completely reduced. However, when substrate was added in a large excess under the same conditions, the signal magnitude remained unchanged, which meant that copper valency was unaffected.

Mondovi et al. (556) also subjected their purified DAO preparations to EPR spectroscopy and obtained results that appeared to be in good agreement with those of Yamada et al. (224). On evaluation of the EPR spectra, Mondovi et al. concluded that the copper atoms that are present in DAO appear to be all in an equivalent environment and in the cupric state. The spectrum is typical of Cu(II) in a complex of tetragonal symmetry (573). The EPR spectrum of DAO does not show the low hyperfine splitting constant $(A_{\|})$ of other copper oxidases, such as ceruloplasmin, laccase, ascorbic acid oxidase, and cytochrome oxidase (221), the first three of these enzymes being of deep-blue color. Mondovi et al. (556) further point out that, whereas the cupric copper of the oxidases, mentioned above, changes its valency in the presence of an excess of substrate, no such change occurs in the valency of the copper of DAO. Similar observations were reported on crystalline beef-plasma amine oxidase, as mentioned above, by Yamada et al. (224) and on crystalline hog-plasma amine oxidase by Buffoni et al (569, 570).

The EPR spectrum of DAO shows superhyperfine lines that presumably arise from nitrogenous ligands (556). Although subtle changes in the EPR spectrum and in the superhyperfine structure are observed on addition of substrates (223), Mondovi and his colleagues (556), on con-

---

* Also personal communication by G. A. Hamilton of data from (572) (1969).

sideration of results obtained with ^{15}N-labeled substrates, deny the possibility that this superfine structure may reflect direct binding between substrate and copper. Mason (574) is indeed of the opinion that the changes which Mondovi and Beinert (223) have observed in the EPR spectrum of hog-kidney DAO in the presence of substrate suggest a change of symmetry rather than a change of ligands. This would indicate a change of conformation of the protein and point toward the substrate binding at a place remote from the copper atom. The fact that the changes in the EPR spectrum differ for different substrates may, however, suggest that the environment of the copper is sensitive to the presence of substrate as well as to the type of substrate added (556). No change in the spin relaxation of the cupric copper was detected by Mondovi et al. (556).

Furthermore, Mondovi et al (554, 555) reported that purified hog-kidney DAO showed in the presence of diethyldithiocarbamate (575) a typical copper protein absorption spectrum and that the copper content of the enzyme preparations increased during purification to about 0.06% (10.57 ng-atoms of copper per milligram of protein). Because of these findings Mondovi and his colleagues (555) suggested that hog-kidney DAO was a copper-containing protein. These workers were also able to demonstrate that a pure homogeneous hog-kidney preparation contained besides pyridoxal phosphate 11 to 12 ng-atoms of copper per milligram of protein, the ratio between pyridoxal phosphate and copper having been found to be 1 : 1 (564).

The essential nature of copper in maintaining enzymic activity was shown by the loss of activity in the diethyldithiocarbamate-treated DAO and its reactivation on addition of suitable amounts of cupric copper. However, it is noteworthy that an excess of cupric copper inhibits the DAO activity (556). Hill and Mann (576) suggest that Cu^{2+} ions exert their inhibitory effect on DAO by reacting with the enzyme-substrate complex.

On the basis of copper analyses a minimum molecular weight of 87,000 was calculated for hog-kidney DAO (556). The molecular weight measured in the ultracentrifuge was, however, higher; this suggests that the hog-kidney DAO is made up of more than one unit, each having a molecular weight 87,000 and containing one atom of copper (556).

Amine oxidases generally contain similar amounts of copper and have molecular weights are that multiples of approximately 90,000. Thus, the molecular weight of crystalline DAO was reported to be 185,000 (258); of pea-seedling DAO, 96,000 (576); of crystalline benzylamine oxidase, 195,000 (146); of crystalline beef-plasma amine oxidase, 255,000 (143); and of amine oxidase of *Aspergillus niger*, 252,000 (577). These findings may suggest that amine oxidases may consist of one or more units of molecular weight approximately 80,000 to 100,000 (556).

Concentrated solutions of hog-kidney DAO are pink yellow and show an absorption maximum at 405 nm (553, 556) and another one at 480 nm (556). The latter absorption maximum was found by Mondovi et al. to decrease on addition of substrate under anaerobic conditions, but was restored on oxygenation (556). Yamada et al. (258) reported that the pink color of solutions of crystalline hog-kidney DAO was associated with an absorption maximum at 470 nm. The pink color was discharged by putrescine and by sodium dithionite and was restored by oxygenation. Analyses of the crystalline DAO by both atomic-absorption spectrophotometry and chemical tests revealed that copper was the only metallic component of the enzyme and that its content amounted to 11.7 ng-atoms/mg DAO.

The copper content of crystalline hog-plasma benzylamine oxidase, as determined by radioactivation analysis, was 3 or 4 g-atoms/mole of enzyme (146).

In this context it should be mentioned that Gorkin (578) reported that beef-plasma amine oxidase was a zinc protein and that there was a direct proportionality between the zinc content and the specific activity of this enzyme. Moreover, Gorkin demonstrated that zinc could be removed by ethylenediamine tetraacetic acid and that the lost enzyme activity was regained on addition of zinc. Yamada and his colleagues (224) were not able to confirm Gorkin's findings, since copper was the only metallic component these workers encountered in crystalline beef-plasma amine oxidase (143, 579, 580).

The function of copper in amine oxidases is, however, not yet fully understood [see Discussion in (221) ; see Chapter III.1.C].

Hamilton (571, 572) has lately reported studies on the mechanisms by which enzymes catalyze reactions of molecular oxygen. This work was carried out on a model system (581) that bears a close similarity to reactions catalyzed by pyridoxal phosphate–dependent amine oxidases. Among enzymes containing pyridoxal phosphate, the amine oxidases seem to be unique in that they require for activity a transition-metal ion (copper) as well as the pyridoxal phosphate coenzyme. Moreover, since amine oxidases are the only pyridoxal phosphate–dependent enzymes that use oxygen as a reactant, copper may be essential for facilitating the transfer of hydrogen to oxygen. Electron-paramagnetic-resonance measurements, carried out on various amine oxidases, have indicated that the copper on the amine oxidases is in the Cu(II) state and does not change its valency during the enzymic catalysis (224, 556, 570).

In his study of a model system for the enzymic reactions, mentioned above, Hamilton (571, 572) observed that many amino acids and amino acid derivatives are readily oxidized by oxygen at room temperature and pH 9, with the formation of α-keto derivatives and ammonia, when catalytic amounts of pyridoxal and certain metal ions, especially Mn(III) or Co(II), are present. From a detailed kinetic study of the mechanisms involved Hamilton (571) concludes that the metal ion does not change its valency during the catalysis, that the reaction does not proceed by a free-radical chain mechanism, and that pyridoxamine is not an intermediate in the oxidation reaction. Hamilton proposes a mechanism for the model reaction consistent with all the experimental information. The main differences between the model and the reactions catalyzed by amine oxidases are pointed out by Hamilton (571) as follows:

1. Whereas simple amines and diamines are substrates oxidized by amine oxidases, the model system oxidizes amino acids more readily than those amines.

2. Cu(II), the metal ion essential for the catalysis by amine oxidases, is ineffective in the model system.

On the basis of the model designed by Hamilton for the oxidation mechanism of amino acids in the presence of pyridoxal and metal ion (Mn^{3+}), Buffoni and Della Corte (569) recently proposed a reaction scheme illustrating the mechanism of the action of pyridoxal phosphate–dependent hog-plasma amine oxidase on its substrate, benzylamine.

It should be also mentioned here that Hamilton has very recently* designed a scheme relating to the mechanism of the oxidation of amines by an oxygen-amine oxidase system which bears great similarity to the mechanism suggested by this author in 1966 for the model reaction of pyridoxal phosphate–dependent amine oxidases (581) as well as to the mechanism proposed by Buffoni and Della Corte (569).

As with other pyridoxal phosphate enzymes, the pyridoxal phosphate is probably attached to the enzyme by a Schiff base linkage (553). For mechanistic reasons Hamilton (571, 572) suggests that the phenolic hydroxy group of the pyridoxal phosphate is bound to the Cu(II) which is attached to the enzyme.

In the scheme shown overleaf, the cupric ion performs two functions that are distinctly different:

1. It catalyzes the formation of *B* and its tautomerization to *C*.
2. It complexes with O_2 and transfers electrons in the conversion of *D* to *E* and to *F*.

---

* Personal communication from the author from (572) (1969).

Mechanism for the oxidation of benzylamine by hog-plasma benzylamine oxidase as suggested by Buffoni and Della Corte (569).

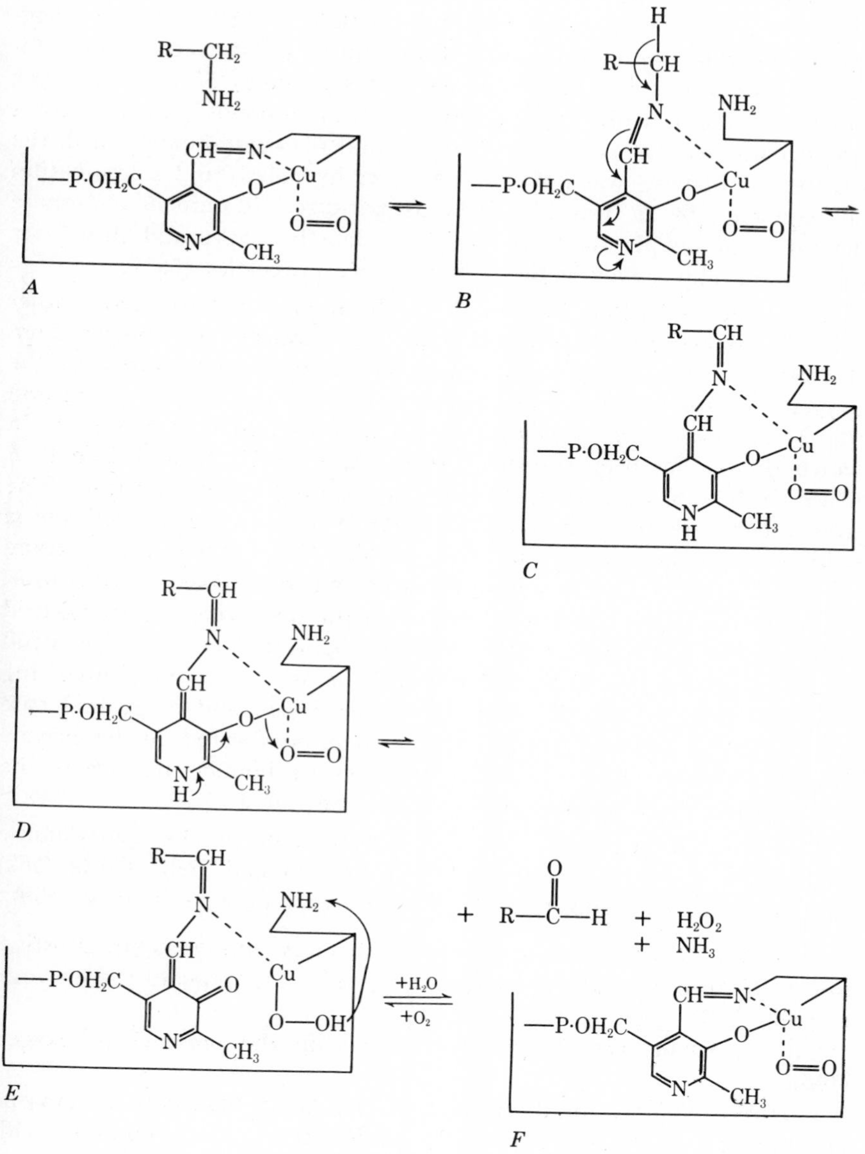

$R—CH_2$
NH_2

$CH=N$ Cu
$—P·OH_2C$ O
$O=O$
N CH_3

A

⇌

$R—CH$ H
N
CH NH_2
$—P·OH_2C$ O Cu
$O=O$
N CH_3

B

⇌

$R—CH$
N NH_2
CH
$—P·OH_2C$ O Cu
$O=O$
N CH_3
H

C

$R—CH$
N NH_2
CH
$—P·OH_2C$ O Cu
$O=O$
N CH_3
H

D

⇌

$R—CH$
N NH_2
CH
$—P·OH_2C$ O Cu
N CH_3 $O—OH$

E

$+H_2O$ / $+O_2$

$+$ $R—C—H$ $+$ H_2O_2
O
$+$ NH_3

$CH=N$ Cu
$—P·OH_2C$ O
$O=O$
N CH_3

F

85

In the first step of the reaction, A, a transamination between the substrate $R-CH_2-NH_2$ and the pyridoxal phosphate form of the enzyme occurs with the formation of the Schiff base B. Direct evidence for the formation of a Schiff base between enzyme and substrate in the first step of the reaction has been obtained by Buffoni (582), who used hog-plasma benzylamine oxidase and ^{14}C-histamine as substrate. Under anaerobic conditions the latter enzyme binds 3 moles of histamine per mole of protein, with the formation of imine bonds. This reaction product was acted upon by borohydride, and the compound formed, the reduced Schiff base, was isolated, and after hydrolysis and fractionation on Sephadex G 25 was identified as pyridoxylhistamine-5′-phosphate derivative of pig-plasma benzylamine oxidase by paper and thin layer chromatography as well as by its fluorescence properties (582).

The tautomerization of B to C has been proposed for many other enzymes involving pyridoxal phosphate. The transfer of a proton from the nitrogen of the pyridoxal phosphate through the solvent to the oxygen, and the transfer of electrons through the conjugated system and the metal ions to the oxygen, as shown in D, would lead first to E, in which the O_2 has been converted to the level of H_2O_2, and then to F in which the rest of the complex has been oxidized by two electrons. The subsequent steps to yield H_2O_2, $R-CHO$, NH_3, and regenerated amine oxidase would be expected to occur rapidly (572). According to Hamilton (572), the oxidation step in this scheme is the conversion of D to E, and this occurs by a mechanism in which the hydrogens are transferred as protons with no intermediate radicals being formed and the metal ion not changing its valency. Hamilton believes that reactions involving proton transfer are usually subject to general acid and/or general base catalysis. Acid and base groups may be therefore located on the enzyme in such a manner that, when the substrates are bound, they are in the correct orientation for catalysis [see (583) ; Section 1.E].

The mechanism of the reaction between hog-plasma benzylamine oxidase and benzylamine, proposed by Buffoni and Della Corte (569) and shown here, is supported by experimental data as follows (569) :

1. Benzylamine oxidase acts only on primary amines (146); substitution of hydrogen on the α-carbon of the substrate impedes the enzymic reaction (131).

2. The enzyme activity greatly depends on the presence of oxygen (146).

3. Direct evidence for the formation of a Schiff base between enzyme and substrate in the first step of the enzymic reaction was obtained with ^{14}C-histamine (582).

4. Electron-spin-resonance studies have shown that the copper remains in the Cu(II) state in the oxidized and reduced enzyme (570). There were, however, changes in the ESR line shape after reduction and in the ESR signal throughout the catalytic cycle, which may indicate a direct participation of cupric copper in the catalytic reaction.

All these experimental data suggest a reaction mechanism for hog-plasma amine oxidase in which the formation of the Schiff base is the rate-determining step of the reaction (571).

It is interesting to note that Yamada and co-workers (224), when trying to assess the concentration of cupric copper in crystalline beef-plasma amine oxidase, found a discrepancy between the results obtained by integration of the ESR signal of the native enzyme (2.7 g-atoms of cupric copper per mole of enzyme) and those determined chemically (3.7 g-atoms per mole of enzyme). Similarly, Buffoni et al. (570) found the copper content of crystalline hog-plasma amine oxidase, obtained by integration of the ESR signal, to be lower than that given by chemical assay. This discrepancy in the results could according to Buffoni et al. indicate either that these enzymes contain cuprous copper or that the cupric atoms are interacting with one another or with some other para-magnetic species. However, Yamada and Yasunobu (580) found no indication of cuprous copper. Equally, there is no ESR evidence for cupric interactions, since according to Broman et al. (584) such interactions would result in broadening of the signal relative to that from cupric EDTA. The reason for this discrepancy between the copper content determined by ESR integration and that obtained by chemical analysis remains unexplained.

In a very recent work involving an ESR technique, Knowles and his colleagues (585) have presented evidence that an ESR signal, observed at pH 10 during the oxidation of substrates by oxygen catalyzed by xanthine oxidase, is due to O_2^-, the superoxide free-radical anion, which these authors found to be stabilized by the alkaline medium, but not to interact with the enzyme molecule. This enzymically produced radical was shown by these workers to be identical with a radical obtained nonenzymically—that is, by oxidation of hydrogen peroxide by periodate ions and by other inorganic reactions. By means of the rapid-freezing technique of Bray as modified by Bray, Knowles, and Meriwether (586), the authors obtained for ESR measurements samples which were frozen after short reaction periods. Measurements of g values (to $\pm$ 0.002) and absolute signal-intensity measurements were carried out according to the work published by Bray et al. (586). It was observed that larger radical signals were obtained when smaller jets were employed, as described by Palmer, Bray, and Beinert (587). This finding agrees with the fact that a rather labile species was being investigated (588). Using this technique, Knowles and his colleagues (585) observed an ESR signal with $g \parallel = 2.08$ and $g_\perp = 2.00$, which they ascribed to the presence of the superoxide ion O_2^-, produced enzymically during the reaction between xanthine (2.0 mM) and O_2 (0.76 mM), catalyzed by xanthine oxidase (0.20 mM), at pH 10.0 and 23°, after a reaction time of 150 msec as well as nonenzymically during the reaction between H_2O_2 (140 mM) and $NaIO_4$ (70 mM), at pH 9.9 (carbonate buffer) after a reaction time of 600 msec. The signal $g_\perp = 2.00$ was apparent only at intermediate reaction times and was considerably stronger at pH 10 than at pH 8.2.

Results of an investigation of the effects of oxygen concentration and of enzyme concentration on the intensity of the signals, due to the radical O_2^-, and on other signals, produced during the enzymic oxidation of xanthine, showed that the relative intensity of the radical was the same whether it was measured as the $g_\perp$ or the $g \parallel$ component, thus confirming that this observation was due to the presence of a single species. On lowering the oxygen concentration, the radical signal decreased, and it approached zero intensity when anaerobic conditions were employed. On the other hand, when the enzyme concentration was decreased approximately by 31% of the original value, the intensity of the radical signal did not decrease proportionately. These experimental data (585) seem to provide strong evidence that the radical signal is associated with oxygen rather than with the enzyme. Integration of the oxygen radical signal, produced in experiments at pH 10, indicated that concentrations of the free radical up to about 0.06 mM could be achieved by using a starting oxygen concentration of 0.76 mM.

The authors confirmed that the free radical could be produced also from oxygen by xanthine oxidase in the presence of substrates other than xanthine as well as, as mentioned above, by mixing, at pH 10, hydrogen peroxide with periodate. On the basis of all the results obtained, Knowles et al. (585) conclude that the ESR signal described, which may be produced either by the enzyme system or by the inorganic system, is due to an oxygen free-radical species that appears identical with the unidentified "alkali-stabilized form of O_2^-," first described by Czapski and Dorfman (589).

To explain the stability of the superoxide ion, O_2^-, in alkaline medium, Knowles and co-workers (585) suggest that it is in rapid equilibrium with a small amount of a dimer and that radical disproportionation occurs via this dimer by a proton-catalyzed process:

$$O_2{}^- + O_2{}^- \underset{k_2}{\overset{k_1}{\rightleftharpoons}} O_4{}^{2-} \tag{1}$$

$$O_4{}^{2-} + H^+ \xrightarrow{\hspace{3cm}} O_2 + HO_2{}^-. \tag{2}$$

Proton attack on the dimer would facilitate the intramolecular electron transfer required for disproportionation (equation 2). At neutral pH values, $k_2 [H^+]$ is assumed to be large enough to make k_1 a rate-limiting factor in the decomposition. In alkaline solution, however, $k_2 [H^+]$ would become rate limiting. This would explain the rapid increase in stability with increasing pH in this region. Therefore, because the equilibrium in equation 1 favors dissociation of the dimer, the species detected by ESR would be the superoxide free radical $O_2{}^-$. For investigations of the significance of the oxygen free radical $O_2{}^-$ in enzymic reactions, Knowles and his co-workers (585) suggest that the ESR technique, which they applied, may serve as a clear-cut diagnostic test for one-electron reduction of oxygen in those enzyme reactions that can be studied at pH 10.

When surveying the information gained on the kinetics of amine oxidases from the application of the ESR technique in various laboratories, one must conclude that our knowledge of these enzymes has considerably advanced. A fully satisfactory understanding, however, of many challenging questions—such as the detailed mechanism of the enzymic reduction of oxygen and the mode of action of the various groups of amine oxidases—must await results of further work on these enzymes which will have to be repeated and extended.

E. Substrate Pattern, Substrate Specificity, and Mechanism

Hog-kidney DAO acts not only on aliphatic diamines and histamine (590) but also on aliphatic monoamines, such as propylamine (591), and on benzylamine (145), tyramine (591), and mescaline (138). Large substrate concentrations (591) as well as considerable amounts of enzyme (592), however, are usually required to demonstrate an oxidative deamination of monoamines by DAO. The K_m values of diamines are about one-tenth those of the corresponding monoamines (591). Zeller and co-workers (590) proposed the following pattern for specific DAO substrates:

$$
\begin{array}{ccccc}
\text{B} & & \text{C} & & \text{A} \\
\overbrace{} & \overbrace{} & & \overbrace{} \\
\text{R}' & \text{R}'' & & \text{R}''' & \\
\bullet & \bullet & & \bullet & \\
\text{R} \bullet \text{N} \bullet \text{C} \bullet \bullet \bullet \bullet \text{C} \bullet \text{CH}_2\text{NH}_2 \\
\bullet\ \bullet & & & \bullet \\
& & \text{H} &
\end{array}
$$

where R represents H or an organic residue.

The two basic residues A and B perform entirely different functions. Group A must be attached to a CH_2 residue and is not replaceable by a secondary amine. This is the NH_2 group that is split off during the enzyme reaction. The structural pattern of group B, which acts as an electron donor, is not so limited. Mono — and dimethyl derivatives of putrescine (C_4) or cadaverine (C_5) are good substrates, but the corresponding trimethyl analogues are hardly affected by DAO. Group B may also represent part of a guanidine residue. The optimal length for α-ω-diamino alkanes is C_5; and for α-amino-ω-guanidino alkanes, C_3 (590).

Furthermore, not only the number of methylene groups between A and B, contained in group C, but also the total length of the aliphatic chain is of importance. Although 1,3-diaminopropane is hardly a substrate (150), 1,3-diaminobutane is easily degraded by DAO (590). Group B can also be formed by aromatic, nitrogen-containing systems, as shown by the aminoethyl derivative of imidazole (histamine) as well as that of pyrazole and of 1, 2, 3-triazole (590). The compounds 2-(2-aminoethyl) pyridine (593, 594) and 4-(2-aminoethyl) pyridine (590) are not, however, substrates of DAO. Group C must contain the α and β H atoms, which are essential for the deamination reaction. Moreover, C is the site of attack for the van der Waals and hydrophobic forces by which the substrate is attached to the enzyme surface. Although methylation of the imidazole ring in the 1-, 2-, or 5-position does not hinder degradation by histaminase (124, 590, 593), methylation of the side chain in the α-position or of the amino residue completely abolishes the enzymic oxidative deamination (595, 596).

Zeller (126, 127) recently enlarged on his theory with regard to the binding of the substrate to the active site of DAO. He suggests that in the interaction between DAO and substrate two types of enzyme-substrate complexes may form, which according to whether the binding is of the "eutopic" or "dystopic" type, may either catalyze the enzyme reaction or prevent it from occurring; for example, 2- (2-aminoethyl) pyridine forms with DAO a dystopic complex with no ensuing enzyme interaction. On the other hand, aminoethyl derivatives of imidazole, pyrazole, or triazole form eutopic complexes, whereby the interaction between DAO and those substrates takes place, although at markedly different rates. This interesting concept requires further elucidation.

Fasella and Manning (583) have lately studied the relationship between substrate structure and reaction specificity in pyridoxal phosphate–dependent enzymes. Since the suggestion that purified hog-kidney histaminase is a pyridoxal phosphate–dependent enzyme (553) has been recently confirmed by Mondovi et al. (556, 564), the interesting work of

Fasella and Manning (583) should be discussed here. Amine substrates bind the pyridoxal enzymes to their amino group forming a Schiff base with the carbon in the 4' position of the coenzyme (see schemes 1 and 2 on pp. 79 and 80). According to Fasella and Manning, substrate specificity depends on the interaction of parts of the substrate molecule with those protein sites that display suitable chemical and geometrical properties. Thus, the geometry of the active site may determine not only "substrate" but also "reaction" specificity. Dunathan (597) has recently pointed out that, of the various bonds that involve the α-carbon of the substrate, bound as a Schiff base to the holoenzyme, only the one oriented in a direction normal to the plane of the coenzyme ring will be labilized.

A large number of the reactions catalyzed by pyridoxal phosphate–dependent enzymes proceed through the initial labilization of the bond between the α-carbon of the substrate and hydrogen, the latter being eliminated as a proton. The coenzyme-substrate Schiff base is thus transformed into an intermediate that is probably of semiquinonoid nature and shows an absorption band at 490 nm (583, 598). The subsequent fate of this intermediate will depend on the structure of the substrate as well as on that of the enzyme.

Fasella and Manning (583) have studied various groups of pyridoxal phosphate–dependent enzymes and suggest that in such systems the enzyme structure may determine which substrate will be bound to its active site, how tight this link will be, and which bonds of the substrate will be labilized. The structure of the substrate will determine, at least in part, the subsequent fate of such an intermediate.

F. Inhibitors

The literature on the inhibitors of histaminase (DAO) has been reviewed by various workers (126, 130, 131).

a. IN VITRO INHIBITION

The most important inhibitors of hog-kidney DAO activity, belonging to the group of carbonyl reagents, are hydroxylamine (150, 505, 514, 594), hydrazine (599), the isonicotinic acid hydrazide isoniazid (135, 142, 145, 600), semicarbazide (150, 594), and aminoguanidine (145, 522, 591, 599). These most widely used inhibitors in general block DAO activity at 10^{-5} to 10^{-7} M concentrations. All these substances fail to inhibit any of the monoamine oxidases (126). On the other hand, the *cis*- and *trans*-2-phenyl-cyclopropylamines, which are the most powerful inhibitors of MAO (251, 344), are without any effect on the DAO of hog kidney (251) or guinea-pig liver (521). The 1,2-disubstituted hydrazine derivative, iproniazid, (see Chapter III. F), although unable to form a hydrazone, is a potent inhibitor of DAO (142, 145, 522, 600–602). Many experimental data indicate that iproniazid is first hydrolyzed before its action on amine oxidases (344). This hypothesis may explain why the hydrolytic product, the isopropylhydrazine, is almost a hundred times as effective as iproniazid (603), and why purified DAO is relatively resistant to iproniazid (518, 603–605).

To the group of basic inhibitors of hog-kidney DAO belong guanidine and its derivatives (145, 594, 606), the amidines and diamidines (606), imidazole (607), thiamine and its pyrimidine components (594), atabrine (605, 608), bulbocapnine (an alkaloid from the plant *Corydalis cava*) (609), and the dyestuffs methylene blue, toluylene blue, and pyocyanine (126). All these compounds in 10^{-3} to $10^{-5}M$ concentrations block hog-kidney DAO by at least 50% (126). Similar results were obtained with cat kidney (522) and guinea-pig liver (126). Kobayashi and Okuyama (610) presented evidence for the quantitative recovery of hog-kidney DAO activity, originally inhibited by isoniazid, by the addition of ferricyanide. The latter compound apparently caused evolution of nitrogen in an amount equivalent to that contained in the hydrazide group of isoniazid and thus prevented loss of enzyme activity.

All guanidine derivatives are competitive inhibitors of hog-kidney DAO. The inhibitory effect of guanidines increases on methylation (130). Aminoguanidine is one of the most potent *in vitro* inhibitors of DAO. In a concentration of $5 \times 10^{-8}M$ aminoguanidine blocks crude DAO preparations up to 50% (599). Other *in vitro* inhibitors are cysteamine (611), diisothiourea derivatives (606), stilbamidine and its methyl derivatives (606), and the long-chain diamines (560). Cysteamine is a short-acting inhibitor of the hog-kidney enzyme, for it is readily transformed into cystamine, which is a substrate of DAO (histaminase) (611). Diisothiourea derivatives exert only a moderate inhibitory effect on hog-kidney histaminase, but stilbamidine and dimethylstilbamidine are strong inhibitors of the enzyme (606). Although the long-chain diamines are not substrates of hog-kidney DAO, they show an affinity for the enzyme that results in a competitive inhibition (560).

In regard to the inhibition of hog kidney DAO by potassium cyanide Zeller, on the basis of results obtained, concludes that cyanide does not act on DAO through its ability to combine with heavy metals but in its capacity as a carbonyl reagent (122, 150).

Of the chelating agents, pyrophosphate and azide do not inhibit crude hog-kidney histaminase preparations (149). Similarly, sodium sulfide and thiourea inhibit the hog-kidney enzyme only to a very small extent (607). Comparatively high concentrations of 8-hydroxyquinoline ($3 \times 10^{-3}M$) and of diethyldithiocarbamate ($10^{-2}M$) are required to block hog-kidney

DAO (612). It is noteworthy that, whereas Werle and Hartung (612) achieved a total inhibition of DAO activity by diethyldithiocarbamate with cadaverine as substrate, with histamine as substrate the chelating agent not only did not act as an inhibitor but proved to be an activator of the hog-kidney DAO. Mondovi and his colleagues (554) confirmed Werle and Hartung's findings when they added $10^{-2} M$ diethyldithiocarbamate to their purified kidney enzyme using cadaverine as substrate. The activity of the purified enzyme was also almost completely abolished. It is regrettable that these authors did not fully repeat Werle and Hartung's experiment, since they omitted to note the effect of their purified enzyme preparation also on histamine as substrate.

b. IN VIVO INHIBITION

There is no parallelism between the potency of DAO inhibitors *in vivo* and *in vitro* (613). Many of the diamidines cause an increase in blood histamine concentration by releasing histamine from the tissues (154, 614), and some of these compounds are also known as potent inhibitors of histaminase. But blockage of histaminase activity and release of histamine do not always exhibit parallel phenomena *in vivo*. The most selective inhibitors of histaminase (DAO), however, both *in vitro* and *in vivo* are semicarbazide, aminoguanidine, and isoniazid (613).

A potentiation of the pharmacological effects of histamine by histaminase inhibitors was first observed by Schild and co-workers (594, 615). They found that such histaminase inhibitors as carbonyl reagents, guanidines, and imidazole derivatives potentiated the response to histamine of the isolated guinea-pig tracheal chain, ileum, and uterus. The effect of histamine on cat blood pressure was not consistently potentiated. The trachea, ileum, and uterus of the guinea pig are known to contain histaminase. There was a significant correlation between the molar concentration of the inhibiting substances and the pharmacological effect. Moreover, the potentiation by histaminase inhibitors concerned only histamine; no potentiation of choline or acetylcholine, or of histamine analogues resistant to histaminase, was obtained. Lindell and Westling (616) found that histaminase inhibitors potentiated the effect of histamine injected into the renal artery of cats, but were without any effect when histamine was injected into the femoral artery or vein. Similar results were obtained in dogs (617). All these findings agree with the fact that in cats and dogs the histaminase activity is especially high in the kidney and very low in other tissues. Blaschko and Kurzepa (618) observed that the effect of histamine on the isolated guinea-pig ileum is potentiated by the three picolylamine isomers, which are all inhibitors of histaminase. The corresponding N-methyl derivatives, which do not inhibit histaminase, did not potentiate the response of the ileum to histamine (618). Mescaline potentiates the hypotensive effect of histamine in rats, anesthetized with urethane, and also potentiates the contractile effect of histamine on guinea-pig ileum (619). Papaverine and bulbocapnine potentiate the effect of histamine on the permeability of the vessels of guinea pigs. Furthermore, papaverine, when injected along with histamine or histidine, induces a catatonic state in mice (620). This effect is probably mediated through an inhibition of histaminase (620). Ivy and his co-workers (621) demonstrated that aminoguanidine potentiated the effect of histamine on gastric secretion, and also the effect of those histamine analogues that are substrates of DAO.

Imidazole acrylylcholine (murexine) and imidazole propionylcholine (dihydromurexine) are inhibitors of DAO (622). This evidence may explain their potentiation of the histamine effect on smooth muscle such as guinea-pig ileum, lung, and trachea (622). It is significant that murexine, dihydromurexine, and aminoguanidine do not potentiate the hypotensive action of intravenous histamine in dogs or the vasodilator effect of intraarterial histamine (623).

G. Other Biological Mechanisms than Histaminase for the Elimination of Histamine *in vivo*

a. ACETYLATION OF HISTAMINE

On oral administration of histamine to a variety of animals, acetylhistamine is excreted in the urine (624). The conversion of histamine to acetylhistamine with the acetate bound to the side chain nitrogen has been also demonstrated *in vitro* with rabbit-liver slices, pigeon-liver slices, or extracts of acetone powder of pigeon liver in the presence of coenzyme A (625). The enzyme that catalyzes the acetylation of histamine has been purified from pigeon-liver acetone powder (626). According to Schayer (627), the *in vivo* acetylation of histamine is largely bacterial, since quantitative analyses of the urines of rats, guinea pigs, rabbits, and cats, that had been injected with ¹⁴C-histamine, hardly indicated an acetylation of the side-chain amino group by mammalian tissues. Livingston and Code (628) infused histamine into the portal vein of dogs, monkeys, and rats and found by analysis of the urine no evidence for acetylation of histamine in dogs and monkeys, but a small increase in the "conjugated histamine" fraction (acetylhistamine) in rats.

A marked decrease in the excretion of acetylhistamine in the urine of patients suffering from dystrophia myotonica has been reported by Sjaastad and Sjaastad (629).

b. METHYLATION OF HISTAMINE

This pathway for the metabolism of histamine *in vivo* involves the methylation of the imidazole ring. Karjala and Turnquest (630) and Karjala, Turnquest, and Schayer (631) isolated from the urine of mice, that had been injected with large quantities of low-activity ^{14}C-histamine, a radioactive crystalline substance that proved to be methylimidazole acetic acid. The probable intermediate is 1-methylimidazole-4-acetaldehyde, which is oxidized to 1-methylimidazole-4-acetic acid (632). It was found by isotope-dilution assays that minute quantities of ^{14}C-histamine are *in vivo* methylated on the imidazole ring nitrogen remote from the side chain, to form 1-methyl-4-(β-aminoethyl) imidazole (632). No methylation of the ring nitrogen adjacent to the side chain—that is, formation of 1-methyl-5-(β-aminoethyl) imidazole—occurs, when small amounts of histamine are injected. If mice are injected with large quantities of histamine, some of the abnormal isomers, 1,5-methylhistamine and 1,5-methylimidazole acetic acid, may be formed (632). It is, therefore, possible that the latter compounds may be produced in the case of a massive release of tissue histamine (633). The existence in various tissues of an enzyme, imidazole-N-methyltransferase, that methylates histamine has been described by Brown and his colleagues (529, 634), Lindahl (635-637), and White (528). It should be noted that if mice, in which methylation is normally the principal pathway of histamine catabolism, are injected with large quantities of histamine, histaminase assumes the main role in the degradation of this drug (631). It is remarkable that in sheep oxidative deamination seems to be the major catabolic pathway for both exogenous and endogenous histamine.* After intravenous injection of ^{14}C-histamine, imidazoleacetic acid, free and conjugated, was found to account for 70 to 80% of the radioactivity excreted in the urine.

As was first shown by Kapeller-Adler and Iggo (596), 1,4-methylhistamine is a very good *in vitro* substrate of hog-kidney histaminase. The same observation was independently and almost simultaneously reported by Lindell and Westling (593). Those authors found, however, that *in vitro* 1,4-methylhistamine was not only a substrate of histaminase, but that this compound was oxidized, although not so rapidly also by MAO. Because of the latter observation by Lindell and Westling, and his own negative results for inhibition *in vivo* with aminoguanidine (633), a known strong inhibitor of DAO, as well as because of his *in vivo* findings of inhibition with the MAO inhibitors, iproniazid and 1-isonicotinyl-2-isobutyl hydrazine (632), Schayer concluded that MAO and not histaminase was responsible for the oxidative degradation of methylhistamine

---

* Personal communication from O. V. Sjaastad (1968).

in vivo (638). This assumption found no support by the present author, who believes that results of *in vivo* inhibition experiments are inadequate for the identification of enzyme activities (562).

In this connection work by Kobayashi and Ivy (519, 520) on histaminase activity in the stomach and intestine of dog and rat should be recalled. These workers reported that *in vivo* histaminase activity, unlike that *in vitro*, was not completely abolished by aminoguanidine (10^{-4} M) or semicarbazide (10^{-3} M). According to Kobayashi and Ivy, this may have been due to these histaminase inhibitors not having penetrated all the cells (519, 520). Moreover, Bjuro, Lindberg, and Westling (640) also failed to obtain any measurable inhibitory *in vivo* effect of aminoguanidine on histamine excretion in the urine of five pregnant and three nonpregnant women. Further elucidation of the problem of the *in vivo* inhibitory effect of aminoguanidine on histaminase is urgently required.

Some experimental evidence seems to point to an enzyme system, playing a predominant role in the oxidation of methylhistamine *in vivo* (641–643), that is insensitive to aminoguanidine, but is sensitive to iproniazid, an inhibitor of both MAO and DAO (632). It is obvious that on this experimental basis it is impossible to decide whether this enzyme system is identical with the classical MAO that acts on catecholamines and indoleamines, or if it represents yet another group of animal amine oxidases. As mentioned above, (Chapter III.2) Kobayashi (148, 644) found that mouse liver as well as cat liver contains an enzyme system that resembles MAO in its sensitivity to inhibitors, but clearly differs from the classical MAO by displaying both MAO and DAO activities. It should be further pointed out that the formation of 1,4-methylimidazole-acetic acid is not affected by the *p*-tolyl ether of choline (632), which is an inhibitor of MAO in intact animals (645). Recently, Snyder and Axelrod (642) studied the *in vivo* action of various drugs on histamine metabolism in the mouse by measuring the effect of various compounds on the disappearance of administered ^{14}C-histamine and the appearance of ^{14}C-methylhistamine in the whole mouse. In the whole mouse, 10 min after an intravenous injection of 10 μg of 2-^{14}C-histamine hydrochloride, 14 to 24% remained as circulating ^{14}C-histamine and 10 to 22% was present as ^{14}C-methylhistamine. In mice, that had been injected intraperitoneally with either chlorpromazine, serotonin, bufotenine, or aminoguanidine 15 to 60 min before the administration of ^{14}C-histamine, the disappearance of histamine was markedly slowed. Whereas the amount of ^{14}C-methylhistamine formed was reduced when chlorpromazine, serotonin, and bufotenine—all known *in vitro* inhibitors of N-methyltransferase (529)—were given, it was found to be elevated after the administration of aminoguanidine. Aminoguanidine would be expected to block the disappearance of injected histamine. The elevation of the methylhistamine level observed after aminoguanidine treatment was considered by Snyder and Axelrod (642) to indicate either a diversion of the histamine metabolism into the methylation pathway, or an inhibition of the methylhistamine catabolism. It should be remembered, however, that, whichever of the amine oxidases may be active *in vivo* on methylhistamine, the latter compound will be oxidized by the amine oxidase only to the stage of 1,4-methylimidazoleacetaldehyde (562). The problem with regard to the enzyme system that will further oxidize the latter compound to the final, *in vivo* identified, metabolite of methylhistamine, the 1,4-methylimidazoleacetic acid, awaits clarification.

In this context an observation made by Lindberg and his colleagues (646) on human embryonic tissues should be mentioned. When tissues of human fetuses, 19 to 24 weeks of age, obtained at abortions induced according to the Swedish law, were incubated with ^{14}C-histamine, the main route of histamine degradation was that of imidazole-N-methyl-transferase. The major end product obtained was, however, ^{14}C-1,4-methylhistamine instead of the expected 1,4-methylimidazoleacetic acid. When ^{14}C-histamine is injected into human beings, the major end product is methylimidazoleacetic acid (647). The reason for this discrepancy is not clearly understood. The authors suggest that the enzyme capable of oxidizing methylhistamine to methylimidazoleacetic acid may not yet be fully active in fetal tissue of 19 to 24 weeks of age.

Attempts to demonstrate formation of 1,4-methylimidazoleacetic acid by direct methylation of imidazoleacetic acid have been unsuccessful (642).

It is established (638) that histamine is metabolized *in vivo* by two main routes, shown in the scheme below, as follows:

1. The classical route of enzymic degradation, with the ultimate excretion of imidazoleacetic acid, which is largely conjugated with ribose to 1-ribosylimidazole-4-acetic acid.

2. By methylation of the imidazole nitrogen remote from the side chain, with the formation of 1-methyl-(4-β-aminoethyl)imidazole (1-methylhistamine). The latter compound is further oxidized by another enzyme system (562, 596, 638) to methylimidazoleacetic acid.

C. FORMATION OF HISTAMINE RIBOSIDE AND RIBOTIDE

In 1955 Tabor and Hayaishi (648) demonstrated that imidazoleacetic acid could be converted *in vivo* to the corresponding riboside (I). Using a purified rabbit-liver preparation in a reaction requiring adenosine

Catabolic Pathways of Histamine

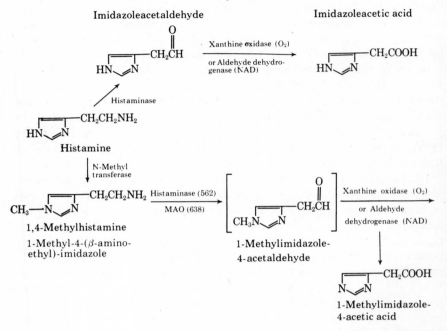

triphosphate and 1-pyrophosphoryl-ribose-5-phosphate, Crowley (649, 650) was able to demonstrate the *in vitro* conversion of imidazoleacetic acid to the corresponding ribotide (**II**). This work was confirmed by Fernandes and co-workers (651), who used a homogenate of the mucosa from dog small intestine and found that their preparation produced the ribotide of histamine as well as the ribotide of imidazoleacetic acid. The ribotides formed could be converted to the corresponding ribosides by a snake-venom nucleotidase.

A most striking observation has been recently reported by Snyder, Axelrod, and Bauer (652). In their study of the fate of ^{14}C-histamine in animal tissues these authors found that relatively large quantities of the imidazoleacetic acid and its riboside were retained in the tissues. This formation and retention of imidazoleacetic acid and its riboside may be of particular interest with regard to a recent finding of Fox and Lasker (653) that imidazoleacetic acid offers a more effective protection against anaphylactic shock in mice than cortisone.

Abdel-Latif and Alivisatos (654) reported that in the presence of an enzyme from bull semen histamine reacted directly with either nicotinamide ribotide or nicotinamide riboside, with the formation of histamine ribotide and histamine riboside, respectively.

d. HISTAMINOPEXIC ACTION OF BLOOD SERUM

Parrot and Laborde (655) observed that a solution containing histamine dihydrochloride (10^{-6} M) and 5% of dialyzed normal human serum, when tested for its biological activity on the guinea-pig ileum, showed a decrease in its histamine content of about 30%. These authors found that the factor responsible for this blood-serum action, named by them "histaminopexic action," was contained in the γ-globulin fraction of the serum. None of the other serum protein fractions displayed this property. The serum γ-globulins seem to form labile complexes with histamine. Furthermore the serum contains a factor, partly linked with the albumin fraction, that inhibits histaminopexy. This inhibitor can be eliminated, however, by prolonged dialysis. There is evidence that in allergic conditions in man and after adrenalectomy in the rat the decrease in histaminopoxy observed is caused by an increase of this inhibitor in the serum.

H. Biological Significance of Classical Histaminase, (DAO)

The identity of the activities of the classical histaminase and DAO on the biogenic amines histamine, putrescine, and cadaverine has not been proved in all tissues. It therefore appears profitable to consider separately the biological significance of the enzymic degradation of those amines.

The function of histamine and its importance in physiological and pathological processes have been discussed in a countless number of publications (656). The endogenous formation of histamine in tissues of human adults has been difficult to demonstrate. Human adult tissues in which histamine formation has been observed are lung tissue (657), skin lesions from patients with *urticaria pigmentosa* (658) and gastric mucosa.*

Radioisotope techniques have materially advanced the knowledge of histamine formation *in vivo* (659). In 1959 Kahlson and Rosengren (660, 661) demonstrated an extremely rapid formation of histamine

---

* S. E. Lindell, H. Westling, and B. Zederfelt, unpublished data (1963).

during pregnancy in the rat and advanced the theory that an increased synthesis of histamine was necessary for fetal development. It could also be shown that the fetuses of the rat contained an extremely active histidine decarboxylase.

These findings have been confirmed by Burkhalter (662) in his recent study on the formation of histamine by fetal rat liver.

West and co-workers (663) have suggested that the function of histamine formed by the rat fetus is to regulate the blood flow through the placenta.

Rosengren (664) has been able to demonstrate that in the pregnant hamster the placenta is the site of the increased rate of histamine synthesis.

Studies on the histamine concentration in whole blood and plasma of the umbilical arteries and vein have produced some evidence that the human fetus may also form histamine *in vivo* (665).

Kahlson, Rosengren, and White (665) found that plasma from fetal arterial blood contained histamine in measurable amounts, whereas no plasma histamine was detectable in the mother. These authors also observed that the histamine content of the umbilical-vein plasma was lower than that of the umbilical-artery plasma, which may suggest a removal of histamine by the placenta. Further evidence that the umbilical-cord blood contains more histamine than the maternal blood during labor was obtained independently by Wicksell (666) and Kapeller-Adler (536). Moreover, Bjuro et al. (667) reported high urinary histamine excretion during some normal pregnancies, which immediately ceased on parturition. The analogous findings in the pregnant rat (668, 669) seem to strengthen the belief that the human fetus can also produce histamine.

Further experiments by Kahlson and Rosengren (661) showed that histamine was also formed in regenerative and reparative growth as well as in certain tumors in the rat. The histamine, which has been related by Kahlson to growth processes and which this author called "nascent histamine", is not bound in the tissues in the conventional way. In growing tissue, when histamine formation is rapid, the active histamine content is generally quite low. Schayer (670) has introduced the concept of intracellular histamine formation as an intrinsic regulator of the microcirculatory system. The evidence for this "induced histamine" is indirect, and is based on the observation that subjection of amino acids to procedures or stimuli likely to increase the requirements for blood in tissues, produced an increased histidine decarboxylase activity, resulting in the formation of "induced histamine." The status of Kahlson's "nascent histamine" in growing tissues is thus similar to that of Schayer's "induced histamine," both being formed in the tissue but not staying there for a long time. The physiological effects of this "sustained histamine action" may well be radically different from those of exogenous or suddenly liberated histamine.

However, no evidence has been presented that would indicate whether Schayer's concept of histamine as a circulatory regulator is applicable to man (671). The methods presently available for experimentation in man are insufficient, and further extension of this theory to man will have to await newer and better techniques.

Animal experiments have shown that the rate of formation and degradation, and the biological activity of histamine is under hormonal control by the anterior pituitary, the adrenals, and the thyroid gland (671). There is some indication that in man the metabolism of histamine may be similarly influenced by the adrenal cortex and the thyroid gland (671). A lower level of histaminase activity has been found in various tissues of adrenalectomized animals, such as rats (672–674), cats (675), and guinea pigs (676). The decrease of histaminase levels observed in the kidney and intestinal mucous membrane of adrenalectomized cats may be due to a release of the enzyme into the thoracic-duct lymph (523). In agreement with this, an increase in the histamine content of tissues of adrenalectomized animals (674, 677–679) and also an increase in their sensitivity to histamine (680, 681) has been reported. In contrast to these observations in tissues, investigations of the catabolism of injected ^{14}C-histamine in adrenalectomized rats have not revealed any hormonal effect on the urinary excretion of imidazoleacetic acid, although the excretion of free histamine tended to be lower in comparison with previous results in the rat (643, 667) and in man (682). A renewed study of these *in vitro* and *in vivo* findings should be made to resolve many obvious discrepancies.

No direct experimental evidence on the effect of the thyroid hormone on the histaminase content of various tissues has so far been obtained, but decrease of intestinal histaminase activity in mice after treatment with thyroxine has been reported (683). Of the sex hormones, the estrogens seem to have a potentiating effect on the activity of placental histaminase, but the chorionic gonadotrophins a reducing effect (509).

Histaminase seems to pass from the kidney to the lymph, from the human placenta to the maternal blood, and from the prostatic gland to the seminal fluid (122). This transfer from the cytoplasm to the body fluids seems to be hormonally controlled (684).

An increase in the blood histaminase level has been demonstrated in various animals during anaphylactic shock, for example, in the rabbit (685), the rat (686–689), and in the guinea pig (690–693). Hahn and his colleagues (692, 693) investigated the histaminase activity in the blood plasma of guinea pigs that had been subjected to various forms of shock, and demonstrated that the liver played an important role in the action of various histaminase liberators, such as anaphylatoxin, heparin, protamine sulfate, and EDTA. Heparin has been found to be an especially active liberator, capable of exhausting the entire histaminase content of the guinea-pig liver by a single injection. The enzyme liberated from the guinea-pig liver by heparin has been shown to be sensitive to carbonyl reagents, to be inactivated by aminoguanidine, and to degrade not only histamine, cadaverine, and putrescine, but also the monoamine benzylamine (692, 693). Ohkura (694) observed during anaphylactic shock in a dog a large increase in histaminase activity in the lymph of the thoracic duct, but not in the blood plasma. Buffoni (131) reports on two incidents of human anaphylactic shock occurring after the injection of penicillin. In both instances an increase in the activity of the blood-plasma histaminase was observed.

In the blood of allergic persons no increase in histaminase activity has been so far encountered. It is even probable that the blood of allergic individuals may contain some histaminase inhibitors (695–697).

Recently, Hansson and his colleagues (698) succeeded in inducing in healthy, nonpregnant persons a rise in plasma DAO by intravenous injections (15,000 units) of heparin. The DAO activity was observed within 30 sec of administration; it reached its maximum at about 60 min, when the plasma DAO became comparable to the DAO plasma content of women in the third trimester of pregnancy, and returned to its normal level within 6 to 12 hr. The DAO activity in the blood plasma was determined by the isotope method of Okuyama and Kobayashi (547), as modified by Tryding (699).

This interesting finding may help to elucidate the recently reported effect of heparin on the cutaneous tests for allergic asthma (700).

Tabor (124) observed that histaminase activity in tissues increases with age, as do the plasma amine oxidases (153). The increase in histaminase activity with age may indicate an adaptive mechanism whereby the enzyme is induced by the substrate (131). Against this assumption, however, is the fact that, whereas animals adapted to histamine showed a lower sensitivity toward histamine than the control animals, the histaminase level in various tissues of the adapted animals remained the same as that in the controls. This was found in guinea pigs (701, 702), rats (703), and dogs (504). These results are, however, not surprising for, as Schayer (659) indicated, exogenous histamine is not incorporated, but is rapidly metabolized.

There are many reports on histaminase activity in various human pathological conditions. An increase in the histaminase level has been observed in the blood serum of patients with bronchial (704, 705) and endometrial carcinoma (706). A higher histaminase activity in the serum of schizophrenic patients as compared with that in normal persons has been reported (707). Analyses of urinary metabolites in animals, previously given small doses of ^{14}C-histamine, labeled in the imidazole ring, have indicated that in various species histaminase and N-methyltransferase are both responsible for the metabolism of the major part of exogenous histamine (638). The relative importance of these pathways is different in different species (638, 708). The oxidative products of histamine catabolism, imidazoleacetic acid and methylimidazoleacetic acid, represent the largest amount of the urinary metabolites, (60-80%) in the rat, mouse, rabbit, guinea pig, cat, and dog, and in man (638, 708). Schayer (638) found that the ratio of imidazoleacetic acid (free as well as conjugated with ribose) to methylimidazoleacetic acid is high in the rat, somewhat lower in the mouse, rabbit, and guinea pig, and very low in the cat and dog, and in man. Beall and Vanarsdel (709) reported that intravenously injected ^{14}C-histamine was metabolized in humans at a surprisingly constant rate, that it was rapidly cleared from the blood, and that the excretion of radioactivity in the urine was usually complete within 30 hr. These authors found, however, that after injection of ^{14}C-histamine more imidazoleacetic acid and less methylimidazoleacetic acid was excreted in the human urine than Schayer and his colleagues had previously reported (647, 710). Thus, a separation of the excreted radioactive metabolites, carried out by Beall and Vanarsdel by means of paper chromatography and radioautography, revealed imidazoleacetic acid to be the principal product of histamine metabolism. Whereas imidazoleacetic acid riboside as well as 1,4-methylimidazoleacetic acid were also consistently recovered in the urine, ^{14}C-histamine was not detected. The above findings were obtained by the authors in a study of ^{14}C-histamine metabolism in seven control subjects, five patients with Laennec's cirrhosis, three patients with bronchial asthma, and three other ones with histamine cephalalgia. No abnormalities in the histamine metabolism, as compared with control subjects, were found. Since injected ^{14}C-histamine is rapidly metabolized and appears in only trace amounts in the urine, Beall and Vanarsdel doubt whether urinary assays of free or conjugated histamine are altogether appropriate as indicators of endogenous histamine release.

These findings of Beall and Vanarsdel (709) were criticized by Kobayashi and Freeman (711), who injected intradermally microgram amounts of ^{14}C-histamine into 12 patients suffering from chronic schizophrenia and into 7 normal subjects. Unlike Beall and Vanarsdel, the latter group of workers did find a difference between the schizophrenic and normal persons with regard to the metabolism of ^{14}C-histamine. Thus, Kobayashi and Freeman observed, that the schizophrenics excreted less radioactive 1,4-methylimidazoleacetic acid and more imidazoleacetic acid riboside in comparison with the normals, subjected to the same treatment. Since a higher histaminase activity was reported in the serum of schizophrenic persons in comparison with that in the serum of normal individuals (707), Kobayashi's results obtained with urines of schizophrenics are hardly surprising. Considering that both groups of workers, Beall and Vanarsdel on the one hand, and Kobayashi and Freeman on the other, have used different techniques for the identification of the metabolites in the urine, further studies on the *in vivo* metabolism of ^{14}C-histamine in man seem to be necessary.

As to the distribution of histaminase in various mammalian tissues it is remarkable that, whereas high levels of histaminase have been commonly encountered in the intestinal mucosa (522), this enzyme has been reported to be absent from the brain of such species as the rabbit, guinea pig, and cat (534, 712–714). Cotzias and Dole (151) found some histaminase activity in the brain of rats and mice, but Burkard et al. (712), using ^{14}C-putrescine as substrate, could not demonstrate any DAO activity in the brain of those animals. The absence of histaminase from the brain seems to be in agreement with the observation of a relative impermeability of the brain tissue to histamine (715–718).

The occurrence of histamine in the central nervous system has been discussed by various workers, and the subject of brain histamine has been reviewed by Green (719). In an interesting study on the distribution of histamine in the central nervous system of the dog Adam (720) has been able to establish that, like norepinephrine (268) and serotonin (272), histamine occurs in the phylogenetically older parts of the brain. Using a very sensitive technique of assaying histamine on the superfused guinea-pig ileum, Adam (721, 722) found in the dog's brain the highest concentration of histamine in the hypothalamus and the area postrema, somewhat lower amounts in the midbrain, and the lowest ones in the cerebral cortex and white matter. In agreement with the findings of Harris and his colleagues (723), the highest histamine values were, however, found in the hypophysis, where, in contrast to the brain, histamine seems to derive at least in part from mast cells (720). The brain does not contain any mast cells (716, 724). According to Michaelson and Dowe (725), histamine is present in the hypothalamus of the dog mainly in particles that sediment with microsomes, but the cellular origin of these particles is not known.

In a recent study on the brain of the cat Adam and Hye (716) obtained the highest histamine values in the corpora mammillaria (1150 ng/g of tissue) and in the ventral part of the hypothalamus (800 ng/g of tissue). Lower histamine concentrations were found in the remainder of the brain stem, and they were lowest in the bulb (<100 ng/g). No histamine was detected in the cerebellum or in the white matter (716).

These investigations have been lately considerably extended in Adam's laboratory by Stephen, who in his study of the concentration of histamine in the brain and hypophysis in different species also included such animals as the rabbit, guinea pig, rat, hamster, mouse, chick, chicken, pigeon, and frog.* The estimation of histamine was carried out in tissue extracts, previously purified by ion-exchange chromatography (Amberlite CG 50-cellulose), by a bioassay using the isolated, superfused guinea-pig ileum (722). Values for whole-brain histamine were lowest in mammals (range 60–130 ng/g), intermediate in birds (range 90–330 ng/g), and highest in the frog (380 ng/g). Mast cells were not detected in the brain of any of the species examined. Mast cells were, however, encountered in connective tissue of the pia-arachnoid and choroid plexus of birds, and were particularly numerous in the chicken. A detailed study was made of the distribution of histamine in rabbit brain; the concentration was highest in the hypothalamus (660 ng/g) and lowest in the cerebellum (60 ng/g). Similar values were obtained for the concentration of histamine in the hypothalamus of the rat (560 ng/g) and guinea pig (400 ng/g). Histamine concentration was also estimated in the hypophysis of the rabbit, guinea pig, and rat. The histamine content was highest in the hypophysis of the guinea pig (960 ng/g), with mast cells occurring mainly in the posterior lobe. No mast cells were found to be present in the hypophysis of the rabbit or rat. In the rabbit, the anterior lobe of the hypophysis contained most of the histamine extractable from the gland.*

In 1964 McGeer (726) reported on the distribution of histamine in the human brain. The highest histamine concentration was found in the hypothalamus (about 2500 ng/g tissue); the thalamus contained 400, the midbrain 250, the pons-medulla 900, the cerebral cortex 200 to 400, the cerebellum 500, and the caudate nucleus 500 ng of histamine per gram of tissue.

Snyder, Glowinski, and Axelrod (727) have recently reported on studies in the rat brain with regard to the uptake, storage, and metabolism of intraventricularly injected ^{3}H-histamine of high specific activity. They found that ^{3}H-histamine is taken up and retained in the rat brain for a prolonged period of time. It disappears from the brain in a multiphasic fashion, with a half-life of 1.6 hr during the first 6 hr and

---

* Personal communications from W. R. G. Stephen and H. Adam (1968).

a half-life of 11 hr from 6 to 24 hr. This is in marked contrast to the fate of exogenous histamine in peripheral tissues, from which it seems to disappear rapidly. Thus, after its intravenous injection, ^{14}C-histamine disappears from the whole mouse so suddenly that within 90 min only about 5% remains. Schayer (659) observed in 1952 that 4 hr after the injection of ^{14}C-histamine, it could not be detected in guinea-pig tissues. The retention of ^{3}H-histamine in the brain seems to suggest that it may be bound in the tissue in a manner that would protect it from rapid enzymic destruction (727). The retention in rat brain of ^{3}H-histamine and its multiphasic disappearance from the brain is reminiscent of the results obtained by Glowinski and Axelrod (382) on the administration of ^{3}H-norepinephrine to the rat brain. These authors found that retained ^{3}H-norepinephrine showed the same subcellular distribution and response to drugs as did endogenous norepinephrine (382, 472).

The major metabolic product of ^{3}H-histamine in the rat brain is ^{3}H-imidazoleacetic acid (727); there was no evidence of the presence of imidazoleacetic acid riboside or ribotide, which have been detected as metabolites of ^{14}C-histamine in peripheral rat tissues (728, 729). It is of interest, however, that after administration of ^{14}C-histidine Robinson and Green (729) detected the formation of the riboside and ribotide of imidazoleacetic acid in the rat brain. Tritiated imidazoleacetic acid disappeared from the rat brain more slowly than tritiated histamine. This slower disappearance of ^{3}H-imidazoleacetic acid may be due to its mixing with a large pool of endogenous compound which may arise from the metabolism of histidine (727). It is known that imidazoleacetic acid may be derived from an enzymic oxidation and decarboxylation of imidazolepyruvic acid, which in turn is formed by the transamination of histidine.

Only small amounts of ^{3}H-methylhistamine were encountered by Snyder et al. (727) as a metabolic product in the rat brain. This seems to be in accordance with the relatively low levels of the enzyme histamine methyltransferase, that are found in the rat brain (529).

In subcellular localization experiments ^{3}H-histamine was present in the "synaptosomal" or pinched-off nerve-ending fraction of the rat brain (727). A similar subcellular localization for ^{3}H-norepinephrine, administered intraventricularly into the rat brain, was previously observed by Snyder et al. (473) and Glowinski et al. (472).

It should be noted here that Carlini and Green (730), when applying in brain-tissue studies the fluorometric method of histamine estimation (731), obtained histamine values eight times higher than those determined by bioassay. Presumably, this may have been due to the presence of substances other than histamine which yield fluorescent compounds, when reacting with o-phthalaldehyde (730). When an ion-exchange resin was used prior to the fluorometric estimation of histamine in rat-brain tissue, Green and Erickson (732) obtained low values, which compared well with the amounts found by Carlini and Green in biological assays.

Carlini and Green (730) examined by bioassay the subcellular localization of histamine in the brain of the rat and guinea pig, and found the highest histamine concentration in the microsomal fraction. The crude mitochondrial fraction, which also included the synaptosomal part, contained about half the amount present in the microsomes.

The accumulation and retention of ^{3}H-histamine by the rat brain was found to be unaffected by reserpine treatment (727). Since the storage of brain norepinephrine and serotonin is markedly impaired by reserpine, it is probable that ^{3}H-histamine is not retained by binding sites that normally store serotonin or norepinephrine. Although reserpine has been known to lower endogenous histamine concentration in the cat brain (716), it does not affect the endogenous content of histamine in the rat brain (727), whether measured by bioassay (722) or by fluorometry after removal of contaminants (732).

The disappearance of ^{3}H-histamine from the rat brain is delayed by pretreatment with pheniprazine, an MAO inhibitor, but not by aminoguanidine, an inhibitor of DAO. Inasmuch as these findings may suggest that in the rat brain ^{3}H-histamine is metabolized predominantly by MAO, it should be recalled here that pheniprazine is an inhibitor of histaminase also (521). Moreover, it is possible that the missing effect of aminoguanidine was due to an inability of the latter drug to penetrate the blood-brain barrier.

The high histaminase activity in the small intestine and liver may be considered a limiting factor for the intestinal absorption of histamine. In sheep, whose rumen contains large amounts of histamine, the liver plays an important role in protecting those animals against intoxication by histamine, absorbed by the alimentary tract (733).

Histamine given orally appears to be less toxic than that given by injection (734). Injected ^{14}C-histamine has been found not to be equally distributed among the tissues (659, 718, 728, 735). In the dog (736), intravenously injected ^{14}C-histamine is removed from the blood by the kidney in the form of its metabolites. In anesthetized dogs with a steady infusion of small quantities of ^{14}C-histamine (736) evidence was obtained that the kidney oxidizes the greater part of the infused histamine to imidazoleacetic acid, which is excreted in the urine, besides much smaller quantities of methylhistamine and methylimidazoleacetic acid. This finding is consistent with the extremely high *in vitro* DAO activity of the kidney reported by Waton (522). However, since an injection of histamine into dogs by other routes led to a preponderance of methylated metabolites (627, 736), it is probable that in the dog the kidney is the most active *in vivo* source of DAO activity (708). On the other hand, the intestine and the liver appear to play a predominant role in the metabolism of orally administered histamine. In mice fed on histamine, the metabolite mainly excreted is imidazoleacetic acid in its free and riboside forms. Mice injected subcutaneously with ^{14}C-histamine, however, excrete in the urine mainly methylhistamine (19%) and imidazole-acetic acid, free and conjugated (21%) (632). It is noteworthy that large doses of exogenous histamine are mainly degraded by histaminase in mice (631) and in rats (737). An important factor to be considered is the evidence that large quantities of injected histamine destroy the ability of rats to metabolize this substance (633). This *in vivo* observation compares well with the *in vitro* results pointing to the fact that in the presence of superoptimal concentrations of histamine a considerable inhibition of histaminase activity takes place (562).

The concept that, in certain animal species at least, histamine may play a role as a chemostimulator of gastric secretion still awaits confirmation (738–740). Some experimental evidence for the above theory was obtained in studies on the distribution of histamine in the gastric mucosa of the rat, from pharmacological studies of the action of histamine on gastric secretion, from determinations of its formation in the mucosa by histidine decarboxylase, and its rate of disappearance from the mucosa during various secretory states (741–745). In all species studied, however, little or no histamine is found in gastric venous effluent (739, 746, 748, 749) or in gastric juice (739, 746, 747). Moreover, it is well established that histamine is rapidly catabolized in animal tissues (728). In recent work, Navert, Code, and their associates (747) studied the fate of ^{14}C-histamine infused intravenously, for 140 min continuously, into conscious dogs with Heidenhain pouches. The data suggested that histamine is actively methylated by the acid-secretory mucosa of the dog and is thus consumed in the process of secretion. This offers an explanation for the failure to detect fluctuating concentrations of histamine in the gastric venous blood or in the gastric juice during secretion, and the lack of correlation between the amounts of histamine in the plasma and the amounts of acid secreted (739, 746, 748, 749). Moreover, the results obtained indicate that the methylated histamine—which may consist of 1,4-methylhistamine, N-methylhistamine, and/or N,N-dimethylhistamine—accumulates in the gastric mucosa during stimulation of gastric secretion by histamine (747). Very recently, H. Navert and C. F. Code succeeded in demonstrating the presence of N,N-dimethyl-histamine and N-methylhistamine in the gastric juice of dogs during stimulation by histamine (747). Their preliminary results are presented in Table I, which shows the distribution of the metabolites of ^{14}C-histamine in the gastric juice of dog after 140 min of continuous intravenous infusion. The major metabolite appears to be N-dimethylhistamine, with much smaller amounts of 1,4-methylhistamine. This interesting work is being continued.

TABLE I.

Distribution of ^{14}C-Histamine Metabolites in Canine Gastric Juice
After 140 min of Continuous Intravenous Infusion (747).

Compound	Percentage of total dpm
Histamine	1.4
N-methylhistamine	4.3
N-dimethylhistamine	48.6
1,4 methylhistamine	21.1
1,4 methylimidazoleacetic acid	2.5
Imidazoleacetic acid	1.31
Imidazoleacetic acid riboside	0
Acetylhistamine	5.3
Undetermined	15.5

Thin-layer chromatography on cellulose-silicic acid (3:1, w/w) horizontal; solvent: methanol-chloroform-ammonia (4:4:1, v/v).

Parrot and his colleagues (750) have recently reported that fasting gastrectomized rats excreted less histamine in the urine than fasting normal rats. Subcutaneous injections on 6 consecutive days of amino-guanidine sulfate (20 mg/kg) to fasting gastrectomized and normal rats further accentuated the difference in the histamine excretion of the two groups of rats. The authors suggest that these findings may point toward the formation of histamine by the gastric mucosa.

It is also noteworthy that the urine of gastrectomized rats, unlike that of normal rats, did not contain any acetylhistamine (750). Whereas the rat gastric mucosa contains histaminase (519), the gastric mucosa of dogs does not show any histaminase activity (520). Unlike Kobayashi and Ivy (519), Gunther and Glick (751) have not encountered any histaminase activity in the gastric mucosa of the rat and suggest that histamine may not be destroyed in the stomach in order to preserve its function in acid secretion. Likewise, Code (746) was unable to demonstrate any histaminase activity in the gastric mucosa of man or dog.

Since N-methyltransferase is present in the gastric mucosa of guinea pigs, cats, rabbits, and mice (529), histaminase activity does not seem to be the limiting factor for the biological response of the gastric mucosa to the released endogenous histamine. Animals treated with amino-guanidine show an increased gastric secretion, which may reflect an increase in the histamine content of the blood, expected to occur in view of the observed increase in the urinary excretion of histamine that follows treatment with aminoguanidine (621).

Green and his colleagues (752) determined the ratio of methylhistamine to histamine excreted in the urine of women in various hormonal conditions. These workers suggested that histamine, probably released by estrogenic disruption of uterine mast cells (753), is catabolized, normally, in women with a relatively low estrogenic response by N-methyltransferase to methylhistamine. With an increased estrogenic response, however, a larger proportion of histamine is released and is degraded by histaminase, with the result of a diminished urinary histamine excretion but an unchanged excretion of methylhistamine.

The occurrence of high histaminase activity in the human placenta, first demonstrated by Danforth and Gorham (530) seems to be in agreement with the fact that histaminase is present in tissues with a high blood supply. The placenta is considered to be the source of the raised plasma histaminase level in human pregnancy, the enzyme being concentrated in the maternal decidua (542) . High plasma histaminase levels were found only in those pregnant animals in which the placenta had a substantial maternal component—such as rats (643) and guinea pigs (754). That the fetus did not play any essential part in the production of placental histaminase was demonstrated by Swanberg (542), who found in a number of cases of experimental deciduoma in rabbits a strong histaminase activity in the nonpregnant uterus. This worker also encountered large amounts of histaminase in the uterus of rabbits with pseudopregnancy as well as in the endometrium of rabbits that had been treated with progesterone.

The metabolism of injected ^{14}C-histamine in human pregnancy was studied by Schayer (638) and Lindberg et al. (646, 755). Whereas in pregnant women imidazoleacetic acid was the predominant metabolite, in nonpregnant women it was methylimidazoleacetic acid (755). During a continuous infusion of ^{14}C-histamine the ^{14}C-plasma histamine level was about 50% lower in pregnant than in nonpregnant women (755). Lindberg concluded that the major route of histamine catabolism in the placenta and myometrium of pregnant women is oxidative deamination by histaminase, and in the myometrium of nonpregnant women, methylation by N-methyltransferase (755). Lindberg and his colleagues also showed that the human fetus possesses mechanisms for the formation and degradation of histamine, the main route of degradation being methylation (646).*

It has been suggested (131) that the metabolic pathway catalyzed by histaminase is concerned with limitation of the biological effects of exogenous histamine as well as of that form of endogenous histamine that, when released in large amounts, would escape into the general circulation. N-Methyltransferase, on the other hand, seems to play the role of a regulator of the levels of endogenous histamine (756). However, more work is required for further elucidation of this concept.

It has been mentioned in Section 1.A that in various tissues the activities of histaminase and DAO do not always run parallel (509). As long ago as 1952 it was demonstrated by the present author (536) that in human pregnancy cadaverine was a much better substrate of the placental and blood-plasma amine oxidase than histamine. In subsequent work (537, 538) this was found also for putrescine, and it has recently been shown (548) that human amniotic fluid also exerts a very large effect on cadaverine and a somewhat smaller one on putrescine—but a very much smaller one on histamine. In this context it should be recalled that Lindahl et al. (757) incubated ^{14}C-histamine with the plasma of pregnant women in the last weeks of gestation, and found a comparatively low histaminolytic activity, which, however, was well above that obtained in the plasma of nonpregnant women. The amine oxidase encountered in the human placenta, pregnancy serum, and amniotic fluid thus appears to be a true diamine oxidase, cadaverine being its best substrate.

For the sake of completeness it has been decided to include in this discussion of the biological significance of the diamines, cadaverine and putrescine, also that of the two putrescine derivatives, spermine and spermidine, although enzymes, specific for the degradation of the latter compounds are described in Sections 2.A and 6.

The biological occurrence and distribution of the polyamines, pentamethylenediamine (cadaverine) and tetramethylenediamine (putrescine), the di-(γ-aminopropyl)putrescine (spermine) and the mono-(γ-aminopropylputrescine (spermidine) have been studied by many investigators. Although these amines have been known for many years as normal biological constituents (758), it is only comparatively recently that considerable knowledge has accumulated with regard to their involvement in various biological reactions.

The prostate gland of sexually mature males of many mammalian species contains large amounts of the polyamines, spermine and spermi-

---

* This is supported by the findings of S. E. Lindell, H. Westling, and B. Zederfeldt (unpublished data 1963).

dine (759–761). In some species spermine and spermidine may appear in the secretions produced by the prostatic epithelial cells (760, 761).

Polyamines occur in every animal tissue and in many microorganisms (156, 758, 760, 762–764), but little is known about their intracellular distribution and biological function. Polyamines are not confined to the nucleus (765) and have been found in ribosomes isolated from *Escherichia coli* (766–768) as well as in those from mammalian pancreas and liver (768–770).

Cohen and Lichtenstein (766) found that the ribosomal fraction of *E. coli* contained putrescine and spermidine in amounts equivalent to approximately 15% of their total content in this bacterium. Furthermore, these authors concluded that ribosomal putrescine, which was the main amine, and spermidine both occurred in a nonexchangeable form and were not derived from the extracting medium. C. W. Tabor and Kellog (771), however, maintain that the polyamine content of *E. coli* ribosomes does depend on the nature and concentration of the ions in the medium used for their preparation.

It has been calculated that the amounts of putrescine and spermidine would suffice to neutralize about 8% of the ribosomal ribonucleic acid in *E. coli* (766). According to Zillig et al. (768), the putrescine and cadaverine content of *E. coli* could neutralize at least one-third of the acidic groups of the ribosomal ribonucleic acid. The presence of cadaverine in the ribosomal particles of *E. coli* has been also observed by Spahr (767). The total diamine and polyamine content amounted in Spahr's study to 0.4% of dry ribosomes, and the amines were found in an approximately molar ratio putrescine:spermidine:cadaverine of 3:2:1. Only traces of spermine were detected.

In contrast to the ribosomes of *E. coli* (766), whose polyamines are at least partly nonexchangeable, the ribosomes isolated from guinea-pig pancreas rapidly take up labeled spermidine from the medium (772). Moreover, Tabor and Tabor (760) have reported that in liver homogenates added labeled polyamines were quickly taken up by particulate fractions of the disintegrated cell.

Raina and Telaranta (773) have recently studied the subcellular distribution of spermine and spermidine in rat liver. In subcellular particles isolated in a medium containing only a low concentration of salt (e.g. 2 mM $CaCl_2$) about 60% of the total cell polyamines was recovered from the microsomal fraction, and the percentage distribution of spermidine and spermine was found to be very close to that of ribonucleic acid. With increasing salt concentrations, especially in the case of divalent cations, the polyamines were partly released from the microsomes into the medium. Added labeled polyamines were taken up by particles and again released by increasing salt concentrations (773).

Lately, the polyamines have attracted much interest because of their physical affinity for nucleic acids (774–777) and their stimulating effect on cell proliferation (778–780). Thus, polyamines have been reported by various authors to occur in high concentrations in actively proliferating tissues, such as human semen (761), developing chick embryos (759, 781, 782), and in germinating plant seeds (783). Furthermore, such workers as Mahler and his colleagues (784–786), Tabor et al. (760, 762, 787), Siekevitz and Palade (772), and Raina and Telaranta (773) have been able to demonstrate an association of the polyamines with the nucleic acids. These authors suggested that polyamines bind tightly to nucleic acids, forming complexes that, if sufficiently concentrated, produce precipitates. These complexes seem to involve a noncovalent linkage between the basic groups of the polyamines and the highly acidic phosphate groups of the nucleic acids. The negatively charged phosphate groups, which normally repel each other, may be thereby neutralized; this may result in an increase in the net strength of various cohesive forces such as van der Waals forces and hydrogen bonding, followed by an increase in the strength of secondary structures, such as, for example, the double helix of DNA.

Much of these authors' evidence for the interaction between DNA and polyamines is based on the ability of these substances to decrease the denaturation of DNA, that occurs when DNA is heated ("melted") by increasing the transition ("melting") temperature (T_m) of DNA. This protective effect of polyamines was most clearly demonstrated by Tabor (787) in solutions of relatively low ionic strength. In their studies on the relative activity of a homologous series of diamines Mahler and his colleagues (784–786) found that the protective effect on DNA was optimal with 1,5-diaminopentane (cadaverine). These observations have been interpreted in terms of a model with structure-independent binding sites for the diamines on a DNA helix. These sites may involve binding between the phosphate residues of one strand and the adenine and/or thymine residues of the other one, with the most stable interaction apparently occurring when chain length (n) of the amine is 5 (785, 786).

Similar observations have been lately made by Stevens (788) in his study on the interaction of spermine and of some of its homologs with DNA. The effect of spermine and its homologs on the T_m of double-stranded DNA does indeed seem to depend on the chain length, maximum stabilization occurring with spermine or with the homolog that has one carbon atom more in its molecule than spermine.

It has been reported (789) that polyamines are involved in the synthesis of messenger RNA, catalyzed by RNA polymerase, when the reaction is primed by native double-stranded DNA. Raina and his colleagues (790) found a strong correlation between RNA and polyamine synthesis in the regenerating rat liver, and in another report Raina (791) showed that spermidine was able to stimulate RNA synthesis in *E. coli*. In a study on the effect of low and high concentrations of spermine on protein synthesis by a subcellular system derived from the L1210 mouse ascites leukemia, Ochoa and Weinstein (792) suggest that polyamines may be involved in a codon-anticodon stabilization mechanism. Martin and Ames (793) relate the polyamine stimulation of protein synthesis in subcellular systems to enhance conversion of 70-S to 100-S ribosomes. Recent results by Raina and Telaranta (773) seem to indicate that the polyamines, bound to subcellular particles are probably associated with RNA. According to the experiments of these authors, the binding of polyamines to RNA in microsomes seems to be ionic, which is in agreement with findings of Siekevitz and Palade (772). Moreover, it follows from the work of Raina and Telaranta (773) that polyamines have a much greater affinity for microsomes than has Mg^{2+}, the cation that is generally considered to act as a stabilizer for ribosomal particles (794).

Results obtained by Raina (759) in a very interesting study on the metabolism of spermine and spermidine in the developing chick embryo have thrown considerable light on the question of the biosynthesis of polyamines. The chick embryo was chosen as the experimental object because, in preliminary experiments, neither spermine nor spermidine were detectable in unincubated eggs or in their hydrolysates. Both these amines, however, were found to be present in embryos, suggesting synthesis during embryonic development. Raina succeeded in demonstrating that the concentration of polyamines in the whole chick embryo was highest at a very early stage of development (759). Another peak was found when the embryo was 15 to 16 days old, with a sharp decrease of the polyamine level thereafter. It is noteworthy that the reported changes in both spermine and spermidine concentrations were fairly parallel, and, further, that these concentrations of the polyamines in whole chick embroys varied only slightly in groups of the same age. On the other hand, the changes in the polyamine contents at different stages seemed to reflect the changes in the metabolism occurring during embryonic development. On the basis of all these findings Raina suggests that the occurrence of spermine and spermidine in relatively large amounts in developing chick embroys may indicate that these amines are in some way connected with growth and protein synthesis. He found marked differences in the polyamine content of different tissues of chick embryos (759). The highest values were encountered in the chick liver, but Raina's results were somewhat lower than those reported by Rosenthal and Tabor (795) for the polyamine contents of mouse, rat, and rabbit livers. The observed decrease in the total concentrations of polyamines in chick embryos after the age of 16 days Raina attributes to a probable increase in enzymic degradation of these amines, since no excretion of polyamines into the yolk sac or the amniotic or allantoic fluids could be demonstrated (759). In further experiments of the same study, with labeled compounds, he was able to show that putrescine, ornithine, and methionine can act as precursors in the biosynthesis of spermine and spermidine in the developing chick embryo.

Work by Tabor and his colleagues (796, 797) with microorganisms has shown that putrescine and ornithine are both incorporated into spermidine and spermine, thus indicating the probable source for the four-carbon chain of these polyamines, and Greene (798) has demonstrated that the source of the three-carbon chain in spermidine is methionine. Experiments with partially purified enzyme systems from *E. coli* have confirmed these results with regard to the enzymic biosynthesis of spermidine from putrescine and methionine (797, 799). Up to the present time, however, no evidence has been presented with regard to the enzymic biosynthesis of spermine. Isotope data obtained with whole cells seem to suggest that the mechanism of spermine biosynthesis is probably comparable with that of spermidine (762).

As to the question of a biosynthesis of spermine and spermidine in animal tissues, Tabor and his co-workers (796) reported in a preliminary communication on a small incorporation of ^{14}C-^{15}N-diaminobutane into the polyamine fraction of minced rat prostate. Results obtained by Raina in his study on the developing chick embryo (759) seem to suggest that the mechanism of the biosynthesis of the polyamines in the chick embryo is similar to that in microorganisms. After administration of labeled putrescine, ornithine, and methionine the incorporation of radioactivity into spermidine was found to be rapid and to start without any clear lag period. In contrast to this observation, 5 hr after the injection of labeled putrescine or ornithine the radioactivity in the spermine fraction hardly exceeded that of the background. The total radioactivity of the spermine fraction started increasing after that lag period and continued increasing for at least 60 hr, but the total radioactivity of the spermidine fraction began to decrease after only 30 hr. This may indicate that spermidine is a precursor of spermine. Raina (759) sees support for this suggestion in his finding that after administration of labeled methionine, which is the probable source of the three-carbon chain of the polyamines, the radioactivities of both spermidine and spermine were found to be more or less parallel from the beginning, indicating that labeled methionine was simultaneously incorporated into both polyamines in about equivalent amounts. It should be noted that C. W. Tabor (799) has reported that in *in vitro* systems, containing purified propylamine transferase from *E. coli*, spermidine was not able to substitute for putrescine as an acceptor of the propylamine moiety.

In continuation of previous work with regard to the stimulation of polyamine synthesis in relation to nucleic acids in regenerating rat liver (790, 791) Raina and his colleagues (800) have recently reported on the effect of growth hormone on the synthesis of polyamines, RNA, and protein in normal rat liver. Although the stimulation of RNA and protein synthesis by growth hormone is well established, the primary site of action of this hormone is not yet known. One of the earliest effects of growth hormone on liver metabolism is the stimulation of RNA polymerase (801). According to Korner (802), the enhancing effect on protein synthesis by growth hormone seems to be mediated through prior stimulation of nuclear RNA synthesis. Growth hormone has been also found to cause an accumulation of spermidine in the liver of hypophysectomized rats (803) and to stimulate the biosynthesis of putrescine and spermidine from their precursors in the normal rat liver (804).

The stimulation of polyamine synthesis by growth hormone is of particular significance, since increasing evidence suggests that these amines function as physiological stabilizers of nucleic acids (805). Furthermore, polyamines seem to be also involved in the biosynthesis of RNA. Thus, polyamines stimulate *in vitro* RNA polymerase of both bacterial and animal origin (789, 806), and *in vivo* they influence the rate of RNA synthesis in *E. coli* (791).

In their recent study, mentioned above, Raina and his co-workers (800) have been able to demonstrate that growth hormone markedly increases the synthesis of liver spermidine from exogenous methionine. This result seems to imply that the increase in spermidine synthesis depends on a primary increase in the concentration of liver putrescine. This hypothesis, they believe, is supported by the following findings. Growth-hormone treatment causes marked stimulation of putrescine synthesis from exogenous ornithine as well as considerable accumulation of putrescine in the liver. Growth hormone also stimulates the synthesis of putrescine from ornithine *in vitro*. Moreover, the accumulation of putrescine and the increase in the synthesis of spermidine could both be prevented by simultaneous administration of actinomycin D or puromycin. Under these experimental conditions, the inhibition by puromycin could be partially reversed by added putrescine. This finding seems to rule out the possibility that puromycin directly inhibits the synthesis of spermidine. Furthermore, after partial hepatectomy, growth-hormone treatment resulted in an early accumulation of hepatic putrescine, with a simultaneous increase in the synthesis of spermidine from ^{14}C-methionine (790, 804). Raina and colleagues (800) have further observed that exogenous putrescine increases *in vivo* the synthesis of spermidine from labeled methionine. Finally, the incorporation of ^{14}C-methionine into spermidine in rat-liver homogenates is considerably enhanced by exogenous putrescine (804).

In this context it should be mentioned that Pegg and Williams-Ashman (807) have very recently reported that putrescine stimulates *in vitro* S-adenosylmethionine decarboxylase, partially purified from rat liver and prostate. In microorganisms the formation of putrescine is catalyzed by two distinct enzymes, ornithine decarboxylase and arginine decarboxylase (808, 809). Results obtained by Raina et al. (800) seem to suggest that in animal tissues putrescine may be formed from ornithine in a direct decarboxylation reaction mediated by ornithine decarboxylase.

Finally, these workers stress the significance of their finding that under conditions characterized by a period of rapid growth there is an increase in the concentrations of putrescine and spermidine (800).

As mentioned above, Pegg and Williams-Ashman (807, 810) have very recently reported on the enzymic synthesis of putrescine by a cell-free preparation of rat ventral prostate. They showed that the biosynthesis of putrescine in rat ventral prostate gland occurs by the direct decarboxylation of L-ornithine by a soluble pyridoxal phosphate–dependent L-ornithine decarboxylase. It is noteworthy that they were unable (810) to find any evidence of the existence in rat prostatic gland of an L-arginine decarboxylase even when homogenates were tested in the presence of pyridoxal phosphate and Mg^{2+} ions, which are needed for the *E. coli* enzyme (808). Pegg and Williams-Ashman (807) have also demonstrated that in rat ventral prostate gland the decarboxylation of (—)-S-adenosyl-L-methionine is markedly and specifically stimulated by putrescine, with the stoichiometric formation of spermidine.

Dykstra and Herbst (811) have lately demonstrated that after partial hepatectomy the content of spermidine increases in rat liver and the rate of polyamine accumulation closely approximates the increased rate of RNA synthesis in the regenerating liver. The uptake by the rat liver of intravenously injected ^{3}H-putrescine and the biosynthesis of spermidine were found by these authors to be accelerated within 2 hr after operation. Dykstra and Herbst also studied the time course of the uptake of ^{3}H-spermidine by normal and regenerating rat liver and showed that the uptake pattern for ^{3}H-spermidine greatly resembled that for ^{3}H-putrescine. The uptake of labeled putrescine by regenerating liver was three to five times the rate of uptake in normal liver. Moreover, the rate of conversion of putrescine to spermidine in regenerating rat liver was approximately doubled as compared with normal liver. The metabolic rate of the amines was found to be increased already two hours after partial hepatectomy and both putrescine uptake and spermidine biosynthesis remained elevated throughout the 24-hr experimental period.

Dykstra and Herbst (811) point out that there is no evidence that the accelerated polyamine utilization by the regenerating liver is specifically linked to either DNA synthesis, which is maximal at 20 to 24 hr, or mitotic activity, which reaches its peak at 22 to 26 hr (812). There is, however, a definite similarity in the time course of the increased rate of putrescine uptake and spermidine synthesis and the sharply elevated rate of RNA synthesis in regenerating liver tissue (812, 813). It should be recalled here that Fujioka and his co-workers (813) were able to demonstrate a sharp increase in the rate of incorporation of ^{14}C-orotate into RNA within 2 hr of partial hepatectomy. Dykstra and Herbst (811) believe that the results obtained in their study suggest a functional relationship between spermidine and RNA synthesis in regenerating liver and, furthermore, that their findings seem to support the data on polyamine stimulation of RNA synthesis by purified preparations of RNA polymerase (789, 814).

In this connection an interesting recent work by Schwimmer (815) should be mentioned. As Tabor and Tabor (760) pointed out, the aliphatic polyamines seem to be involved in a diversity of both *in vivo* and *in vitro* biochemical and biophysical effects. Among the varied properties of the polyamines is their ability to influence the synthesis of DNA *in vitro*. Whereas Brewer and Rusch (816) have reported stimulation by spermine of DNA polymerase from isolated nuclei of *Physarum polycephalum* supplied with an exogenous source of DNA and deoxynucleoside triphosphate, O'Brien and his colleagues (817) have found that spermine inhibits the activity of partially purified DNA polymerase in the presence of a native DNA primer. Schwimmer (815) has studied the differential effects of putrescine, cadaverine, and glyoxalbis(guanylhydrazone) on DNA-dependent and nucleohistone-supported DNA synthesis.

It should be mentioned here that glyoxalbis(guanylhydrazone) and its methyl derivatives are considered to be anticancer agents (818). Schwimmer (815) found that whereas putrescine and cadaverine, but not glyoxalbis(guanylhydrazone), inhibit DNA-dependent DNA synthesis by *E. coli* polymerase, all these three polyamines stimulate DNA-polymerase activity with nucleohistones or chromatin from various sources as primers. Glyoxalbis(guanylhydrazone) proved to be the most effective stimulator of nucleohistone-primed DNA polymerase activity. Moreover, cadaverine and putrescine were found to inhibit DNA polymerase in the presence of thymus and liver DNA as well as in the presence of pea-embryo DNA. On the other hand, these diamines brought about an apparent augmentation in the rate of DNA-polymerase activity in the presence of pea and thymus nucleohistone, as well as in the presence of pea and liver chromatin. The reported inhibition of DNA-primed DNA-polymerase by simple diamines (815) agrees well with findings of O'Brien and his co-workers (817), who observed an inhibition by spermine of DNA-primed DNA polymerase. The stimulation by both aliphatic diamines and glyoxalbis(guanylhydrazone) of nucleohistone- and chromatin-primed DNA polymerase found by Schwimmer (815) is quantitatively similar to the stimulation by spermine of the DNA-polymerase activity of nuclei of *Ph. polycephalum,* demonstrated by Brewer and Rusch (816). This stimulation is also reminiscent of that obtained on addition of an endonuclease to a nucleohistone-dependent DNA-polymerase reaction mixture after maximum DNA synthesis has occurred (819). Schwimmer (815) believes that the simplest explanation of this stimulation by aliphatic diamines is that these amines partially displace histone from the DNA, and that histones exert a greater inhibitory effect than do the amines. The observation that stimulation by the aliphatic amines is not as great as that exerted by glyoxalbis(guanylhydrazone) is in agreement with Schwimmer's idea of competition between histone and diamine for DNA.

According to Schwimmer (815) a reversal of the stimulating effect of diamines on the nucleohistone-supported DNA reaction would be expected at sufficiently high concentrations of diamine. No such reversal was, however, observed, *not* even at a concentration as high as 2×10^{-3} M, at which a 70 to 80% inhibition of the DNA-primed reaction did take place. When trying to interpret the kinetics and action of endo-nuclease on nucleohistone-dependent DNA polymerase, Schwimmer and his colleagues (819, 820) suggested that the polymerase may be reversibly bound in a complex with product DNA in such a way that it may be no longer available for further catalysis. One possible function of the diamines may be to uncouple this kind of complex and thus to allow DNA synthesis to proceed. The overall effect would be an increase in the activity of the DNA polymerase. Thus, data obtained by Brewer and Rusch (816), which indicate a release or activation by spermine of a "latent" DNA polymerase in Ph. *polycephalum* nuclei may be explained on the basis of this hypothesis of Schwimmer (815). It is a surprising fact that glyoxalbis(guanylhydrazone), the methyl derivative of which has been shown to complex with sarcoma DNA (821), does not inhibit DNA-supported DNA-polymerase activity, but still stimulates the nucleohistone-supported reaction. Schwimmer tries to account for this fact by suggesting that internal hydrogen bonding may shape this amine in such a manner that its complexing with DNA does not interfere with the activity of the polymerase (815). Binding of glyoxalbis(guanylhydrazone) to nucleohistone, possibly by complexing with histone-bound RNA (822), may, however, prevent histone from interfering with DNA-polymerase activity. The substitution of glyoxal hydrogen by a methyl group to yield methylglyoxalbis(guanylhydrazone) may thus hinder this hydrogen bonding. The methyl derivative could then bind to DNA as do other polyamines and so interfere with DNA synthesis (821).

Caldarera, Barbiroli, and Moruzzi (782) have been able to enlarge upon Raina's investigations of the metabolism of spermine and spermidine in the developing chick embryo (759). In their study on interactions between nucleic acids and polyamines in chick embryos Caldarera and his colleagues (782) found that at a very early stage of development spermine, spermidine, cadaverine, and putrescine were all synthesized in considerable amounts by the embryonic cells. Thus, whereas in preliminary experiments, in agreement with Raina (759), none of the polyamines was detected in unincubated eggs, as early as 48 hr after incubation the polyamines were all present in appreciable quantities in the blastoderms of fertilized eggs. The concentrations of these compounds progressively increased, reached maximum values between the fifth and tenth day of incubation, and then steadily decreased. Both RNA and DNA, which were determined under the same experimental conditions, showed in each stage of embryonic development the same behavior as the polyamines. In this context these authors point out that it is particularly the period of early embryonic development during which the most obvious morphological changes and the most active protein synthesis become evident. In the course of their study these workers investigated the effect on chick embryos of injections of spermine and of iproniazid into the yolk sacs of embryos at the tenth day of incubation. Injection of spermine into the yolk sac resulted in an increased activity of amine oxidases (spermine oxidase and spermidine oxidase) and, consequently, in an increased oxidation of spermine and spermidine. Spermine, by inducing an increase of amine oxidase activity, apparently, leads to a decrease of the concentration of polyamines in the embryo. Correspondingly, the concentrations of nucleic acids also decrease. Under the experimental conditions of the authors (782) the concentration of the total free acid-soluble nucleotides was increased, mainly due to the triphosphates, precursors of the nucleic acids. It may be that the increase in the free acid-soluble nucleotide fraction is partly caused by the enzymic oxidation of spermine and spermidine to the corresponding aminoaldehydes, which are toxic (823).

Reference should be made here to recent important findings by Bachrach and his co-workers (824–828; see Section 6) with regard to the complex, that oxidized spermine forms with the DNA of coliphage T_5.

An increase in the concentrations of spermine and spermidine was obtained by Caldarera et al. (782) on injection into the yolk sac of iproniazid, an inhibitor of the amine oxidases. A simultaneous increase in the content of RNA (18%) and DNA (27%) and a decrease in the concentration of total free nucleotides, particularly of the triphosphates, was recorded. Finally, an injection into the yolk sac of a mixture of spermine and iproniazid brought about an increase in the concentration of spermine and a greater increase in the contents of RNA (30%) and DNA (33%). At the same time the concentration of total free nucleotides decreased.

On account of these biochemical findings in the developing chick embryo, Caldarera, Barbiroli, and Moruzzi (782) conclude that a significant relationship seems to exist between polyamines and nucleic acids. Accordingly, polyamines may play an important role also *in vivo* by stimulating the biosynthesis of nucleic acids, especially of DNA during polymerization, controlled by RNA polymerase and DNA polymerase. This concept would possibly strengthen the significance of the occurrence of biogenic polyamines, particularly in organs and tissues characterized by a rapid cellular proliferation (759, 829, 830).

In a very recent paper Moruzzi, Barbiroli, and Caldarera (831) report on further investigations with regard to a possible relationship between polyamines and nucleic acids. These authors have studied the incorporation of labeled precursors (e.g. ^{3}H-orotic acid) into the nucleic acids of chick-embryo subcellular fractions, the polyribosomal profiles, and DNA-dependent RNA polymerase activity in conditions in which polyamine contents were experimentally modified.

Thus, these authors observed that an increase in polyamine concentration, brought about by inhibiting the amine oxidase activities with iproniazid, caused an increase in the incorporation of ^{3}H-orotic acid into chick embryo RNA and DNA. On the other hand, a decrease in polyamine concentration, obtained by causing an increase in amine oxidase activities, decreased ^{3}H-orotic acid incorporation into nucleic acids. This was particularly evident for nuclear DNA and ribosomal RNA. Moreover, the activity of DNA-dependent RNA polymerase, assayed under the same experimental conditions, also varied in the same manner with changes in polyamine concentration. These authors conclude that the variation of polyamine concentration, resulting from different physiological requirements of the cell, seem to indicate that spermine and spermidine may have a role in the control of protein synthesis (831).

In this context an interesting study on the effect of polyamines on the stability of brain-cortex ribosomes by Datta and his colleagues (1092) should be mentioned. These authors examined the stability of ribosomes isolated from the cortex of goat brain in the presence of such polyamines as spermidine, spermine, cadaverine, and putrescine, and found that the pH-dependent degradation of ribosomes into protein, RNA and acid-soluble nucleotides was markedly decreased by polyamines when present in the suspension medium. On an equimolar basis, spermidine and spermine exerted higher stabilizing effects than did cadaverine and putrescine.

No direct information is available as to the question of biosynthesis of polyamines in man. Indirect evidence, however, suggests that the human fetus may be capable of producing besides histamine (646) also cadaverine and putrescine.* The interesting possibility that these biogenic amines may play an important role in man in mechanisms of cell growth and tissue differentiation has, in the present author's opinion, gained considerably in significance in relation to the teratogenic action of the drug thalidomide [$(\pm)$-α-phthalimidoglutarimide].

As is well known, this compound, nontoxic to adult tissues, has been found to exert disastrous effects on the early development of the human fetus, resulting in serious malformations. According to its chemical configuration, thalidomide consists of two groups, a highly reactive phthalimide group and a glutarimide group (832). In very interesting studies on penetration by ^{14}C-thalidomide into the rabbit embryo Keberle and his colleagues (832) found that neither the endometrium, nor the blastocyst membranes, nor the placenta appeared to prevent the thalidomide from reaching the embryo. In this connection it should be noted that thalidomide is a nonpolar compound (833), which may explain its easy transport *in vivo* across various membranes (833). On consideration of the great reactivity of the phthalimide nucleus of thalidomide on the one hand and the apparent importance of the biogenic polyamines (histamine, cadaverine, putrescine, spermidine, spermine) for cell growth and development on the other, the idea suggested itself that these amines may in the early human embryo react with the phthalimide group of thalidomide with the formation of stable condensation products. This suggestion (834) was borne out in a communication by Williams et al. (833) who reported that thalidomide reacted rapidly with spermidine, putrescine, cadaverine, and spermine, whereby very stable, non-dissociable compounds were formed. This reaction of thalidomide with the biogenic polyamines seems to be one of acylation and involves splitting of the phthalimide group of thalidomide (833). It is well known that substituted phthalimides are able to acylate hydrazine and that this reaction

---

* R. Kapeller-Adler, unpublished data (1965).

is in fact used to split substituted phthalimides (833). Hence, it appears to be feasible that thalidomide, when consumed by the mother at a very early stage of pregnancy, would pass unchanged into the embryo, and by condensing with some biogenic amines would deprive the human fetus of vital growth factors at an early critical stage of morphogenesis, and may thus induce disastrous teratogenic effects (834).

It should be mentioned here that Bignami and his colleagues (835) have recently reported on the embryotoxic effects of several phthalimide derivatives, and Champy-Hatem (836) has demonstrated *in vitro* the formation of a condensation product between phthalimide and imidazole. Referring to the work of Kahlson et al. (669) with regard to the vital importance of histaminogenesis for the embryo, Champy-Hatem suggests that a condensation reaction between the phthalimide nucleus of thalidomide and histamine *in vivo* may have a detrimental effect on fetal development.

Considering these most interesting and important aspects of the metabolic function of the biogenic amines, it may be assumed that the classical DAO is not restricted only to the control of the levels of the biogenic amines, cadaverine and putrescine in cells and body fluids, but that this enzyme may play a more essential role as a regulator of many vital reactions in which those diamines take part. A support for this concept may be seen, for example, in the fact that, on simultaneous examination, the fetal blood does not show any DAO activity, whereas the maternal blood during labor still contains a very active enzyme. In agreement with this, the histamine content of the fetal blood has been found to be higher than that of the maternal blood (562).

Investigations of the DAO activity of amniotic fluids of a large number of women at various stages of pregnancy showed that in the first weeks of gestation the DAO values with all three substrates, cadaverine, putrescine, and histamine were rather low. A spectacular increase in the enzymic activity took place at about the 20th week, reached a peak at about the 34th week of gestation, and declined somewhat steeply toward term (548)*

All these findings seem to point toward the significant role which the biogenic diamines seem to play in the early development of the human fetus and may also stress the importance of a regulatory function of DAO *in vivo*. The striking increase in DAO activity seems to concern only the protection of the mother. The fetus does not appear to require it since, in agreement with Kahlson's work (837), a degradation of the biogenic amines may interfere with the normal growth and development of the fetus, especially in early pregnancy (562, 834).

---

* The author has unpublished data (1965).

2. PLASMA AMINE OXIDASES (EXTRACELLULAR ENZYMES)

Amine oxidases occur in the blood plasma of many mammals, and two main types may be encountered in different species (153). One of these enzyme types was discovered by Hirsch (140) in the plasma of sheep and cattle and was described as spermine oxidase because of its high activity on spermine and spermidine. The other type of enzyme was found in the blood plasma of the horse, pig, and many other mammals and was given the name of benzylamine oxidase, since it readily oxidized that substance (144, 153, 154). So far only one difference has been found between these two kinds of plasma amine oxidases. Whereas plasma spermine oxidase acts not only on spermine and spermidine, but also on all the amines that are attacked by benzylamine oxidase, the latter enzyme has no effect on spermine or spermidine (153). Both types of plasma amine oxidases are inhibited by cyanide and by carbonyl reagents and both are unable to oxidize N-methyl secondary amines (153). In their substrate specificity both types of plasma amine oxidases differ from histaminase (DAO) and are more similar to the intracellular MAO (153).

A. Ruminant-Plasma Amine Oxidase, Plasma Spermine Oxidase*

a. OCCURRENCE

The soluble enzyme spermine oxidase was first found in the blood plasma of the sheep (140) and of the ox (141). Subsequently, Blaschko and his co-workers (128, 145, 152, 153) reported the presence of this enzyme in the plasma of a number of animals that possess a rumen— such as the camel, llama, giraffe, fallow deer, ox, sheep, and goat (152). Spermine oxidase activity was not found in the sera of nonruminants (152, 838), with the exception of the hippopotamus and of two members of the order Hyracoidea (153). The amount of spermine oxidase present in the plasma varies with age. Thus, in the newborn goat, when the rumen is not yet functioning, the plasma shows no spermine oxidase activity, or very little, but the level of this enzyme increases gradually during the first few months (153). It should be stressed that, whereas the plasma of all the animal species mentioned above showed not only spermine oxidase but also benzylamine oxidase activity (153), the plasma of many other species contained only benzylamine oxidase activity and did not act on spermine or spermidine at all (145, 153, 154 162).

b. CHEMICAL PROPERTIES

Spermine oxidase has been purified (141, 839–841) and crystallized (143) from beef plasma. The oxidation of spermine and spermidine by

---

* E.C. 1.5.3.3. Spermine: O_2 oxidoreductase (donor cleaving).

purified beef-plasma amine oxidase may be represented by the following equations:

$$NH_2(CH_2)_3NH(CH_2)_4NH(CH_2)_3NH_2 + 2O_2 + 2H_2O \qquad \rightarrow$$
$$\text{spermine}$$

$$\underset{H}{\overset{O}{\diagdown}}C(CH_2)_2NH(CH_2)_4NH(CH_2)_2C\underset{H}{\overset{O}{\diagup}} + 2NH_3 + 2H_2O_2.$$

$$NH_2(CH_2)_3NH(CH_2)_4NH_2 + O_2 + H_2O \qquad \rightarrow$$
$$\text{spermidine}$$

$$\underset{H}{\overset{O}{\diagdown}}C(CH_2)_2NH(CH_2)_4NH_2 + NH_3 + H_2O_2.$$

In experiments with radioactive compounds Tabor, Tabor, and Bachrach (842) presented evidence for the validity of the reactions, shown above. Thus, they were able to demonstrate that spermine is oxidized by purified plasma amine oxidase in the terminal positions with the formation of an dialdehyde containing all of the carbon atoms of the original spermine. The dialdehyde formed was identified by reduction with $NaBH_4$ to the corresponding dialcohol, N,N′-bis(3-hydroxypropyl)-1,4-diaminobutane.

If this oxidation of spermine were to occur at a secondary amino group, a variety of other products would be expected, including 1,3-diaminopropane, 1,4-diaminobutane (putrescine), spermidine, and several short-chain aldehydes. No evidence, however, was found in the reduced enzymic mixture for the presence of such substances or of 3-aminopropanol-1, 4-aminobutanol-1, or pyrrolidine. The findings of Tabor and his colleagues (842) that the oxidation of spermidine by the purified plasma enzyme does not lead to the formation of either spermidine or 1,4-diaminobutane contrasts with previous results, obtained when less strict techniques were applied to cruder enzyme preparations (141, 841, 843, 844).

In the oxidation of spermidine by plasma amine oxidase the mono-aldehyde $NH_2(CH_2)_4NHCH_2CH_2CHO$ is obtained as the major reaction product, which indicates that the oxidative deamination takes place at the primary amino group of the aminopropyl moiety. There was no evidence for oxidation at the other primary amino group or at the secondary amino group.

Yamada and his co-workers (845) have reported that the molecular weight of crystalline plasma spermine oxidase is about 250,000. Copper is a constituent of the crystalline enzyme (224, 579, 580; see Section 1.D), but it has been also claimed that zinc is a prosthetic group, although a less purified enzyme preparation was used (578). Some evidence has been presented by Yamada and Yasunobu (579, 846) that suggests that plasma spermine oxidase is a pyridoxal enzyme, containing 2 moles of pyridoxal phosphate per mole of enzyme. According to these authors, the beef-plasma amine oxidase could be represented as a protein-$(Cu^{2+})_4$-(pyridoxal phosphate)$_2$ (579). Spectral studies on purified beef-plasma amine oxidase by Gorkin showed no evidence for the presence of pyridoxal phosphate in this enzyme (578). Even though results of previous investigations of Hirsch (140) and Tabor et al. (141) on the inhibitory action of carbonyl reagents on the beef-plasma amine oxidase would be consistent with a role of pyridoxal phosphate as a cofactor of this enzyme, this compound has been so far neither isolated nor identified in any work on beef-plasma amine oxidase. The report of Yamada and Yasunobu (579) on this problem requires further study and confirmation.

c. SUBSTRATE SPECIFICITY

Although beef-plasma amine oxidase attacks spermine and spermidine most rapidly, other amines are also oxidatively degraded—for example, benzylamine, homosulfanilamide, phenylethylamine, and kynuramine as well as some aliphatic amines, amylamine and heptylamine, and the aliphatic diamine decamethylenediamine (141, 143, 838). The relative rates of reaction with various amines are essentially the same for both crude and purified enzyme preparations at the optimum pH, 7.2 (143). No significant oxygen uptake is obtained with tryptamine, serotonin, norepinephrine, epinephrine, histamine, putrescine, or cadaverine (141, 143). Using the crystalline preparation of beef-plasma amine oxidase Yamada and Yasunobu (143) did not obtain any oxygen uptake with tyramine or mescaline. The substrate specificity of beef-plasma amine oxidase, thus, differs from that of tissue MAO as well as from that of histaminase (DAO).

d. INHIBITORS

The oxidation of spermine and spermidine by the purified beef-plasma enzyme has been reported to be inhibited by cyanide (1×10^{-3} M), hydroxylamine (4×10^{-4} M), semicarbazide (4×10^{-3} M), tripelenamine (2×10^{-3} M), diphenhydramine (2×10^{-3} M), dibenamine (2×10^{-3} M), quinacrine (1×10^{-3} M), ephedrine (2×10^{-2} M), and several amidines (141, 394). Studies with isoniazid and iproniazid showed that the inhibitory effect of these drugs was not immediate, but required preincubation of the enzyme and the inhibitor (141). Beef-plasma amine oxidase seems to differ also in inhibitor specificity from the particulate MAO described in liver (141).

e. BIOLOGICAL SIGNIFICANCE OF RUMINANT-PLASMA AMINE OXIDASE

It is now recognized that spermine is a growth factor for a number of microorganisms (847), and it is also known that the rumen contains a large population of microorganisms. Because of the close association between the presence of a rumen and the ability of the blood plasma to oxidize spermine and spermidine Blaschko and Hawes (152) suggested that spermine or a related compound may be formed in the rumen by some of the microorganisms, and pass into the general circulation where it is acted on by plasma amine oxidase. The aldehydes produced by the reaction are unstable, and attempts to isolate them directly have been unsuccessful (141, 841). These aldehydes are of particular interest because of their toxicity for various bacteria, bacteriophages, and cells in tissue culture. Thus, concentrations of spermine that do not damage the cells or organisms may become very toxic when mixed with beef-plasma amine oxidase, presumably due to the toxicity of the metabolites. This observation was first made by Hirsch and Dubos in their studies on *Mycobacterium tuberculosis* (140, 848–850). Subsequently, similar effects were observed with *E. coli* (843), *Staphylococcus Aureus* (843) *Trypanosoma equiperdum* (843), the T-odd series of *E. coli* bacteriophages (824), mammalian *spermatozoa* (843), and a variety of mammalian and chick cell lines in cell culture (851, 852). Similarly, oxidized spermidine is much more toxic than spermidine to spermatozoa, bacteria, and bacteriophages (843).

In this context an interesting observation by Holtz and his co-workers (853) should be mentioned. Those authors found that dopamine (I), which is usually a pressor agent in the cat, is transformed on incubation with MAO into a compound that lowers the cat's blood pressure. In a careful study these authors found subsequently that in the action of MAO on dopamine the 3,4-dihydroxyphenylacetaldehyde (II) that is formed condenses with an unchanged molecule of dopamine to yield tetrahydropapaveroline (III). By means of thin-layer chromatography they showed that the latter compound is indeed formed during the action of MAO on dopamine. They were also able to demonstrate that both the enzymically formed papaveroline as well as the authentic compound exert a depressor action on the cat's blood pressure.

In 1952 Rosenthal et al. (854) showed that spermine had also a marked renal toxicity. Thus, parenteral administration of 0.075 to 0.15 millimoles of spermine per kilogram of body weight to a variety of animals (mouse, rat, guinea pig, rabbit, dog) caused first only relatively mild acute effects. However, proteinuria and retention of nonprotein nitrogen both gradually developed, followed within a week by death from renal failure. Animals injected with considerably smaller amounts of spermine showed an early diuresis and a moderate proteinuria but did not develop permanent renal damage. Spermidine is much less toxic than spermine, and 1,4-diaminobutane is not toxic at comparable doses.

In contrast to the effect of parenteral spermine, administration of large amounts of spermine in the diet causes only a slight transient proteinuria (843).

In man, vomiting, albuminuria, hematuria, acetonuria, azotemia, and hyperglycemia are observed after intramuscular injection of spermine (0.033 millimoles/kg) (855).

The kidneys of animals that die after parenteral administration of a toxic dose of spermine show marked necrosis of the epithelial lining of the proximal convoluted tubules without any inflammatory changes (854, 856), whereas the glomeruli, distal convoluted tubules, collecting tubules, and blood vessels remain normal. Direct administration of spermine into the renal artery of rabbits brings about extensive glomerular and tubular necrosis of the corresponding kidney, and eventually extensive scarring. Numerous juxtaglomerular cells appear in the injected kidney, and hypertension develops. The uninjected kidney escapes damage and undergoes compensatory hypertrophy (857).

It should be pointed out that although spermidine does not cause as much renal damage as spermine, when injected subcutaneously or intraperitoneally, both polyamines are equally toxic when they are injected into the renal artery. Rosenthal and Tabor (795) have demonstrated that after intravenous administration of spermine to rabbits approximately one-fourth of the injected dose is present in the blood after 1 hr, two-thirds of this amount in the formed elements, and one-third in the plasma. Five hours after the injection, the plasma contains practically no spermine, whereas the spermine content of the cells is still twice the normal value.

The mechanism of the renal toxicity induced by spermine is not known. One may speculate that spermine is concentrated within the renal cells and binds nucleic acids, phospholipids, or other acidic constituents. Alternatively, it is possible that spermine may be converted within the kidney to a toxic metabolite.

It is not known, whether spermine is degraded by the kidney, but by analogy with other enzymic reactions for spermine oxidation one may suggest the formation of an aminoaldehyde that is either toxic itself or is converted to a toxic product, such as acrolein. In this connection a very interesting finding by Tabor et al. (823) should be mentioned. These workers were able to demonstrate that the dialdehyde, formed from spermine by beef-plasma amine oxidase, decomposes upon heating, probably by a nonenzymic β-elimination reaction, to 1,4-diaminobutane; at the same time spermidine is formed by a similar elimination reaction from the intermediately occuring monoaldehyde. The other product of the spermine decomposition, described above, may be acrolein. This would help to explain recent findings of Alarcon that acrolein can be detected after heating and distillation of the oxidation products of spermine (858). There is no evidence, however, that any spermidine occurs as an intermediate in the enzymic oxidation of spermine by purified beef-plasma amine oxidase, or that any significant amounts of 1,4-diaminobutane or 1,3-diaminopropane are formed as primary products of spermine, or spermidine oxidation (842). Some evidence has been presented (795) for the oxidation of spermine to spermidine *in vivo*. Thus, when spermine is parenterally administered to rats, mice, or rabbits, up to 20% of the dose injected is excreted unchanged in the urine. Spermidine is also excreted in amounts corresponding to 4 to 8% of the injected spermine. When, on the other hand, spermidine is injected into mice, 18% of the given spermidine dose appears unchanged in the urine, but no spermine is excreted. To render these findings more conclusive, Tabor and Tabor (760) suggest that experiments with labeled spermine and spermidine be carried out to show definitely whether the spermidine originates from the injected spermine, or whether it is released from tissue stores. All the observations, recorded above, seem to indicate that the biological significance of spermine oxidase, occurring in the blood plasma of ruminants and a few other herbivores, is to be ascribed to a function of this enzyme other than that of a mere catalyst, responsible only for the removal of an excess of extracellular polyamines. More specific research work will have still to be carried out before this problem of great biological interest can be solved.

B. Nonruminant-Plasma Amine Oxidase, Benzylamine Oxidase

a. OCCURRENCE

In 1957 Bergeret, Blaschko, and Hawes (154), in a study on the enzymic oxidation of the antimalarial compound, primaquine, in the blood of certain mammals, discovered that horse serum contained an enzyme which showed most of the properties characteristic of spermine oxidase, with the one important exception that it did not possess the rapid action of the latter enzyme on spermine and spermidine. Since benzylamine was found to be rapidly oxidized, the provisional name "benzylamine oxidase" was proposed by the authors for this enzyme.

Plasma enzymes of the benzylamine oxidase type are widely distributed in mammals (129, 145, 153). They occur not only in ungulates and carnivores but also in some primates. This enzyme was, however, not found in the blood plasma of newborn piglets, and its activity in newborn foals is rather low (153). The ability of human plasma and serum to convert benzylamine to benzaldehyde has been recently demonstrated (859). The benzylamine oxidase level in the plasma of pregnant women does not differ from that in nonpregnant women (544).

b. CHEMICAL PROPERTIES

The enzyme benzylamine oxidase has been purified from hog plasma by Buffoni (508) and obtained in crystalline form from the same source by Buffoni and Blaschko (146). It has likewise been recently purified from human plasma (226).

Both hog-plasma benzylamine oxidase and human-plasma benzylamine oxidase catalyze the general reaction of amine oxidase (see Chapter II.2).

The crystalline hog-plasma benzylamine oxidase was homogeneous, as shown by starch-gel electrophoresis and ultracentrifugation. Its molecular weight was found to be 196,000 (146). The enzyme is stable at 0 to 4° under ammonium sulfate in the dark for months (146, 215). The spectrum shows a maximum at 280 nm an a shoulder at 480 nm. The enzyme contains 4 moles of phosphate per mole, 15.12% of nitrogen (215), and 3 to 4 g-atoms of copper per mole, as determined by radioactivation analysis (146).

Acid hydrolysis of the crystalline enzyme and subsequent treatment with 7 M urea (860, 861) yielded diffusible material showing spectroscopic and fluorescence properties of pyridoxal. By the use of bacterial apodecarboxylase, traces of pyridoxal phosphate and rather large amounts of free pyridoxal were detected (215). It was calculated that about 3 or 4 moles of pyridoxal were contained in a mole of enzyme. Considering results obtained in phosphate estimations these observations seemed to indicate the presence of 3 to 4 moles of stably bound pyridoxal phosphate per mole of hog-plasma benzylamine oxidase (215).

Blaschko and Buffoni (215) consider hog-plasma benzylamine oxidase to be a copper protein. To judge from the sensitivity of this enzyme toward inhibitors both copper and pyridoxal phosphate seem to be essential for the action of hog-plasma benzylamine oxidase (215; see Section 1.D and Chapter VI.4.B.a).

In this context it should be mentioned that Buffoni (582) has recently stated that, although benzylamine oxidase is a copper-pyridoxal phosphate-containing protein, its spectrum is not typical of a pyridoxal enzyme.

More fundamental work is still required for elucidation of the crucial question of the prosthetic groups of this and other amine oxidases.

The pH optimum of the purified enzyme with benzylamine as substrate at 25° is 8.0 (146). Blaschko and colleagues (145) obtained the same result for the pH optimum of crude hog-serum benzylamine oxidase with β-phenylethylamine as substrate.

It has been recently suggested (862) that in many pyridoxal enzymes the aldehyde group is not free but is combined with the ϵ-amino group of a lysine residue. It is not known how the pyridoxal group is attached to hog-plasma amine oxidase, but Blaschko and Buffoni (215) tentatively suggest that the pyridoxal group may be linked with a lysine residue. According to these workers their hypothesis seems to be supported by the results they obtained on amino acid analyses showing that the peak that contained the pyridoxal also contained most of the lysine of the enzyme protein (215). This observation will have to be confirmed and further elucidated.

Concentrated solutions of the purified hog-plasma enzyme have a pink color, which disappears when benzylamine is added under anaerobic conditions.

McEwen (226) has succeeded in isolating and purifying about 5000-fold benzylamine oxidase from human plasma. The specific activity of the latter preparation with benzylamine as substrate was, however, only one-thirtieth of the specific activity of the crystalline hog-plasma enzyme (131). The purest enzyme preparations obtained from human plasma were colorless, and the absorption spectra of concentrated enzyme solutions usually showed an absorption maximum at 404 nm and a small broad shoulder at 500 nm.

Since neither of these spectral features altered on anaerobic addition of benzylamine, McEwen is inclined to attribute these characteristics of the visible spectrum to a possible contamination of his purified plasma enzyme with catalase, which was present even in his most purified preparations (226).

In phosphate and pyrophosphate buffers 3.3 mM concentrations of benzylamine are optimally oxidized at neutral pH values (226). Unlike benzylamine oxidase of hog plasma, this amine oxidase of human plasma does not seem to require any cofactor for maximal activity (859).

C. SUBSTRATE SPECIFICITY

The amines that were first tested as potential substrates of pure crystalline hog-plasma amine oxidase were benzylamine, histamine, mescaline, and 4-picolylamine (146). All these substances were previously known to be oxidized by crude preparations of hog-plasma amine oxidase (145). As was to be expected, the substrate most actively oxidized by the pure hog-plasma enzyme was benzylamine. The affinity of the enzyme for this substrate was more than 100 times that for histamine. However, since the relative rate of oxidation of histamine by the crystalline enzyme was of the same order of magnitude as that by the crude enzyme, Buffoni and Blaschko concluded that the degradation of histamine by the hog-plasma enzyme is an intrinsic property of the enzyme that acts on benzylamine and other monoamines (146). The crystalline hog-plasma enzyme does not act on short aliphatic diamines such as putrescine and cadaverine and in this respect clearly differs from hog-kidney DAO. The plasma enzyme does not act on catecholamines nor on spermine and spermidine. It is active, however, on tyramine, tryptamine, 4-picolylamine, and mescaline.

The best substrate of human-plasma oxidase was found to be also benzylamine (226, 859). This enzyme specifically oxidizes primary aliphatic amines that lack substituents on the α-carbon atom, at a rate comparable to that found for benzylamine under the same conditions (226). The optimal chain length for oxidation of the primary aliphatic amines appeared to be that of butylamine. Secondary and tertiary amines, and amines with α-substituents, for example, amino acids, are not affected. Of the amines of pharmacological and biological interest, tyramine, dopamine, tryptamine, kynuramine, ethanolamine, and β-mercaptoethylamine were deaminated by the human-plasma amine oxidase. Typical substrates of histaminase (DAO), such as histamine or 1,4-diaminobutane, were poor substrates of human-plasma amine oxidase.

Whereas the short-chain diamines were hardly attacked by this enzyme, one of the long-chain homologs, 1,8-diaminooctane, was actively oxidized. On the other hand, the long-chain polyamines, spermine and spermidine, were not attacked (226).

The substrate specificity of the human-plasma benzylamine oxidase seems to differ from that reported for the amine oxidases from plasma of any other species. Thus, whereas the plasma amine oxidases of the ox and other ruminants oxidize spermine and spermidine as actively as simple aliphatic amines (128, 141), these polyamines are not substrates of the human-plasma enzyme. Unlike the hog-plasma amine oxidase (146) or the DAO, found in human plasma during normal pregnancy (122, 536), the human-plasma benzylamine oxidase does not oxidize histamine or 1,4-diaminobutane at a significant rate (226). Whereas the human-plasma amine oxidase deaminates tryptamine but not serotonin, the amine oxidase of horse plasma is reported to oxidize both tryptamine and serotonin (145). The hallucinogenic amine, mescaline, which is a substrate of plasma amine oxidases of many lower primates (128), is not deaminated by the human-plasma enzyme (226).

d. INHIBITORS

Hog-plasma benzylamine oxidase and the human-plasma enzyme are both sensitive to carbonyl reagents and to semicarbazide (215, 226). They are also inhibited by cyanide, which helps to distinguish these plasma enzymes from intracellular MAO (131). Isoniazid and iproniazid are effective inhibitors of the mammalian plasma oxidases (128, 141). As for beef-plasma amine oxidase (141), both isoniazid and iproniazid inhibited the human-plasma enzyme only after a preliminary period of incubation (226). Whereas the sensitivity of the human-plasma enzyme to isoniazid inhibition was considerably less than that reported for the beef-plasma enzyme under similar conditions (141), the sensitivity of the human-plasma amine oxidase to inhibition by iproniazid was more marked than that of the beef-plasma enzyme (226). In contrast to the beef-plasma amine oxidase (580) and to the hog-plasma enzyme (215), the human-plasma amine oxidase is not markedly inhibited by sodium diethyldithiocarbamate (226). It is only slightly inhibited by 2,2'-bipyridyl and by o-phenanthroline, which contrasts with the behavior of the beef-plasma enzyme (580), but resembles that of hog-plasma amine oxidase (215). However, the human-plasma enzyme is strongly inhibited by cuprizone (biscyclohexanone oxaldihydrazone) which, in analogy to the behavior of the hog-plasma enzyme (131), may be suggestive of the presence of copper in the human-plasma amine oxidase. According to Buffoni (131), the different sensitivity to chelating agents of the amine oxidases of hog plasma and human plasma may depend on a different affinity for copper of the respective apoenzymes, as well as on the steric accessibility of copper to the chelating agents (228). Thus, diethyldithiocarbamate can remove copper from hog-plasma benzylamine oxidase. The copper-free enzyme is inactive, but the activity can be regenerated by dialysis against cupric sulfate (131). In agreement with Buffoni (131), McEwen (226) considers it possible that human-plasma amine oxidase is a copper protein.

It is remarkable that n-octyl alcohol, a specific inhibitor of the mitochondrial MAO (123), is also an effective inhibitor of human-plasma amine oxidase (226). Its inhibition of benzylamine oxidation is instantaneous and competitive in nature.

Because of all these findings, McEwen has emphasized that the behavior of human-plasma amine oxidase in comparison with that of any other amine oxidase thus far recognized in animal tissues is unique (226). Thus, unlike the classical histaminase (DAO), the human-plasma amine oxidase does not actively degrade histamine or putrescine. Unlike the mitochondrial MAO, it does not attack secondary and tertiary amines. Moreover, it is easily inhibited by carbonyl reagents and semicarbazide. On the other hand, the intracellular MAO actively oxidizes all substrates of human-plasma amine oxidase.

As already mentioned (see Section 2.B.c), McEwen (544) has pointed out that blood plasma from pregnant women seems to contain two independent amine oxidases, an MAO (human-plasma benzylamine oxidase), whose activity remains unaltered throughout pregnancy, and a DAO (histaminase), the enzymic activity of which steeply increases during the first trimester of gestation and reaches a peak early in the third trimester.

e. BIOLOGICAL SIGNIFICANCE OF NONRUMINANT PLASMA AMINE OXIDASE

Since the biological substrates of both hog-plasma and human-plasma amine oxidase have so far not been detected, the biological significance of these enzymes is unknown.

In his very interesting comparative study on amine oxidases of mammalian blood plasma Blaschko (128) considers the possibility of an evolution of spermine oxidase having taken place from a "benzylamine oxidase."

3. RABBIT-SERUM AMINE OXIDASE

Recently, McEwen and his co-workers (863) have identified in the rabbit serum an amine oxidase, which they purified considerably. The purified serum enzyme oxidizes only primary amines, with the stoichiometric formation of the corresponding aldehydes, hydrogen peroxide, and ammonia. Substrates most actively oxidized by this enzyme are benzylamine, dopamine, tyramine, and mescaline, and long-chain aliphatic amines. The ability of rabbit serum to oxidize benzylamine was found to be tenfold greater than that of human serum (863). Less actively oxidized substrates include tryptamine, serotonin, histamine, and simple aliphatic amines. With respect to its general substrate specificity, the rabbit-serum enzyme resembles the plasma amine oxidases of other mammals. Accordingly, the rabbit-serum amine oxidase deaminates only primary amines; secondary and tertiary amines, as well as amines with α-substituents (e.g., α-amino acids) are not attacked. The rabbit-serum amine oxidase differs, however, from the bovine-plasma enzyme (141, 143) in that it does not degrade spermine and spermidine, and likewise from bovine-plasma (141, 143) and human-plasma amine oxidases (226) in that it does not attack kynuramine but, on the other hand, deaminates mescaline as actively as it does long-chain aliphatic amines (863). Although similar in many respects to that of the human-plasma amine oxidase, the substrate specificity of the rabbit-serum resembles more closely that of the hog-plasma enzyme (128, 145, 146).

The purified rabbit-serum amine oxidase has an absorption maximum at 470 nm which is decreased in the presence of substrate (863). Despite the lack of direct evidence that this enzyme is a pyridoxal protein, the marked similarity of the latter to the hog-plasma amine oxidase in substrate specificity and absorption spectrum makes this probable (863).

4. RABBIT-LIVER AMINE OXIDASE, MESCALINE OXIDASE

The hallucinogenic drug mescaline is the active substance of a mexican cactus *(Lophophora Williamsii)*. When fed to various animals, this compound was shown by Slotta and Müller (864) to be excreted in the urine in the form of 3,4,5-trimethoxyphenylacetic acid.

In their classical study on the oxidation of mescaline Bernheim and Bernheim (155) were able to demonstrate that rabbit-liver preparations effectively inactivated mescaline by oxidative deamination. In their extensive investigations these authors were surprised to find that the mescaline-oxidizing enzyme differed significantly from the classical intracellular MAO, that oxidized tyramine. Whereas rabbit liver rapidly oxidized mescaline, rat and guinea-pig liver and rabbit kidney degraded this compound very slowly, and rat and guinea-pig kidney, and cat and dog liver, and kidney did not act on mescaline at all (155). Moreover, it was subsequently demonstrated that neither octanol nor choline-*p*-tolyl ether, both specific inhibitors of MAO, had any effect on the oxidation of mescaline by rabbit-liver preparations (138, 865). All these findings thus suggested that the rabbit liver contained in addition to MAO another enzyme that was responsible for the oxidation of mescaline (155). This has been confirmed by workers in different laboratories (137–139, 865).

In its inhibitory specificity rabbit-liver mescaline oxidase seems to resemble hog-plasma amine oxidase (129). In this context it should be recalled that mescaline is rapidly oxidized by hog-plasma benzylamine oxidase, crude and crystalline alike (145, 146). However, the highly purified plasma enzymes, spermine oxidase (143) and human-plasma amine oxidase (226), do not attack mescaline (see Sections 2.A.c, and 2.B.c).

Thus, the introduction of methoxy groups into the benzene ring of tyramine seems to modify the molecule with regard to its enzymic degradation by tissues as well as in its pharmacological action (155). Nothing is known about the inactivation of mescaline *in vivo*.

5. PLANT AMINE OXIDASE, PEA-SEEDLING DAO*

A. Occurrence and Properties

In 1948 Werle and co-workers (866, 867) reported the occurrence in extracts of some leguminous plants as well as of sage and lavender of an enzyme that catalyzes the degradation of 1,4-diaminobutane, 1,5-diaminopentane, and histamine, and suggested that this enzyme corresponds to the DAO of animal tissues. Subsequently, Werle and Pechmann (516) compared the DAO content of resting and germinating seeds in the pea, red clover, and lucerne (alfalfa), and found that the enzyme activity becomes apparent in these seeds on the onset of germination and, 5 days later attains values that are about 10 times higher than the highest values obtained for DAO in animal tissues. Kenten and Mann (868) and Mann (592) were able to show that the plant enzyme also catalyzes the oxidation of phenylalkylamines and aliphatic monoamines, and suggested the name "plant amine oxidase" for this enzyme. This name was subsequently used by Mann and co-workers (869–872) for the enzyme that they studied mainly in pea seedlings. Because of the work by Fouts et al. (591), who showed that the animal DAO catalyzes also the oxidation of monoamines and phenylalkylamines, Hill and Mann (873) thought that, by analogy with the animal DAO, a distinction between MAO and DAO in plants was not justifiable merely on the basis of a substrate specificity; hence, in their subsequent investigations of the plant enzymes they reverted to the name of plant diamine oxidase.

---

* E.C. 1.4.3.6. Diamine: O_2, oxidoreductase, deaminating.

B. Prosthetic Group(s)

Mann (872) devised a method by which a highly purified pea-seedling DAO was obtained. The final preparations were pink and the absorption spectrum showed a band with a maximum at 500 nm. The pink color of the pure enzyme preparations was discharged by sodium dithionite and restored by oxygenation. The pink color was also discharged by the addition of the substrate 1,4-diaminobutane under anaerobic conditions and was restored by subsequent oxygenation. Moreover, on addition of hydrazine, which is an inhibitor of the enzyme, the pink color of the pure enzyme solution changed to yellow. On standing in air the yellow color slowly faded, but the pink color did not reappear. All these results seem to suggest that the pink color is a property of the enzyme (872). The pure enzyme preparation contained 0.08 to 0.09% of copper, which could be removed by sodium diethyldithiocarbamate. The copper-free preparations were catalytically inactive, but regained most of their original activity on the addition of Cu^{2+} ions. Mann (872) suggested that the copper is present in the pea-seedling enzyme as a complex with a carbonyl compound and that this complex forms the prosthetic group of the enzyme.

Further evidence that the plant DAO is a metalloenzyme, containing cupric copper, was provided by Hill and Mann (873), who studied the inhibitory effect of various metal chelating agents on the pea-seedling enzyme. Werle, Trautschold, and Aures (517) also obtained from pea seedlings highly purified preparations of DAO that were rose red in concentrated solutions and contained 0.12% of copper. The latter authors also suggested that the color is a property of the enzyme. Later results by Hill and Mann (576) seem to add more evidence to the theory of Mann (872) that the prosthetic group of plant DAO contains copper in a complex with a carbonyl compound.

In 1949 Werle and Pechmann (516) observed that the oxygen uptake of dialyzed clover-seedling extracts was increased in the presence of cadaverine by the addition of pyridoxal. Since hydrogen peroxide was produced during the enzymic reaction, these authors suggested that both pyridoxal phosphate and FAD were prosthetic groups of the plant DAO. Likewise, Goryachenkova (560) reported increases in both oxygen uptake and ammonia formation, when pyridoxal phosphate and FAD were added to dialyzed clover-seedling extracts in the presence of 1,6-diaminohexane. These claims with regard to pyridoxal phosphate and FAD as prosthetic groups of plant DAO have so far not been substantiated. Werle (517), working with a highly purified pea-seedling enzyme preparation, was unable to produce any evidence for his previous suggestion (516) with regard to the presence of pyridoxal phosphate or/and FAD in the molecule of plant DAO.

Similarly, recent attempts of Hill (874) to demonstrate unequivocally the existence of pyridoxal phosphate in plant DAO have not met with success.

C. Substrate Specificity

With regard to substrate specificity of the highly purified plant DAO, the 1,4- and 1,5-diamines are readily oxidized by the pea-seedling enzyme, but not the 1,2- and 1,3-diamines (517, 576). Histamine is oxidized much less readily than 1,4-diaminobutane, the rate of the catalyzed oxidation of the latter being about 20 times that of histamine (576). Werle et al. (517) found that the purified pea-seedling enzymes oxidized cadaverine at a rate 23 times that of histamine. Whereas spermidine is rapidly oxidized, spermine is only very slowly degraded (517, 576). With the aliphatic monoamines the rates of oxygen uptake have been found to be very much lower than with the aliphatic diamines (576). Diethylamine and isopropylamine were not oxidized at all. Of the phenylalkylamines tested, 2-phenylethylamine was oxidized faster than benzylamine, but 1-methylbenzylamine was not oxidized. In agreement with Werle et al. (517), Hill and Mann (576) found that the plant DAO catalyzed the oxidation of L-norepinephrine but not that of L-epinephrine.

In its substrate specificity the plant enzyme (576) thus resembles the DAO of animal tissues, which, as shown by Fouts et al. (591), catalyzes the oxidation of both monoamines and diamines. As with the animal enzyme (122), the rates of oxidation of aliphatic diamines by the plant enzyme depend on the chain length; but, although both enzymes attack preferentially the short-chain diamines, 1,4-diaminobutane, and 1,5-diaminopentane, the oxidation of 1,3-diaminopropane is catalyzed only by the animal enzyme. Secondary amines are not oxidized by either enzyme (576), in contrast to MAO which oxidizes both primary and secondary amines (161, 243).

D. Inhibitors

Several workers have reported that the DAO of pea and clover seedlings is inhibited by chelating agents (517, 875) and have, therefore, suggested that the enzyme may be a metalloprotein. The purified enzyme is inhibited by sodium diethyldithiocarbamate, salicylaldoxime, potassium ethylxanthate, diphenylthiocarbazone, 8-hydroxyquinoline, 2,2'-bipyridyl, and o-phenanthroline (517, 612, 873, 875). Hill and Mann (873) suggest that the primary reaction, involved in the inhibition of the plant DAO by the chelating agents, is the combination of these reagents with enzyme-bound copper to form inactive enzyme-inhibitor complexes. These complexes are, however, dissociable and usually break down on dialysis, or on addition of metal ions with which the inhibitor preferentially forms a complex and thus liberates the enzyme. As already mentioned, the pea-seedling enzyme belongs to the group of amine oxidases that are inhibited by carbonyl reagents. By analogy with the inhibition of animal DAO by cyanide (607), Kenten and Mann (868) proposed the theory that the plant enzyme is inhibited by cyanide, because this compound acts as a carbonyl reagent. Later findings by Hill and Mann (576), pointing to the behavior of plant DAO as a metalloenzyme that contains cupric copper, suggest the possibility that the inhibition by cyanide may be due, at least in part, to this. Werle et al. (876) report that the pure pea-seedling enzyme is completely inactivated by a 10^{-4} M solution of aminoguanidine.

E. Biological Significance of Plant DAO

Little is known about the significance of plant DAO for the process of germination. Werle and his colleagues (517) suggested that the formation of enormous amounts of DAO during germination is caused by the embryo, when this enzyme is apparently resynthesized along with other enzymes (876). However, the effect of DAO on germination is a matter of speculation.

6. BACTERIAL AMINE OXIDASES

There are reports on the enzymic oxidation of diamines in a great number of microorganisms, such as *Mycobacterium smegmatis, M. tuberculosis, Pseudomonas aeruginosa, Serratia marcescens, Neisseria perflava, Ps. pyocyanea, Escherichia coli,* and many others (511, 512, 514, 516, 877–880).

Mycobacterium smegmatis is able to catalyze the oxidative deamination of aliphatic diamines, histamine, and spermine in a manner similar to that of the hog kidney (514). However, the bacterial enzyme shows preference for C_3 and C_4 compounds; thus, putrescine is much more readily attacked by this enzyme than are cadaverine or histamine. The enzymic metabolism of diamines seems to be "inducible," since it is increased by the addition of the substrate to the culture medium (879). The first step in the metabolism of the diamine by bacterial preparations is assumed to be an oxidative deamination of the "DAO type." However, Kim and Tchen (881) have recently shown that in extracts from 1,4-diaminobutane-adapted *E. coli* the first step was transamination to a ketoacid, catalyzed by pyridoxal phosphate. Further work with extracts of other bacteria is necessary to demonstrate which of the two reactions—transamination or oxidative deamination—is responsible for the metabolism of diamines by a bacterial system. Maggio et al. (882) incubated for 2 hr various bacterial suspensions with the simultaneous addition of two substrates, histamine and cadaverine. They found an additive formation of ammonia with cultures of *Staphylococcus aureus, Proteus* Ox_{19}, *Ps. aeruginosa, Brucella melitensis,* and *Shigella dysenteriae,* indicating that each of the two substrates induced enzyme formation. *Streptococcus pyogenes* showed, however, preference for histamine, and *Klebsiella pneumoniae,* for cadaverine as substrates (882). Moreover, Satake, Ando, and Fujita (883) demonstrated that in cultures of *Achromobacter sp.* the addition of histamine, putrescine, and amylamine induced substrate-specific enzymes. The amine oxidases of *Aspergillus niger, Penicillium chrysogenum, Monascus anca,* and *Fusarium bulbigenum* catalyze the oxidative deamination of histamine, although their best substrate is the monoamine butylamine (577). The amine oxidase of *M. anca* does not act on diamines or polyamines (577).

An oxidative degradation of spermidine and spermine by bacterial preparations was first observed by Silverman and Evans (884), who used whole cells or lyophilized preparations of *Ps. pyocyaneae.* Oxidation of the polyamines has since been observed in intact cells or crude extracts of *M. smegmatis, N. perflava, Ps. aeruginosa,* and *S. marcescens* (156, 514, 879, 885, 886). The oxidative activities for these amines usually increase when the organisms are grown with the amines added to the growth media, which may indicate that the relevant enzymes are adaptive in nature (156, 879).

In their studies on the oxidation of polyamines by *N. perflava,* Herbst and Weaver (156) demonstrated that the bacterial enzyme oxidatively deaminates spermine as well as spermidine with the consumption of 1 mole of oxygen and the production of 1 mole of 1,3-propanediamine, 1 mole of hydrogen peroxide, and 1 mole of the appropriate aldehyde, according to the equation

$$RCH_2NH(CH_2)_3NH_2 + O_2 + H_2O \rightarrow RCHO$$
$$+ NH_2(CH_2)_3NH_2 + H_2O_2.$$

It follows from the above equation that γ-aminobutyraldehyde should be formed in the oxidation of spermidine. This compound, which is also a product of the action of DAO on putrescine, undergoes spontaneous cyclization under physiological conditions to form Δ^1-pyrroline (551, 869, 887). Since the reaction mixture after spermidine oxidation showed no aldehyde by the 2,4-dinitrophenylhydrazine method of Friedmann and Haugen (888), Weaver and Herbst concluded that a cyclization might take place in the enzymic oxidation of spermidine by *N. perflava* (156). The *Neisseria* polyamine oxidase has an optimal pH of 6.8 to 7.0 and is inhibited by cyanide, hydroxylamine, and semicarbazide in concentrations similar to those that act on plasma oxidases (156); it has no MAO or DAO activity and does not attack benzylamine. It thus differs from beef-plasma spermine oxidase, which oxidizes a number of monoamines besides spermine and spermidine (141).

A different attack on spermine and spermidine is brought about by the amine oxidase of *Ps. aeruginosa* (879) and by the amine oxidase isolated from *M. smegmatis* (886). 1,3-Diaminopropane, β-alanine, and γ-aminobutyric acid, as well as spermidine, have been identified among the oxidation products of spermine. The amine oxidase from *M. smegmatis* also differs from the *N. perflava* enzyme in that it oxidizes 1,4-diaminobutane (886). Like the *Neisseria* polyamine oxidase, the *Mycobacterium* enzyme is, however, inhibited by carbonyl reagents; this, as mentioned above, is reminiscent of plasma amine oxidases (156, 886).

In his recent studies on *S. marcescens* Bachrach (513, 889) isolated and partially purified from *Serratia* an enzyme that oxidizes spermidine and its homolog bis(3-aminopropyl)amine, but not spermine or any of the diamines or monoamines tested. This partly purified enzyme catalyzes the following reaction:

$$NH_2(CH_2)_3NH(CH_2)_4NH_2 \quad + \quad O_2 \quad + \quad H_2O \quad \longrightarrow$$

$$[NH_2CH_2CH_2CH_2CHO] \quad + \quad NH_2(CH_2)_3NH_2 \quad + \quad H_2O_2$$
$$1,3\text{-Diaminopropane}$$

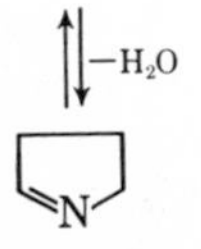

Δ^1-Pyrroline

Δ^1-Pyrroline, which is formed by the oxidation of spermidine but not by the oxidation of bis(3-aminopropyl)amine, reacts with *o*-aminobenzaldehyde to give a yellow color (601, 890). On these reactions is based a specific enzymic assay for spermidine (207).

Campello, Tabor, and Tabor (891) further purified the polyamine oxidase from *S. marcescens* and found that the oxidation of spermidine by that enzyme requires FAD as well as an additional electron carrier, such as phenazine methosulphate. In this behavior the *Serratia* enzyme differs strikingly from MAO and DAO preparations (126, 128), because the latter enzymes do not require any added electron carriers to react with molecular oxygen. Campello, Tabor, and Tabor (891) regard the spermidine-oxidizing enzyme from *S. marcescens,* which is linked to an electron-transport system, as a spermidine dehydrogenase, with FAD serving as its cofactor.

In this context it should be mentioned that Yamada and co-workers (880) have recently reported on a bacterial putrescine oxidase, a DAO requiring FAD as a prosthetic group. From cells of *Micrococcus rubens,* grown on a nutrient broth, these authors prepared a homogeneous protein fraction (sedimentation analysis, $S^0_{20, w} = 5.5$) which catalyzes the oxidation of putrescine as follows:

$$putrescine + O_2 + H_2O \rightarrow \Delta^1\text{-pyrroline} + H_2O_2 + NH_3$$

Whereas cadaverine and spermidine are also attacked, but at very low rates, other monoamines and polyamines are not oxidized. This enzyme, purified by ammonium sulfate fractionation, chromatography on DEAE-cellulose, DEAE-Sephadex, and hydroxylapatite, shows an absorption spectrum typical of a flavoprotein, with absorption maxima at 380 and 460 nm which bleach on addition of putrescine as well as of sodium dithionite. The flavin component has been identified as FAD. The molecular weight of putrescine oxidase, calculated by the equilibrium method, was found to be about 80,000 and the flavin content, assessed from the spectral data, was 1 mole FAD per mole of enzyme. This enzyme is markedly inhibited by hydroxylamine, cyanide, and iproniazid—but not by phenylhydrazine, semicarbazide, or isoniazid.

The functional significance of these microbial amine oxidases is still unknown. However, it is interesting that, in most bacteria enzymes, that show strict substrate specificity, can be induced (883). As to the name "adaptive enzymes," the classification of microbial enzymes by Karström (892) may be recalled here. Karström distinguishes between two classes of microbial enzymes. The first includes enzymes that are always present when the determining gene is present in the organism. These enzymes he calls constitutive enzymes. The second class comprises enzymes that are formed when bacteria are exposed during growth to specific substrates; these Karström calls adaptive enzymes.

Recent studies by Bachrach and his co-workers (824–828) may have thrown considerable light on the biological significance of microbial amine oxidases. These authors were able to demonstrate that oxidized polyamines are toxic for bacteria (825), bacteriophages (824), plant viruses (826), and Ehrlich ascites cells (827). In their very careful studies Bachrach and Leibovici (893) have found that oxidized spermine forms a complex with the DNA of coliphage T_5, and that upon infection this complex is injected into the bacterial host. The injected complex does not significantly interfere with the metabolic activity of the host (894) nor does it induce the formation of complementary RNA (893). In a recent study Bachrach and Eilon (828) have presented evidence for an interaction of oxidized polyamines with DNA. As discussed in Section 2.A.b, Tabor, Tabor, and Bachrach (842) have demonstrated that the naturally occurring polyamines, spermine and spermidine, are oxidized by purified plasma amine oxidase to the corresponding aminoaldehydes. Thus, spermine is oxidized in the terminal positions to a dialdehyde that contains all the carbon atoms of the original spermine. The dialdehyde can be identified by reduction with sodium borohydride to the corresponding dialcohol. Spermidine is also oxidatively deaminated by plasma amine oxidase but with the formation of a monoaldehyde only (842).

As shown by Tabor (787), nucleic acids, being polyelectrolytes, readily form complexes with basic polyelectrolytes, such as spermine and spermidine. Dissociation can normally be achieved by the addition of cations, which apparently reverse the electrostatic binding of the basic electrolytes. Bachrach and Eilon (828) have obtained evidence that the amine dialdehyde produced by the enzymic oxidation of spermine interacts with DNA. This interaction has been found to involve two classes of bonds: electrostatic binding and covalent cross-links. Gel-filtration experiments show that the binding by electrostatic forces is reversible and that complexes between DNA and oxidized spermine are partially dissociated by 3 M sodium chloride.

Urea has no effect on DNA-oxidized spermine complexes. The occurrence of covalent cross-links between the paired strands of bihelical DNA is demonstrated by thermal denaturation profiles and by sucrose-gradient centrifugation of complexes between DNA and oxidized spermine. Thermal-denaturation profiles also confirm the importance of electrostatic bonds, which bring about a shift of the melting temperature of the DNA. On the other hand, the covalent cross-links prevent strand separation and thus reduce the hyperchromic effect. Whereas the partial binding by electrostatic forces may be explained by the cationic nature of oxidized spermine because of the presence in its molecule of secondary amino groups, the binding of oxidized spermine to DNA by nonelectrostatic bonds seems to be due to the two carbonyl groups in the molecule of oxidized spermine. Thus, reduction of the aldehyde groups of oxidized spermine by sodium borohydride abolishes the induction of cross-links (828). According to Bachrach and Eilon, oxidized spermine seems to behave like a bifunctional alkylating agent that induces bridges between the complementary strands of the DNA (895). It is noteworthy that oxidized spermidine, which possesses only one carbonyl group, does not form cross-links in DNA (828). Bachrach et al. have previously shown that oxidized spermine prevents DNA replication (893) and transcription (896). These effects may be now explained by the irreversible binding by nonelectrostatic links of oxidized spermine to DNA (828).

Yamada and co-workers have reported (897) that the oxidation products of spermine as well as of two homologues of spermine, obtained with bovine-plasma amine oxidase, react with phage DNA and interfere with the maturation of the phages in infected cells.

All these recent findings seem to add further evidence for the important regulatory function that the amine oxidases may have *in vivo*.

CHAPTER V

Differentiation of Amine Oxidases

It has been known for a long time that enzymes from different species that catalyze the same reaction and show the same substrate specificity may be in many other respects significantly different. It has also been long recognized that the technique of identifying an enzyme by the species and tissue or organ from which it originated is not sufficient for the specification of an enzyme. Moreover, brilliant work on ribonuclease, published in 1951 and 1953 (898–900), revealed that crystallized bovine pancreatic ribonuclease is not a single protein, but can be resolved by partition and ion-exchange chromatography into several components. It soon became evident that this heterogeneity of ribonuclease is not a unique property of the bovine enzyme but is shared by the ribonuclease from the pancreas of the sheep (901).

In 1937 Blaschko, Richter, and Schlossmann (161) were the first ones to point out that differences in the specificity of substrates of mitochondrial monoamine oxidase (MAO) of various origin may be due to the activity of more than one MAO. Subsequently, Alles and Heegard (164), in testing a large number of monoamine substrates with liver preparations from different species, observed marked variations in the relative rates of oxidation and also suggested that more than one MAO seems to occur in the livers of various species. The latter suggestion found considerable support in the work of many researchers in different laboratories (189, 198, 234, 254, 902–908). All these workers emphasized that MAO exists in multiple forms. The differences in the behavior of MAO observed by the authors, quoted above, prompted Blaschko (128) to suggest that either the MAO from each tissue is an entirely different protein, or that each

145

organ contains a mixture of different amine oxidases in differing proportions.

Further evidence on the problem of the heterogeneity of MAO has been recently provided by Oswald and Strittmatter (194), who in their comparative studies on substrate specificity and Michaelis constants of MAO preparations from the liver, kidney, and heart of the rat and guinea pig found marked tissue differences. These authors concluded that each tissue contains a single distinctive major MAO. Their findings also suggest a dual intracellular localization of MAO with the major part in the mitochondria and a smaller, but significant, portion sedimenting with the microsomal fraction. In the course of many investigations of the substrate and inhibition pattern Zeller (122, 344) found qualitative and quantitative differences among amine oxidases of different origin, and concluded that the name monoamine oxidase does not represent a single entity, but an extremely large group of closely related enzymes. It is characteristic of members of this group, called by Zeller the homologous group, that they all catalyze the same classical reaction of oxidative deamination, all are insensitive to semicarbazide, and all show preference for certain alkylmonoamines and aralkylamines (344).

In 1963 Gorkin (236) achieved a partial separation of rat-liver mitochondrial MAO into two components, each attacking a different amine; that is, 3-nitro-4-hydroxybenzylamine and 4-nitrophenylethylamine. In more recent experiments (237) with density-gradient electrophoresis of mitochondrial sonicates, treated with urea or hydroxyquinoline he confirmed and extended these data. According to this author, the particles of mitochondrial membranes may dissociate and the subfragments formed may acquire a substrate specificity distinct from that of parent particles. This hypothesis of the multiplicity of MAO suggests the possibility of more or less selective inhibition of deamination of various monoamines by different inhibitors. Several compounds exhibit this interesting property both *in vitro* and *in vivo* (222, 909). Thus, harmine is 1000 times more potent as an inhibitor of the deamination of serotonin than of tyramine or tryptamine (910).

Youdim and Sandler (911) recently prepared a highly purified MAO from rat liver as well as from human-placenta mitochondria, and subjected these preparations to polyacrylamide-gel electrophoresis. The purified enzyme from rat liver showed three bands of enzyme activity, and the human-placenta enzyme preparation showed two bands. These isoenzymes vary with regard to heat inactivation, pH activity curve, and electrophoretic mobility (912). Lately Youdim and Collins (1093) have obtained evidence that the multiplicity of rat liver MAO is due to conformational differences.

With regard to histaminase it was Zeller's finding in 1938 that preparations of hog kidney act on 1,4-diaminobutane and 1,5-diaminopentane as well as on histamine, that induced him to suggest that the name histaminase be replaced by diamine oxidase (DAO) (505). As mentioned in Chapter IV.1.A, Zeller's opinion that the activities of histaminase and DAO always run together has been contested by the present author (509). Moreover, Zeller's view is not compatible with findings that hog-plasma amine oxidase (146) acts on histamine but does not attack the short-chain diamines, such as 1,4-diaminobutane and 1,5-diaminopentane. Work carried out by Blaschko and his associates (145, 506) has led to the suggestion that there exist at least two kinds of histaminase, both present in the hog. The enzyme occurring in the hog kidney acts, according to Blaschko (129), on the dicationic form of histamine, in which not only the terminal amino group but also the imidazole ring carries a positive charge. The hog-plasma amine oxidase, however, attacks the monocationic form of histamine, in which the basic group in the imidazole ring carries no charge. Thus, hog-plasma amine oxidase behaves like an MAO and indeed shares many substrates with the intracellular MAO. However, hog-plasma amine oxidase is, like DAO, inhibited by cyanide and carbonyl reagents as well as by semicarbazide and aminoguanidine and differs in this property fundamentally from mitochondrial MAO (129, 146). Furthermore, the unique behavior of human-plasma amine oxidase toward substrates and inhibitors, which differs from that of any animal-tissue amine oxidase, should be recalled here (226; also see Chapter IV.2.B.d). It should also be remembered here that mouse-liver histaminase displays both MAO and DAO activities (148; see Chapter III.2), and that pea-seedling amine oxidase represents an enzyme that is sensitive to both cyanide and carbonyl reagents but attacks diamines as well as monoamines almost identically (517, 576; see Chapter IV.5.C).

In 1962 Burkard, Gey, and Pletscher (913) reported very interesting results, obtained in the *in vivo* study of the effects of inhibitors of MAO and DAO on the activity of these enzymes. In the kidney of cats, pretreated with inhibitors of MAO and DAO, the activity of these enzymes was measured simultaneously, using typical cell fractions and substrates. After iproniazid pretreatment, inhibition of DAO occurred somewhat earlier and was of shorter duration than inhibition of MAO. Both amine oxidases may be inhibited independently *in vivo*. Thus, *dl-trans*-phenylcyclopropylamine and pargyline inhibit mainly MAO, whereas aminoguanidine interferes only with DAO. In confirmation of earlier findings (527, 613) Burkard and his colleagues concluded that *in vivo* the time course of inhibition is different for MAO and DAO, and that these enzymes also differ in their inhibitor spectra. These differences between MAO and DAO suggest that both these enzymes may represent isoenzymes (913).

On chromatography with DEAE-cellulose and hydroxylapatite Yamada and Yasunobu (143) found that highly purified plasma amine oxidase can be resolved into two or three discrete enzymically active fractions. The appearance of multiple peaks in the chromatography of plasma amine oxidase the authors attributed to a possible occurrence of this enzyme as an equilibrium mixture of different forms. It is noteworthy that McEwen (226, 859) suggested in his studies on human-plasma amine oxidase that this enzyme may be identical with one of the components of mitochondrial MAO.

Chromatographic experiments with DEAE-cellulose and electrophoretic investigations revealed that partly purified hog-kidney histaminase can be resolved into two discrete enzymically active fractions.* Electrophoresis on cellulose acetate gave two bands, one located in the albumin region and another, slower moving band between the α_2- and β-globulin fractions of normal human serum. Only the slow-moving fraction was further purified, and a rather stable preparation of histaminase was obtained, which acted only on histamine and not on cadaverine or putrescine, and which appeared to be homogeneous by the criteria of chromatography and analytical electrophoresis (553). Since the appearance of the fast-moving peak, showing a lower specific activity toward histamine as substrate than the slower moving one, was attributed by Kapeller-Adler and MacFarlane* to an association of histaminase with other enzyme proteins and/or inert proteins, this enzyme fraction was, unfortunately, not further purified. However, this observation of a separation of hog-kidney histaminase into two active fractions with somewhat different substrate specificities points to a behavior of histaminase as an isoenzyme, and requires for its clarification further experiments. The long debated question whether hog-kidney histaminase is identical with hog-kidney DAO may perhaps find a simple explanation on the basis of isoenzymes.

A very recent observation by Kobayashi and his colleagues (914) should be quoted in this connection. These authors purified DAO from hog intestines and from human amniotic fluid by ammonium sulfate fractionation and column chromatography. Whereas they were able to concentrate by this procedure the DAO from hog intestines 1000 times, the enzyme from human amniotic fluid was concentrated only about tenfold. Moreover, the DAO prepared from human amniotic fluid showed differ-

* R. Kapeller-Adler and H. MacFarlane, unpublished data (1960).

* Unpublished data (1960).

ent pH optima for the destruction of putrescine and histamine. Furthermore, the putrescine/histamine ratio varied from 4 to 16 during the purification of DAO from the amniotic fluid. These workers conclude that DAO isolated from human amniotic fluid is, like the DAO obtained from hog intestine, a heterogeneous enzyme.

Finally, the interesting properties of the bacterial enzyme spermidine dehydrogenase from *Serratia marcescens* (891) should be recalled here. As mentioned in Chapter IV.6, this enzyme, when acting on spermidine, follows the classical course of oxidative deamination (see Chapter II.2). It differs strikingly from MAO and DAO in that it requires an added electron carrier to react with molecular oxygen.

In view of all these resemblances and differences among amine oxidases of different origin, which all catalyze the fundamental equation of oxidative deamination, it becomes evident that amine oxidases belong to a group of enzymes with closely related multiple forms. In 1959 Markert and Möller (915) proposed to use the term "isozymes" to describe the different molecular forms in which a protein may exist with the same enzymic specificity. In 1962 Blaschko (129) in his comparative study on amine oxidases of mammalian blood plasma suggested that underlying the differences in the behavior of plasma enzymes of different origin must be variations in the enzyme proteins. These differences in the configuration of the apoenzymes of amine oxidases are apparently responsible for variations in their substrate specificity. According to present ideas on protein evolution (916), the amino acid sequence of a protein determines its steric structure and its specific biological activity. In this context it should be mentioned that Blaschko (129) suggested that benzylamine oxidase, spermine oxidase, histaminase, and mescaline oxidase as well as plant and microbial enzymes might represent members of the same family in evolution.

In a recently published study *On the Evolution of an Enzyme* (917) Bonner and his associates suggest that mutations leading to alterations in the specificity or in the stability of proteins are preserved or eliminated by natural selection. This hypothesis is based on studies on gene-enzyme relationships, indicating that mutations can lead to alterations in substrate affinities, thermolability, sensitivity to metals, and antigenic structure of specific proteins or enzymes.

In this context recent work by Aures and colleagues (918) on the microanalysis of tissue amines and of enzymes involved in their metabolism should be mentioned. These authors have developed simple and sensitive methods for the separation and detection of biogenic amines by thin-layer chromatography and have adapted them for microradiometric assays of amine oxidases with a high degree of sensitivity and specificity (see Chapter VI.2.D.m).

It is the present author's opinion that a consideration of the results of biochemical, genetic, and immunochemical studies on amine oxidases of different origin will facilitate the differentiation and classification of these enzymes.

CHAPTER VI

Analytical Procedures

1. INTRODUCTION

Until fairly recently there has been no standard method for measuring or expressing reaction velocities of enzymes. A large number of arbitrary units came into being, often several different ones for the same enzyme activity. The conditions of enzymic assays were often inadequately defined, and uniformity in the mode of presentation of results was not maintained. Thus, in some instances the enzyme activity, measured in terms of the amounts of substrate changed, is given in milligrams, micromoles, or microliters of equivalent gas. In some other cases the activity of the same enzyme is expressed in the form of a change in absorption of light, or of fluorescence, or of pH, and no efforts were made to correlate the results obtained and to compare them with each other. The time interval observed appears to have been even more inconsistent, variations from a few minutes to several hours not being uncommon. The conditions of measurement were also very diverse. Often no optimal substrate concentration appears to have been ascertained so that in some assays an enzyme was subjected to high substrate concentrations and in others to very low ones. The same observations hold true for the pH conditions. It is obvious that, due to this somewhat chaotic state, the comparison of data from different laboratories is often rather difficult, if not impossible.

In 1961 the first *Report of the Commission on Enzymes of the International Union of Biochemistry* was published and contained strong recommendations to enzymologists with regard to the standardization of enzyme assays (919).

151

A standard unit was proposed that is applicable to all enzymes and expresses their activity in absolute terms, thus allowing a comparison between the activities of different enzymes. One standard unit (U) of any enzyme is to indicate the amount that under defined assay conditions catalyzes the transformation of 1 micromole of the substrate per minute or, where more than one bond of each substrate molecule is attacked, one microequivalent of the group concerned per minute. Where two identical molecules react together, the unit is the amount that catalyzes the transformation of 2 micromoles per minute. Since different enzymes have different ratios of activities, some very large and others very small, results may be expressed in milliunits (mU) or kilounits (kU).

In cases of enzymes whose activity is not usually measured in terms of a chemical reaction, but in terms of some physical change, for example, an alteration in viscosity, a conversion factor, quantitatively relating the physical changes to the chemical ones, or more directly to the number of enzyme units present, should be determined and used. If the physical and chemical changes are not directly proportional, curves should be constructed from which the number of units could be read off.

With regard to the conditions of the enzyme assay, the temperature should be stated and, where practicable, should be 25°. The other conditions of the enzymic measurement should also be defined and, whenever possible, be optimal, especially the pH and substrate concentration.

Moreover, the Commission recommends that enzyme assays should be based, wherever possible, on measurements of initial rates of reaction and not on amounts of substrate changed by the end of a certain period of time, unless it is known that the velocity of the enzymic reaction remains constant throughout this period. To facilitate measurement of the initial velocity the substrate concentration should be high enough to saturate the enzyme. The kinetics of the standard assay will then approach zero order. If high concentrations of the standard cannot be used—for example, because of limited solubility of a substrate or because of a low affinity of the enzyme for the substrate—the Michaelis constant K_m should be determined and the observed rate converted into that which would be obtained on saturation of the enzyme with the substrate. If the observed velocity at substrate concentration s is v, the velocity at saturation is given by $V = v(1 + K_m/s)$. When s is very small in comparison with the K_m, v is related to V by the equation $v = (V/K_m)s$. In such cases it has been sometimes the practice to express the activities in terms of the first-order velocity constant k. As $v = ks$, V will be obtained by multiplying k by K_m.

The use of measurements of initial rates of reaction will avoid complications due, for instance, to reversibility of reactions or to formation of inhibitory products. The concentration of an enzyme is normally to be given as units per milligram of protein. Molecular activity is defined as units per micromole of enzyme at optimal substrate concentration, that is, as the number of molecules of substrate transformed per minute per molecule of enzyme. When the enzyme has a distinguishing prosthetic group, or a catalytic center, the concentration of which can be measured, the catalytic power can be expressed as catalytic center activity, which is defined as the number of molecules of substarte transformed per minute per catalytic center. This gives the term "turnover number" another sense. The molecular activity and the catalytic center activity will be equal, if the enzyme molecule contains one active center. If there are n centers per enzyme molecule, the molecular activity will be n times the catalytic center activity. It is recommended that the use of the term "turnover number" be abandoned and be replaced either by the term "molecular activity" or "catalytic center activity" in accordance with the meaning intended. In the past, as long at the term "turnover number" was used with only one meaning, it served a useful purpose, but when it began to be used with several different meanings it merely introduced misunderstanding and confusion.

When critically investigated in the light of the recommendations quoted above, many of the methods used in the past for the determination of the activities of amine oxidases fall short of the necessary requirements. However, most of the well-established techniques, to be described in detail, will, when adjusted to conditions of modern enzymology, help to throw more light on some so far unexplained problems of the amine oxidases.

2. METHODS OF ISOLATION, SOLUBILIZATION, PURIFICATION, AND ESTIMATION OF MAO

Since its discovery in 1928 (1) MAO has withstood until comparatively recently many attempts at its purification. This behavior of MAO found its explanation in the demonstration by Cotzias and Dole in 1951 (191) and by Hawkins in 1952 (192) of the intimate adherence of this enzyme in most organ tissues to the insoluble structures of mitochondrial membranes (see Chapter III.1.B).

It should be mentioned, however, that Weissbach, Redfield, and Udenfriend (920) were able to show that guinea-pig liver unlike other organs and other species contains MAO in a soluble form. This unique finding was confirmed by Oswald and Strittmatter (194).

Attempts to solubilize mitochondrial MAO have until recently not met with much success. However, partially purified MAO was prepared by various workers by treatment of the mitochondrial membranes, mostly of rat livers, with detergents or with ultrasonic waves or by disruption of the mitochondria by homogenization and subsequent treatment with the detergent Triton X–100.

A. Procedures for Isolation of Tissue Mitochondria

a. PREPARATION OF MITOCHONDRIA FROM RAT-LIVER HOMOGENATES
   ACCORDING TO SCHNEIDER AND HOGEBOOM (921), HAWKINS (192),
   AND KOBAYASHI AND SCHAYER (400)

Adult male (Sprague-Dawley) rats, fasted overnight to remove liver glycogen, are killed by decapitation (233, 241). Immediately upon removal by excision, the livers are chilled over cracked ice and forced through a masher to remove connective tissue. The liver pulp is weighed and homogenized in a Potter-Elvehjem homogenizer with a Teflon pestle in 9 volumes of ice-cold isotonic (0.25 M) or hypertonic (0.88 M) sucrose in distilled water at the low speed of 600 g for about 5 min. The first super-natant is centrifuged at 22,000 g for 45 min on the refrigerated centrifuge to separate the soluble, enzymically inactive protein from the sediment of formed elements, the mitochondria, and the microsomes, which contain all the enzymic activity of the first supernatant. The sediment is re-suspended in 3 to 4 ml/g of original liver weight of isotonic or hypertonic sucrose solution and is again centrifuged at 22,000 g for 45 min. All operations are carried out at 4°. The mitochondrial enzyme preparation can be stored frozen without noticeable loss of activity over a period of 14 days.

b. PREPARATION OF MITOCHONDRIA OF RAT LIVER, KIDNEY, AND HEART
   ACCORDING TO OSWALD AND STRITTMATTER (194)

Freshly excised tissues from adult male Wistar rats are homogenized in ice-cold 0.25 M sucrose-0.001M EDTA, pH 7.0, (sucrose-EDTA), with a Potter-Elvehjem homogenizer. The "total particulate preparation," which contains essentially all the MAO activity of the homogenate (using tyramine as substrate), is obtained by centrifuging a 20% homogenate at 100,000 g for 1 hr and then washing the sediment once by resuspension in the original volume of sucrose-EDTA and recentrifugation. The sedi-ment is resuspended in sucrose-EDTA and stored at −10°. Monoamine oxidase activity is stable in the frozen state for several weeks.

c. PREPARATION OF RAT-BRAIN MITOCHONDRIA ACCORDING TO SEIDEN AND
   WESTLEY (201)

Male albino rats (Holtzman, Sprague-Dawley line), between 60 and 75 days old, are killed by decapitation. After removal, each brain is transferred to a Tenbroek tissue grinder containing 9 volumes of 0.25 M sucrose. The brain is homogenized with the aid of a motor operating at 1750 rpm. The homogenate is centrifuged at 1500 g for 6 min, and the sediment of nuclei and cell debris is discarded. The supernatant fluid containing suspended mitochondria and microsomes as well as the "white fluffy layer" (613) is layered over 15 ml of 0.88 M sucrose and centrifuged for 20 min at 18,000 g. This results in retention of the white fluffy layer at the interface of the 0.25 M and 0.88 M sucrose solutions, while the mitochondria sediment to the bottom of the tube. The supernatant fluid as well as the white fluffy layer are removed by aspiration. The mitochondria are lysed when resuspended by gentle agitation in a glass homogenizer with a volume of 0.05 M phosphate buffer (pH 7.4) equal to three times the weight of the intact brain. The suspended, lysed mitochondria may be frozen and stored at −30° for at least 2 months without significant loss of activity.

d. PREPARATION OF BEEF-LIVER MITOCHONDRIA ACCORDING TO BARBATO AND ABOOD (234)

About 400 g of liver are homogenized in 3 volumes of 0.25 M sucrose, using a Servall Omnimixer at maximum speed for 90 sec. The thick suspension is diluted with doubly distilled water to 2 liters, and the pH is adjusted to 7.0. After sedimenting twice at 800 g for 20 min to separate nuclei and cellular debris, the mitochondria are sedimented at 8000 g, homogenized in doubly distilled water, and brought to a final volume of 300 ml. The suspension is immediately frozen. No change in MAO activity occurs within several weeks. All operations described above are performed at 0 to 5°.

e. PREPARATION OF LYOPHILIZED KIDNEY MITOCHONDRIA OF THE DOG OR RABBIT ACCORDING TO GIORDANO, BLOOM, AND MERRILL (922)

Rabbit or dog kidney is removed surgically, sliced, weighed, and then homogenized with 4 volumes of 0.25 M sucrose. The pH is adjusted to 7.0 with potassium hydroxide. The kidney brei is then placed in a 50-ml Vitrex tube, and the preparation is further homogenized for 1 min with a motor-driven pestle. The brei is then centrifuged at 700 g for 10 min, and the sediment is discarded. The supernatant is recentrifuged at 700 g for 10 min, and the sediment is again discarded. The resulting supernatant is centrifuged at 11,000 g for 20 min. The supernatant fluid is discarded, and the sediment is washed with 10 to 20 ml of 0.25 M sucrose and centrifuged at 11,000 g for an additional 20 min. All these operations are performed in a coldroom. The final precipitate of mitochondria is weighed and suspended in 0.067 M

phosphate buffer, pH 7.4, using 1.0 ml of buffer for each 50 mg of wet sediment. This suspension is quick-frozen and lyophilized, and the resulting powder is stored at $-10°$.

In the hands of Giordano and his colleagues (922) this method has proved to be a rapid, reliable, and simple one to perform. The lyophilized mitochondria may serve as a stable source of MAO, since its MAO activity remains after 6 months and even after 13 months essentially unchanged from the initial enzymic values.

B. Procedures for the Solubilization and Purification of Mitochondrial MAO

a. PRODUCTION BY SONICATION OF A SOLUBLE MAO PREPARATION FROM RAT-LIVER MITOCHONDRIA ACCORDING TO KOBAYASHI AND SCHAYER (400)

Isolated rat-liver mitochondria, suspended in 0.25 M sucrose (3–4 ml/g of original liver weight), are subjected to sonication for 60 min at 2 to 3° in a Raytheon 9 kHz magnetostriction oscillator, model S–102 A. The vibrated preparation is centrifuged at 21,000 g for 2 hr. The clear enzyme preparation is decanted and used as such for MAO activity estimations with tyramine as substrate. There is no increase in specific activity of MAO after sonication as compared with the original mitochondrial preparation before sonication. Nor has dialysis of the sonicated MAO preparation against 0.25 M sucrose overnight in the coldroom any effect on the specific activity of MAO, which remains unchanged. The sonicated preparation can be stored frozen without noticeable loss of activity over a period of 2 months. Although by the technique described above no purification of mitochondrial MAO can be achieved, this procedure yields a clear, stable, and uniform solution of MAO.

b. METHOD OF PURIFICATION OF BEEF-LIVER MITOCHONDRIAL MAO ACCORDING TO BARBATO AND ABOOD (234)

The original mitochondrial preparation obtained with the isolation technique described in Section 2.A.d is first treated with the detergent Triton X–100, and the soluble enzyme solution is purified on calcium phosphate gel.

Treatment with Triton X–100. The following procedure is used:

1. The original mitochondrial suspension (300 ml) (see Section 2.A.d) is diluted to twice its volume with distilled water, and enough 10% Triton X–100 is added to yield a final concentration of 0.5%. After adjusting the pH to 7.0 with 0.1 N sodium hydroxide, the suspension is kept in the cold for 30 min, with occasional stirring, and then centrifuged at 80,000 g for 20 min. The reddish, opaque supernatant, which contains

little MAO activity, is discarded.

2. The residue is brought to the same volume as in step 1 with 0.5% Triton X–100 and the pH is adjusted to 7.5. After standing in the cold-room for 30 min, the suspension is centrifuged for 20 min at 8000 g, and the supernatant is discarded.

3. The residue of the above centrifugation is treated exactly as described in step 2, except that the pH is adjusted to 8.0. After standing in the coldroom for 30 min, the suspension is centrifuged for 20 min at 8000 g, and the supernatant is discarded.

4. The residue is suspended in half the volume of 0.5% Triton X–100 used in step 2, and the pH is brought to 8.5. After standing in the cold-room for 30 min, the suspension is centrifuged at 100,000 g for 30 min.

5. On adjustment to pH 7.0 of the supernatant the latter is again centrifuged at 100,000 g for 2 hr. The residue is discarded, and the supernatant is subjected to further purification.

Treatment with Calcium Phosphate Gel. The clear, slightly yellow supernatant of the last centrifugation is adjusted to pH 8.5 and calcium phosphate gel is added in an amount equal to four times the weight of protein in the supernatant. After standing at 0° for 20 min, the sus-pension is centrifuged at 5000 g for 15 min; the supernatant is adjusted to pH 7.0, and calcium phosphate gel is added as before. The suspension is centrifuged at 5000 g for 20 min, and the protein-free supernatant is discarded. The residue is washed with two 5.0-ml portions of 0.2 M potassium dihydrogen phosphate and centrifuged at 50,000 g. It is then extracted three times with 5.0-ml portions of the 0.2 M potassium dibasic phosphate containing 0.5% Triton X–100. Although by the procedure described no solubilization of MAO can be achieved, this method yields, according to the authors, an about twenty fold purified enzyme preparation suitable for spectrophotometric studies.

C. SOLUBILIZATION AND PURIFICATION OF MAO ACCORDING TO GUHA AND KIRSHNA MURTI (923)

This is a comparatively simple method for preparing a 350-fold purified MAO from rat-liver mitochondria. Rat-liver mitochondria, prepared in 0.25 M (isotonic) sucrose solution according to the method of Schneider and Hogeboom (921), are washed once with 0.01 M phos-phate buffer, pH 7.6, resuspended in 0.01 M phosphate buffer, pH 7.6, and sonicated at 25 kHz (output amperage, 2.5) for 25 min in a Mullard magnetostrictor generator. The suspension is placed in a stainless-steel flask, fitted to the transducer element. During the exposure to sound waves the contents of the steel flask are kept chilled, and thereafter the suspension is centrifuged in a Servall refrigerated centrifuge at maximum speed, at 2°. The resulting supernatant contains over 90% of the original mitochondrial MAO activity.

For further purification, aliquots of the supernatant containing 40 mg of protein are applied directly to a DEAE-cellulose column (30 × 2.5 cm), previously equilibrated with 0.01 M phosphate buffer, pH 7.6, containing sodium chloride in a 0.01 M concentration. The MAO is eluted by stepwise addition of four 100-ml portions of 0.01 M phosphate buffer, pH 7.6, containing 0.01, 0.05, 0.1, and 0.5 M sodium chloride, respectively. The stepwise elution is carried out at 8°, 10-ml fractions being collected. The active eluates are water-clear and pale yellow, and the entire MAO activity is recovered in the early fractions at low sodium chloride concentration. The active fractions are pooled, lyophilized, and rechromatographed on DEAE-cellulose. Fractions that on rechromatography emerge at a 0.01 M sodium chloride concentration usually contain the enzyme, whose specific activity may indicate a 350-fold overall purification.

d. PREPARATION OF SOLUBLE BEEF-LIVER MITOCHONDRIAL MAO AND ITS
  PURIFICATION ACCORDING TO NARA, GOMES, AND YASUNOBU (220)

Beef-liver mitochondrial MAO was purified by Nara, Gomes, and Yasunobu (220) as much as 58-fold by a procedure that involves a disruption of the mitochondria by homogenization and Triton X–100 treatment, ammonium sulfate fractionation, Cγ-alumina gel treatment, and chromatography on DEAE-cellulose. Beef-liver mitochondria were prepared according to the method of Schneider and Hogeboom (921).

About 120 g of purified mitochondria were suspended in 500 ml of 0.1 M phosphate buffer, pH 7.4, and were homogenized in batches, for 2 min each, in a Potter Elvehjem homogenizer (25 × 200 mm) using a stirring motor operating at 3,000 rpm. The homogenate was adjusted to a protein concentration of 30 mg/ml by the addition of 0.1 M phosphate buffer, pH 7.4 (fraction I). In order to obtain a better extraction of the enzyme, 73 ml of 20% Triton X–100, adjusted to pH 7.4, were added to a 900-ml aliquot of the homogenate so that the final Triton X–100 concentration was 1.5%. The mixture was stirred gently for 3 hr. The homogenate was centrifuged at 20,000 rpm for 45 min in the No. 21 rotor of the Spinco model L ultracentrifuge. A second homogenate batch of 900 ml was treated in the same fashion, and the supernatants of both treated homogenates were combined, yielding 1530 ml of fraction II. Then 162 g of solid ammonium sulfate were added to the clear, reddish-brown solution to achieve 20% ammonium sulfate concentration, and the solution was centrifuged. A further addition of 219 g of solid ammonium sulfate made the solution 45% saturated with regard to ammonium sulfate. The solution was then centrifuged at 20,000 rpm in the ultracentrifuge for 25 min. The insoluble enzyme protein rose to the top and was collected by removing the solution with a hypodermic syringe. The enzyme (7.43 g) was dissolved in 450 ml of 0.1 M phosphate buffer, pH 7.4, yielding fraction III. At this stage the enzyme was rather stable in the frozen state and could be stored frozen for several days. In the next step the concentration of the protein was adjusted to 10 mg/ml by the addition of 0.1 M phosphate buffer, pH 7.4. To this solution 20.8 ml of 20% sodium cholate (0.56 mg of cholate per milligram of protein) and about 137 g of solid ammonium sulfate were added to provide a final concentration of ammonium sulfate of 25%. The concentration of ammonium sulfate, already present in the solution, was first determined (924), and then the solid ammonium sulfate was added to a final concentration of 25%. The solution was centrifuged at 8,000 rpm for 15 min in the Servall centrifuge, and an additional 115 g of ammonium sulfate were added per liter of solution to 45% saturation. The enzyme floated to the top, and the solution was again removed with a hypodermic syringe. The enzyme precipitate was dissolved in 240 ml of 0.1 M phosphate buffer, pH 7.4 (fraction IV). This solution was then dialyzed against 2 liters of 0.01 M phosphate buffer, pH 7.4, for 2 hr, whereupon the buffer was changed, and the solution was dialyzed for an additional 2 hr against 2 liters of 0.01 M phosphate buffer, pH 7.4. The enzyme solution was adjusted to a protein concentration of 20 mg/ml by adding 52.8 ml of 0.01 M phosphate buffer, pH 7.4. In the next step, 162 ml of Cγ-alumina (18 mg/ml, dry weight), prepared according to Willstätter and Kraut (925) were added to the solution so that the protein-to-alumina ratio was 2 : 1. The solution was then stirred gently for 5 min, after which the enzyme was centrifuged at 8000 rpm for 15 min, and the precipitate was discarded. An additional 486 ml of Cγ-alumina gel were added to make the protein-to-alumina ratio 1 : 2, and the mixture was allowed to stand overnight. The gel obtained by centrifugation was washed with approximately 200 ml of 0.01 M phosphate buffer, pH 7.4. The elution process was repeated two or three times, depending on the yield of the enzyme. The brown eluates were combined (470 ml) and concentrated by the addition of 120 g of solid ammonium sulfate (45% saturation). The floating, reddish-brown precipitate, obtained after centrifugation, was dissolved in 53 ml of 0.1 M phosphate buffer, pH 7.4 (fraction V). The enzyme solution was then passed through a Sephadex G–25 column (30 $\times$ 3.5 cm) in order to remove the ammonium sulfate. Thereupon the enzyme solution was applied to a DEAE-cellulose column (2 $\times$ 45 cm), previously equilibrated with 0.01 M phosphate buffer, pH 7.4. Elution was started with 0.01 M

phosphate buffer, pH 7.4, and fractions of 14 ml were collected. At tube 20, the buffer concentration was increased to 0.1 M, pH 7.4, and from tube 40 onward gradient elution was used by mixing 1000 ml of a 1.0% Triton X–100 solution in 0.1 M potassium phosphate buffer, pH 7.4, into 1000 ml of 0.1 M potassium phosphate buffer, pH 7.4. At tube 175, the buffer concentration was increased to 0.3 M potassium phosphate buffer, pH 7.4, containing 1% Triton X–100.

It was estimated that the active enzyme was eluted at a final Triton X–100 concentration of 0.2 to 0.3%. The slightly brownish solution was concentrated by the addition of solid ammonium sulfate to 45% saturation. After centrifugation, the solid phase was dissolved in 15 ml of 0.1 M potassium phosphate buffer, pH 7.4, yielding fraction VI. For some of their studies Nara and his colleagues stored the enzyme in the frozen state. There was a 20% loss in enzyme activity within a 2-week period, after which the activity remained constant. The authors emphasize that they obtained different results with different batches of mitochondria. In about half their experiments the enzyme appeared to be solubilized, as was shown by the facts that the enzyme precipitated to the bottom after the addition of ammonium sulfate, and that the enzyme could be dialyzed against buffer that did not contain any detergent. These enzyme preparations were stable for indefinite periods, when stored in the freezer. When the purity of the enzyme was tested in the ultracentrifuge, preparations with a specific activity of 4760 showed a single peak at pH 7.4. Determination of the sedimentation coefficient of an 0.5% solution of the beef-liver enzyme in 0.1 M phosphate buffer, pH 7.4, yielded a value of 10.5S (uncorrected). The sedimentation coefficient obtained for beef-plasma amine oxidase, when corrected to 20° and water, was 9.2S (845), which may suggest that the two enzymes may have similar molecular weights. Until further detailed criteria of the purity of mitochondrial MAO are put forward, it may be taken that this method yields a highly purified enzyme preparation.

e. PREPARATION OF SOLUBLE, PURIFIED MITOCHONDRIAL MAO FROM RAT
    LIVER ACCORDING TO YOUDIM AND SOURKES (241)

Mitochondria are prepared from the livers of male (Sprague-Dawley) rats by the method of Hawkins (192) (see Section 2.A.a). The mitochondria are suspended in 0.0125 M phosphate buffer, pH 7.4, containing 0.003 M benzylamine. After sonication for 100 min at 20 kHz, cholic acid is added to a final concentration of 1%. The material is allowed to stand for 30 min and is then centrifuged at 160,000 g for 90 min. The supernatant contains up to 85% of the enzyme activity, three-quarters of which is precipitated between 30 and 55% saturation with ammonium sulfate. The precipitate redissolves readily on addition of water. For further treatment this partially purified enzyme is placed on a Sephadex G–200 column and is eluted with 0.05 M phosphate buffer, pH 7.4. The fractions containing active MAO are pooled and subjected to fractionation with ammonium sulfate as before. The enzyme is then chromatographed on a column of DEAE–Sephadex, A–50, in the presence of 0.05 M phosphate buffer, pH 7.4, and is eluted with an increasing gradient of concentration of sodium chloride. The active eluate is then chromatographed on a column of hydroxylapatite with elution by phosphate buffer, pH 6.8, in concentrations increasing from 0.005 M to 0.5 M. In this last stage of purification four protein peaks are obtained of which only the last one contains MAO activity. Fractions containing this peak are pooled and are used for further study.

By this method mitochondrial MAO of the rat liver can be purified up to 208-fold as compared with the crude homogenate. About 25% of the original activity may be recovered by this technique. In this pure form the enzyme is yellow in color. It can be stored in 0.05 M phosphate buffer, pH 7.4, at 4° with little loss of activity for at least 2 weeks. It is stable in the pH range 5.5 to 9.5, but outside this range the rate of inactivation is high. Once precipitated, the active protein can easily be redissolved in an aqueous medium without use of dispersing agents. It is interesting that the estimated molecular weight of MAO from the rat liver, about 290,000, is of the same order as that reported for the plasma enzyme (143) and the fungal enzyme (577).

f. SOLUBLE MAO FROM HUMAN-PLACENTA MITOCHONDRIA ACCORDING TO
    YOUDIM, COLLINS, AND SANDLER (912)

A highly purified MAO can be prepared from human-placenta mitochondria by means of a modification of the method of Youdim and Sourkes (241) (see Section 2.B.e). Mitochondria are subjected to sonication, followed by treatment with Triton X–100, ammonium sulfate fractionation, and column chromatography on Sephadex G–200 and DEAE-Sephadex. A final product with a specific activity of 3500, showing a 400-fold purification over the original homogenate, may be obtained by this method. When subjected to polyacrylamide-gel electrophoresis, the human-placenta MAO showed two bands of enzyme activity, whereas solubilized and partly purified MAO from the rat liver (241) shows in these circumstances three bands of enzyme activity.

g. SIMPLE METHOD FOR THE EXTRACTION AND PURIFICATION OF HOG-BRAIN
    MITOCHONDRIAL MAO ACCORDING TO TIPTON (213)

There are only very few reports dealing with the purification of brain MAO. Nagatsu (926) was the first to demonstrate that MAO can be extracted from beef-brain mitochondria by sonication in the presence of a detergent. Tipton has developed a procedure by which hog-brain mitochondrial MAO can be extracted and purified about 1000 times by using in the absence of a detergent a simple, if rather tedious, technique of repeated sonication, freezing, and thawing (213).

Brains were removed from freshly killed hogs, defatted, suspended in 9 volumes (w/v) of 0.25 M sucrose, adjusted to pH 7.6 with 1.0 M K_2HPO_4 and homogenized in a Waring Blender at medium speed for 1 min. The pH was then readjusted to 7.6 with 1.0 M K_2HPO_4. Mitochondria from this homogenate were prepared by a method similar to that described by Brody and Bain (927). The homogenate was centrifuged at 1,500 g for 20 min and the supernatant was carefully decanted and centrifuged at 18,000 g for 20 min. The residue was resuspended in half the original volume of 0.25 M sucrose-K_2HPO_4, pH 7.6, and again centrifuged at 18,000 g for 20 min and resuspended in the sucrose-phosphate medium.

For the liberation of the MAO from the mitochondria the above suspension was sedimented at 18,000 g for 20 min and then suspended in about one-tenth of the original homogenate volume in ice-cold water. The suspension was homogenized by hand in a Potter-Elvehjem homogenizer with an-all nylon pestle and centrifuged at 18,000 g for 45 min. The residue was taken up in a similar volume of 0.01 M phosphate buffer, pH 7.6, and stored frozen for not less than 3 days. The preparation was then allowed to thaw out and was diluted with 0.01 M phosphate buffer, pH 7.6, to give a final protein concentration of 10 to 15 mg/ml.

The suspension was then sonicated for 45 min in a Dawe Soniprobe, fitted with $1/2$-in. probe at an output amperage between 6 and 7. During sonication the suspension was cooled in ice. The sonicate was then centrifuged at 108,000 g for 120 min, and the supernatant, which should be a cream-colored opalescent solution, was carefully decanted. If the color of the supernatant was a distinct brown, the solution was recentrifuged. The residue was resuspended in 0.01 M phosphate buffer, pH 7.6, homogenized, diluted to give a protein concentration of 10 to 15 mg/ml, and sonicated as before. The sonicated suspension was stored frozen, thawed out at room temperature, and centrifuged as before. This procedure of sonication, freezing, and thawing was continued six times, whereby 60 to 80% of MAO activity was liberated into the supernatants. An excellent purification was obtained with only two sonication steps. The active supernatants were then pooled.

The MAO activity was determined by a modification of the method of Creasey, adapted for use with an oxygen electrode (928). For description of the method see Section 2.D.c.

The combined active supernatants from the sonications were then exposed to low pH. After cooling in ice, 1.0 N HCl was added with continuous stirring until the pH fell to 3.0. The pH of the solution was then immediately brought back with 1.0 N NaOH to 7.0. A white precipitate formed during this procedure. After standing for 20 min at 4°, the suspension was centrifuged at 35,000 g for 20 min. The residue was discarded, and the supernatant, adjusted to pH 7.6, was treated with DEAE-cellulose (Whatman DE–52). For this purpose DEAE-cellulose was equilibrated with 0.01 M phosphate buffer, pH 7.6, and the excess liquid was removed by using a sintered glass funnel and applying gentle suction with a water pump. The cake of DEAE-cellulose, that formed, was added to the supernatant (0.5 g DEAE-cellulose per milligram of protein). The suspension was stirred and allowed to stand at 4° for 20 min. The DEAE-cellulose was then removed by centrifugation at 35,000 g for 30 min.

The clear supernatant, which should contain between 0.5 and 1.5 mg protein per milliliter, was then subjected to alcohol fractionation. After cooling in ice, an equal volume of absolute ethanol was added slowly within 60 min, with continuous stirring. The mixture was then stirred in ice for another 2 hrs. The fine precipitate, that formed, was discarded after centrifugation at 35,000 g for 45 mins at —5°. Half the original volume of absolute alcohol was then added slowly to the cooled supernatant, and the mixture was allowed to stand at —10° overnight. The fine precipitate formed, which contained the active enzyme, was separated by centrifugation at 35,000 g for 45 min at —5° and was taken up in the minimum volume of ice-cold 0.01 M phosphate buffer, pH 7.6. Insoluble material was removed by centrifugation, and the solution was stored frozen. A further yield of MAO was obtained by adding the original volume of absolute ethanol to the alcoholic supernatant, allowing the mixture to stand overnight at —10°, and centrifuging as before. The residue was again taken up in 0.01 M phosphate buffer, pH 7.6, and the insoluble material was removed as before. The two active solutions were pooled and stored in the frozen state for 2 days. The precipitate, that formed, was discarded after centrifugation at 35,000 g for 45 min. The supernatant contained MAO that was found to be purified about 1000-fold as compared with the original mitochondrial activity.

The purified enzyme appears to be homogeneous by cellulose-acetate electrophoresis and has a molecular weight of approximately 102,000. It is inhibited by iproniazid, which is characteristic of MAO (160), and it is also inhibited by p-chloromercuribenzoate. The latter finding compares well with results of Lagnado and Sourkes (233) and may indicate that a sulfhydryl group is involved in the enzyme activity. Inhibition of the purified enzyme by metal-binding agents is also in agreement with results quoted in the literature (200,239). It has been reported that MAO contains copper as an essential cofactor (160, 239). Preliminary results obtained by Tipton (212) with purified brain mitochondrial MAO have suggested that this enzyme contains FAD as a prosthetic group. Recent findings of Yasunobu et al. (219) indicate that purified beef-liver MAO contains FAD covalently attached to the enzyme (see Chapter III.1.C).

The pH optimum of the purified brain mitochondrial MAO was found by Tipton (213) to be at about 7.2, which compares well with values reported by Guha (929) and Severina and Sheremetevskaya (930) for the purified rat-liver mitochondrial MAO.

Preparation of Apomonoamine Oxidase. As mentioned above (see Chapter III.1.C), Swoboda (217) has shown that it is possible to remove the FAD from glucose oxidase without irreversibly inactivating the enzyme. The technique designed by Swoboda was adapted by Tipton (214) for the production of apomonoamine oxidase, as follows. An 0.5-ml sample of the purified enzyme was added rapidly from a syringe to 5 ml of a saturated solution of ammonium sulfate in 0.1 M HCl, which was cooled in ice. The mixture was allowed to stand for 30 min; the precipitate was then removed by centrifugation and taken up in 0.5 ml of 0.01 M phosphate buffer, pH 7.6. The precipitation procedure was repeated three more times, and the final precipitate was taken up in 0.5 ml 0.01 M phosphate buffer, pH 7.6, and was assayed for enzyme activity. The best preparations of apomonoamine oxidase produced by this method had no detectable activity and showed negligible flavin fluorescence. The inactive apoenzyme could be partly reactivated by incubation with authentic FAD, but not with FMN.

Determination of Copper. Copper was assayed in MAO preparations by two methods:

1. Reaction with cuprizone (931).
2. Reaction with bathocuproine (932).

The enzyme preparation was dialyzed for 36 hr against 0.01 M phosphate buffer, pH 7.6, before being assayed for its copper content. A 1-ml aliquot of the dialyzed enzyme solution was made 2 M with respect to HCl, 0.5 ml of 20% (w/v) trichloroacetic acid was added, and the reaction mixture was centrifuged. The supernatant was then submitted to copper analyses. Glassware used in copper determinations was soaked in concentrated nitric acid for 14 days and was exhaustively washed with deionized water before use.

Protein Determination. Protein concentration was estimated by the microbiuret method of *Goa* (933), using bovine serum albumin as a protein standard.

C. Method of Purification of Soluble MAO from Guinea-Pig Liver According to Weissbach, Redfield, and Udenfriend (920)

Although MAO has been found to be localized primarily in the mitochondria in most organ tissues, appreciable activity is present in soluble form in some tissues. Weissbach and his colleagues (920) have described a procedure for the purification of soluble MAO from guinea-pig liver.

Male guinea pigs (Hartley- or National Institute of Health-strain) weighing about 250 g were used. Tissues were homogenized in water (1 part of tissue to 5 parts of water) in a Waring Blendor. The tissue homogenate was first centrifuged at 8,000 g for 20 min in a refrigerated International angle head centrifuge, and the residue was discarded. The supernatant fluid was then centrifuged in a Spinco preparative ultracentrifuge at 100,000 g for 30 min. The supernatant fluid of this high speed centrifugation was made 25% saturated with respect to ammonium sulfate by the addition of saturated ammonium sulfate solution, adjusted to pH 8.0, and the precipitate was discarded. The supernatant fluid was then adjusted to 40% saturation with ammonium sulfate. The resulting precipitate was dissolved in one-third the volume of the initial high-speed supernatant with 0.01 M phosphate buffer, pH 7.4. The pH was then adjusted to 5.1 by dropwise addition of 0.5 N acetic acid, and the precipitate formed was quickly isolated by centrifugation and dissolved in 0.05 M phosphate buffer, pH 7.4. This precipitate contained the MAO. The fractions could be dialyzed at any stage for 5 hr against phosphate buffer, pH 7.4, without any loss of activity, but dialysis against water resulted in a rapid loss of activity in a few hours. The MAO could be also kept at 3° for several days with little loss of activity.

D. Methods for the Quantitation of MAO Activity

a. INTRODUCTION

Methods available for the assay of MAO activity were reviewed by Werle (130) in 1964. They include a variety of techniques—manometric, diffusion, and spectrophotometric—based on the consumption of oxygen (934), production of ammonia (935), concomitant analyses of oxygen uptake and ammonia evolution (936), estimation of substrate disappearance under the action of MAO (937), determination of the MAO effect on kynuramine (938), and production of a dinitrophenylhydrazine derivative, intensely colored in alkaline solution, of the aldehyde formed in the reaction between MAO and the substrate (939).

Until recently the methods most commonly used for the estimation of MAO depended on the measurement of oxygen (934) or the determination of substrate disappearance (938, 940).

According to Slater (941), the manometric technique, if properly used under correctly chosen conditions, is still the most accurate one for measuring the oxygen uptake. He emphasizes, however, that it is essential to measure the rate of oxygen consumption during steady-state conditions, since the manometer responds slowly to changes of the steady state. When a constant rate of oxygen consumption prevails for at least 15 min after the addition of mitochondria, the total oxygen consumption can be accurately calculated.

The application of Warburg's manometric techniques for estimation of MAO has been severely criticized by many workers. Zeller (942) has very recently pointed out that about 40 to 50 times as much tissue homogenate is required for the manometric procedures as for a sensitive spectrophotometric technique. For this reason, according to Zeller, it is impossible to analyze the distribution of MAO in organs of low enzymic activity, obtained from small laboratory animals, for example, mouse brain. Further, Zeller emphasizes that the manometric technique does not lend itself to investigations of MAO inhibitors. The large amounts of homogenate, required for the Warburg technique, overcome most of the inhibitory effects of such drugs, and thus render screening studies of MAO inhibitors futile. Moreover, with preparations of low MAO activity an apparent time lag is often noticed so that readings become erratic and the determination of the initial velocity cannot be precisely carried out. Furthermore, the oxygen consumption may be caused not only by the MAO reaction but also by other reactions with endogenous substrates. Although it is possible to avoid this extraneous oxygen consumption by adding cyanide and semicarbazide to the sample (161, 943), the latter agents may considerably complicate the evaluation of the results obtained. As an example Zeller (942) quotes the strong effect of cyanide on the inhibition of MAO by hydrazine derivatives (402, 944). In 1938 Petering and Daniells (945) were the first to measure the respiration of a variety of cell suspensions and rat-liver homogenates by means of a dropping-mercury electrode, and this polarographic method has been used since to measure oxygen uptake by microorganisms, fibroblasts, and mitochondria (946–948). In 1957 Connelly (949) published a brief statement on the theory and evaluation of the oxygen electrode. According to this author the oxygen electrode is a polarographic device for measuring the concentration of oxygen dissolved in the medium in which the tip of the electrode is placed. Connelly considers the oxygen electrode to be a part of an electrochemical system in which oxygen is caused to be electrolytically reduced. The rate of reduction is measured by the electrode current and is determined by, among other things, the concentration of dissolved oxygen. A platinum wire, the tip of which is immersed in an aqueous medium, is connected to the negative terminal of a polarograph. The electrical circuit is completed through a galvanometer to a nonpolarizable reference electrode such as a silver–silver chloride electrode. When the platinum is several tenths of a volt negative to the reference electrode, oxygen dissolved in the medium, is electrolytically reduced at the surface of the platinum. The disappearance of oxygen at the electrode surface causes oxygen to diffuse toward the surface and establishes a concentration gradient. When a potential difference of 0.6 to 0.7 volt is applied across the platinum and the reference electrode, immersed in an aqueous medium, the magnitude of the current flowing is proportional to the concentration of oxygen in the solution and the area of platinum exposed to the solution. The current is increased when the medium is stirred.

In early experiments with stationary electrodes in direct contact with a stirred mitochondrial suspension a progressive decline in the current was noticed, when the electrodes were immersed in a fluid in contact with room air or, when the incubation media contained sucrose, phosphate, and magnesium chloride. But by far the greatest difficulties arose, when it was found that the electrodes were inactivated on the addition of mitochondria to the incubation medium. There was a rapid fall in the current, simulating respiration, which continued for about 5 min (948–950). Presumably, the inactivation is due to deposition of protein on the electrodes. These difficulties were overcome by ensuring that the electrodes have a constant ionic environment and by shielding the electrodes from the incubation medium with thin membranes (polyethylene films) that are permeable to gases but not to solutions (951). This modification of the oxygen electrode (the Clark-type electrode) has been since used by many workers (950, 952–955). Tipton and Dawson (928) have very recently applied the Clark oxygen electrode to the estimation of MAO activities in the hog brain (see Section 2.D.c).

Connelly (949) has emphasized that two points must be considered in specific applications of the oxygen electrode. One is that measurements with an oxygen electrode are largely empirical; the electrode must be calibrated under conditions essentially the same as those used experimentally. The second point is that the oxygen electrode does not operate as a reversible potentiometric device for measuring the activity of a chemical component, as does the hydrogen electrode or the chloride electrode, but acts rather as a kind of amperometric device, which irreversibly consumes the material whose concentration it is measuring. Some authors, therefore, call this device an oxygen cathode in preference to oxyfen electrode.

In the analytical procedures based on the estimation of substrate disappearance (170, 940), serotonin or tyramine have been used as substrates, since both these substances can be readily measured. However, this technique of MAO estimation is very cumbersome, since it involves multiple extraction procedures (956).

Most of the methods, mentioned above, have proved to be too insensitive for determinations of MAO in organs with low enzyme activity, such as dog heart, or in small amounts of tissue, such as a single sympathetic ganglion. Since attention is being especially focused on the pharmacological actions of MAO inhibitors in such sites, it has been obvious for some time to many scientists that more sensitive and accurate methods would have to be developed to facilitate a direct correlation of biochemical and pharmacological effects. Modern radiometric, fluorometric, spectrophotometric, and histochemical procedures have been recently published for the estimation of MAO, some of which are described here in detail. Moreover, the present author has decided to retain in this section a few of the older, well-known techniques of MAO estimation, although they may have been superseded by some present-day methods.

b. STANDARD WARBURG TECHNIQUE FOR THE MEASUREMENT OF
   OXYGEN UPTAKE ACCORDING TO CREASEY (934)

Monoamine oxidase activity is measured manometrically at $38°$ by following the oxygen uptake during the deamination of tyramine in the presence of 10^{-2} M semicarbazide and 10^{-3} M cyanide. Under these conditions only MAO and catalase appear to be active, spontaneous oxidations are inhibited, and oxygen is absorbed at the rate of one atom per molecule of tyramine deaminated. Tyramine is rapidly oxidized; oxygen uptake is linear over the first 30 min of the reaction and is also directly proportional to the MAO concentration.

Reagents. The following reagents are required:

0.24 M sodium phosphate buffer, pH 7.0,
0.01 M potassium cyanide solution,
0.1 M semicarbazide,
0.1 M tyramine,
2.0 M potassium cyanide.

The composition of the recommended reaction mixture is as follows:

1. Main compartment: enzyme preparation (1 ml); 0.24 M sodium phosphate buffer, pH 7.0 (0.2 ml); 0.01 M potassium cyanide (0.2 ml); 0.1 M semicarbazide (0.2 ml); water (0.2 ml).
2. Side arm: 0.1 M tyramine (0.2 ml).
3. Center well: 2 M potassium cyanide (0.1 ml), filter paper strip.

All solutions except the 2 M potassium cyanide are adjusted to pH 7.0. The flasks are gassed with oxygen for 2 min and equilibrated for 10 min before tipping. Readings are being taken at 0, 10, 20, and 30 min for an activity determination and, when total oxygen uptake is to be measured, at 30-min intervals and thereafter, until oxygen ceases to be absorbed. A control reaction mixture containing no substrate is also prepared, and any oxygen absorbed by it is subtracted from that absorbed in the test reaction mixtures.

c. DETERMINATION OF MAO ACTIVITY BY A POLAROGRAPHIC METHOD FOR MEASURING OXYGEN UPTAKE ACCORDING TO TIPTON AND DAWSON (928)

MAO activities were determined in homogenates and in the mito-chondrial fraction prepared from individual regions of hog brain by measuring oxygen uptake with a Clark oxygen electrode (Yellow Springs Instrument Co., Inc., Yellow Springs, Ohio), connected via a voltage divider to a Honeywell-Brown 1-mV strip-chart recorder. The apparatus was similar to that described by Dixon and Kleppe (955). All assays were carried out at 30°, and the apparatus was standardized with distilled water saturated with air at 30°. Proportionality of reaction velocity and concentration of enzyme was verified in each assay. The MAO activity is expressed as millimicrogram-atoms of oxygen consumed per minute and per milligram of protein. The authors used six samples in each of the investigated cases and expressed the MAO activities as means ±S.E.M. Tyramine was used as substrate, and its oxidation was followed by measuring the oxygen consumption with the Clark oxygen electrode. The assay mixture contained in a total volume of 2.4 ml 200 micromoles of sodium phosphate buffer, pH 7.0, 50 units of catalase (British Drug Houses, Poole, England), 20 micromoles of semicarbazide, 2 micromoles of potassium cyanide, and enzyme. The mixture was equilibrated with air at 30°, and the reaction was started by the addition of 100 μl of 0.2 M tyramine hydrochloride (British Drug Houses). Separate blanks were determined in the absence of enzyme.

Protein concentrations were determined by the microbiuret method of Goa (933).

d. DETERMINATION OF MAO ACTIVITY BY MICROESTIMATION OF AMMONIA, USING THE WARBURG MANOMETRIC APPARATUS ACCORDING TO BRAGANCA, QUASTEL, AND SCHUCHER (957)

Measurement of the ammonia liberated in the action of MAO on its substrate (see Chapter II,2) may provide a reliable, sensitive, and not very laborious method when applied in the form proposed by Braganca, Quastel, and Schucher (957).

The principle of this method is essentially the same as that of the microdiffusion procedure of Conway and Byrne (958), in which the ammonia liberated from an alkaline solution is allowed to diffuse within a closed compartment into a vessel containing acid. In the technique of Braganca and his colleagues the enzymic reaction as well as the diffusion of the ammonia formed is allowed to take place in Warburg manometric flasks.

Procedure. A small roll of filter paper soaked with 0.2 ml of 1 N sulfuric acid is placed in the center well of a Warburg manometric vessel, and 0.3 ml of a saturated potassium carbonate solution is placed in the side tube. In the main compartment the enzyme activity of the homogenate is assayed (923) in the presence of 0.05 M phosphate buffer, pH 7.6, and 30 micromoles of tyramine hydrochloride, neutralized before use, total volume 3 ml. Incubation is carried out for 30 min at 37°. The potassium carbonate solution present in the side arm is then tipped into the main compartment; this raises the pH of its contents to 10.5 and arrests the course of the enzymic reaction. The ammonia formed diffuses into the center well, where it is absorbed by the acid on the filter paper. The vessel is shaken at 37° for 3 hr to permit complete diffusion and absorption of quantities of ammonia, ranging from 10 to 500 μg. At the end of the 3-hr diffusion period the vessels are removed from the bath, and the small rolls of filter paper are taken out from the center well with forceps and placed in graduated tubes. The contents of the center well are quantitatively removed and added to the graduated tube by washing five times with distilled water. This manipulation is easily accomplished with the aid of a micropipet. The content of the tube is made up to a definite volume, and aliquots corresponding to 20–30 μg NH_4^+ are taken for the assay by nesslerization. The aliquots are made up to a final volume of 10 ml with distilled water, 1 ml of Nessler solution, and 2 ml of 2 N NaOH. The intensity of color produced is estimated at 425 nm. All values are corrected for blanks.

Recently, this method has been successfully used by Guha and Krishna Murti (923) for MAO estimations in rat-liver mitochondria.

e. SPECTROPHOTOMETRIC ASSAY OF MAO BASED ON THE RATE OF
   DISAPPEARANCE OF KNURAMINE ACCORDING TO WEISSBACH,
   SMITH, DALY, WITKOP, AND UDENFRIEND (938)

This assay for MAO activity utilizes a direct spectrophotometric method which allows a rapid determination of the disappearance of kynuramine, or the appearance of 4-hydroxyquinoline. Makino and his colleagues (959) showed in 1955 that mouse-liver homogenates convert kynuramine, the decarboxylation product of kynurenine, to 4-hydroxyquinoline, which these authors identified by its paper chromatographic and spectral properties.

Kynuramine → 4-Hydroxyquinoline

According to Weissbach et al. (938) MAO probably deaminates kynuramine **(I)** to the corresponding aldehyde **(II)**, which may either condense to 4-hydroxyquinoline **(III)** or undergo further oxidation to the corresponding acid **(IV)**.

Intramolecular (nonenzymic) condensation of the amino aldehyde **(II)** proved, however, to be faster than further oxidation of the aldehyde to the corresponding acid **(IV)**.

Procedure.   Tissues were homogenized by Weissbach et al. (938) in 5 volumes of cold distilled water and passed through a cheese cloth. The homogenate was centrifuged at 500 rpm for 15 min to remove cellular debris.

Incubations with kynuramine were performed by these authors in 3-ml silica experimental cuvettes. A cuvette contained 0.1 or 0.2 ml of tissue homogenate, about 0.3 micromoles of kynuramine, 0.3 ml of 0.5 M phosphate buffer, pH 7.4; distilled water was added to a total volume of 3 ml. The initial absorbancy of this reaction mixture at 360 nm was approximately 0.5. Since the incubations were run at room temperature, only the enzyme solutions were kept cold. A blank cuvette was prepared in which the kynuramine was replaced with water. After the final addition, the mixing was achieved by inversion and an initial reading was made at 360 nm. Further readings were then recorded at suitable time intervals, depending on the activity of the enzyme preparation. With crude tissue preparations, the readings observed in the first minute or two may be erratic, because of settling of particles in the cuvette. But after this period there is no difficulty, and the assay can be employed for tissue homogenates. Activity is expressed as the change in absorbancy at 360 nm per unit of time.

Kynuramine disappearance was linear with time until the absorbancy fell below 0.150. Enzyme activity was found to be proportional to enzyme concentration. Moreover, the reaction was found to be extremely sensitive to iproniazid, a very good inhibitor of MAO (160, 402). Rat-liver mito-chondria, which are an excellent source of MAO, rapidly metabolized kynuramine, as did a partially purified MAO preparation from guinea-pig liver (920). Although the authors mention that their spectrophoto-metric method also allows a rapid and exact determination of the appearance of 4-hydroxyquinoline by measuring the absorbancy between 310 and 335 nm, no experimental data for such procedure are given, which is regrettable.

Enzyme experiments were performed either with a Beckman model DU spectrophotometer or the Cary recording spectrophotometer, model 14.

Kynuramine can now be obtained from the Regis Chemical Company, Chicago.

The spectrophotometric assay of MAO by Weissbach et al. (938) has been lately applied by Barbato and Abood (234) in their kinetic studies on the effect of various inhibitors on purified beef-liver mitochondrial MAO, and the results obtained have been compared with those found with the spectrophotometric technique of Tabor and co-workers (141), which utilizes benzylamine as substrate. Tabor and his associates were the first investigators who followed the action of a plasma amine oxidase (spermine oxidase) by a spectrophotometric method, using benzylamine as a substrate. The product of the reaction, benzaldehyde, displays a very high extinction coefficient at 250 nm and can easily be detected spectro-photometrically. Barbato and Abood (234) found that kynuramine was a better substrate for a spectrophotometric MAO estimation than benzyl-amine, particularly when working with carbonyl reagents, such as potassium cyanide, which can react with the benzaldehyde formed, or when using the inhibitor phenanthroline, whose extinction coefficient at 250 nm is extremely high. With 10^{-4} M kynuramine as substrate the absorbancy is measured at 360 nm (938) at which wavelength the absorption caused by phenanthroline is negligible. Furthermore, carbonyl reagents have no effect when kynuramine is used, evidently because the intramolecular reaction between the aldehyde group and the amino group of kynuramine is much more rapid than the reaction of the aldehyde group with the carbonyl reagents.

The spectrophotometric method of Tabor et al. (141), utilizing benzylamine as substrate, has been recently used by Nara and his associates (220) for estimations of the activity of purified beef-liver mitochondrial MAO.

Zeller and his colleagues (942) have adversely criticized the spectrophotometric kynuramine method of Weissbach et al. (938) for the estimation of MAO for the following reasons. Upon the action of MAO, kynuramine is deaminated, and the resulting aldehyde condenses intramolecularly to form 4-hydroxyquinoline. The latter substance has a much lower molar extinction coefficient at 360 nm than the starting product. Since the initial optical readings with 0.1 mM kynuramine are relatively high, small differences in optical density may be of doubtful value. This difficulty could be, at least partly, avoided by following the increase in optical density between 310 and 335 nm, due to the formation of 4-hydroxyquinoline. But, as mentioned above, no experimental data were given for the latter reaction.

f. A RAPID SPECTROPHOTOMETRIC METHOD FOR THE DETERMINATION OF THE
    ACTIVITY OF MAO AND MAO INHIBITION ACCORDING TO
    ZELLER, RAMACHANDER AND ZELLER (942)

This method is based on the high degradation rate of m-iodobenzylamine and on the strong absorbancy at 253 nm of the m-iodobenzaldehyde formed during the action of MAO on m-iodobenzylamine. This primary reaction product does not appear to undergo further oxidation by oxidoreductases. When solubilized homogenates were tested, well-defined reaction curves were obtained by the authors during the first minute of incubation. They found a linear relationship within a wide range between enzyme concentration and increase in absorbancy, and between the reaction rate and the reciprocal of the substrate concentrations (Lineweaver-Burke relationship). This procedure has been used by these authors for the determination of MAO levels in the liver and brain of several species and has facilitated a study on the effects of age, sex, and castration on the MAO levels of mouse liver. Furthermore, since this technique requires very little time for its performance, Zeller and his colleagues found it to be very useful for the study of rapid processes, such as analysis of progressive inhibition of MAO during the incubation of the latter enzyme with pargyline and o-chloropargyline.

The compound m-iodobenzylamine was chosen by the authors as a substrate of MAO because of E. A. Zeller's studies (343) on the occurrence of eutopic and dystopic complexes in the MAO reaction, which indicated that, with the exception of fluorine, the introduction of any substituent in the meta position led to much better substrates than benzylamine itself. By far the highest reaction rates, as expressed by maximum velocity (V), were obtained with m-iodobenzylamine. In beef and mouse liver a value for V was obtained that was higher than that for tyramine, the classical substrate of MAO.

Synthesis of m-Iodobenzylamine. m-Iodotoluene (Eastman White Label Chemical) yielded on photobromination (960) m-iodobenzylbromide (m.p. 50°), which was converted to m-iodobenzylamine by means of hexamethylenetetramine (961). m-Iodobenzylamine hydrochloride was purified by recrystallization from absolute ethanol (m.p. 193°).

At 253 nm the molar extinction coefficient in 0.067 M phosphate buffer, pH 7.4, was 780. For a 0.33 $\times$ 10^{-3} M solution, which according to Zeller et al. (942) represents the final substrate concentration in the standard test solution of this method, the absorbancy was 0.26.

Treatment of Tissues with a Detergent. When tissues are homogenized in the presence of a suitable detergent, transparent solutions are obtained (942); they show the same activity as homogenates, prepared by a conventional method (138, 151), and display much steadier light absorption. Without the use of a detergent, homogenates render the reading erratic for the first few minutes, because of settling of particles in the cuvette (938).

The tissue or mitochondria were homogenized by Zeller and his colleagues (942) by using a cold solution of 1.92 ml of 0.067 M phosphate buffer, pH 7.2, and 0.08 ml of "Cutscum" (isooctylphenoxypolyethoxyethanol, Fisher Scientific Co.) per 100 mg of tissue. These workers homogenized the material with 50 up-and-down strokes in a Teflon-pestle tissue grinder (Thomas, Philadelphia) and centrifuged the product for 30 min at 12,000 g. The thin lipid layer, that topped the liver preparation, was gently pierced so that 1 to 2 ml of the clear aqueous layer could be withdrawn.

This solution was diluted with an equal volume of cold phosphate buffer, pH 7.2, but for tissues poor in MAO, (e.g., brain) the enzyme solution was used without further dilution. Under all circumstances the enzyme solution was kept in an ice bath prior to the actual start of the assay.

Assay Procedure. The standard test solution consists of 0.2 ml of the enzyme solution and 2.8 ml of 0.36 $\times$ 10^{-3} M m-iodobenzylamine in phosphate buffer, 0.067 M, pH 7.2 (giving a final substrate concentration of 0.33 $\times$ 10^{-3} M). The control contains 0.2 ml of enzyme in phosphate buffer only. Whereas the vessels containing the substrate in buffer solution are shaken under oxygen at 38° for at least 10 min, and are then sealed with Parafilm, the enzyme solution, as mentioned above, is kept in an ice bath and is not flushed with oxygen. After breaking the seal of the tube containing the buffered substrate solution, the enzyme (0.2 ml) is pipetted into 2.8 ml of the latter solution, and the reaction velocity for the first minute or two from the start of recording is measured. The recording is started usually 10 to 20 sec from the time the enzyme is pipetted into the buffered substrate solution. In very active liver preparations the reaction rate decreases after 2 min and becomes nonlinear after this period.

The change in absorbancy—that is, increase in optical density at 253 nm as a function of time—that takes place upon the addition of the enzyme solution to the substrate-buffer solution, is measured in a model 14 Cary recording spectrophotometer at a compartment temperature of 38°.

Rates are expressed as differences in optical density per gram of enzyme per minute and refer to milligrams or grams of enzyme present in 1 ml.

The Beckman ultraviolet spectrophotometer and Gilford four-channel photometer have been found by Zeller et al. (942) to be also instruments suitable for the above measurements.

Finally, Zeller et al. emphasize that the rate of degradation of *m*-iodobenzylamine is much slower in air than in pure oxygen.

This modern, rather refined microspectrophotometric method of MAO estimation seems to ensure reliable results in kinetic studies on MAO activity in various tissues, especially in the presence of inhibitors. A more lucid and coherent presentation of the author's experimental data is, however, highly desirable.

g. A SENSITIVE FLUOROMETRIC ASSAY OF MAO ACTIVITY IN VITRO BASED ON
   THE RATE OF INDOLEACETIC ACID FORMATION ACCORDING TO
   LOVENBERG, LEVINE, AND SJOERDSMA (962)

Tryptamine is known to be a good substrate for MAO (123). On incubation of tissue homogenates with tryptamine, indoleacetaldehyde is first formed, and on addition of aldehyde dehydrogenase and nicotinamide-adenine dinucleotide NAD (DPN), is oxidized to indoleacetic acid (IAA). Whereas it is often not possible to quantify the disappearance of small amounts of tryptamine, corresponding amounts of IAA formed can be easily measured, since endogenous IAA is below the level of detection, and since the fluorescent properties of the latter compound allow the assay of microgram quantities.

As is well known, the average fluorescence measuring instrument will permit the determination of quantities of fluorescent material as small as 0.1 to 0.001 $\mu g/ml$. With the more sensitive instruments and with compounds having a high absorption and a high quantum yield of fluorescence, a millimicrogram (nanogram) of a fluorophor can emit sufficient fluorescence to allow measurement (963). Such concentrations are much smaller than are needed for other assay techniques. Moreover, fluorometry offers an added means of specificity.

Preparation of Tissues.   Tissues were frozen on removal and thawed prior to homogenization in 0.25 M sucrose using a motor-driven glass homogenizer. In studies on small samples the homogenate was frequently diluted 50 to 100 times.

For the preparation of aldehyde dehydrogenase guinea pigs were pretreated intraperitoneally, 18 to 24 hr before the experiment, with the irreversible MAO inhibitor, pheniprazine (JB–516, Catron, Lakeside Laboratories, Milwaukee, Wisc.), using 5 to 10 mg/kg. The pretreated guinea pigs were then killed, and a 20% kidney homogenate was centrifuged in a Spinco model L centrifuge for 1 hr at 78,000 g. The high-speed supernatant fraction showed a high aldehyde dehydrogenase activity and a negligible MAO effect. It could be stored at $-5°$ for periods up to 2 weeks without deterioration. The activity of the aldehyde dehydrogenase preparation is checked by measuring the rate of $NADH_2$ formation during incubation with acetaldehyde (964). The addition of this aldehyde dehydrogenase preparation to tissues already containing this enzyme did not reduce their MAO activity, which indicated the absence of significant amounts of residual pheniprazine.

Of the reagents required, stock solutions of tryptamine HCl (5.6 mg of free base per milliliter), nicotinamide (24.4 mg/ml), and IAA (100 $\mu g/ml$) were stored at $2°$ and used for several weeks. NAD (23.2 mg/ml) was prepared fresh daily.

Assay Procedure.   Portions of homogenate equivalent to 10 to 200 mg of tissue were pipetted into 20-ml beakers containing 60 micromoles nicotinamide (0.3 ml); 14 micromoles NAD (0.4 ml); 250 micromoles phosphate buffer, pH 7.4; 0.2 ml of the aldehyde dehydrogenase prepration; and distilled water to a total volume of 2.8 ml. Incubations were carried out in air at $37°$ in a Dubnoff metabolic shaking incubator. After a few minutes of equilibration, the reaction was initiated by addition of 7 micromoles of tryptamine (0.2 ml). At various intervals, 0.5-ml aliquots were removed and placed in tapered, glass-stoppered 50-ml centrifuge tubes, containing 3 ml of 0.5 N HCl, to stop the reaction. After addition of 15 ml of toluene, the tube was agitated for 5 min in a mechanical shaker. After centrifugation, 10 ml of the organic layer, which now contained the IAA formed, were transferred to another glass-stoppered tube, containing 1.5 ml of 0.5 M phosphate buffer, pH 7.0. The tube was shaken again for 5 min to extract the IAA into the aqueous phase. The fluorescence of the IAA in the buffer was then measured directly in an Aminco Bowman spectrophotofluorometer, at 280 nm excitation and 370 nm fluorescence (uncorrected wavelengths). Appropriate blanks and internal standards were carried through the procedure to permit direct calculation of results. In this procedure IAA was recovered quantitatively (95–100%) in the presence of tissue, as compared with internal aqueous standards. In all experiments a control assay was performed on the aldehyde dehydrogenase preparation to correct for traces of MAO activity.

In studies on the left ventricle of dog heart the authors could easily detect MAO activity with the fluorometric technique, just described. The reaction was shown to proceed at a linear rate for at least 30 min, and this rate was proportional to the amount of tissue used. In other experiments they proved that all the components of the reaction were present in amounts adequate to ensure a maximal rate of reaction. Furthermore, in still another set of experiments, they found that the formation of IAA could be completely prevented by the presence of 10^{-4} M pargyline, whereas up to 10^{-4} M concentrations of pargyline did not interfere with the activity of aldehyde dehydrogenase. The stoichiometry of the above fluorometric reaction was tested by these workers by measuring simultaneously the rate of tryptamine disappearance (965) and IAA formation. Working with a more active source of MAO (i.e., guinea-pig liver) Lovenberg and co-workers (962) were able to demonstrate that on a 20-min incubation of the liver extract with 3.5 micromoles of tryptamine virtually all of the amine that had disappeared could be recovered as IAA. These authors have demonstrated, however, that, whereas dog heart contained almost saturating amounts of endogenous aldehyde dehydrogenase, in the case of the sympathetic ganglion, it was apparent that the rate of the reaction was strongly dependent on the addition of exogenous aldehyde dehydrogenase. This dependence on another enzyme, besides MAO, is a disadvantage of this fluorometric technique.

This fluorometric assay for MAO determination seems according to Lovenberg and his colleagues (962) to be particularly suited to a great variety of problems, such as detailed mapping of MAO activity in specific parts of the central and peripheral nervous system, action of cardiovascular drugs on heart MAO, study of needle-biopsy specimens in man, and in elucidating relationships between pharmacological and biochemical effects of MAO inhibitors. Thus, the fluorometric assay was used by Lovenberg et al. to clarify the question concerning the inhibition of sympathetic ganglionic transmission by various MAO inhibitors, raised by Gertner (966), and Goldberg and Da Costa (967). In preliminary studies, Lovenberg et al. (962) were able to demonstrate by means of their fluorometric technique that potent inhibitors cause marked inhibition of MAO in the sympathetic ganglia of dogs and cats within a few minutes after injection.

When studying the effects of MAO inhibitors with this fluorometric method, certain precautions must be taken (962). It must be first established that the effects obtained are due solely to inhibition of MAO and not to inhibition of aldehyde dehydrogenase, or possibly to a reaction with the immediate metabolite, indoleacetaldehyde. It seems to the authors, however, to be unlikely that in *in vivo* experiments any of the above problems will be encountered, since the tissue concentrations of various injected inhibitors are rather low. The authors did not experience any difficulty in *in vivo* or *in vitro* experiments with β-phenylisopropylhydrazine (JB–516), pargyline, and iproniazid. However, they warn that, since small samples of tissue require considerable dilution for handling, the possibility must not be disregarded that in the case of reversible inhibitors, such as harmine, the results obtained might not, because of the very considerable dilution, accurately reflect the degree of *in vivo* inhibition.

h.   A RAPID MICROFLUOROMETRIC DETERMINATION OF MAO ACCORDING TO KRAJL (968)

This method is a fluorometric adaptation of the MAO assay of Weissbach et al. (938), (see this Section e). However, instead of determining the disappearance of kynuramine, Krajl (968) monitors fluorometrically the appearance of 4-hydroxyquinoline, which arises from the spontaneous cyclization of the intermediate aldehyde, formed by the enzymic oxidative deamination of kynuramine (938). The activation and fluorescence spectra for kynuramine and 4HOQ were obtained by the author in 1 N NaOH, employing an Aminco Bowman spectrofluorometer, coupled to an Electro Instruments model 101 X-Y recorder. All wavelengths reported are uncorrected values. On activation at 315 nm, 4HOQ exhibits an intense fluorescence maximum at 380 nm. Under these conditions a tenfold excess of kynuramine has no appreciable fluorescence. At the sensitivity settings employed, slit system 3, meter multiplier 0.03, sensitivity 50 (American Instrument Co. Service Manual No. 768 A, 1960, p. 15), full-scale galvanometer deflection is obtained with between 0.1 and 0.2 μg of 4 HOQ per milliliter. At the highest sensitivity settings, as little as 1 $m\mu$g (1 ng) of 4HOQ could be detected. This extremely sensitive fluorometric method has been applied to various tissues (e.g., rat brain, guinea-pig atria, and cat ganglion).

Procedure. To 1.0 ml of enzyme (an aqueous homogenate of tissue containing the desired amount of wet weight of tissue) were added 0.5 ml of kynuramine (100 μg kynuramine dihydrobromide), 0.5 ml of phosphate buffer (0.5 M, pH 7.4), and distilled water to 3.0 ml. After incubation at 37° for 30 min with air as the gas phase, 2.0 ml of 10% trichloroacetic acid (TCA) were added. The precipitated proteins were spun down, and 1.0 ml of supernatant was added to 2.0 ml of 1 N NaOH in a silica cuvette. The solution was activated at 315 nm and the fluorescence measured at 380 nm, or was recorded by scanning from 320 to 450 nm. Appropriate blanks and 4HOQ standards were carried through the entire procedure. A 4HOQ standard of 10 millimicromoles (1.99 μg) carried through the entire procedure gave an arbitrary fluorescence value of 77 $\pm$ 2 units, and under these conditions the blank values are 5 $\pm$ 1 units; 4HOQ carried through the entire procedure gives only one-quarter the fluorescence obtainable in 1 N NaOH. This is not due to adsorption losses on precipitated proteins, or to quenching by tissue extracts, or kynuramine, these effects being minimal, but to quenching by TCA. In the concentration range applied, a 10% change in TCA concentration causes a 5% change in 4HOQ fluorescence. Since the 4HOQ standard has never varied more than 2 units from the average value of 77, fluctuations caused by TCA are in practice minimal. TCA has been retained as deproteinizing agent because it is convenient to use and because, despite quenching, the fluorescence yield was sufficient to permit MAO determination in all tissues so far tested by the author.

The agreement between duplicate analyses is very good, the average error being less than 2%. The manipulations involved are very simple and consume very little time.

Moreover, this microfluorometric assay does not depend on the tissue sample containing aldehyde dehydrogenase. It should be recalled here that MAO is responsible only for the oxidative deamination of a given amine to the corresponding aldehyde. Many methods for the determination of MAO are based, however, on the assay of the acid produced by further oxidation of the relevant aldehyde by a second enzyme, an aldehyde dehydrogenase (962, 969, 970). Whereas liver tissue contains sufficient aldehyde dehydrogenase to make MAO rate limiting, this does not hold true for all tissues (969, 970). Thus, in ganglia Lovenberg et al. (962) had to add exogenous aldehyde dehydrogenase to obtain true MAO levels (see this Section, g). In Krajl's method the aldehyde produced by the oxidative deamination of kynuramine undergoes, as mentioned above, spontaneous and complete cyclization to 4HOQ. Thus, 4HOQ production depends only on MAO activity.

This simple, highly sensitive, and rapid microfluorometric technique of MAO estimation will, no doubt, prove to be valuable for various studies on MAO activity.

i. A SENSITIVE AND SPECIFIC RADIOMETRIC ASSAY FOR THE ESTIMATION OF MAO ACCORDING TO WURTMAN AND AXELROD (969)

Fluorometric methods of MAO estimation, although more sensitive than the spectrophotometric ones, still require milligram quantities of tissue and, besides, may be subject to error because of potential variation in endogenous levels of fluorescent material. Radiometric methods have been recently designed to allow quantitative estimations of MAO in microgram amounts of tissue.

Wurtman and Axelrod (969) have developed a simple, sensitive, and specific assay of MAO, utilizing the measurement of deaminated ^{14}C-metabolites of ^{14}C-tryptamine. Sixty or more assays, each requiring as little as 5 μg of tissue, can be carried out in 3 hr.

Procedure. Tissues are homogenized in chilled isotonic KCl, and 1 to 100 μl (10 μg to 1 mg) of the homogenate are used for assays. ^{14}C-Tryptamine (2-^{14}C-tryptamine hydrochloride, New England Nuclear Co., 1.3 mC/millimole) is dissolved in water and stored at —4°. A typical assay contained 25 μl of enzyme preparation, 25 μl (6.25 nanomoles, 10,000 cpm) of ^{14}C-tryptamine, and 250 μl of 0.5 M phosphate buffer, pH 7.4, in a 15-ml glass-stoppered centrifuge tube. The mixture was incubated at 37° for 20 min; the reaction was then stopped by the addition of 0.2 ml of 2 N HCl, and the deaminated radioactive material was extracted by shaking into 6 ml of toluene. After centrifugation, a 4-ml aliquot of the organic layer was transferred to a vial, containing 10 ml of phosphor [0.4% 2,5-diphenyloxazole and 0.005% 1,4-di-(2,5-phenyloxazolyl)benzene in toluene] and counted for 1 to 5 min in a liquid scintillation spectrometer. A small amount of ^{14}C-tryptamine (less than 0.3%) is extracted by this procedure. A correction is made for the blank value (about 30–50 cpm) by incubating ^{14}C-tryptamine with boiled enzyme.

The reaction was linear with time for at least 20 min and with enzyme concentration over a range of 5 to 1000 μg of liver. Duplicate determinations of the MAO activity of 250 μg of many liver specimens differed by less than 2%

The authors identified by ascending paper chromatography the radioactive metabolite of ^{14}C-tryptamine, formed by incubation with rat heart or liver, as ^{14}C-indoleacetic acid. This confirmed the findings of Lovenberg et al. (962), who used nonradioactive material in their assays. Indoleacetic acid was separated from tryptamine in a butanol : acetic acid

: water (4 : 1 : 1) system, and from indoleacetaldehyde in an isopropyl alcohol : ammonia : water (8 : 1 : 1) system. Moreover, since Wurtman and Axelrod (969) found that the intermediate metabolite, indoleacetaldehyde, has solubility characteristics similar to those of indoleacetic acid in an acid-toluene system, they concluded that their method can be used in the presence as well as in the absence of aldehyde dehydrogenase. The enzyme specificity of this method was investigated by the authors by determining the effect on hepatic MAO activity of pretreatemnt *in vivo* or preincubation *in vitro* with tranylcypromine, this compound being a potent inhibitor of MAO, at doses that are without effect on DAO (521). Both *in vivo* and *and vitro*, tranylcypromine produced a decrease of 97 to 99% in MAO activity.

k. A RADIOISOTOPIC ASSAY FOR MAO DETERMINATIONS ACCORDING TO OTSUKA AND KOBAYASHI (971)

This method is based on the formation of a radioactive anisole-soluble end product from the substrate, ^{14}C-tyramine. After the enzyme incubation, the end product is extracted into anisole, the aqueous phase is frozen, and the anisole, containing both the radioactive end product and phosphor, is poured into a counting vial for assay in a liquid scintillation spectrometer. This method can detect the metabolism of 1.4 ng of tyramine.

^{14}C-Tyramine, 4.6 mC/millimole was purchased from Calbiochem. This was diluted with nonisotopic tyramine to give a solution with an activity of 50,000 dpm per 4 μg tyramine, per 0.1 ml solution. MAO was prepared from hog kidney by the method of Satake (972). Hogkidney acetone powder (65 g) was extracted twice with 30 ml of 0.3 M phosphate buffer, pH 7.2. The residue was suspended in 20 ml of water inside a Visking cellulose casing bag and dialyzed for 4 hr against tap water, followed by 4-hr dialysis against distilled water. This dialyzed suspension, containing 1.12 mg nitrogen ($\equiv$7.00 mg protein) per milliliter, has a specific activity of 44 μl O_2 uptake per hour, per milligram of protein, with tyramine as substrate.

Liquid Scintillation Counting Method. A Packard Tri-Carb liquid scintillation spectrometer, model 314 EX, was used in this technique. The extraction solvent consisted of reagent-grade anisole, containing 0.6% 2,5-diphenyloxazole (PPO). This solution (anisole–PPO) gave 53% counting efficiency, when used in the extraction of end products. Toluene containing 0.4% PPO (toluene–PPO) was also tested in preliminary studies as a possible useful solvent.

^{14}C-Tyramine was assayed at a counting efficiency of 50% by using a toluene–ethanol mixture (70 : 25, v/v) containing 0.37% PPO and

0.01% 1,4-bis-2-(4-methyl-5-phenyloxazolyl)benzene (dimethyl-POPOP).

For the assay of radioactivity in protein solutions, 1 ml of methanol containing 1 M hyamine hydroxide (Packard Instrument Co.) was added to 0.2 ml of sample and heated for 4 hr at 60° to dissolve the protein, and assayed with the addition of 10 ml of toluene–PPO with an efficiency of 32%.

Counting efficiencies of various solvents were determined by the internal-standard method using a standard solution of ^{14}C-toluene containing 10,000 dpm/0.1 ml. All counting was done in low-potassium 5-dram glass vials (Wheaton Glass Co.).

Incubation and Extraction Procedure. All enzyme assays were performed in a screw-cap culture tube (10 $\times$ 1.5 cm) in a final volume of 2.0 ml, in air at 37°, in a Dubnoff metabolic shaker for 60 min. At the end of the incubation period, 0.4 ml of 2 M citric acid was added to the culture tube, followed by 10 ml of toluene–PPO or anisole–PPO solution. The mixture was shaken vigorously for about 1 min, centrifuged at 1,000 rpm in an International centrifuge, model PR–1, and allowed to stand at −20° until the lower aqueous phase was frozen. The upper layer was then poured into a counting vial and assayed in a liquid scintillation counter.

Standard Test Solution for Extraction Studies. One milliliter of kidney MAO preparation was incubated with 40 μg of ^{14}C-tyramine and 10^{-3} M EDTA in 20 ml of 0.1 M sodium phosphate buffer, pH 7.5, at 38° for 60 min. After that time the reaction mixture was heated in a boiling water bath for 5 min and centrifuged. Aliquots of the supernatant part were used in the extraction studies. EDTA was found to accelerate the enzymic reaction and was also used to prevent further oxidation of the reaction product.

In extraction studies the authors demonstrated that the anisole–PPO system was superior to that of toluene–PPO for the extraction of the reaction products of MAO, using tyramine as substrate. Hence, they adopted the anisole–PPO system as the extraction solvent for the isotopic MAO assay. Under the conditions of their assay the authors found the formation of the end product of the MAO reaction to be roughly proportional to the incubation time during the first hour, but to deviate from linearity with larger incubation periods. Only a small increase in extractable radioactivity was found after a 240-min incubation, indicating the virtual completion of the reaction.

The stoichiometry of the MAO reaction was determined by these workers by estimating the remaining tyramine, the end product formed, and oxygen consumption during the incubation. Four micrograms ^{14}C-tyramine, 10 micromoles of nonisotopic tyramine, 0.25 or 0.50 ml of enzyme, and 10^{-3} M EDTA were made up in 0.1 M sodium phosphate buffer, pH 7.5, to a total volume of 2.0 ml. After 1-hr incubation at 37° in air, the product formed was extracted eight times with 10-ml portions of anisole–PPO. The amount of product was calculated from the total radioactivity of the combined extracts. The stoichiometry of the MAO reaction was found experimentally to average 1.88 moles of product, formed per mole of oxygen consumed. This approximates the expected 2 : 1 ratio for MAO in the absence of aldehyde oxidase and the presence of catalase in the enzyme preparation (123). Because of their results Otsuka and Kobayashi (971) assume that in their assay procedure MAO activity is being measured in the presence of catalase and that aldehyde oxidase was absent from their enzyme preparation.

The sensitivity of this method is apparently a function of the specific activity of the tyramine available. In their diluted preparation the authors could easily measure the metabolism of 17 ng of tyramine. Without dilution, the metabolism of 1.4 ng of tyramine (100 cpm at 50% efficiency) could be detected by the authors.

According to Otsuka and Kobayashi (971), their radioisotopic assay for the determination of MAO offers also a simple means of evaluating *in vivo* the effectiveness of MAO inhibitors in the circulation as well as a method for determining the biological half-life of these drugs.

1. AN ISOTOPIC METHOD FOR THE MICRODETERMINATION OF MAO ACTIVITY IN MICROGRAM QUANTITIES OF TISSUE ACCORDING TO McCAMAN, McCAMAN, HUNT, AND SMITH (970)

A procedure is described for the quantitative estimation of MAO activity in samples of nervous tissue weighing as little as 2 to 5 μg (dry weight), using ^{14}C-labeled substrates, such as serotonin, 3-hydroxytyramine, and tyramine, showing different specific activities.

The buffer substrate solution, containing 0.1 M potassium phosphate buffer, pH 7.2, and 0.8 mM 3-^{14}C-serotonin was stored at $-20°$ until needed. In a pointed microtube (2.5 mm $\times$ 4 cm) 10 μl of ice-cold buffer-substrate solution were added to the sample, which was equivalent to 1 to 20 μg of dry brain or 1 μl of homogenate (1 g tissue per 25 ml H_2O). All the tubes were placed in a tray of ice water; the contents of each tube were mixed without warming and incubated at 38° for 30 min. The reaction was then stopped by the addition of 1 μl of 3 N HCl. Then 50 μl of ethyl acetate were added to each tube, and the samples thoroughly mixed and centrifuged in order to separate the phases. A portion (40 μl) of the ethyl acetate layer was removed from each tube and transferred to another tube containing 30 μl of 0.3 N HCl. This "washing step" served to remove the last trace of the radioactive substrate. After thorough mixing and centrifugation, 35 μl of the ethyl acetate layer were transferred to a counting vial (20 ml) and mixed with 1 ml of absolute methanol, followed by 15 ml of scintillator-toluene solution [4 g of 2,5-diphenyloxazole and 0.1 g of 1,4-bis-2-(5-phenyloxazolyl)benzene per liter of toluene]. Radioactivity was determined in a Packard scintillation counter, and the enzyme activity calculated from the known specific activity of the substrate. The same protocol was used with ^{14}C-3-hydroxy-tyramine (2.5 mM in 0.1 M phosphate buffer, pH 6.75), and ^{14}C-tyramine (3.0 mM in 0.1 M phosphate buffer, pH 7.95). The radioactivity extracted into ethyl acetate from the blanks was equivalent to less than 0.5% of the amount of serotonin or tyramine contained in the buffer-substrate, and with one backwash it was reduced to less than 0.005% (40 cpm to no detectable activity, depending on specific activity). Tubes containing the buffer–substrate solution alone or in conjunction with tissue homogenate, added after incubation, or boiled homogenate, gave similar blank values. For every nanomole of serotonin (specific activity, 1 mC/millimole) metabolized by brain in the above procedure, a delta of approximately 2000 cpm with a blank of 5 cpm was obtained. The isotopic measurements of the compounds extracted into ethyl acetate after incubation with ^{14}C-tyramine indicated that the ratio of acid to aldehyde was 4 : 1. Similarly, measurements of the end products of the brain-enzyme reaction with ^{14}C-serotonin showed that the indoleacetaldehyde and indoleacetic acid were present in approximately equal quantities. The optimum concentration during incubation was 0.8 mM for serotonin and 2.5 mM for 3-hydroxytyramine. A marked inhibition was observed for these two substrates at higher concentrations. Thus, a 50% MAO inhibition was found with a 8 mM concentration of serotonin and a 14 mM concentration of 3-hydroxytyramine as substrates. With tyramine as substrate, essentially the same activity was observed with concentrations between 2.5 and 10 mM. The amount of radioactive product formed during the MAO reaction was proportional to the amount of brain present, over the range of 5 to 170 μg of wet brain, at a 30-min incubation at 38°. The activity of MAO in brain decreased somewhat with time after 15 min incubation, so that values at 30 min fell off by 15% and those at 60 min by 30%.

McCaman et al. (970) tested some well-known inhibitors of MAO activity for their effect on the assay with rabbit-brain homogenates. Three noncompetitive inhibitors of MAO activity, tranylcypromine, pheniprazine, and iproniazid, were preincubated for 15 min at 38° with the buffered homogenate prior to the addition of the substrate. Each of these compounds was found to be without effect on the extraction of the labeled products of the enzyme reaction. The simple unsubstituted amines (tryptamine and phenylethylamine), added to the buffer-substrate solution, were the most potent inhibitors of tyramine and 3-hydroxy-tyramine deamination, whereas α-methyl- or N-methyl-substituted compounds were more potent inhibitors of serotonin deamination than nonsubstituted analogs.

McCaman et al. (970) point out that their isotopic micromethod for the quantitative determination of MAO activity permits investigations of discrete histological areas in laminar structures found within the central nervous system (i.e., cerebellar and cerebral cortex). Results obtained by these authors seem to suggest the presence of MAO in high concentration in synaptic structures.

m. A SIMPLE, RAPID AND SENSITIVE METHOD FOR THE MICRORADIOMETRIC
     ASSAY OF MAO IN SUBMICROGRAM AMOUNTS OF TISSUE ACCORDING TO
     AURES, FLEMING, AND HAKANSON (918)

Current interest in the function of biogenic amines and in the enzymes involved in their metabolism has been stimulated by the recent development of simple and sensitive methods for the chemical detection of these amines. Studies on the cellular localization and turnover of biogenic amines have been impeded by the lack of methods for the adequate separation and discrimination of these amines as well as of their precursors and metabolites. Recently, thin-layer chromatography has been successfully applied to the rapid isolation and identification of minute amounts of tissue biogenic amines. Nanogram amounts of these amines can be detected by exposing the thin layer to an o-phthalaldehyde spray or paraformaldehyde gas, whereby the biogenic amines are converted into highly fluorescent derivatives (918, 973). These methods have served as a basis for the development of simple radiometric assays of amine oxidase activity.

Assay of MAO in the Pineal Gland of the Rat. One rat pineal gland was homogenized in 50 μl of 0.1 M phosphate buffer, pH 6.8. The homogenate, corresponding to approximately 1 mg of pineal tissue, was incubated with 2 μg ^{14}C-serotonin (free base; 39.7 mC/millimole; Radiochemical Center, Amersham, England), in a total volume of 60 μl, under oxygen at 37°. Incubation was stopped after 1 hr by the addition of 1 ml of the solvent system ethyl acetate–acetic acid (95 : 5). Precipitated proteins were spun down, and 0.5 ml. of the clear supernatant was evaporated to dryness under reduced pressure. The dry material was extracted with a drop of the ethyl acetate–acetic acid (95 : 5) mixture, which quantitatively extracts 5-hydroxyindoleacetic acid but only traces of serotonin. The extract was then submitted to thin-layer chromatography on silica gel. A microscope slide (2.5 $\times$ 7.5 cm) was coated with a thin layer (250 μ) of silica gel (Kieselgel G and Kieselgel H, Merck, Darmstadt, Germany). The layer was applied as a slurry, consisting of 30 g of silica gel suspended in 70 ml of redistilled water. The chromatoslide was then activated by drying in an oven at 100° for 1 to 2 hr, and the ethyl acetate–acetic acid extract was applied to it by means of a glass capillary. The reaction product, 5-hydroxyindoleacetic acid, was separated from residual serotonin in less than 15 min by means of the solvent system n-butanol–acetic acid (1 : 1). The thin-layer material containing 5-hydroxyindoleacetic acid was scraped off and transferred to a counting vial containing 0.1 ml of water as eluant. The radioactivity was measured by liquid scintillation counting.

Estimation of MAO Activity in Rat-Brain Caudate Nucleus Slices. The corpus striatum has a high dopamine content (331). The uptake and metabolism of dopamine in the latter tissue was studied by Aures et al. (918) by incubating slices of rat caudate nucleus, approximately 50 mg, with 0.1 μC of ^{14}C-dopamine (22 mC/millimole, Nuclear-Chicago) in 2 ml of Krebs-Ringer bicarbonate solution, fortified with 1% D-glucose, pH 7.4, at 37° for 30 min in an atmosphere of 95% O_2 and 5% CO_2. The slices were removed from the incubation medium, rinsed, and homogenized in acidified methanol. The protein precipitate was spun down, and an aliquot of the supernatant was chromatographed on a cellulose thin layer along with reference compounds. A slurry was made by suspending 15 g of cellulose powder (MN–300–HR, Machery, Nagel, and Co., Düren, Germany) in 90 ml of redistilled water. After two-dimensional development in the solvent systems consisting of (a) n-butanol saturated with 0.1 N HCl and (b) isopropanol–5 N NH$_4$OH–H$_2$O (8 : 1 : 1), the compounds were visualized by exposure to paraformaldehyde gas and subsequent scanning in ultraviolet light. The spots were scraped off, and the radioactivity was determined by liquid scintillation counting. Dopacetic acid was found to be the major dopamine metabolite in the rat caudate nucleus. The results obtained compare well with earlier observations by Goldstein and co-workers (974).

At present, work is in progress in the laboratory of Aures and her colleagues (918) on the development of microtechniques, so urgently required for the assay of enzymes involved in the formation and degradation of biogenic amines.

n. THE STUDY OF MAO ACTIVITY BY HISTOCHEMICAL PROCEDURES

Since Zeller's discovery (142, 342) of the potent inhibitory effect of iproniazid on MAO an extremely large number of new MAO inhibitors has been prepared and their use in pharmacotherapy has correspondingly increased. The importance of MAO in the metabolism of serotonin, epinephrine, and norepinephrine suggests that an investigation of the inhibitory effects on MAO activity of different substances—such as psychotropic drugs, vasodilators, and antihypertensives—might be of interest for pharmacological evaluation (976). Because of technical simplicity histochemical experiments with MAO inhibitors may help to elucidate the physiological role of MAO in brain.

Francis (977, 978), applying tyramine as substrate, showed that MAO can be localized histochemically by the use of a tetrazolium compound as hydrogen acceptor. As is well known, hydrogen peroxide is formed during the oxidation of tyramine by MAO, indicating that in this reaction dehydrogenation takes place (see Chapter II.2). Instead of molecular oxygen, neotetrazolium [pp'-diphenylene-bis-2-(3,5-diphenyl) tetrazolium chloride] can be applied as hydrogen acceptor, and this compound, on being reduced will be precipitated as blue formazan. This is the principle of the histochemical demonstration of MAO activity.

Koelle and Valk (979) treated tissue sections with hydrazine and incubated these in the presence of tryptamine and 3-hydroxy-2-naphthoic acid hydrazide in phosphate buffer, containing a high concentration of sodium sulfate. The aldehyde formed by the action of MAO condenses with the hydrazide, and the condensation product is then converted to a bluish-purple pigment by coupling with tetrazotized o-dianisidine. The specificity of the method was demonstrated by the use of a selective inhibitor of MAO, iproniazid. Despite its validity, this method has not been frequently used because it is so laborious.

In 1957 Glenner, Burtner, and Brown, Jr. (975) modified the histochemical technique of Francis (977, 978) by using tryptamine as substrate rather than tyramine in the presence of tetrazolium salts. Under these conditions there was a distinct localization of the formazan at tissue sites during an hour of incubation. Until the present, the latter technique has remained the most satisfactory method for general use in the histochemical demonstration of MAO activity (980).

o. STANDARD TRYPTAMINE-TETRAZOLIUM METHOD FOR THE HISTOCHEMICAL DEMONSTRATION OF MAO ACTIVITY ACCORDING TO GLENNER, BURTNER, AND BROWN, JR. (975)

Selected tissues of young adult guinea pigs, rabbits, and rats were utilized for the localization of the formazan, precipitated during incubation with tryptamine. The kidneys of young adult female guinea pigs were used throughout all the experimental studies on substrate specificity, effects of inhibitors, pH optimum, and relative reactivity of various tetrazolium compounds. All tissues were freshly removed, and 20 μ thick frozen sections were cut on the sliding microtome, using the Adamstone-

Taylor technique (981).

Recommended Procedure. In a 20 ml polyethylene jar place 25 mg of tryptamine hydrochloride; 4 mg of sodium sulfate; 5 mg of nitro-blue tetrazolium; 5 ml of 0.1 M phosphate buffer solution, pH 7.6; and 15 ml of distilled water. Warm to 37° and incubate tissue sections in this mixture for 30 to 45 min at 37°. Wash in running water, fix in buffered 4% formaldehyde for 24 hr, dehydrate and clear in graded acetone–xylene solutions, and mount in Permount.

Control. A modified Koelle-Valk (979) technique was used for the controls to compare sites of staining with the tryptamine–tetrazolium method. In this modified technique, the previously prepared hydrochloride of 3-hydroxy-2-naphthoic acid hydrazide, a compound with greater solubility, was utilized and the addition of sodium sulfate to the incubation mixture was eliminated. This modification reduced the incubation time to 1 hr and obviated the need for constant aeration of the incubation medium.

Several commercially available tetrazolium salts—including blue tetrazolium, neotetrazolium, 3-(p-iodophenyl)-2-(p-nitrophenyl)-5-phenyl tetrazolium chloride (INT), and tetrazolium violet—as well as several specially synthesized tetrazolium compounds—including 2,5-diphenyl-3-(2,4-dinitrophenyl)tetrazolium chloride and nitro-blue tetrazolium (nitro-BT)—were used as the tetrazolium component in the tryptamine–tetrazolium reaction.

Inhibitor Studies. Potential inhibitors in 0.01 M concentrations were added to a pH 7.6 phosphate buffer solution, and mounted tissue sections were preincubated in these solutions for 15 min at 37°. The slides were then thoroughly washed in running water and placed in a tryptamine–tetrazolium (INT) incubating solution for 30 min, at 37°. For concomitant incubation inhibition experiments, various potential inhibitors in 0.1 M concentration were incorporated in the tryptamine–tetrazolium (INT) incubating solution, with immersion of slides for 30 min at 37°. Controls without potential inhibitors were utilized throughout all assays.

Reaction Optimum with the Tryptamine–Tetrazolium Method. Optimal activity was obtained between pH 7.2 and 7.6, at 37°. No reaction occurred below pH 6.8 and above pH 8.0. Inhibition of the reaction was also noted on incubation of slides for 20 min at 60°. Incubation under anaerobic conditions failed to increase the speed of the reaction, or to change the localization of formazan precipitation. A tryptamine concentration of 0.005 M was found to produce the most rapid reaction, a lesser concentration requiring a longer incubation time. When a 20% sodium sulfate solution was incorporated into the incubating solution, there was a moderate reduction in staining intensity, but a slightly sharper localization of the formazan dye.

Substrate Specificity. Experiments on substrate specificity were carried out by replacing tryptamine with various compounds in 0.005 M concentration in the standard incubation medium. Incubations using INT, neotetrazolium, and nitro-BT as the tetrazolium components were carried out for periods from 30 min to 2 hr, at 37°.

Furthermore, various compounds known to be degraded by MAO and DAO, were substituted for tryptamine as substrates in the incubating solution. Several indole derivatives were also used as substrates. Visible reactions with similar localization during incubation from 30 min to 2 hr occurred only with tryptamine and serotonin. No visible reaction occurred with 5-methoxy-1-methyltryptamine, 5-methoxy-2-methyltryptamine, gramine, bufotenine, indole-3-acetic acid, DL-tryptophan, indole, skatole, tyramine, histamine, spermine, and acetaldehyde. DOPA, dopamine, and ephinephrine produced spontaneous dye reduction in the absence of tissue sections.

In a simultaneously published communication (982), Weissbach et al. extended their histochemical observations to *in vitro* investigations. Their aim was to determine whether the reduction of tetrazolium salts was a valid measure of MAO activity as well as to try to explain the biochemical basis of the reduction of tetrazolium salts in histochemical assays. Data obtained by these workers showed that, although the reduction of the tetrazolium salts is the basis of a rapid and simple histochemical method of localization of MAO in tissues, there are limitations to the application of this reaction to *in vitro* studies. It should be pointed out, however, that the *in vitro* experiments of Weissbach et al. (982) indicated that MAO and the system that transfers electrons from the indole carbonyl intermediate to the tetrazolium salt are present in the same particulate fraction of the cell. Until pure 5-hydroxyindoleacetaldehyde is made available, it will not be possible to ascertain whether the aldehyde reduces the dye directly, or not. However, the data of Weissbach et al. seem to point to the involvement of a diaphorase-like system that is present in the mitochondria along with MAO. Whether this is associated with the enzymic or nonenzymic oxidation of the aldehyde has yet to be determined.

The development in 1962 of a controlled temperature freeze-sectioning technique by Cunningham et al. (983), by means of which unfixed tissue can be sectioned at 8 microns, made it possible to achieve the degree of preservation of cellular structure that is required for accurate histochemistry. Cohen, Bitensky, and Chayen (984), in a study of MAO activity with the histochemical tetrazolium technique, have been able to show in sections of human endometrium, frozen and sectioned according to the method of Cunningham et al. (983), that MAO occurred either in discrete particles or in diffuse form, depending on the stage of the menstrual cycle.

Finally, Graham and Karnovsky (980) have succeeded in demonstrating histochemically MAO activity in the liver and kidney of rats and guinea pigs by a coupled peroxidatic oxidation technique. In the presence of peroxidase, its substrate 3-amino-9-ethylcarbazole has been found to be oxidized by the hydrogen peroxide, generated during the oxidation of tryptamine, catalyzed by MAO. The resulting insoluble, red oxidation product is apparently deposited at sites of MAO activity. This histochemical coupled oxidation method of MAO estimation may be useful as an alternative technique to the tetrazolium method, since it depends on an entirely different type of chemical reaction (980).

p. DIRECT MEASUREMENT OF MAO INHIBITION IN HUMANS ACCORDING TO LEVINE AND SJOERDSMA (985)

Most methods in current use for determining the drug-induced inhibition of MAO activity in man depend on indirect approaches, such as measurement of the urinary excretion of a substrate amine, for which there are no alternative routes of degradation. Certain errors are inherent in the indirect approach. Urinary levels of the excreted amines may be altered by variations in the intake of their precursors as well as in their rate of formation or in their rate of release from tissue. Moreover, the degree of enzyme inhibition required to produce any given increment in the urinary excretion of an MAO substrate is not known.

Levine and Sjoerdsma (986) recently demonstrated the feasibility of accurate assay of MAO activity in specimens of the mucosa of the human jejunum obtained by peroral biopsy. This observation forms the basis for a direct approach to the measurement of MAO inhibition in man. The authors determined mucosal levels of MAO activity before, during, and after the administration of two different MAO inhibitors. Tissue MAO levels were then correlated with urinary levels of excreted tryptamine, and the relative efficacy of oral and parenteral routes of administration of the drug was compared. Finally, the relationship of MAO inhibition in the intestine to that in other organs was studied in laboratory animals.

Procedure. Nine adults who were hospitalized with primary diagnoses of essential hypertension were the subjects of these studies. In five subjects specimens of jejunal mucosa were obtained by peroral biopsy (987), before treatment began and assayed for MAO activity by the fluorometric method of Lovenberg, Levine, and Sjoerdsma (962); (see this Section, g).

Treatment was then started in two subjects with the MAO inhibitor, isocarboxazid, a hydrazine type of MAO inhibitor, and in three subjects with pargyline hydrochloride, a nonhydrazine MAO inhibitor. The drugs were administered once daily in doses that were adjusted to produce postural hypotension. Between the 10th and the 60th day of treatment specimens of jejunal mucosa were obtained again and assayed for MAO activity. To obviate possible errors, due to local concentrations of drug in the intestine, biopsies during treatment were done from 20 to 24 hr after drug administration. The results obtained in these experiments have shown that pargyline and isocarboxazid in doses commonly used in clinical practice produce marked inhibition of MAO activity in the mucosa of human jejunum. Furthermore, this study indicates that the assay of MAO activity in specimens of jejunal mucosa, obtained by biopsy before and after the administration of a drug, is a satisfactory means for direct demonstration of MAO-inhibiting properties of a drug in human subjects. Moreover, the findings of the authors support the validity of one of the indirect methods—that is, the assay of urinary excretion of tryptamine. Urinary tryptamine levels reached a maximum 2 or 3 days after maximum MAO inhibition in the gut.

There are certain limitations of the direct method of estimating MAO inhibition. This method is useful only for the study of drugs that produce irreversible inhibition of MAO. Tissue specimens as small as those obtained by jejunal biopsy (986) must be homogenized at dilutions so high that the concentrations of reversible inhibitors might be decreased below effective levels. Since MAO inhibitors may produce different degrees of inhibition in various tissues, the authors recommend that new inhibitors be tested first in experimental animals in order to determine the relationship of MAO inhibition in the intestinal mucosa to that in other organs. It appears that man is much more sensitive to MAO-inhibiting drugs than are several other mammalian species.

3. METHODS OF ISOLATION, PURIFICATION, CRYSTALLIZATION, AND ESTIMATION OF HISTAMINASE (DAO)

The first attempts at purification of histaminase were made in 1932 by McHenry and Gavin (149), who prepared from the hog-kidney cortex an acetone powder that was stable at room temperature for about 12 months. Subsequently, partial purification of hog-kidney extracts was achieved by fractionation with sodium and ammonium sulfate (988, 989), by thermal denaturation of the concentrated inert proteins at 60 to 62° (990), or by isoelectric precipitation (991). Laskowski, Lemley, and Keith (990) as well as Leloir and Green (991), by combining various techniques, accomplished a purification of histaminase of about 100-fold. In 1951 Tabor (551) published a method, consisting of fractionation of hog-kidney cortex by acetone, sodium sulfate, heating at 60°, adsorption on C γ-alumina, and pH precipitation, and with this technique he was able to achieve a 300-fold purification of hog-kidney DAO. Arvidsson, Pernow, and Swedin (992) replaced Tabor's Cγ-alumina batch procedure with Dowex–50 column chromatography, and Swedin (993) with alumina column chromatography. Both groups of workers accomplished by means of column chromatography a further considerable degree of purification of the hog-kidney enzyme. Likewise Uspenskaya and Goryachenkova (552), when applying starch-block electrophoresis, obtained a 700-fold purification of hog-kidney DAO.

A. Procedures for the Isolation, Purification, and Crystallization of Hog-Kidney DAO

a. PREPARATION OF PURIFIED DAO FROM HOG KIDNEY ACCORDING TO TABOR (551)

Acetone Powder Extract. In a Waring Blendor 100-g portions of fresh hog kidney were homogenized with 500 ml of acetone, previously chilled to −10° for 1 min, and quickly filtered with suction. The filter cake was homogenized in another 500 ml of cold acetone, again filtered with suction, quickly dried at room temperature, and sifted through wire mesh. The powder obtained could be stored at 2° for at least 6 months without loss of activity.

Protein concentrations were determined by total nitrogen (Kjeldahl) estimations. Enzyme activity was estimated either by measuring oxygen consumption or by a colorimetric microtechnique determining histamine disappearance (994). 150 g of acetone powder were extracted with 3000 ml of 0.2 M phosphate buffer, pH 7.2, at room temperature, with occasional stirring for 20 min; the mixture was then centrifuged, and the residue was discarded.

Sodium Sulfate Fractionation. To each portion of 100 ml of the acetone powder extract 14.5 g of anhydrous sodium sulfate were added, and the solution was filtered. The filtrate was treated with 7.0 g of sodium sulfate per 100 ml, and the solution was again filtered. The precipitate, dissolved in 275 ml of 0.1 M phosphate buffer, pH 6.8, was dialyzed against running tap water for 1 hr.

Heating. The dialyzed solution, placed in 100-ml test tubes, was heated in a 60° water bath, with mechanical stirring, for 20 min and then centrifuged; the precipitate was washed with 100 ml of water.

Gel Adsorption and Elution.   The combined supernatants were treated with 109 ml of C γ-alumina gel for 10 min and centrifuged. The supernatant was discarded, and the gel washed with 60- and 30-ml portions of 0.2 M phosphate buffer, pH 7.2. Elution was carried out by treating the C γ-alumina gel with successive portions of 200, 100, and 100 ml of 1 M K_2HPO_4 and collecting each eluate by centrifugation. The combined supernatants were added to 40 ml of 0.5 M KH_2PO_4 and dialyzed overnight against 4 liters of cold distilled water. All subsequent operations were performed in the coldroom at 2°. The enzyme was precipitated by adding 40 g of ammonium sulfate per 100 ml of the dialyzed solution. After centrifugation at 12,000 r.p.m. for 10 min, the precipitate was extracted with 20 ml of 0.2 M phosphate buffer, pH 7.2, and dialyzed against cold running 0.04 M sodium acetate for 4 hr.

First pH Precipitation.   A 6-ml volume of 0.2 M sodium acetate buffer, pH 5.0, was added to the solution; after 30 min, the precipitate was collected by centrifugation and dissolved in 6.5 ml of 0.1 M phosphate buffer, pH 7.2.

Second pH Precipitation.   This solution was treated with 3 ml of 0.2 M sodium acetate buffer, pH 5.0; after 40 min, the precipitate was collected by centrifugation and dissolved in 6.5 ml of 0.1 M phosphate buffer, pH 7.2. A similar degree of purification was obtained, when the dissolved precipitate from the first pH precipitation instead of being reprecipitated with 0.2 M sodium acetate buffer (second pH precipitation) remained at 0° for 1 to 2 days, and the precipitate that formed was collected by centrifugation.

The overall yield was 6 to 16% and the product had 240 to 300 times the activity of the acetone powder extract. The enzyme precipitate at the first and second pH precipitations was white or gray. The enzyme solutions were usually colorless, although occasionally they still contained a faint yellowish tinge. The purified preparations were free from catalase activity and showed no oxygen uptake in the absence of substrate. When stored at −10°, there was no loss of enzymic activity for at least several weeks.

With histamine, putrescine, and cadaverine as substrates the purified DAO preparation yielded results indicating that in the attack of this enzyme on any of the substrates, mentioned above, one molecule of aldehyde, ammonia, and hydrogen peroxide are formed from each molecule of substrate. Whereas the product of the histamine–enzyme reaction, presumably imidazoleacetaldehyde, could be further oxidized by oxygen in the presence of xanthine oxidase or by NAD in the presence of aldehyde oxidase to imidazoleacetic acid, the products of the reaction of putrescine or cadaverine and DAO, that is, the corresponding aliphatic amine aldehydes, seemed to undergo cyclization to Δ^1-pyrroline or Δ^1-piperideine, respectively.

b. PURIFICATION OF HOG-KIDNEY HISTAMINASE ACCORDING TO
   KAPELLER-ADLER AND MACFARLANE (553)

Chromatographic, electrophoretic, and spectrophotometric techniques have been combined and applied to extracts of hog-kidney cortex, prepurified by fractionation with ammonium sulfate and heat. A moderately stable preparation of histaminase has been isolated which has appeared to be homogeneous by the criteria of chromatography and analytical electrophoresis.

Analytical Methods. For all dialyzing procedures the Visking dialyzing tubings were pretreated with a 10^{-3} M solution of EDTA (ethylenediaminetetraacetic acid disodium salt, dihydrate, British Drug Houses, Analar) (995).

Protein. Protein was estimated in the initial stages of enzyme purification by the biuret method, as described by Gornall et al. (996), and in the later stages of purification by the method of Lowry et al. (997).

Enzyme Assays. Enzyme activity was determined by a simple microvolumetric technique (998) involving a coupled oxidation reaction in which indigo disulfonate is oxidized by hydrogen peroxide, formed in the primary reaction. The enzyme activity was expressed in units, one of which corresponds to the amount of enzyme that will under standard assay conditions catalyze the degradation of 6.9 $\times$ 10^{-5} micromoles of histamine or cadaverine per minute. The specific activity of the enzyme was defined as the number of enzyme units per milligram of protein.

This method of estimation of histaminase or DAO activity has been recently greatly improved by the replacement of the microvolumetric determination of the residual indigo disulfonate by a spectrophotometric method (548; see Section, 3.D.f).*

Analytical Electrophoresis on Cellulose Acetate Strips. Electrophoretic assays were carried out in horizontal tanks (999) using 2.5 $\times$ 10 cm cellulose acetate strips (1000), various buffer solutions (ionic strength 0.05–0.1) of a pH range extending from 4.6 to 8.6, with a current of 0.6 mA/cm width of cellulose acetate strip, for 4 hr. The strips were stained with 0.2% Ponceau S in 6% salicylsulfonic acid. Enzyme solutions containing as little as 0.4 mg of protein per milliliter could be successfully electrophoresed. Cellulose acetate strips were supplied by Oxoid, Ltd., London.

Partial Purification of Crude Histaminase by Fractionation with Ammonium Sulfate and Heat Treatment. Unless otherwise stated, all procedures were carried out at room temperature. The cortices of 24

---

* Also unpublished data by the author (1964).

hog kidneys were minced to a fine brei in an electric mincer, yielding a wet pulp of approximately 2000 g. A 200-ml volume of 1% NaCl per 100 g of wet tissue was added, and the mixture was allowed to stand for 60 min with intermittent gentle shaking and then filtered through muslin. The pulp was returned to the original flask and the extraction procedure repeated, using the same volume of 1% NaCl as before. The combined extracts were then precipitated with finely powdered ammonium sulfate (A.R.) to 0.6 saturation by adding slowly with mechanical stirring 42.4 g of ammonium sulfate per 100 ml of solution. The precipitate was immediately filtered through Whatman No. 1 fluted filters and the reddish filtrate discarded. The precipitate was dissolved in small portions, using approximately 200 ml of 0.02 M phosphate buffer, pH 6.8, for each portion. Aliquots of 50 ml of the dark-brown solution obtained were dialyzed for 60 min at 4° against 2000 ml of 0.02 M phosphate buffer, pH 6.8, with two changes of the buffer, then freed by centrifugation at 4° from the precipitate formed. The clear, yellowish-brown supernatant was then heated with mechanical stirring at 60° for 20 min in a constant-temperature water bath. After rapid cooling, the turbid solution was centrifuged at 4°, the precipitate of inert protein washed three times with small amounts of 0.02 M phosphate buffer, pH 6.8, and after recentrifugation discarded. The combined yellowish-brown, clear supernatants were precipitated with ammonium sulfate to 0.6 saturation.

This partially purified enzyme fraction served as a starting material for chromatography on DEAE-cellulose columns. In each fractionation step the protein content as well as the enzyme activity with histamine and cadaverine as substrates were determined in parallel.

It should be mentioned here that on heating the enzyme fractions, which had been partially purified by ammonium sulfate fractionation, at 60° for 20 min, the specific activities of histaminase greatly differed with the substrate used. With histamine as substrate, the specific activity of histaminase was found to be considerably increased by the thermal purification step, indicating in some assays up to a four fold purification of the enzyme. With cadaverine as substrate, a sharp decrease in the specific activity of the enzyme occurred in most of the assays (553).* As discussed in Section 3.B, this loss of enzyme activity with cadaverine as substrate was found to be so substantial, when placental extracts, partially purified by ammonium sulfate fractionation, were heated at 60° for 20 min, that this further purification step by thermal denaturation of inert proteins had to be abandoned in the case of placental extracts (538).

Chromatography of Histaminase on DEAE-cellulose Columns, Followed by Electrophoresis of Chromatographic Effluents on Cellulose Acetate

* R. Kapeller-Adler and M. Fletcher, unpublished results (1958).

Strips. Swedin (993) was the first to attempt a purification of histaminase by chromatography on DEAE-cellulose columns. The standard conditions of his assay were, however, not very well outlined and very few details as to the specific activity of the chromatographed enzyme were given.

In the work now being described, many preliminary experiments had shown that histaminase was quantitatively adsorbed on a DEAE-cellulose column in the presence of 0.005 M borate–HCl buffer, pH 8.6, and that on simple elution with 0.1 M acetate buffer, pH 5.5, the enzyme emerged from the column with virtually no loss of activity. Attempts to elute the adsorbed enzyme from DEAE-cellulose columns by gradient elution were not very successful. At pH 8.6, histaminase is apparently so tightly bound to the cationic groupings of the resin that it can only be eluted from the column by a change of the pH to a lower range in the eluting agent.

Sufficient DEAE-cellulose powder (Whatman DE–50) to pack a column (2.0 $\times$ 23.0 cm), was first thoroughly washed with 0.1 N NaOH, then with distilled water to a neutral pH, and finally equilibrated with 0.005 M borate–HCl buffer, pH 8.6. An aliquot of the partially purified enzyme preparation, obtained as described above, after heating at 60°, and after the second ammonium sulfate precipitation, was dissolved in 30 ml of 0.02 M phosphate buffer, pH 6.8, and dialyzed for 3 hr against 2000 ml of 0.005 M borate–HCl buffer, pH 8.6, at 4°, with two changes of the buffer. The slight precipitate that usually formed after dialysis, was removed by centrifugation, and the dialyzed enzyme solution, containing about 300 mg of protein, was placed on the column; the column was developed with 75 ml of 0.005 M borate–HCl buffer, pH 8.6. Enzyme elution was accomplished with 200 ml of 0.1 M acetate buffer, pH 5.5. The rate of elution was approximately 16 ml/hr, and 15-min samples of the eluate were collected by a time-regulated fraction collector. Fractions were collected until no more protein was eluted from the column. It was noticed during each elution procedure that a distinct brownish band remained firmly adsorbed to the top of every column, while soon after the commencement of the elution a very pale yellow band moved down the column. The effluents emerging from such a column showed a pH of 7.0. Each effluent was assayed for its content of protein and for its enzymic action on histamine and cadaverine as substrates.

Effluents with a high specific histaminase activity were subjected to electrophoresis on cellulose acetate strips, applying to each strip 5 to 8 μl of each effluent and 5μl of normal human serum, the latter to serve as a marker. One-half the width of the strip was used for the effluent and the other half for the serum. Effluents from several individual columns showing the same high degree of purity, as indicated by their specific activity as well as by their electrophoretic patterns, were pooled and were subjected for maximal purification either to rechromatography on DEAE-cellulose or to a precipitation with ammonium sulfate to 0.6 saturation.

Rechromatography. The method of rechromatography of histaminase on DEAE-cellulose columns was essentially the same as for chromatography, as described above. Five to ten chromatographic effluents of a similar degree of purity, usually obtained from three DEAE-cellulose columns, were combined into one fraction (approximately 35 ml) and dialyzed for 3 hr against 2,000 ml of 0.005 M borate–HCl buffer, pH 8.6, at 4°. The slight precipitate which formed was separated by centrifugation, and the clear, yellowish enzyme solution was applied to the DEAE-cellulose column.

For precipitation with ammonium sulfate, effluents from several columns with a similar degree of purity were pooled and solid ammonium sulfate was slowly added under mechanical stirring to give an 0.6 saturation. The precipitate was separated by centrifugation, dissolved in a small amount of 0.02 M phosphate buffer, pH 6.8, and the solution obtained was freed by centrifugation from the small insoluble residue, which consisted mainly of inert protein. The clear, pale-yellow supernatant was dialyzed against 0.02 M phosphate buffer, pH 6.8, for 30 min, at 4°, and after dilution, was investigated for its histaminase and protein content.

Concentration of Chromatographic Effluents. The effluents were concentrated by osmosis using a simple rapid technique, described by Hsiao and Putnam (1001). A dialysis casing, containing the chromatographic eluate, was covered with dry powdered sucrose for 1 hr. The sucrose was then removed from the dialyzing tubing by washing with water, and dialysis of the concentrated effluent against 0.02 M phosphate buffer, pH 6.8. The three fold to four fold concentration of each chromatographic effluent thus achieved, gave a final concentration of about 1 to 2 mg of protein per milliliter of solution, suitable for most analytical purposes.

Discussion of Results Obtained at Various Steps of Histaminase Purification. In the fractionation procedure with ammonium sulfate, followed by heat treatment at 60° for 20 min, a purification of about 60 times of the enzymic effect on histamine as substrate, as compared with the crude extract, was achieved. However, with cadaverine as substrate only 22-fold purification was obtained by the same procedure. Moreover, the ratio of the specific activity on histamine in relation to that on cadaverine was not constant at all, but differed significantly from one purification step to another. This absence of relative constancy at various stages of purification of the enzymic effect on histamine to that on cadaverine confirms previous findings (537).* Y. Kobayashi, in his recent attempts to purify hog-intestine DAO, has isolated several fractions that metabolized both histamine and cadaverine.* This author also found that the ratio of the enzymic effect on histamine to that on cadaverine varied considerably among the different purification fractions.

The method designed in this work for the purification of histaminase by chromatography on DEAE-cellulose was used in 55 independent assays, each of them comprising 3 to 4 individual columns. In a typical chromatographic assay (Fig. 1), an enzyme preparation, partially purified by ammonium sulfate fractionation and heat treatment, and containing 315 mg of protein and 22,500 enzyme units (corresponding to a specific activity of 71.4), using histamine as substrate, and with a trace activity on cadaverine, was subjected to further purification on DEAE-cellulose, as described above.

Figure 1 represents a typical elution diagram. At the tube, marked by an arrow, simple elution with 0.1 M acetate buffer, pH 5.5, was started. On the basis of the elution pattern, this diagram has been divided into three parts, marked A, B, and C. Part A contained mainly inert protein and negligible amounts of histaminase activity, part B contained comparatively little protein but 80 to 90% of the total histaminase activity applied to the column, part C contained small amounts of histaminase activity but large quantities of protein. It can be seen from this diagram that soon after the commencement of elution with acetate buffer, pH 5.5, the entire histaminase activity was eluted in a single symmetrical peak (part B). The peak effluents of this part showed a specific histaminase activity of 1400, indicating a twentyfold purification as compared with the enzyme material before chromatography. The effluents of part B displayed a distinct greenish-yellow fluorescence, when examined under the ultraviolet lamp.

Each effluent of parts A, B, and C was subjected to electrophoresis on cellulose acetate strips at pH 6.8, and the results were compared with the results of electrophoresis of the partially purified enzyme preparation before chromatography on DEAE-cellulose. When compared with normal human serum, the partially purified enzyme showed two diffuse bands in the globulin fraction and one well-defined band in the albumin region, with some protein remaining at the point of application. At pH 6.8, the peak effluent of section B moved as a single, distinct band toward the anode, and was located in the region between the α_2- and β-globulin fractions of normal human serum. Part A, which contained a negligible amount of histaminase and consisted mainly of inert protein, showed on

---

* Also unpublished data by R. Kapeller-Adler and M. Fletcher (1958).

* Personal communication from the author.

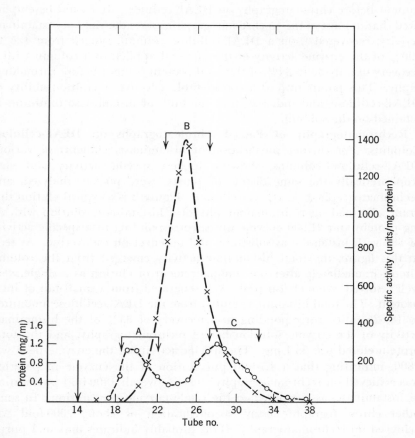

Figure 1. Elution diagram of hog-kidney histaminase chromatographed on DEAE-cellulose columns at pH 8.6. Simple elution commenced at the arrow with 0.1 M acetate buffer, pH 5.5. Parts A and C represent effluents with the lowest specific activity. Part B is representative of effluents with the highest specific histaminase activity.

electrophoresis at pH 6.8 a broad diffuse band in the globulin fraction, with some protein remaining at the point of application. Part C, containing small amounts of histaminase activity but larger quantities of protein, as compared with part B, showed two bands, one, a slower moving band located in the region between the α_2- and β-globulin fractions, the second, a fast-moving distinct band with the mobility of normal human serum albumin. The total histaminase activity eluted from the column amounted to 34,594 units and the total protein to 101.65 mg. This corresponds to a recovery of 154% of histaminase activity and only 32.4% of the protein content before chromatography on DEAE-cellulose. It should be emphasized that in most of the 55 chromatographic assays the yield of histaminase activity, recovered from a DEAE-cellulose column, ranged from 125 to 190% of the enzyme activity of the material applied to a column with a recovery of only 30 to 34% of the total protein content before chromatography. This points toward a considerable selective separation ability of DEAE-cellulose, and indicates that the bulk of histaminase inhibitors is retained on the column.

Rechromatography of Pooled Chromatographs on DEAE-cellulose Columns. For further purification of histaminase, effluents of various DEAE-cellulose columns, showing highest specific activity and electrophoretically the same degree of purity, were pooled, dialyzed, and rechromatographed on DEAE-cellulose. Figure 2 is a typical elution diagram, obtained on rechromatography of a histaminase solution with 57 mg protein and 21,900 enzyme units, corresponding to a specific activity of 384 with histamine as substrate, and no effect on cadaverine. As seen in this figure, the total histaminase activity emerged from the column almost immediately after the commencement of elution as a single, very well resolved, symmetrical peak, well separated from a small tail of inert protein. The total histaminase eluted from the DEAE-cellulose amounted to 18,625 units, corresponding to a recovery of 85% of the histaminase activity of the enzyme solution before rechromatography, and the total protein eluted was 25.1 mg. The specific activity of the enzyme peak was 1800, indicating that a 4.7-fold purification of the enzyme preparation was achieved on rechromatography with an overall 1800-fold purification of histaminase, as compared to the crude enzyme preparation. In some other chromatographic assays a purification of about 2000-fold was achieved on rechromatography. This probably indicates maximal purity of the enzyme, since the specific activity of the enzyme did not change on further rechromatography.

When pooled effluents of DEAE-cellulose columns with a very high specific histaminase activity are precipitated with ammonium sulfate to 0.6 saturation, a further almost tenfold purification of the enzyme can be accomplished. Thus, in one case the purified enzyme showed a specific activity of 2700 and was recovered in a yield of 89.7%.

The results obtained in the work, described here, on the purification of histaminase by fractionation with ammonium sulfate, followed by thermal denaturation of inert proteins at 60°, and chromatography on DEAE-cellulose columns, seem to substantiate previous suggestions (144, 537, 883) by demonstrating that it is possible to isolate from hog-kidney extracts histaminase fractions that display a very high activity on histamine but are without effect on cadaverine.

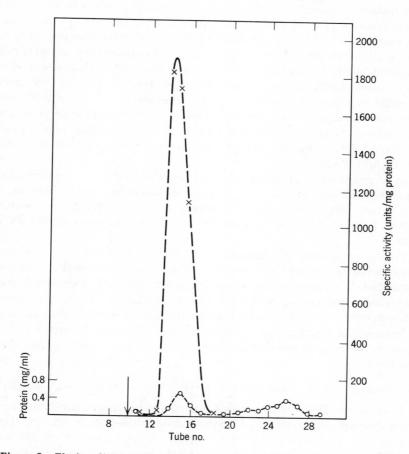

Figure 2.   Elution diagram of a histaminase preparation after rechromatography on a DEAE-cellulose column. Effluents of a chromatogram, similar to effluents of Part *B*, Figure 1, of highest specific activity were pooled and rechromatographed under conditions as in Figure 1. Elution was started at the arrow. 0 - - - - - 0 mg protein per milliliter. X- - - -X specific activity (units/mg protein).

Mondovi and his colleagues (555), who extensively purified hog-kidney histaminase, have recently reported that their purest preparation acted on both cadaverine and histamine and that the ratio of cadaverine rate to histamine rate remained practically constant during the purification procedure. In this connection it should be mentioned that these workers heated the crude hog-kidney extract at 60° for 10 min and not, as in the present work, the enzyme fraction partially purified by ammonium sulfate fractionation. Moreover, the latter preparation was subjected to the heat treatment for 20 min. Since the effect of partially purified histaminase on cadaverine, in contrast to that on histamine, has proved to be very sensitive to application of heat (553),* the discrepancy in the results obtained by Mondovi et al. (555) and those reported in the present work may possibly be explained on the basis that Mondovi and his colleagues, by subjecting their crude hog-kidney homogenates to 60° for only 10 min, were not able to achieve the separation of the enzymic effects on histamine and cadaverine which has been demonstrated to be, at least partly, possible by subjecting the partially purified enzyme to heat treatment at 60° for 20 min (553).

In recent work, Uozumi et al. (1002) reported on an almost complete separation of the effects of histaminase on histamine and cadaverine in hog-kidney preparations that had been partially purified by ammonium sulfate fractionation, and were subjected to heat treatment at 60° for 15 min.

Finally, as quoted in Chapter V, the finding that hog-kidney histaminase can be be resolved into two discrete enzymically active fractions with somewhat different substrate specificities, demonstrated in electrophoretic assays as a slow-moving peak with a very high activity on histamine and a fast-moving one with a much lower activity on histamine, may point to a behavior of histaminase as an isoenzyme.* As already mentioned in Chapter V, this observation requires further experiments for its clarification.

C. PURIFICATION OF HOG-KIDNEY DAO ACCORDING TO MONDOVI ET AL. (555)

Hog-kidney DAO was purified by controlled heat denaturation of crude homogenates, pH precipitation, ammonium sulfate fractionation, column electrophoresis, and column chromatography on DEAE-cellulose. A 1700-fold purification was achieved by the authors.

Analytical Methods. Estimation of DAO and histaminase activities was performed in parallel by two methods: oxygen uptake and oxyhemoglobin oxidation.

Oxygen Uptake. Oxygen uptake was followed in conventional Warburg manometers in the presence of cadaverine or histamine, as described by Mondovi et al. (554). One unit of DAO is defined as the amount of enzyme that will catalyze the transformation of 1 micromole of substrate per minute under standard conditions at 38°. Air was used as gas phase. The standard conditions were: 100 micromoles of substrate were dissolved

---

* Also unpublshed results by R. Kapeller-Adler and M. Fletcher (1958).
* R. Kapeller-Adler and H. MacFarlane, unpublished data (1960).

in 0.1 ml of water and placed in the side arm; the main compartment contained the enzyme, 1 ml of 0.3 M potassium phosphate buffer, pH 7.4, 100 μg of catalase, 100 micromoles of ethanol, 10μg of pyridoxal-5-phosphate, water to 3 ml. The center well contained 0.2 ml of 30% NaOH.

The Oxyhemoglobin Test. The procedure for the oxyhemoglobin test was as follows: Human oxyhemoglobin solution (0.5%) was prepared according to Rossi-Fanelli and Antonini (1003). Solutions were prepared to contain 0.3 ml of the enzyme preparation; 1 ml of 0.3 M potassium phosphate buffer, pH 7.4; 0.1 ml of oxyhemoglobin solution; 25 micromoles of substrate; and water to give a total volume of 3 ml. The mixture was incubated at 38° for 30 min. In the presence of DAO the solution turned from red to light brown, due to the formation of oxidation products of hemoglobin by hydrogen peroxide produced by the action of DAO on the substrate. This method is only semiquantitative, but very sensitive. Extinctions were measured with a Beckman model DU spectrophotometer.

Column Electrophoresis. Column electrophoresis was performed according to the method of Porath and Hjerten (1004). The column contained a 3.2 $\times$ 60 cm cellulose bed, equilibrated with the buffer in a coldroom. The cellulose was pretreated with borohydride, as described by Porath and Hjerten; 10-ml fractions were collected. Rate of elution varied from 2 to 4 ml/min.

Protein. Protein was determined by the biuret method of Goa (933). Crystalline albumin from bovine serum (Sigma Chemical Co.) was used as the standard.

Extraction and Preliminary Purification. Frozen hog-kidney cortex (1000 g) was homogenized in 1200 ml of water for 10 min at room temperature in a Waring Blendor (fraction A1). The crude homogenate was placed in a beaker and stirred in a water bath at 60° for 10 min. The beaker was then immersed in an ice-water bath to cool the contents quickly to about 25°. The cooled homogenate was centrifuged at 6000 g for 1 hr, and the precipitate was discarded. To the supernatant fluid (fraction A2) solid ammonium sulfate was added to 0.33 saturation; the mixture was allowed to equilibrate for 30 min and was then centrifuged at 6000 g for 20 min. The supernatant was brought to 0.6 saturation with ammonium sulfate, and after 30 min the precipitate was collected by centrifugation. It was dissolved in 80 ml of 0.01 M sodium acetate buffer, pH 5.3, to give fraction A3.

All the operations described were performed at 0 to 4°.

Precipitation at pH 5.3. The solution A3 from the preceding stage was divided into two portions and each was first dialyzed with vigorous mechanical stirring against 10 liters of 0.01 M sodium acetate buffer, pH

5.3, for 10 hr and then, with two changes, against 5 liters of 0.1 M sodium acetate buffer, pH 5.3, for 24 hr. The dialyzed enzyme solution was centrifuged at 20,000 g for 40 min. The precipitate was suspended in 80 ml of water, dissolved by adjusting the pH between 7.2 and 7.4 with 1.0 N ammonia and lyophilized to give fraction A4.

Column Electrophoresis.   Two grams of the freeze-dried powder (A4), corresponding to about 700 mg of soluble protein, were dissolved in 25 ml of 0.01 M potassium phosphate buffer, pH 6.5, and dialyzed with stirring against 2 liters of the same buffer for 12 hr. After the elimination of the insoluble material by centrifugation, the enzyme solution was placed on the top of an electrophoresis column of cellulose previously equilibrated with the same buffer (1004). The top of the column was connected to the cathode and the bottom to the anode. The electrophoretic run was made in 0.01 M potassium phosphate buffer, pH 6.5, for 36 hr, with a current of 18 mA at 620 volts. After the run, the column was eluted with 0.01 M potassium phosphate buffer, pH 6.5, and 0.5 ml of each fraction was tested for DAO by the oxyhemoglobin test. The enzyme activity of 1 ml of every second active fraction was estimated in the Warburg apparatus. Fractions with a specific activity (units/mg of protein) higher than 0.085 were combined and freeze-dried to give fraction A5.

Column Chromatography on DEAE-Cellulose.   The freeze-dried preparations from two electrophoresis columns, corresponding to about 100 mg of protein, were dissolved in 6 ml of 0.05 M potassium phosphate buffer, pH 7.2, dialyzed against 2 liters of this buffer for 12 hr, and placed on a column of DEAE-cellulose (2.2 $\times$ 35 cm) that had been previously equilibrated with the same buffer. Three chambers of the Varigrad device were charged, in order, with 150 ml of potassium phosphate buffer, pH 7.2, of the following concentrations: 0.05, 0.15, and 0.3 M. Of the three protein peaks obtained on column chromatography only the first peak contained the enzymic activity. All fractions of this protein peak that were positive in the oxyhemoglobin test were pooled and freeze-dried. The freeze-dried powder was dissolved in 4 ml of water, dialyzed against 0.1 M potassium phosphate buffer, pH 7.4, and concentrated again by lyophilizing to give fraction A6.

When comparing the specific activity of the purified DAO (fraction A6) with that of the crude homogenate (fraction A1), it follows that Mondovi et al. achieved an overall 1700-fold purification of hog-kidney DAO with cadaverine as substrate, and a 1300-fold purification with histamine as substrate. The yield of the enzyme obtained with the purification technique, just described, was a very poor one indeed (4.5% with cadaverine as substrate). The main loss of enzyme activity (41%) apparently occurred at the stage of precipitation at pH 5.3 (fraction

A4). No figures for the yield of purified enzyme with histamine as substrate were given.

It might possibly have been more profitable to have subjected the enzyme preparation, partly purified by ammonium sulfate fractionation, first to column chromatography on DEAE-cellulose, and to have applied column electrophoresis as well as the precipitation procedure at pH 5.3 at the later stages of purification.

Recently, Mondovi and his colleagues (556) slightly modified the method described above and obtained from hog kidney a DAO preparation that appeared to be homogeneous by the criteria of gel electrophoresis and ultracentrifugation, and had approximately twice the specific activity that was previously reported (555). The purified enzyme (556) was pink-yellow and showed two absorption bands, at 405 and 480 nm.

Determination of Copper in Purified Hog-Kidney DAO Preparations. For the estimation of the copper content of the purified enzyme the method of McFarlane (1005) as modified by Smith and Krueger (1006) was applied. Copper was also determined by atomic absorption in a Hilger and Watts Uvispek H 700.308 spectrophotometer, equipped with the atomic absorption attachment H 1100. A hollow-cathode copper lamp was used as light source, and pure propane as fuel. The spectral line chosen was at 342.7 nm. The copper content increased during the purification of hog-kidney DAO, and in the final purified preparation the average value was 11.6 ng-atoms of copper per milligram protein (556).

Effect of Diethyldithiocarbamate on Hog-Kidney DAO (556). Seven milligrams of purified DAO were dissolved in 7 ml of 0.1 M potassium phosphate buffer, pH 7.4. 1 ml of this solution was analyzed in the atomic absorption spectrophotometer and was found to contain 0.725 μg of copper per milligram of protein. The residual solution was then dialyzed for 12 hr against 2 liters of 0.1 M potassium phosphate buffer, pH 6.95, containing 0.05 M diethyldithiocarbamate. Absorption spectra recorded after the latter treatment showed a peak at 442 nm, which is characteristic of the behavior of copper proteins in the presence of diethyldithiocarbamate (554). The enzyme solution treated with diethyldithiocarbamate was centrifuged at 105,400 g for 2 hr. The precipitate dissolved in carbon tetrachloride showed the spectrum of the diethyldithiocarbamate–copper complex. The supernatant was dialyzed against 2 liters of 0.01 M Tris buffer, pH 7, for 48 hr to remove the excess of diethyldithiocarbamate. The amount of copper, determined by atomic absorption, indicated that about 70% of the copper was removed. The enzyme activity, tested in the presence of 1.7 mM cadaverine as substrate, was completely abolished by the treatment with diethyldithiocarbamate, but was restored to about 40% of its original value on addition of 1 to

10 nanomoles of copper sulfate per 100 μg of enzyme.

It is noteworthy that interaction between diethyldithiocarbamate and hog-kidney DAO does not seem to lead immediately to the removal of the enzyme copper, since the diethyldithiocarbamate–copper binding could be partially reversed, when the copper–diethyldithiocarbamate complex was not removed by high-speed centrifugation. Thus, in such an experiment the enzyme solution was, as described above, first dialyzed against diethyldithiocarbamate and then against Tris buffer; the diethyldithiocarbamate–copper complex was, however, not separated by centrifugation. Although the enzyme activity was found to be completely abolished, 70% of its original value could be restored by the addition of 10 nanomoles of Cu^{2+} per 100 μg of enzyme. This seems to indicate that the copper–diethyldithiocarbamate complex remains attached to the enzyme, when high-speed centrifugation is omitted, and, moreover, that the added copper ions remove the diethyldithiocarbamate bound to the enzyme copper, thereby partially restoring enzyme activity. Greater concentrations of Cu^{2+} diminish the extent of enzyme reactivation, and complete inhibition is observed when 300 nanomoles of Cu^{2+} are added per 100 μg of enzyme.

Mondovi et al. (556) emphasize that the addition of Zn^{2+}, Fe^{2+}, and Mg^{2+} in the form of their sulfates, and in concentration ranges in which Cu^{2+} is able to restore enzyme activity, is without any effect at all.

Electron Paramagnetic-Resonance Spectrum of Hog-Kidney DAO. Mondovi and colleagues (556) investigated the electron-paramagnetic-resonance (EPR) spectrum of hog-kidney DAO in the presence and absence of putrescine. Enzyme of 90% purity was dissolved at 0.4 mM concentration in 0.020 ml of 0.05 M phosphate buffer, pH 7.4, in an anaerobic cell. The spectra were examined before the addition and 1 min after the addition of 0.01 ml of a 0.5 M neutralized putrescine dihydrochloride solution. The spectra were recorded at a microwave power of 10 mW, a modulation amplitude of 6 gauss, a scanning rate of 200 gauss/min, a time constant of 0.3 sec, and a temperature of $-160°$. Values for g_m, $g_{||}$, and $A_{||}$ were determined as described by Malmström and Vänngard (573), and were not confirmed by calculation of spectra. For resolution of weak, hyperfine lines, a Varian C–1024 time-averaging computer was used. For generation of the second derivative spectrum, the signal from the computer was fed into a differentiating operational amplifier and presented on an x-y plotter with the same sweep of the x-axis as the corresponding first-derivative signal.

For the evaluation of the EPR spectra obtained by Mondovi and his colleagues (556) see Chapter IV.1.D.

d. CRYSTALLIZATION OF DAO FROM HOG KIDNEY, ACCORDING TO
   YAMADA ET AL. (258)

All operations throughout the entire purification procedure were performed at 5°.

Step 1. Cortices of fresh hog kidneys (1.1 kg) were suspended in 1100 ml of 0.03 M phosphate buffer, pH 7.0, homogenized in a Waring Blendor, and centrifuged.

Step 2. The supernatant solution was fractionated with ammonium sulfate (30–60% saturation); the precipitate obtained at 60% saturation was dissolved and dialyzed against 0.03 M phosphate buffer, pH 7.0.

Step 3. The dialyzed solution was applied to a DEAE-Sephadex A–50 column (6 × 60 cm) that had been equilibrated with 0.03 M phosphate buffer, pH 7.0. The column was washed with 0.03 M phosphate buffer, pH 7.0, and the enzyme was desorbed from the column by stepwise elution with 0.07 and 0.1 M phosphate buffer, pH 7.0. The active fractions were combined and concentrated by precipitation with ammonium sulfate at 60% saturation. The precipitate was collected and dialyzed against 0.1 M phosphate buffer, pH 7.0.

Step 4. The dialyzed solution was fractionated with ammonium sulfate at 35 to 55% saturation. The precipitate obtained at the latter ammonium sulfate saturation was dissolved and dialyzed against 0.1 M phosphate buffer, pH 7.0.

Step 5. The dialyzed solution was applied to a hydroxylapatite column (5 × 15 cm) that had been equilibrated with 0.1 M phosphate buffer, pH 7.0. The column was then washed with 0.1 M phosphate buffer, pH 7.0, containing 0.2 M ammonium sulfate, and the enzyme was eluted from the column with 0.1 M phosphate buffer, pH 7.0, containing 0.5 M ammonium sulfate. The active fractions with a specific DAO activity greater than 0.200 were combined and concentrated by the addition of ammonium sulfate at 60% saturation.

Step 6. The precipitate was dissolved in 0.1 M phosphate buffer, pH 7.0, and passed through a Sephadex G–150 column (2 × 100 cm), equilibrated with 0.1 M phosphate buffer, pH 7.0. Enzyme fractions showing a specific activity greater than 0.500 were pooled and concentrated by precipitation with ammonium sulfate at 60% saturation. The collected precipitate was dissolved in 0.1 M phosphate buffer, pH 7.0.

Step 7. Finely powdered ammonium sulfate was carefully added to the enzyme solution until it became slightly turbid, and the mixture was then placed in an ice bath. Crystallization began after about 3 hr and was virtually completed within a week. Crystals appeared as fine, highly refractive needles showing a faint pink color. Recrystallization was carried out by repeating the last step.

The enzyme activity was determined (a) manometrically by measuring the oxygen uptake at 38° in a reaction mixture containing 100 μg of the enzyme, 50 μg of catalase, 10 micromoles of substrate, and 100 micromoles of potassium phosphate buffer, ph 7.4, in a total volume of 3 ml, and (b) by a colorimetric method, described by Yamada et al. (1007), measuring the formation of Δ^1-pyrroline, an oxidation product of putrescine. The latter method was in the present work of the authors modified in that the enzymic assay was carried out at pH 7.4 in preference to pH 8.0. A unit was defined as the amount of enzyme that under standard conditions of the assay catalyzed the formation of 1 micromole of Δ^1-pyrroline per minute. The protein concentration was determined spectrophotometrically by measuring the absorbancy at 280 nm. An E value of 1.63 for 1 mg of protein per ml, and a 1-cm light path was used by the authors throughout their present work. This value was arrived at by determinations of absorbancy and dry weight. The specific activity of the purest crystals as compared with that of the crude homogenate indicated a 1350-fold purification.

On electrophoresis according to Tiselius—which was carried out at 4°, at pH 7.4, and an ionic strength of the sodium phosphate—sodium chloride buffer of 0.1—the crystalline DAO preparation migrated as a single band. In the ultracentrifuge this enzyme preparation sedimented as a single symmetrical peak. When using 0.1 M potassium phosphate buffer, pH 7.4, 16°, a sedimentation constant of $\overset{o}{S}_{20\,w}$ of 9.90×10^{-13} (cm/sec) was determined on extrapolation to zero protein concentrations of the data obtained from four runs in the ultracentrifuge. The sedimentation constant decreased by 0.03×10^{-13} cm/sec, per mg with increasing protein concentration. A diffusion constant, $D_{20,\,w}$ of 5.14×10^{-7} (cm²/sec) was determined for a solution containing 6.18 mg of protein per milliliter. Assuming a partial specific volume of 0.75, a value of 185,000 was calculated for the molecular weight of the enzyme. Spectrophotometric investigations of the crystalline DAO preparation revealed that its pink color was associated with the absorption maximum observed at 470 nm. The pink color was discharged by putrescine as well as by sodium dithionite and was restored by oxygenation. This finding compares well with recent observations made by Goryachenkova (552) and Mondovi et al. (556). Analyses for metals in the molecule of crystalline DAO by both the atomic absorption spectrophotometry as well as chemical analysis showed that copper was the only metal present and that its content amounted to 11.7 ng-atoms per mg of enzyme. This value corresponds to 2.17 g-atoms of copper per mole of enzyme.

The crystalline enzyme oxidized cadaverine, putrescine, 1,6-diaminohexane, histamine, agmatine, and 1,3-diaminopropane at relative rates of 100, 97, 61, 59, 40, and 17 respectively. Using the manometric method under standard conditions, the authors found that 1 mg of crystalline enzyme oxidized 2.04 micromoles of cadaverine per minute.

B. Method for the Isolation and Purification of Histaminase, DAO, from Human Placentae (538)*

Attempts at purification and identification of histaminase from human placentae have been hampered by the fact that, despite intensive efforts for complete exsanguination of placentae by perfusing the vascular bed with normal saline, the placental extracts or even the partially purified enzymic fractions contained hemoglobin and its derivatives in overwhelming quantities compared to the amount of histaminase present (536, 538, 542, 558, 600).†

In the work to be discussed*, a technique is described which by the use of DEAE-cellulose at various hydrogen ion concentrations allows not only a complete separation of hemoglobin from histaminase but also a considerable purification of the latter. In 1962 Hennessey and his co-workers (557) published a method for the separation of erythrocyte enzymes from hemoglobin by treatment of the hemolysate at pH 7.0 with DEAE-cellulose; under these conditions proteins having an isoelectric point below 7.0 are adsorbed onto DEAE-cellulose, whereas hemoglobin, having an isoelectric point near neutrality, is not adsorbed. The hemoglobin-free enzyme protein fraction thus obtained may serve as a useful starting material for further purification.

Kirkman (1008) was the first to use DEAE-cellulose for the successful extraction of glucose-6-phosphate dehydrogenase from diluted hemolysates of human erythrocytes. On adsorption of the hemolysate on DEAE-cellulose at pH 6.3, subsequent washing, elution, precipitation with ammonium sulfate, and dialysis, Kirkman obtained an 80-fold purified glucose-6-phosphate dehydrogenase in a yield of 20 to 40% that was essentially free of hemoglobin. However, no experimental details have been published by this author.

The technique of Hennessey et al. (557) has recently been applied with good results by Hatfield and Wyngaarden (1009) to the separation of beef-erythrocyte pyrophosphorylase from hemoglobin.

Since the isoelectric point of hog-kidney histaminase had been found to lie between pH 5.0 and 5.15 (553), the technique of Hennessey et al.

---

* Based partly on unpublished results by the author (1964).

† Also unpublished data by R. Kapeller-Adler and M. Fletcher (1958) and R. Kapeller-Adler and MacFarlane (1960).

(557) was adopted for the separation of histaminase from hemoglobin and its derivatives in extracts of human placentae.

Analytical Methods.

Protein. Protein concentrations of the enzymic fractions were determined by spectrophotometric measurements of the absorption at 280 nm on the basis that a solution containing 1 mg of protein shows in a 1-cm cuvette at 280 nm an optical density of 1 (6).

Enzyme Assays. Enzyme activity was determined by a simple spectrophotometric technique (548) involving a coupled oxidation reaction in which indigo disulfonate is oxidized by the hydrogen peroxide formed in the primary reaction, and the excess of indigo disulfonate is measured, at 613 nm. The enzyme activity is expressed in international milliunits, one milliunit corresponding to a degradation of 1 nanomole of substrate per minute. The specific activity indicates milliunits of enzyme per milligram of protein. For detailed description of this method see Section 3.D.f.

Preparation of Hemoglobin-Free Placental Extracts. Two fresh human placentae were freed from membranes and the umbilical cord, and were then rinsed with cold water to eliminate the excess blood. The placental tissue was freed from the vessels by dissection, was then subjected to another rinse with cold water and weighed, and after addition of 200 ml of 1% NaCl to each 100-g portion of placenta, homogenized in a Waring Blendor at high speed for 1 min. The homogenate was filtered, using a large Buchner funnel with a layer of glass wool; after filtration, the placental pulp was resuspended in 200 ml of 1% NaCl and homogenized again in a Waring Blendor at high speed for 1 min. The homogenate was filtered again, using the same technique as before. The pulp was then discarded, the filtrates were pooled and measured and, for the separation from hemoglobin, added to the same volume of a suspension of DEAE-cellulose, prepared according to Hennessey et al. (557) and Hatfield and Wyngaarden (1009).

Preparation of DEAE-Cellulose. A 40-g volume of commercial DEAE-cellulose was suspended in 1000 ml of 1 M K_2HPO_4 (pH 9.1), stirred mechanically for 1 hr, and filtered with suction. This step was repeated seven times until the filtrate was colorless. The adsorbent was washed further with water until the pH of the filtrate was approximately 7.0. An aqueous suspension of the resin (approximately 8 g/100 ml) was adjusted to pH 7.0 and stored at room temperature. Immediately before it was used, DEAE-cellulose was washed by filtration with 1 volume of 1 M potassium phosphate buffer, pH 7.4, followed by 4 volumes of distilled water. DEAE-cellulose was then suspended in water so that the final solution contained 8 g of DEAE-cellulose (dry weight) per 100 ml of water.

Purification of Crude Placental Extracts on the DEAE-Cellulose Prepared as Described Above. A 500-ml volume of DEAE-cellulose solution of approximately pH 7.0 (8 g dry weight per 100 ml) was added to 500 ml of the crude placental extract. (It was previously demonstrated* that 90% of histaminase was adsorbed on DEAE-cellulose at pH values between 6.5 and 7.5.) This mixture was stirred manually every 4 to 5 min for 45 min and then poured into a Buchner funnel (13.5 $\times$ 6 cm) with glass wool serving as a filter pad. Two to three liters of 0.01 M potassium phosphate buffer, pH 7.4, were passed through the DEAE-cellulose in 200- to 300-ml portions in order to remove hemoglobin. When the eluate became colorless, the DEAE-cellulose was stirred gently with a final passage of 500 ml of the buffer.

The enzyme was desorbed from the DEAE-cellulose by placing the latter in a 1000-ml beaker containing 500 ml of 0.5 M potassium phosphate buffer, pH 7.4. The mixture was stirred manually every 5 min for 30 min and then poured into a Buchner funnel containing a glass-wool filter pad. After most of the solution had passed from the DEAE-cellulose, the latter was gently pressed with a flat glass surface to remove residual fluid. This elution step was repeated once more with an additional 500 ml of 0.5 M potassium phosphate buffer, pH 7.4. The two eluates were combined and measured and then poured onto solid ammonium sulfate (using 45.6 g per 100 ml of eluate) for precipitation at 0.6 saturation, and the mixture was stirred gently to minimize foaming. After the ammonium sulfate had fully dissolved, the precipitate formed was collected by centrifugation at 10,000 rpm for 15 min and redissolved in a minimum of cold water. The clear solution obtained was dialyzed for 9 hr against 6 liters of 0.002 M potassium phosphate buffer, pH 7.4, containing 0.001 M EDTA. It should be recalled here that placental extracts were previously found to contain compounds that strongly inhibit the enzymic effect on cadaverine and on putrescine but not that on histamine (538). Since this inhibitory effect on the placental enzyme activity with cadaverine and putrescine as substrate was completely abolished by the addition of 0.001 M EDTA, the latter was incorporated into all buffers used throughout the entire purification procedure as well as for the estimation of the enzymic activity.

First Fractionation with Ammonium Sulfate. To the dialyzed enzyme solution from the preceding step 22.8 g of ammonium sulfate per 100 ml of solution were slowly added with stirring to achieve 0.3 saturation. The pH of the mixture was adjusted to pH 7.4. After removal of the precipitate by centrifugation, the clear supernatant was brought,

---

* R. Kapeller-Adler and H. MacFarlane, unpublished data (1960).

at pH 7.4, to 0.75 saturation with ammonium sulfate. The resulting precipitate was collected by centrifugation, dissolved in a small amount of cold water, and the solution was dialyzed for 3 hr against 4.5 liters of 0.002 M potassium phosphate buffer, containing 0.001 M EDTA.

As already mentioned (Section 3.A.b), the specific activity of hog-kidney histaminase was found to be considerably increased on heating at 60° for 20 min the enzyme fractions, previously partly purified by ammonium sulfate precipitation. The loss of enzyme activity, especially with cadaverine as substrate, that occurred, when placental extracts purified by ammonium sulfate fractionation were subjected to heat treatment at 60°, was so substantial that this purification step had to be abandoned in the case of placental extracts.

Second Fractionation with Ammonium Sulfate. The dialyzed solution obtained at the preceding step from the precipitate formed at 0.75 ammonium sulfate saturation was subjected to a further fractionation with ammonium sulfate, and the solution containing the precipitate which formed between 0.3 and 0.75 saturation was collected, and the suspension stored at 4°. This enzyme fraction was stable at 4° for many months without any loss of activity, and was used as starting material for further purification of the placental enzyme by chromatography on DEAE-cellulose in the presence of 0.005 M borate–HCl buffer, pH 8.6 (538; see Section 3.A.b).

Chromatography on DEAE-Cellulose. An aliquot of the suspension obtained after the second ammonium sulfate fractionation was filtered; the precipitate was dissolved in cold water and dialyzed for 3 hr against 4.5 liters of 0.005 M borate buffer, pH 8.6, with two changes of the buffer. The dialyzed enzyme solution was applied to a column (28 $\times$ 1.5 cm) packed with DEAE-cellulose (Whatman, Column Chromedia, DE–11) previously equilibrated with 0.005 M borate–HCl buffer, pH 8.6, and was washed into the column with five 1-ml fractions of 0.005 M borate–HCl buffer. The column was developed with 900 ml of 0.005 M borate–HCl buffer, pH 8.6. Elution of the enzyme was accomplished by using a gradient of from 0 to 2.5 M NaCl in 0.005 M borate–HCl buffer. The effluents were collected every 10 min by a time-controlled fraction collector, and analyzed for their protein and enzyme contents.

Removal of histaminase from the DEAE-cellulose column could also be accomplished by substituting the rather steep gradient of 0 to 2.5 M NaCl in 0.005 M borate–HCl buffer, pH 8.6, by a simple elution with 0.1 M acetate buffer, pH 5.5. As mentioned in Section 3.A.b, at pH 8.6, which is the optimal hydrogen ion concentration for the quantitative adsorption of histaminase onto DEAE-cellulose, this enzyme seems to be very tightly bound to the cationic groupings of the resin. Complete desorption of histaminase from DEAE-cellulose columns can apparently be achieved only by using either a steep gradient of the eluting agent or by a change of the eluant to a lower pH range.

Effluents of chromatographic runs with highest specific activity were pooled and concentrated by osmosis using the simple technique of Hsiao and Putnam (1001). A dialysis casing containing the pooled eluates is covered with dry powdered sucrose for 1 hr. The sucrose is then removed from the dialysis tubing by rinsing with water and the contents dialyzed against 0.02 M phosphate buffer, pH 6.8, for 20 min. A threefold to fourfold concentration of the enzymic contents may be achieved.

The pooled concentrated enzyme fractions were subjected for further purification, either to recycling on DEAE-cellulose columns or to repeated fractionation with ammonium sulfate.

Reconditioning of DEAE-Cellulose. The resin is washed with 1000 ml of 1 M potassium phosphate buffer, pH 7.4, and then four times with 1000 ml of distilled water. The suspensions are filtered through Buchner funnels; 1000 ml of distilled water are added to the reconditioned DEAE-cellulose, and the suspension is stored at 4°.

Substrate Specificity. Throughout all the investigations of the various purification stages of placental histaminase the enzymic effect has been found to be highest on cadaverine with decreasing intensity on putrescine and histamine as substrates. This confirms previous findings (535–538, 548; see Chapter IV.1.E).

Taking into consideration all the properties of purified placental histaminase so far studied, it appears that the placental enzyme resembles more the DAO of pea seedlings than hog-kidney histaminase (538).

The method of purification, described above, results in a good yield of an 80 to 100 times concentrated placental histaminase (DAO) which is essentially free from hemoglobin and its derivatives, and which seems to be appropriate for further kinetic studies of this enzyme.

C. Methods for the Histochemical Detection of Histaminase (DAO) Activity in Tissue Sections

a. PROCEDURE ACCORDING TO VALETTE AND COHEN (1010).

It is said that aldehydes can be histochemically identified by the Feulgen reaction (1011), a staining method extensively employed by histologists and cytologists for the staining of cell nuclei and chromosomes. The Feulgen reaction depends on the fact that aldehydes restore the intense magneta color of basic fuchsin, previously discharged by means of sulfurous acid to form Schiff's reagent.

Oster and Schlossman (1012) were the first to use the Feulgen reaction for the detection of aldehyde formation during the MAO action on tyramine as substrate. Valette and Cohen (1010) likewise utilized this reaction for the histochemical detection of histaminase activity in various animal tissues. Although these workers were able to demonstrate that fresh animal tissues, when stained with the Feulgen reagent, displayed a color completely different from that formed with the same reagent during the action of histaminase on histamine, they nevertheless treated the tissue sections before subjecting them to the histaminase-histamine reaction with a 2% (w/v) solution of sodium bisulfite in order to block all the free aldehyde groups (1012).

Procedure. Fresh, frozen sections of various animal tissues, obtained immediately after the death of the animal were placed in a 2% (w/v) solution of sodium bisulfite and kept for 24 hr at 37°. After thorough rinsing with distilled water, the tissue sections were immersed at 37° for 24 hr in a 0.5% (w/v) solution of histamine dihydrochloride that had been previously adjusted to pH 7.2. For controls, other tissue sections were likewise incubated either in histamine-free buffer solutions, pH 7.2, or with histamine dihydrochloride in the presence of the histaminase inhibitor, semicarbazide hydrochloride (0.01%, w/v). After incubation all the tissue sections were washed with boiled distilled water and then immersed for 1 hr in the solution of Feulgen reagent. They were then washed twice with an 0.2% solution of sodium bisulfite and finally with freshly boiled distilled water. These tissue sections were then mounted in glycerine. Whereas the control sections displayed a uniformly distributed rose color, the tissue sections containing histaminase presented plaques of cells of deep violet color, surrounded by rose-colored tissue free of histaminase.

Valette and Cohen (1010) have studied with this technique the localization of histaminase in various organs of different animals. In guinea pigs they detected histaminase activity in the proximal convoluted renal tubules. The rat kidney, which does not contain histaminase, gave a negative test. The presence of histaminase was, however, revealed in the intestinal epithelium of the guinea pig as well as in the bronchioles of these animals. In human placenta, a positive Feulgen reaction was found to be restricted to the decidual cells, a completely negative result having been obtained in fetal membranes. This observation confirms the findings of Swanberg (542).

Preparation of Feulgen Reagent (1013). Dissolve 1 g of basic fuchsin in 200 ml of boiling water; filter, cool, and add 2 g of potassium metabisulfite ($K_2S_2O_5$), and 10 ml of 1 N hydrochloric acid. Let bleach for 24 hr, and then add 0.5 g of activated carbon (Norit), shake for about 1 min, and filter through coarse paper. The filtrate should be colorless.

b. PROCEDURE FOR THE QUANTITATIVE ESTIMATION OF HISTAMINASE
   LOCALIZED IN THE HUMAN UTERUS AND ATTACHED INTACT PLACENTA,
   AT TERM, ACCORDING TO GUNTHER AND GLICK (751)

These authors modified the spectrophotometric method of Aarsen and Kemp (1014) for the determination of histaminase, based on the measurement of the hydrogen peroxide that is formed during the enzymic oxidation of histamine (see Chapter II.2). Aarsen and Kemp employed o-dianisidine as the chromogen for hydrogen peroxide in a reaction in which the peroxide is decomposed by added peroxidase and the resulting oxygen converts the colorless o-dianisidine into an amber-colored product. For the detailed description of this modified technique of histaminase estimation see Section 3.D.e.

Gunther and Glick (751) adapted the modified technique of Aarsen and Kemp (1014) to their studies on histaminase in microtome sections of the human uterus and attached intact placenta, at term.

Preparation of Tissue Sections. A human uterus with unseparated placenta from a term pregnancy was obtained immediately on Caesarean hysterectomy. Blocks of tissue were cut from the uterine wall and the attached placenta, quickly frozen with solid carbon dioxide, and stored at —20°. Blocks of human placentae from vaginal deliveries were obtained and stored similarly. Sampling of fresh frozen cylinders of the tissue and their microtomy in a cryostat at —15°, to obtain circular sections 16 microns thick of 4mm in diameter and 0.2 μl in volume, was carried out by a conventional procedure used in Gunther and Glick's laboratory (1015).

Analytical Procedure. To obtain serial samples throughout the tissue, place one section on a slide for hematoxylin-eosin staining for histologic examination, the next five sections in a 27-mm reaction tube for the no-substrate blank, the next one on a slide for histologic study, and the next five in a tube for enzyme reaction, and repeat this pattern.

Constriction pipets are used for all volumes of 200 μl or less. The procedure is as follows:

1. Pipet 50 μl of 0.1 M sodium phosphate buffer, pH 6.2, freshly prepared before use, into each of the reaction tubes into which the tissue sections for the no-substrate blank were placed.

2. Pipet 40 μl of the same buffer into each of the tubes containing the five sections for the enzyme reaction.

3. Mix well, using a vibration mixer to break up the tissue.

4. In a 5-ml Pyrex tube mix 1.6 ml of 0.1 M sodium phosphate buffer, pH 6.2, 200 μl of horseradish peroxidase stock solution (Boehringer, No. POD 11, 15302, containing 0.4 mg peroxidase per milliliter of distilled

water, stored at $-20°$ up to a week), and 200 μl of o-dianisidine (Sigma) stock solution (containing 5 mg/ml of 95% ethanol, stored at $-20°$ up to 3 days). Pipet 50 μl of the above mixture into each of the tubes containing tissue and mix.

5. Pipet 10 μl of 0.3 M histamine solution (110 mg histamine dihydrochloride per milliliter of distilled water, containing 70 μl 5 N NaOH for neutralization and diluted twice with an equal volume of water) into each of the tubes for enzyme reaction, but not into the blank tubes.

6. Cap tubes, mix, and incubate at $37.5°$ for 8 hr.

7. Stop reaction by adding 50 μl of 50% sulfuric acid (v/v) to each tube, mix, and centrifuge at 1000 g for 2 min in an International clinical centrifuge, model CL, or in a Misco microcentrifuge (Microchemical Specialties, Berkeley, Calif.).

8. Measure absorbancy of the clear supernatant at 530 nm in a Beckman model DU spectrophotometer.

9. Subtract the absorbancies of no-substrate blanks from those of corresponding enzyme reactions to obtain activity, and express activity as the corrected change in absorbancy per unit time.

With a number of samples from different individuals a linear relationship was established between the number of placental tissue sections (up to nine sections tested) and enzyme activity.

With regard to the quantitative determination of histaminase activity in a human uterus with attached intact placenta, at term, the highest activity was encountered in the decidua and in Nitabuch's membrane. These findings compare well with conclusions drawn by Swanberg (542) concerning the localization of histaminase in the placenta, and suggest that the decidua is a significant source of the increased plasma histaminase in human pregnancy (542).

D. Methods for the Quantitation of Histaminase (DAO)

a. INTRODUCTION

The most commonly used methods for the estimation of histaminase (DAO) have been reviewed by Zeller (126) and Werle (130).

All the components of the fundamental histaminase–substrate reaction (150); see Chapter II.2),

$$RCH_2NH_2 + O_2 + H_2O \rightarrow RCHO + NH_3 + H_2O_2,$$

with the exception of water, have been utilized as a basis for the determination of the activity of this enzyme. Many different enzymic units have been proposed by various workers and are still used to express the results of activity measurements. They should be, however, brought up to date and replaced by units that comply with the recommendations of the International Union of Biochemists (919).

The methods published for the estimation of histaminase (DAO) activity include those that measure the amount of substrate disappearance or the oxygen consumption during the enzymic reaction. Other methods have been devised that are based on the determination of aldehyde, hydrogen peroxide, or ammonia. Two types of highly sensitive enzymic methods based on measurements of substrate disappearance have so far been published. However, both kinds refer only to estimations of the amount of histamine that disappears during the action of histaminase on this substrate. One of these methods involves a biological determination, and the other a spectrofluorometric estimation, of histamine.

To the manometric method of histaminase estimation, measuring the oxygen consumption during the enzymic reaction, the same criticism applies that was mentioned in the discussion of techniques for the estimation of MAO (see Section 2.D.a). The manometric method is not sensitive enough to measure adequately the low activities of histaminase occurring in many biological materials. Moreover, in some instances part of the oxygen consumption may be due not only to histaminase, but to further oxidation of the primary enzymic product by other enzymes; for example, the oxidation of imidazoleacetaldehyde to imidazoleacetic acid. Whereas in experiments with MAO this extraneous oxygen uptake can be prevented by means of added cyanide or semicarbazide, this technique cannot be applied to histaminase (DAO), which itself is inhibited by these substances. The appearance of an aldehyde as a consequence of the histaminase action on histamine was ascertained by the isolation of imidazoleacetaldehyde dinitrophenylhydrazone (133). In the oxidative deamination of aliphatic diamines by DAO, the aldehyde metabolite spontaneously reacts with the remaining amino group to form a cyclic Schiff base (azomethine), Δ^1-pyrroline being formed from putrescine (551, 601, 602, 869, 887) and Δ^1-piperideine from cadaverine (869, 887). The production of these compounds can be demonstrated by means of o-aminobenzaldehyde, which condenses with the cyclic azomethine to form yellow or orange 1,2-dihydroquinazolinium derivatives (551, 602, 890).

Ammonia determinations have been carried out in histaminase (DAO) systems by various methods that are not fully specific for ammonia, and are carried out in a strongly alkaline medium so that sensitive intermediate products may undergo degradation (122, 126). Both disadvantages can be overcome by an enzymic method based on the application of the ammonia, formed, to the amination of α-ketoglutaric acid by glutamic dehydrogenase in the presence of $NADH_2$ (1016).

The hydrogen peroxide, formed in the histaminase (DAO) reaction, can act by coupled oxidation on compounds such as ethanol (1017), indigo disulfonate (505, 607), o-dianisidine (751, 1014, 1018), cerous hydroxide (516), and many other substances.

b. BIOLOGICAL METHOD FOR THE ESTIMATION OF HISTAMINASE ACTIVITY
   BY MEASURING HISTAMINE DISAPPEARANCE, ACCORDING TO SPENCER (1019)

The effects of enzyme and substrate concentrations on the rate of inactivation of histamine by histaminase of rat ileum has been investigated by this author. After having demonstrated that the time necessary for one-half of the histamine to be destroyed is inversely proportional to the histaminase content of the incubation mixture, Spencer has used this fact as a basis for a new method for the estimation of histaminase activity in tissue extracts.

Procedure. Male Wistar albino rats weighing 120 to 160 g were used. The small intestine was removed from groups of four or five rats immediately after killing the animals by a blow on the head. After discarding the first 6 and the last 2 in., the tissue was washed with normal saline solution and weighed. The pooled tissue was then ground in a glass mortar with a little silver sand. Tyrode solution (5ml/g) was added, and the extraction was carried out for 5 min; the suspension was then centrifuged at 2800 rpm for 10 min. The separated supernatant contained the expected enzyme. More prolonged grinding, extraction, and centrifugation did not increase the yield of the enzyme. To 95 ml of Tyrode solution in a 250-ml conical flask, containing a known amount of histamine and equilibrated to 37°, 5 ml of the histaminase extract was added. After shaking the flasks to mix the content, a first sample was immediately removed and the remainder incubated with frequent shaking at 37°. Further samples were removed at 10-min intervals. Each sample was brought to the boil to arrest enzyme activity, cooled, and stored at 4° until it was assayed for its histamine content. In experiments in which the effect of changes in enzyme or substrate concentration were investigated, different volumes or quantities of extract and histamine were used, and the volume of Tyrode solution varied so that all incubation mixtures were initially 100 ml in volume. Throughout the first 90 min of incubation, the mixture remained within the limits pH 7.0 to 7.4 For the estimation of histamine the samples were assayed directly on the atropinized ileum of the guinea pig. All estimations were in duplicate and the mean values were used for the calculation of results.

The results obtained showed that an exponential relationship exists between histamine concentration and length of incubation.

The following method is proposed by Spencer for the estimation of histaminase activity. Histamine concentrations are determined at 0 time and at intervals after and a plot of $\log_{10}(x_0/x_t)$ against t is made. x_0 is the initial histamine concentration and x_t is the histamine concentration after incubation for time t respectively. A linear relationship should prevail over most of the reaction, at least as far as 70 to 80% of histamine inactivation. The time taken for 50% of the histamine to be destroyed (or another fixed percentage within the linear part of the graph) is read; this is the DT_{50} value. The ratio of enzyme activity (E) in different tissues is then

$$\frac{E_1}{E_2} = \frac{DT_{50}\ (2)}{DT_{50}\ (1)}$$

To express the percentage of control tissue activity the following equation is used:

$$E\ \text{(experimental)} = \frac{DT_{50}\ \text{(control)}}{DT_{50}\ \text{(experimental)}} \times 100.$$

With this method maximum errors of $\pm 7\%$ have been obtained by the author, using known amounts of tissue extract in different incubation mixtures. If an initial histamine concentration within the range of 1.50 to 2.50 $\mu g/ml$ is applied, then dilution before assay is possible throughout the range of samples obtained, and this excludes difficulties in assay due to interference by tissue proteins. Spencer carried out pilot experiments investigating histaminase activity from rat lung by this new method. An exponential relationship between histamine concentration and length of incubation similar to that obtained with the rat ileum was found. The results obtained showed the lung to possess about one-fifth the activity of the ileum. Mouse ileum and lung show similar exponential curves, and the ratio of activity is approximately the same as in the rat. Thus, by converting the exponential relationship between histamine concentration and incubation time to the linear one of $\log_{10}(x_0/x_t)$ against t, a constant (the slope of the new graph) for a particular incubation mixture is obtained, and the time taken to achieve a certain fraction of inactivation depends on the enzyme content of that mixture.

Spencer points out the advantages of his new method as follows:

1. The DT_{50} time depends on a constant and does not change for a given concentration of enzyme.

2. The value is obtained from data from all the samples assayed for histamine in a given mixture. Thus, one false assay only slightly affects the result, although the initial sample is perhaps the most important for any mixture.

3. Incubation mixtures containing different initial amounts of histamine can be compared, providing they fall within the range 75 to 150% of the mean initial histamine concentration. Varying amounts of endogenous histamine will not alter the result.

4. A maximum experimental error of less than 10% can be expected.

Preparation of Tyrode Solution (1020). The following stock solutions are made up:

Salt	Quantity (g/l)
Sodium bicarbonate, $NaHCO_3$	40.00
Magnesium chloride, $MgCl_2 \cdot 6H_2O$	8.54
Sodium acid phosphate, $NaH_2PO_4 \cdot H_2O$	2.00
Calcium chloride, $CaCl_2$	8.00
Potassium chloride, KCl	8.00
Sodium chloride, NaCl	80.00

For the histamine estimation on the guinea-pig ileum, place in a 1000-ml measuring flask 25 ml each of the first five stock solutions, and add 100 ml of the sodium chloride stock solution. After adding 1 g of glucose and 1 ml of an 0.01% (w/v) aqueous solution of atropine, make up to volume with distilled water.

As to the usefulness of the highly sensitive methods of histaminase estimation based on measurements of histamine disappearance, Zeller (126) points out that complementary experiments must be carried out in order to determine that histaminase alone is responsible for the disappearance of histamine.

As has been reported, histamine undergoes chemical changes and biological inactivation not only through histaminase but also through other metabolic processes, such as ring methylation (638), acetylation (626), and incorporation into proteins (1021) and into NAD-like dinucleotides (654). Hence, if histamine serves as the sole substrate and if no tests with specific inhibitors are carried out, it is often difficult to conclude whether and to what degree the results obtained refer to histaminase and/or to other histamine-degrading systems.

C. CHEMICAL METHOD FOR THE ESTIMATION OF HISTAMINASE ACTIVITY BY MEASURING THE DISAPPEARANCE OF HISTAMINE BY A SPECTROFLUOROMETRIC TECHNIQUE, ACCORDING TO SHORE AND COHN (521) AND SHORE, BURKHALTER, AND COHN (731)

Histaminase activity was estimated by measuring by a simple fluorometric method the rate of disappearance of histamine added to tissue homogenates.

This very sensitive method involves extraction of histamine into *n*-butanol from alkalinized solutions, return of the histamine to an aqueous solution and its condensation with *o*-phthalaldehyde (OPT) to yield a product with strong and stable fluorescence, measured in a spectrofluorometer.

Procedure. Tissues are homogenized in 9 volumes of 0.2 M phosphate buffer, pH 7.2, and after 10 min the homogenate is centrifuged. After preincubation in a Dubnoff metabolic shaker for 15 min, at 37°, in an atmosphere of air, histamine is added to make a final concentration of 7.5 μg/ml, and the mixture is incubated at 37° for 60 min. Samples are then removed for the estimation of histamine. For the extraction of the residual histamine to be assayed add to 4 ml of a sample in a 25-ml glass-stoppered shaking tube, 0.5 ml of 5 N NaOH, 1.5 g of solid NaCl, and 10 ml of *n*-butanol. The tube is shaken for 5 min to extract the histamine into the butanol, and, after centrifugation, the aqueous phase is removed by aspiration. The organic phase is then shaken for about 1 min with 5 ml of salt-saturated 0.1 N NaOH. This wash removes any residual amounts of histidine that may be present. The mixture is then centrifuged and an 8-ml aliquot of the butanol is transferred to a 40-ml glass-stoppered shaking tube, containing 2.5 to 4.5 ml of 0.1 N HCl and 15 ml of *n*-heptane. After shaking for about 1 min, the tube is centrifuged, and the histamine in the aqueous phase is assayed fluorometrically.

Fluorometric Assay. To estimate histamine in the acid extract, a 2-ml aliquot containing 0.005 to 0.5 μg histamine per milliliter is transferred to a test tube and 0.4 ml of 1 N NaOH is added, followed by 0.1 ml of OPT reagent [1% (w/v) of *o*-phthalaldehyde, Mann Research Laboratories Inc., New York, in reagent-grade absolute methanol]. After 4 min, 0.2 ml of 3 N HCl is added. The contents of the tube are thoroughly mixed after each addition. The solution is then transferred to a cuvette, and the fluorescence at 450 nm resulting from activation at 360 nm is measured in a spectrofluorometer (wavelengths uncalibrated). The fluorescence of the acidified solution is stable for at least 30 min. The fluorescence intensity is proportional to histamine concentration over the range of 0.005 to 0.5 μg/ml. The small fluorescence blanks in tissues or reagents may be corrected by omission of the condensation step. This is achieved by adding all the reagents to a separate aliquot of tissue extract, but reversing the order of addition of OPT and 3 N HCl. The resulting solution contains tissue blanks and reagent blanks, but not the fluorophore resulting from condensation of OPT and histamine, since the condensation reaction does not proceed in an acid medium.

Histidine and histidyl histidine produce fluorophores spectrally similar to that of histamine, when treated with OPT, but do not interfere with the histamine estimation, since these compounds are not extracted into butanol from alkaline solution. Ammonia at concentrations greater than 4 μg/ml produces some fluorescence, but is poorly extracted into butanol and consequently does not interfere in the analysis of tissue or blood. Fluorescence readings not exceeding those of reagent blanks were obtained, when as much as 150 μg of ammonia were run through the entire procedure.

In urine, however, there is sufficient ammonia present to interfere with the procedure described above.

This condensation reaction of histamine with o-phthalaldehyde seems to be remarkably specific for N-unsubstituted imidazoleethylamines. By this sensitive method amounts of histamine as low as 0.005 μg/ml can be assayed.

Kremzner and Wilson (1022) have published a modification of the spectrofluorometric method, just described. According to Kremzner and Wilson, the modified method is simple and accurate, and, in contrast to other techniques, yields complete recovery of histamine. This technique is based on the fact that amino acids exist in moderately alkaline solutions partly as anions, whereas histamine is present in such solutions in the form of neutral and positively charged molecules. Using a strongly basic anion-exchange resin such as Dowex–I–X8, BioRad–I–X8, or IRA–410 (50–100 mesh) converted on a large column to the acetate–hydroxide form with several volumes of a mixture of 0.75 M sodium acetate and 0.05 M NaOH (15 : 1 molar ratio), the authors have been able to demonstrate a complete separation of histidine (and other interfering substances that remained on the column) from histamine, collected in the effluents of the column. The histamine eluted from the column is estimated by the fluorometric method of Shore et al. (731).

d. SPECTROPHOTOMETRIC METHOD FOR THE DETERMINATION OF DAO, BASED ON THE ALDEHYDE FORMED WHEN PUTRESCINE IS OXIDIZED BY DAO, ACCORDING TO HOLMSTEDT AND THAM (601) AND HOLMSTEDT, LARSSON, AND THAM (602).

This method is based on the formation of γ-aminobutyraldehyde during the action of DAO on putrescine. This aldehyde undergoes a rapid cyclization to Δ^1-pyrroline, which compound is allowed to react with o-aminobenzaldehyde; a yellow-colored product, 2,3-trimethylene-1,2-dihydroquinazolinium hydroxide, is formed (551, 602, 890; see Section 3.D.a). This colored reaction product shows a strong absorption maximum at 430 nm. To determine its molar extinction coefficient, o-aminobenzaldehyde was allowed to react with varying concentrations of γ-aminobutyraldehyde diethylacetal for 3.5 hr, whereupon the reaction mixture was diluted to 10 ml with phosphate buffer, pH 6.8, and read at 430 nm. A standard calibration curve was obtained by plotting the absorbancy against the concentration of γ-aminobutyraldehyde diethylacetal, which is expected to be equivalent to the concentration of the colored product, mentioned above. Its molar extinction coefficient was found to be 1.86 $\times$ 10^3 mole $^{-1}$ cm $^{-1}$.

Method. The following reagents are required:

1. o-Aminobenzaldehyde, 0.005 M, synthesized according to Bamberger and Demuth (1023, 1024), dissolved in phosphate buffer 0.067 M, pH 6.8, and stored at $-8°$ in ampoules containing nitrogen.

2. Putrescine dihydrochloride, 0.1 M, dissolved in phosphate buffer, pH 6.8 (Light and Co., Ltd., London).

3. Phosphate buffer (Sörensen) 0.067 M, pH 6.8.

4. Trichloroacetic acid (10%, w/v).

5. Enzyme preparation: a catalase- and peroxidase-free DAO preparation of high specific activity was obtained from the hog kidney according to Arvidsson et al. (992).

Procedure. In a test tube 2.5 ml of a solution of 0.005 M o-aminobenzaldehyde in 0.067 M phosphate buffer, pH 6.8, (stable at 4° for a week), and the enzyme preparation (2 to 20 mg) dissolved in phosphate buffer, pH 6.8, and made up in the same buffer to 4.5 ml were shaken at 37°. After temperature equilibration, 0.5 ml of putrescine dihydrochloride (0.1 M freshly prepared in phosphate buffer, pH 6.8) was added. The mixture was incubated at 37° for 3 hr. As controls, assays without the addition of substrate were incubated along with the tests. At the end of incubation, 1 ml of 10% trichloroacetic acid was added to both tubes, containing the test and the control, to stop the enzyme reaction (at this stage the tubes can be kept for 24 hr). After centrifugation, the absorbancy of both supernatants was read in a Beckman model DU spectrophotometer, at 430 nm, in 1-cm cuvettes. When the absorbancy exceeded 0.8, the solution was diluted with distilled water as in ordinary photometric techniques.

The standard calibration curve makes it possible to convert the extinction readings of the dye, formed in the o-aminobenzaldehyde–putrescine–

DAO reaction, into micromoles of putrescine destroyed per milligram of enzyme in 60 min, which units express the DAO activity.

According to Holmstedt and his colleagues (602), this method is simple to perform, requires little apparatus, and many determinations can be carried out simultaneously. Moreover, it seems to lend itself also to the study of the effect of inhibitors on DAO. Furthermore, in tissue homogenates and partly purified enzymes the spectrophotometric measurements are uninfluenced by the occurrence of other oxidative enzymes, such as catalase and peroxidase.

e. MICROSPECTROPHOTOMETRIC DETERMINATION OF HISTAMINASE
ACTIVITY IN TISSUE HOMOGENATES BASED, ON MEASUREMENTS OF THE
HYDROGEN PEROXIDE FORMED BY THE ENZYMIC OXIDATION OF
HISTAMINE. METHOD OF AARSEN AND KEMP (1014), AS
MODIFIED BY GUNTHER AND GLICK (751)

Aarsen and Kemp (1014) devised a microspectrophotometric method for the estimation of histaminase in minute amounts of plasma (0.05–0.1 ml) of small animals. They employed o-dianisidine as the chromogen for the hydrogen peroxide in a reaction in which the peroxide is decomposed by added peroxidase, and the resulting oxygen converts the colorless o-dianisidine into an amber-colored reaction product, measured in a spectrophotometer at 470 nm (see Section 3.C.b).

Gunther and Glick (751) adapted the technique of Aarsen and Kemp to their studies in homogenates of human placental tissue after having modified the method and established optimal conditions for their assays.

Preparation of Tissue Homogenates. Human placentae from women with normal pregnancies were obtained immediately after delivery. The cord blood was allowed to drain prior to delivery of the placenta. The maternal surface was gently rinsed with cold saline and blotted with a paper towel to remove excess blood. To remove fetal blood, the umbilical arteries were gently perfused with cold saline until the return fluid was clear. Small shallow pieces of placenta, including attached decidua, were cut from the maternal surface. These pieces were minced with scissors, weighed, and homogenized with an ice-cold Teflon homogenizer, distilled water being added to give a final pulp concentration of 200 mg/ml. The homogenate was stored in 0.5-ml tubes at −20°.

Rat-stomach homogenates were prepared and stored in a similar fashion.

For use, the frozen material was brought to room temperature, centrifuged briefly, and the residue was discarded.

A suspension of isolated rat peritoneal mast cells in Hank's solution was obtained by methods, previously described by Glick and co-workers (1025,

1026) and 0.5-ml (2.4 $\times$ 10^6 cells) portions were used directly as well as after freeze thawing three times, to disrupt cells.

Analytical Procedure.

Apparatus. Homogenates were prepared in glass tubes with an electrically driven Teflon pestle. Constriction pipets were used for all volumes from 200 μl downward. Mixing was effected by a vibration mixer. Centrifugations were carried out with an International clinical centrifuge, model CL, or a Misco centrifuge (Microchemical Specialties, Berkeley, Calif.) with variable transformer for speed control. Incubations were carried out in a constant-temperature water bath at 37.5°. Absorbancy was measured with a Beckman spectrophotometer, model DU, using microcells with a sample channel of 1.5 $\times$ 10 mm.

Reagents. The following reagents are required:

1. Horseradish peroxidase (Boehringer, No. POD–11, 15302), stock solution: 0.4 mg/ml in distilled water, which may be stored at —20° up to 1 week.

2. *o*-Dianisidine (Sigma) stock solution: 5 mg/ml in 95% ethanol, which may be stored at —20° up to 3 days.

3. Histamine dihydrochloride solution (0.6 M): 110 mg/ml of distilled water containing 70 μl 5 N NaOH for neutralization.

4. Sodium phosphate buffer 0.1 M, pH 6.2.

5. Sulfuric acid 50% (v/v), (or 3 N hydrochloric acid, used in some preliminary tests).

6. Aminoguanidine sulfate, 0.6 M stock solution: 74 mg/ml of distilled water, which may be stored at 4°.

Procedure. In a 2-ml reaction tube pipet into 0.75 ml of buffer 50 μl each of the reagent solutions of peroxidase, *o*-dianisidine, and histamine. Prepare a no-substrate blank by replacing the histamine solution with buffer. Place the tubes in a water bath at 37.5° for 2 min, add 100 μl of the unknown sample solution to each tube, mix, seal tubes with Parafilm, and place in the 37.5° water bath. At intervals from 0 to 18 hr transfer 100 μl reaction mixture from each tube into a 27-mm tube and add 100 μl of 50% sulfuric acid (10 μl of 3 N HCl were used in some preliminary tests instead of 100 μl of 50% H_2SO_4), mix, and spin in the microcentrifuge at 1000 g for 2 min. Measure absorbancy of the clear supernatant at 530 nm (at 400 nm in assays in which 3 N HCl was used instead of 50% H_2SO_4). Correct the absorbancies of the unknowns for those of the no-substrate blanks and express histaminase activity as the corrected change in absorbancy per unit time. The absorbancy in enzyme blanks (i.e., in solutions, in which the homogenate or histaminase preparation have been replaced by an equal volume of water) was found to be negligible. However, it was essential in placental homogenates, to run a no-substrate blank with each determination, since significant absorbancy was obtained at 530 nm with homogenates as well as enzyme preparations as such.

When estimating D-*glucose* by means of glucose oxidase in the o-dianisidine–peroxidase method, McComb and Yushok (1027) found that the use of sulfuric acid for the terminal acidification in the glucose oxidase assay increased the sensitivity of the method and changed the color of the reaction product from amber to deep pink. In the presence of 30% sulfuric acid a sharp absorption peak was obtained at 520 nm.

According to Saifer and Gerstenfeld (1028), the addition of sulfuric acid is necessary to stabilize the spectrophotometric readings, and serves both to stop the enzymic reaction and to eliminate any turbidity formed during the reaction.

Messer and Dahlquist (1029), who developed an ultramicromethod for the assay of intestinal disaccharidases, employed in the glucose oxidase reaction a final sulfuric acid concentration of 16% and found that the absorbancy had a peak value at 530 nm.

In the Gunther and Glick histaminase method, just described, adding 50% sulfuric acid to the reaction mixture on terminating incubation, gave a final acid concentration of 25% and maximum absorption at 530 nm, thus confirming the results of McComb and Yushok (1027) and of Messer and Dahlquist (1029).

Whereas plasma assays remained clear during the incubation procedure, Gunther and Glick (751) observed that with homogenates cloudiness developed during incubation in some reaction mixtures. Although the cloudiness could be cleared by centrifugation, this procedure did not appear to be profitable because of losses of color from the supernatant due to aggregation and insolubility of reaction products.

Guidotti and colleagues (1030) have demonstrated that, when hydrochloric acid is used for the terminal acidification below pH 2, the formation of cloudiness could be prevented because of solubilization of the material on addition of 3 N HCl. In the presence of 3 N HCl the color of the reaction changed from amber to a deep yellow with a shift of the absorption peak to 400 nm. Gunther and Glick (751) confirmed the observation of Guidotti et al. On comparison of the effects of sulfuric acid and hydrochloric acid on the absorption peak they found that a greater absorbancy was obtained with sulfuric acid than with hydrochloric acid, and, therefore, they changed from hydrochloric acid to sulfuric acid.

Although most of the histaminase assays, including that of Aarsen and Kemp (1014), have been carried out at a pH of about 7.0 (126), Gunther and Glick (571) demonstrated that under the conditions of their assay the pH optimum was at 6.2. Furthermore, whereas Aarsen and Kemp used in their assay the *o*-dianisidine reagent at a final concentration of 0.05 mg/ml, Gunther and Glick achieved in placental homogenates maximal enzyme activity only when employing 0.25 mg/ml. The optimal peroxidase concentration was found to be 0.02 mg/ml of reaction mixture. At twice this concentration the apparent histaminase activity decreased by 15 to 30% over an incubation period of 26 hr.

Inhibitor studies carried out by Gunther and Glick (571) in placental extracts with various concentrations of aminoguanidine sulfate have indicated an inhibition of histaminase of 97% in the presence of equimolar amounts of substrate and inhibitor, and an inhibition of 99% when the concentration of aminoguanidine sulfate was twice that of histamine. A linear relationship between activity and time at all concentrations of inhibitor, tested over an 18-hr incubation period, was maintained.

In a recent paper Möller and Ottolenghi (1031) report on the nature and properties of the reaction product, obtained on the oxidation of *o*-dianisidine by peroxide and peroxidase.

f. MICROSPECTROPHOTOMETRIC ESTIMATION OF HISTAMINASE (DAO),
    IN BIOLOGICAL FLUIDS, INVOLVING A PEROXIDATIC OXIDATION OF
    INDIGO DISULFONATE. METHOD OF KAPELLER-ADLER (548)*

The method to be described is a modification of a microvolumetric technique that was designed in 1951 for the determination of histaminase activity in biological media. It involves a coupled oxidation reaction in which indigo disulfonate is oxidized by the hydrogen peroxide, formed in the primary reaction (553, 998, 1032). The excess of indigo disulfonate not oxidized by the hydrogen peroxide is determined by titration with 0.002 N KMnO$_4$ solution. The amount of 0.002 N KMnO$_4$ used for the blank (complete assay without the addition of substrate) minus that used for the assay indicated histaminase activity, expressed in units, one unit representing the amount of enzyme that under standard assay conditions produced the amount of hydrogen peroxide equivalent to 0.1 ml of 0.002 N KMnO$_4$. Since in this stoichiometric reaction one molecule of hydrogen peroxide is formed for each molecule of substrate oxidized (998, 1032; see Section 3.D.a), the enzymic unit, mentioned above, corresponds to the destruction of 0.436 μg of histamine per hour or of 0.425 μg of cadaverine per hour.

The microvolumetric technique, just described, has been criticized, especially by Zeller (126), on two grounds. First, Zeller maintains that the oxidation of indigo disulfonate by potassium permanganate is, in spite of its apparent simplicity, a complex reaction (135). Second, in an especially developed test system (containing phosphate buffer, pH 7.2, 0.2

---

* Also unpublished data by the author (1964).

mM indigo disulfonate, 0.01 M hydrogen peroxide, and millimolar amounts of copper sulfate) Zeller and Buerki (1033) were able to demonstrate that an addition of millimolar quantities of histamine to their system enhanced the catalytic power of copper sulfate.

This remarkable peroxidase-like power of histamine was found to be absent in other substrates of DAO, e.g., cadaverine (1033). Hence, according to Zeller (1033, 1034), histamine acts not only as a substrate of histaminase, but also as a catalyst of the coupled oxidation reaction of indigo disulfonate.

Zeller's objection to the estimation of the residual indigo disulfonate by titration with potassium permanganate is avoided in the spectrophotometric modification of the original indigometric method, discussed below. Zeller's concept that, in contrast to cadaverine and putrescine, imidazole derivatives such as histidine or histamine together with Cu^{2+} ions exert a remarkable peroxidase-like effect on the indigo disulfonate–hydrogen peroxide system cannot be supported by the present author, since in human placental extracts, pregnancy sera, and amniotic fluids, throughout various purification procedures the DAO effect remained highest on cadaverine as substrate and continued to be very small indeed when histamine served as substrate in the indigo disulfonate reaction (548).*

The modified technique of histaminase (DAO) estimation has been greatly improved by a considerable reduction of the end volume of the assay (i.e., from 10 to 1.5 ml), further by reducing the incubation time at 37° from 24 hr to 60 min, and by replacing the microvolumetric estimation method of the residual indigo disulfonate with 0.002 N KMnO$_4$ by a spectrophotometric one, carried out at 613 nm, the wavelength of maximum absorption of indigo disulfonate. The value of ϵ, the molar absorption coefficient ($\times 10^{-4}$), is for indigo disulfonate, at pH 7.0, and 613 nm, 2.22 (1035). Finally, the enzyme activity is expressed in international milliunits (ImU), one milliunit corresponding to the degradation of 1 nanomole of substrate per minute.

The theory underlying the spectrophotometric estimation method of histaminase (DAO), may be described as follows:

In the primary reaction between enzyme and substrate, expressed by the equation

$$RCH_2NH_2 + O_2 + H_2O \rightarrow RCHO + NH_3 + H_2O_2,$$

one molecule of hydrogen peroxide is formed from each molecule of substrate (1017, 1036).

In the subsequent coupled oxidation reaction the hydrogen peroxide thus generated oxidizes the added indigo disulfonate. It is well known

---

* The author also has unpublished data (1964) on this subject.

that indigo is oxidized by oxidizing agents to isatin (1037) according to the equation

$$C_{16}H_{10}N_2O_2 + 2\,0 = 2C_8H_5NO_2.$$

On the basis of these equations it is assumed that two molecules of hydrogen peroxide formed from two molecules of substrate in the enzyme–substrate reaction will oxidize one molecule of indigo disulfonate.

The molecular weight of indigo disulfonate is 466; thus, half a micromole of this compound will amount to 233 μg. Furthermore, since one molecule of indigo disulfonate requires two atoms of oxygen for its oxidation, 1 ml of a 0.002 N indigo disulfonate solution contains 233 μg of indigo disulfonate, which corresponds to half a micromole of this compound. Hence, it is concluded that 1 ml of 0.002 N indigo disulfonate oxidized by hydrogen peroxide in the coupled oxidation reaction described, corresponds to a destruction of 1 micromole of substrate and, if determined per time unit of 1 min, forms the basis for the international unit of histaminase (DAO). Thus, the consumption of 1 ml of 0.002 N indigo disulfonate indicates the destruction of 1 micromole of histamine (111 μg), cadaverine (102 μg), or putrescine (88 μg) per minute. Equally, the consumption of 1 μl of 0.002 N indigo disulfonate indicates the destruction of 1 nanomole of histamine (0.111 μg), cadaverine (0.102 μg), or putrescine (0.088 μg) per minute, and is equivalent to 1 international milliunit.

Protein was determined by a modification of the method of Lowry et al. by Miller (997).

The specific activity is defined as enzyme units per milligram of protein.

Materials and Methods. Human placental extracts, amniotic fluids, and pregnancy sera were examined by the spectrophotometric method for their enzymic effects on histamine, cadaverine, and putrescine. The placental extracts were prepared according to the method outlined in Section 3.B, the amniotic fluids were obtained either by amniocentesis or puncture of the amniotic sac, and pregnancy sera were obtained from women at various stages of pregnancy.

Reagents. The following reagents are required:

1. Buffers: (a) 0.175 M sodium phosphate buffer, pH 6.8; (b) EDTA–phosphate buffer: 0.175 M sodium phosphate buffer, containing 0.001 M EDTA, pH 6.8.

2. Indigo disulfonate solution. Dissolve 23.3 mg of indigo carmine (Analar, British Drug Houses, $C_{16}H_8O_8N_2S_2Na_2$, molecular weight 466.37) in 100 ml of distilled water. The solution keeps well at 4° for 2 to 3 weeks. This 0.002 N indigo disulfonate solution contains 23.3 μg of indigo disulfonate per 0.1 ml.

3. Substrates. The following substrates are used:

Histamine dihydrochloride (British Drug Houses). Dissolve 27.6 mg of histamine dihydrochloride in 10 ml of phosphate buffer, pH 6.8; 0.1 ml contains 0.276 mg of histamine dihydrochloride. Add a few drops of chloroform for preservation. The solution keeps at 4° for a few days.

Cadaverine dihydrochloride (1,5-diaminopentane dihydrochloride, purissimum, Fluka A. G., Buchs, Switzerland). Dissolve 525 mg of cadaverine dihydrochloride in 20 ml of EDTA–phosphate buffer, pH 6.8, and add a few drops of chloroform for preservation. Keeps at 4° for 2 to 3 weeks; 0.2 ml contains 5.25 mg of cadaverine dihydrochloride.

Putrescine dihydrochloride (1,4-diaminobutane dihydrochloride, purum, Fluka A. G.). Dissolve 483 mg of putrescine dihydrochloride in 20 ml of EDTA–phosphate buffer, pH 6.8, and add a few drops of chloroform. Keeps at 4° for 2 to 3 weeks; 0.2 ml contains 4.83 mg of putrescine dihydrochloride.

Procedure. As mentioned in Section 3.B, placental extracts were found to contain substances that strongly inhibited the enzymic effect on cadaverine and putrescine but not that on histamine (538). Since this inhibitory effect on the enzyme activity with cadaverine and putrescine as substrates, present also in other biological media, was completely abolished by the addition of 0.001 M EDTA, the latter compound was incorporated into all buffers used for the estimation of enzyme activity.

Dialysis. About 10 ml of the biological fluid to be examined (i.e., partly purified placental extract—see Section 3.B—amniotic fluids, or pregnancy sera, completely free of hemolysis) are first dialyzed for 3 hr against 5 liters of 0.175 M EDTA–phosphate buffer. The dialyzed fluids are then centrifuged, and the supernatant must be water-clear for the enzymic estimations.

Optimal Substrate Concentrations. Optimal substrate concentrations were found to be 20 mM cadaverine dihydrochloride (5.25 mg of cadaverine dihydrocholoride, 30 micromoles/1.5 ml end volume), 20 mM putrescine dihydrochloride (4.83 mg of putrescine dihydrochloride, 30 micromoles/1.5 ml end volume), and 1 mM histamine dihydrochloride (0.276 mg of histamine dihydrochloride, 1.5 micromoles/1.5 ml end volume).

In all assays the gas phase was air.

Assay Using Dialyzed Placental Extracts or Amniotic Fluids. In a Pyrex test tube (2 $\times$ 15 cm), marked "a", place 0.2 ml of dialyzed enzyme solution and add 0.6 ml of the 0.002 N indigo disulfonate solution. In another test tube, marked "b", place multiples of a mixture consisting of 0.2 ml of cadaverine dihydrochloride or putrescine dihydrochloride solution and 0.5 ml of EDTA-phosphate buffer. When histamine is to serve as substrate, place in test tube "b" multiples of a mixture consisting of 0.1 ml of histamine standard solution and 0.6 ml of phosphate buffer, pH 6.8.

Preincubate all the tubes marked "a" and "b" in a thermostatically controlled water bath at 37° for 15 min. Then pipet carefully into each "a" tube 0.7 ml of the substrate–buffer mixture, present in the "b" tube, mix very gently by inversion, and stopper the tubes. Incubate for 60 min., at 37°. Blanks are set up in the same fashion as tests, with the exception, that for the blanks test tube "b" contains only buffer without the addition of the substrate. End volume in each blank and test is 1.5 ml.

To stop the enzymic reaction, when incubation is completed, place all the tubes containing tests and blanks in a beaker of ice-cold water and rapidly dilute the fluid in each tube to 10 ml by adding 8.5 ml of distilled water. Stopper the tubes, mix gently by inversion, and read blanks and tests in a spectrophotometer (Optica CF4) at 613 nm against a dilute phosphate buffer solution, pH 6.8 (1.5 ml of buffer and 8.5 ml of distilled water), using cuvettes of 1-cm light path.

Calculation. Subtract the extinction reading obtained for the test from that of the blank and calculate for each substrate separately the enzyme activity per milliliter of fluid and per minute, and express it in international milliunits. It is convenient to remember that 1 ml of 0.002 N indigo disulfonate solution diluted to 10 ml with distilled water shows at 613 nm in a 1-cm cuvette an optical density of 1.0.

Assay for Dialyzed Pregnancy Serum. In the Pyrex test tube (2 × 15 cm) marked "a" place 0.5 ml of clear dialyzed pregnancy serum and add 0.6 ml of 0.002 N indigo disulfonate solution. In another test tube marked "b" place multiples of a mixture consisting of 0.2 ml of the stock solution of cadaverine dihydrochloride or putrescine dihydrochloride, and 0.2 ml of EDTA–phosphate buffer. When the enzymic effect on histamine is to be measured, place in the test tube marked "b" multiples of a mixture consisting of 0.1 ml of the stock solution of histamine dihydrochloride and add 0.3 ml of phosphate buffer. Continue as described above for assays of amniotic fluids and placental extracts. In some pregnancy sera the histaminase (DAO) activity may be very low. In such cases the assay should be repeated and the incubation time should be extended from 60 to 120 min. Calculate the results as shown above and express them in international milliunits and in the form of specific activity.

The advantage of this simple spectrophotometric test is that blanks and tests can easily be examined, in replicate, and that the enzymic effect of various biological media on the three substrates cadaverine, putrescine, and histamine can be studied simultaneously.

Previous findings (538) suggesting that placental DAO, due to its significant behavior toward histamine, cadaverine, and putrescine resembled the DAO of pea seedlings more than it did the DAO of hog kidney, have been confirmed with the spectrophotometric method, just described (548).*

This modified technique of histaminase (DAO) estimation has been applied to simultaneous determinations of the enzymic effect on cadaverine, putrescine, and histamine in sera and amniotic fluids from about 70 women at various stages of pregnancy. Throughout all these investigations the DAO activity was highest with cadaverine as substrate and lowest with histamine. It was very interesting to find that amniotic fluids displayed with all three substrates a much higher DAO activity than the simultaneously examined maternal blood. Moreover, because of a very low protein content of the amniotic fluid (averaging 0.2%, as compared with the average protein content of the maternal serum of 5%), the specific activity of DAO in the amniotic fluid with any of the three substrates appeared to be about 100 times that of the simultaneously tested maternal serum.

The results shown in Table II of a simultaneous investigation of the DAO activity in the serum and amniotic fluid of a woman in the 36th week of pregnancy are representative of observations made in the study, mentioned above.

Table II.
Correlation of DAO Levels in the Maternal Serum and Amniotic Fluid

	DAO levels			
	Maternal serum		Amniotic fluid	
Substrate	Activity (ImU/ml)	Specific activity	Activity (ImU/ml)	Specific activity
Cadaverine	0.634	0.0095	2.330	1.080
Putrescine	0.467	0.0070	1.750	0.814
Histamine	0.367	0.0055	1.160	0.543

* The author has unpublished data (1964).

Table III.

Correlation of Specific Activity of DAO in the Placental Extract and
Amniotic Fluid of a Woman at Delivery [a]

	Specific activity	
Substrate	Placental extract	Amniotic fluid
Cadaverine	0.308	1.040
Putrescine	0.131	0.352
Histamine	0.123	0.327

[a] Full term.

Table III shows a correlation of DAO levels in the extract of the
placenta and in the amniotic fluid of a woman at delivery. The results
obtained for the specific DAO activity with all three substrates were
much higher in the amniotic fluid than those determined in the placental
extract.

g. MICROSPECTROPHOTOMETRIC ESTIMATION OF DAO, BASED ON AN
   ENZYMIC DETERMINATION OF AMMONIA WITH THE SYSTEM
   GLUTAMATE DEHYDROGENASE, α-OXOGLUTARATE, AND NADH$_2$,
   ACCORDING TO KIRSTEN, GEREZ, AND KIRSTEN (1016)

This micromethod is based on the application of the ammonia, formed
in the primary DAO–substrate reaction, as substrate of glutamate dehydrogenase, according to the equation

$$\alpha\text{-oxoglutarate} + NH_4^+ + NADH + H^+ \underset{\substack{\text{dehydro-}\\\text{genase}}}{\overset{\text{glutamate}}{\rightleftarrows}} \text{glutamate} + NAD^+ + H_2O.$$

This reaction occurs at a neutral pH and thus avoids the strong alkalin-
ization, necessary for the Conway method (958). With the large amounts
of α-oxoglutarate used, ammonia is converted quantitatively into L-glutamate.

Analytical Method. *Reagents.* The following reagents are required:
1. Glutamate dehydrogenase (Boehringer, Mannheim, Germany) free
from ammonium ions, three times recrystallized, and dialyzed against

sodium phosphate buffer; dissolved in 50% glycerol, pH 7.3. (Boehringer preparation).

2. α-oxoglutaric acid (α-ketoglutaric acid, Boehringer). A 5-ml volume of 0.08 M α-oxoglutaric acid in water is neutralized with 2 N NaOH to pH 7.0 (indicator paper). The neutral solution keeps at 4° for a few days.

3. NADH$_2$ solution (Boehringer) in 0.2 M phosphate buffer, pH 7.6; contains 10 mg of NADH$_2$ per milliliter of buffer.

4. Sodium phosphate buffer 0.2 M (Sörensen), pH 7.6.

The reagent mixture required for the determination of NH$_4$+, formed in the DAO–substrate primary reaction, consists of 29 ml of 0.2 M sodium phosphate buffer, pH 7.0, and 0.6 ml of NADH$_2$ solution (10 mg NADH$_2$ in 1 ml of 0.2 M phosphate buffer, pH 7.6).

Assay. The assay is carried out at 25°. Place in a cuvette (glass cell with lid, 1-cm light path) 1.50 ml of the reagent mixture and mix gently. Read extinction at 366 nm in a spectrophotometer; then start the reaction by stirring into the solution 20 μl of glutamate dehydrogenase, cover the cuvette with the lid, and follow the conversion of the ammonia, formed in the primary enzymic reaction, into L-glutamate by measuring the decrease in extinction at 366 nm within 25 min, when the reaction is usually completed. Express the result in micromoles of NH$_4$+ formed and calculate from the latter the DAO activity. Blanks (1.5 ml of reagent mixture and 0.5 ml of water) must be set up along with every test reaction.

According to Kirsten et al. (1016), 1 to 100 $\times$ 10^{-9} moles of ammonia can be estimated with an accuracy of $\pm$1%.

It is regrettable that the presentation of the above method is not very clear.

h. RAPID METHOD FOR THE DETERMINATION OF DAO ACTIVITY BY LIQUID
  SCINTILLATION COUNTING, ACCORDING TO OKUYAMA AND KOBAYASHI
  (547) , AND MEASUREMENT OF PLASMA DAO DURING NORMAL HUMAN
  PREGNANCY BY AN IMPROVED RADIOASSAY PROCEDURE, ACCORDING TO
  SOUTHREN, KOBAYASHI, ET AL. (549, 1038)

The method of DAO estimation by means of liquid scintillation counting is based on the formation of radioactive, toluene-extractable end products from the action of DAO on ^{14}C-cadaverine. The end products are extracted directly into toluene and assayed in a liquid scintillation spectrometer. This method is also applicable to radioactive putrescine as substrate, but does not work with radioactive histamine. This radioassay method has provided evidence to indicate the quantitative production of Δ^1-pyrroline and Δ^1-piperideine (or polymers thereof) as the end products of DAO action on putrescine and cadaverine, as suggested by Tabor (551).

Materials and Methods (547). The following radioactive compounds were used:

1. ^{14}C-Putrescine (New England Nuclear Corp.) was diluted with non-isotopic putrescine to give approximately 130,000 dpm per 5 μg of base.

2. ^{14}C-Cadaverine picrate, synthesized by Schayer and co-workers (1039) was converted on a micro Dowex-1 column to the dihydrochloride and gave between 26,000 and 40,000 dpm per 5 μg of base.

This range of specific activities of ^{14}C-cadaverine reflected the varying degree of efficiency with which the cadaverine dipicrate was converted, to the dihydrochloride. The specific activity of each batch of ^{14}C-cadaverine was calculated on the basis of 100% conversion of the dipicrate to the dihydrochloride. This variation in specific activity of ^{14}C-cadaverine did not influence the results obtained or conclusions drawn.

Enzyme Preparations. Hog-kidney DAO was used. Frozen hog kidneys (233 g) were minced and extracted with 1060 ml of 0.1 M phosphate buffer, pH 7.6, for 20 min at 59 to 60°, with mechanical stirring. The extract was filtered through cheesecloth and fractionated with ammonium sulfate. The ammonium sulfate precipitate, obtained at 40 to 55% saturation, was dissolved in 0.1 M phosphate buffer, pH 7.6, and dialyzed against demineralized water overnight at 4°. The preparation containing 13.48 mg of nitrogen per milliliter had a specific activity of 2.59 μl oxygen uptake per minute, per milligram of nitrogen, with cadaverine as substrate, and also contained some catalase activity (1040). DAO from hog intestines wes prepared in a similar manner. Commercially prepared hog-kidney DAO (Pentex, Inc.) was used at a concentration of 10 mg of protein per milliliter.

Liquid Scintillation Counting Method. A Packard Tri-Carb liquid scintillation counter, model 314 X, was used by the authors. The extraction solution consisted of reagent-grade toluene containing 0.35% of 2,5-diphenyloxazole (PPO). The solution gave 59% counting efficiency after being used in the extraction of the end products. For the radioactive assay of ^{14}C-cadaverine and ^{14}C-putrescine an ethanol–toluene mixture (25 : 70, v/v) containing 0.35% PPO was used. The counting efficiency of this solution was 36%. All the countings were performed in low-potassium 5-dram glass vials (Wheaton Glass Co.).

General Extraction Procedure. Into a screw-cap culture tube (10 $\times$ 1.5 cm) containing approximately 200 mg of sodium bicarbonate, 2 ml of the enzyme reaction mixture was pipetted. A 10-ml volume of the PPO solution was then added with a rapid, constant-volume dispenser, the mixture was shaken vigorously for about 1 min, and then was allowed to stand at −10° until the lower aqueous phase was frozen. The upper layer was then poured into a counting vial and assayed in the liquid scintillation spectrometer. Whereas a single extraction procedure gave satisfactory results for semiquantitative experiments, for quantitative purposes the sample had to be extracted two to three times with 10-ml portions of scintillation (PPO) solution.

Standard Test Solution. A standard test solution for extraction was prepared as follows: 1.0 ml of the kidney enzyme preparation was incubated with 25 μg of ^{14}C-cadaverine in 20 ml of 0.1 M sodium phosphate buffer, pH 7.6, for 60 min at 38°. Aliquots of this mixture were used to determine the optimal extracting conditions. The mixture was stored frozen.

Results. From chromatographic investigations Okuyama and Kobayashi (547) concluded that the action of DAO on a diamine may yield more than one product. The multiplicity of the end products is probably due to the nonenzymic transformation of the primary end product of DAO. All those products, however, appear to be soluble in toluene and thus lend themselves to quantitative estimations.

With a large amount of enzyme, cadaverine was oxidized completely within a short time. Further incubation at 38° for 4 hr had little effect on the amount of extractable reaction product formed.

With putrescine as substrate, results similar to those for cadaverine were obtained. The reaction product was extracted effectively under the standard conditions, and putrescine itself was not extracted. Moreover, this radioassay was found to be applicable to inhibition studies on DAO. Thus, aminoguanidine, a specific inhibitor of DAO, was preincubated with the hog-kidney enzyme for 15 min at a final concentration of 10^{-5} M before the addition of ^{14}C-putrescine. After an incubation of 60 min at 38°, the enzyme activity was found to be completely inhibited; that is, the extracted radioactivity was at the same level as that of the zero time value. When aminoguanidine was added to the reaction mixture after the completion of the enzymic reaction, a slight but insignificant effect on the extraction efficiency was noticed.

The experimental data presented by Okuyama and Kobayashi (547) seem to indicate that this rapid quantitative technique for the estimation of DAO activity not only compares well in terms of sensitivity with other relevant methods, but has also the advantage of simplicity and speed. The specificity of the radioactive extraction method is said to be superior to that of the bioassay method of histaminase estimation (666), and to be comparable to that of the spectrofluorometric method of histaminase estimation by Shore et al. (731). The extraction method is also claimed to be superior to the colorimetric method of DAO estimation by Holmstedt et al. (602), the latter technique having the lowest sensitivity of all the methods discussed.

However, Okuyama and Kobayashi (547) have to admit that the greatest errors of their extraction method are introduced during the extraction and decantation steps. According to the authors, these errors are within the limits of statistical acceptability, as shown by a standard deviation of less than 4% for 10 replicates, routinely run for DAO activity.

This rapid assay method has been successfully applied in Kobayashi's laboratory for the purification of DAO from hog intestine as well as in the screening of DAO inhibitors.

Measurement of Plasma DAO During Normal Human Pregnancy. In 1963 Kobayashi reported on the application of this highly specific and sensitive radioassay for DAO estimation to the plasma of pregnant women (1041). The method Kobayashi used was as follows: 10 ml of blood is withdrawn from a pregnant woman with an oxalated Vacutainer (Becton, Dickinson & Co., Rutherford, N. J.). The blood is centrifuged and the plasma transferred into a clean test tube. The plasma is heated at 50° for 15 min to activate DAO (990). The assay is carried out in a screw-cap culture tube (10 $\times$ 1.5 cm) containing 2 ml of heated plasma and 5μg of ^{14}C-putrescine, having an activity of approximately 20,000 dpm per microgram. The incubation mixture is shaken in a Dubnoff metabolic shaker for 2 hr in air at 37°. Approximately 200 mg of sodium bicarbonate is then added to the reaction mixture and the final product is extracted into 10 ml of toluene containing 0.35% diphenyloxazole and 0.01% 1,4-bis-2-(5-phenyloxazolyl)benzene. The reaction mixture is placed in a freezer at $-15°$ until the aqueous phase is frozen, and the toluene extract is poured into a counting vial for assay in a liquid scintillation counter. The extraction is repeated once. The counting efficiency is approximately 50%.

The plasma of 17 pregnant women up to the 17th week of pregnancy was tested for DAO activity by the radioassay. The average normal DAO level in 32 nonpregnant women was found to be 249 cpm $\pm$ 110 S.D. The plasma titers of DAO were highly elevated—for example, a value of 36,700 cpm was obtained in the plasma of women in the 12th week of gestation.

In continuation of this work Southren, Kobayashi, et al. (545) observed that, whereas during the first trimester of human pregnancy the plasma DAO level rose almost in a linear fashion, the pattern during the remaining two trimesters appeared to be a plateau with a relatively narrow range. The authors themselves suggested thereupon that this may have been due to the application of too small amounts of substrate in relation to a very high plasma enzyme activity. In order to quantitate their measurements, they established a new unit of DAO, which they defined as the amount of DAO, contained in 1 ml of plasma, that metabolizes 0.01 μg of ^{14}C-putrescine in 2 hr at 37°, using 2 ml of plasma and 5 μg of ^{14}C-putrescine in a final volume of 2.1 ml. Whereas in 34 nonpregnant individuals a plasma DAO level of 3.4 units $\pm$1.5 S.D. per milliliter of plasma was obtained, in the plasma of a pregnant woman in the 12th week of gestation 425 units/ml were found. In subsequent work this technique was modified to use 50 μg instead of 5 μg of ^{14}C-putrescine as substrate. When 50 μg ^{14}C-putrescine was used as substrate, a linear relationship was found between plasma DAO activity and incubation time up to 150 min. Moreover, the increased substrate concentration extended the range of the assay to 5000 DAO units, thus providing a means by which the plasma DAO could be conveniently studied in the second and third trimester of pregnancy.

The modified DAO unit represents the amount of enzyme, contained in 1 ml of plasma, that metabolizes 0.01 μg of putrescine in 2 hr at 37° in air, using 50 μg of ^{14}C-putrescine as substrate in a final volume of 2.1 ml.

Estimation of DAO Activity in Tissues (549). Tissue specimens (trophoblastic and decidual tissues) obtained from patients undergoing caesarean section or caesarean hysterectomy were washed with saline, blotted dry, and weighed. A sample of 1 g of tissue was homogenized in an Omnimixer in 3.0 ml of 0.1 M phosphate buffer, pH 7.8, and then centrifuged at 10.000 g for 30 min. The supernatant solution was decanted and then diluted 1 : 10 with phosphate buffer to a final dilution of 1 : 40. Two milliliters of this solution were taken for the analysis, as described above for plasma. The activity of each tissue was compared to the activity of fetal plasma, 1 ml of plasma having been equated to 1 g of tissue.

The tissues assayed included decidua, cross sections of placenta, cord, fetal membranes, consisting of both amnion and chorion, and amniotic fluid. In 48 women at parturition after full-term pregnancies DAO activity was estimated by the radioassay technique in paired maternal and fetal plasma samples. The mean DAO level in the maternal plasma was 1027 units and in the fetal plasma 5.7 units, with an overall mean maternal-to-fetal ratio of 166 : 1. The amniotic fluid was found to possess approximately two to three times the DAO activity of an equivalent of maternal plasma. When relative DAO activities of the tissues and fluids were computed taking the mean foetal plasma value as unity, the relative DAO content of the tissues and fluids was as follows: decidua = 5200, fetal membranes = 4100, placenta = 1400, amniotic fluid = 100, maternal plasma $= 42$, cord $= 14$, and fetal plasma $= 1$.

On the basis of the above results Southren, Kobayashi, and co-workers (549) suggest that the decidua is the probable source of DAO encountered in human pregnancy. However, they find it difficult to explain on this basis the occurrence of a highly active DAO in the amniotic fluid, especially when taking into account the fact of an extremely low DAO level in the fetal circulation. In this connection it should be mentioned that the present author, when discussing the origin of the large amounts of DAO in the amniotic fluid, suggested that the human amnion itself may be the site of formation of DAO (548; see Chapter IV.1.B).* From detailed studies of the histological structure of the amniotic epithelium Keiffer (1042) and Danforth and Hull (1043) seem to have found cytologic evidence for a secretory activity in the amnion with the amniotic epithelium secreting fluid into the amniotic cavity.

In a recent paper Southren, Kobayashi, et al. (1038) seem to have provided evidence suggesting the presence of a circulating inhibitor of plasma amine oxidase. It should be recalled here that the present author reported in a work on human placental DAO (538) the occurrence in crude placental extracts of compounds that strongly inhibited the enzymic effect on cadaverine and putrescine, but not that on histamine. This inhibitory effect on the placental enzyme activity with cadaverine and putrescine as substrates was found to be completely abolished by the addition of 0.001 M EDTA to the extracting solutions and buffers (538; see Section 3.B).

Tryding (699) has recently slightly modified the radioassay technique for the determination of plasma DAO, discussed above, by increasing approximately three fold the concentration of ^{14}C-putrescine to 0.001 M. Since only 0.05 ml of plasma is required for this test, the analysis can be performed on capillary blood plasma.

It will be, however, necessary to investigate whether the 0.001 M concentration of the substrate is sufficient to saturate the very active plasma DAO in human pregnancy.

---

* The author also has unpublished data (1965).

4. METHODS OF ISOLATION, PURIFICATION, AND ESTIMATION OF PLASMA OR SERUM AMINE OXIDASES (EXTRACELLULAR ENZYMES), OF PLANT AMINE OXIDASE AND OF BACTERIAL AMINE OXIDASES

A. Isolation, Purification, and Crystallization of Plasma Spermine Oxidase (Ruminant-Plasma Amine Oxidase)

a. EXTRACTION, CONCENTRATION, AND ESTIMATION OF SPERMINE OXIDASE FROM BEEF PLASMA, ACCORDING TO TABOR, TABOR, AND ROSENTHAL (141)

In the work to be discussed soluble amine oxidase has been purified 150 to 200 times its original activity from steer plasma. The enzyme oxidatively deaminates a variety of amines, with the stoichiometric formation of the corresponding aldehydes, ammonia, and hydrogen peroxide (see Chapter IV.2.A.b).

Purification of Beef-Plasma Amine Oxidase. The following procedure was used in the purification of beef-plasma amine oxidase:

Step 1. Steer blood was collected at the slaughterhouse and immediately treated with one-sixth volume of a citrate solution, containing 8 g of citric acid and 26.7 g of sodium citrate ($Na_3CH_5O_7 \cdot 5 \cdot 5H_2O$) per liter. The citrated blood was centrifuged at 2000 rpm for 30 min, and the plasma was collected. In other preparations serum was used instead of plasma with essentially the same results. All the subsequent procedures were carried out at 0 to 3°.

Step 2. To 13,600 ml of plasma was added 7,300 ml of saturated ammonium sulfate (final concentration was approximately 1.4 M); the mixture was cooled to 0 to 3° and the precipitate was removed by filtration through fluted filter paper. To the filtrate were added 13,100 ml of saturated ammonium sulfate solution (final concentration approximately 2.4 M). The precipitate was collected by filtration and dissolved in water (final volume 3100 ml). This solution was then refractionated with ammonium sulfate. Conductivity measurements indicated that the solution was 26% saturated (approximately 1 M). An additional 1000 ml of saturated ammonium sulfate were added to obtain a final ammonium sulfate concentration of 1.8 M. The precipitate was discarded after filtration, and 2000 ml of saturated ammonium sulfate were added to yield a final concentration of approximately 2.5 M. The precipitate was dissolved in 900 ml of water and the solution was dialyzed against three changes of 0.01 M sodium acetate (total volume 18 liters), to reduce the ammonium sulfate concentration to <0.02 M. Since inadequate dialysis caused poor fractionation in the subsequent steps, the efficiency of the dialysis was checked by conductivity measurements.

Step 3. The dialyzed solution was divided into 50-ml portions, each of which was heated rapidly (within 1–2 min) to 65° with stirring and maintained at this temperature for 10 min. Usually no precipitate formed, and the solution was immediately cooled to 0°.

Aliquots of 400 ml of the dialyzed and heated solution were then fractionated with ethanol at 0° after 40 ml of 0.1 M manganese chloride had been added to each portion. Then 200 ml of cooled ($-10°$) 25% ethanol were added with mechanical stirring over a period of 5 to 10 min at 0°. The precipitate was collected by centrifugation at 4000 g for 15 min in a refrigerated centrifuge (precipitate 1).

Additional precipitates were collected after successive additions of alcohol in the following manner: 100 ml of 25% ethanol (precipitate 2), 100 ml of 25% ethanol and 20 ml of absolute ethanol (precipitate 3), 100 ml of absolute ethanol (precipitate 4), 80 ml of absolute ethanol (precipitate 5), and 110 ml of absolute ethanol (precipitate 6). Each precipitate was dissolved in approximately 35 ml of cold water. The fractions showing the highest specific activity (usually precipitates 4 and 5) were then pooled. Aliquots (57 ml) of these pooled fractions were refractionated with ethanol at 0° after the addition of 5.7 ml of 0.1 M MnCl$_2$.

Precipitates were collected by centrifugation at 4000 g for 15 min after the following successive additions of cooled absolute ethanol : 3.0 ml (precipitate 1), 2.0 ml (precipitate 2), 1.8 ml (precipitate 3), 2.5 ml (precipitate 4), and 2.5 ml (precipitate 5). The various precipitates were each dissolved in approximately 20 ml of cold water and assayed. The fractions showing the highest specific activity (usually precipitates 3 and 4) were used for the next step of the enzyme purification on calcium phosphate gel.

Step 4. Material in 6-ml portions from the preceding step was treated with 21 ml calcium phosphate gel (calcium phosphate gel was prepared according to Keilin and Hartree (1044) and aged for 4 to 10 months). It contained 14 mg of dry weight per milliliter. The gel was collected by centrifugation and eluted with 42 ml of 0.01 M K$_2$HPO$_4$ buffer, pH 7.2. Most of these authors' studies were carried out with enzyme fractions from the two last steps of purification, i.e. those of alcohol fractionation and adsorption on calcium phosphate gel and desorption with potassium phosphate buffer, pH 7.2. Frequently, the material obtained on alcohol fractionation was lyophilized and stored as the dry powder. Although the enzyme preparations were stable at all stages at 0 to 4° in the refrigerator, prolonged storage was usually carried out at -15 to $-20°$. No substantial loss occurred during a period of 1 to 4 months. Moreover, no activity was sedimented when the enzyme, obtained on alcohol fractionation, was centrifuged in the Spinco preparative centrifuge at 100,000 g for 1 hr.

Enzyme Assay. Two methods were used to determine enzyme activity, the manometric and the spectrophotometric.

Manometric Method. Experiments were carried out in a conventional Warburg apparatus at 37.3°; the gas phase was air. The incubation mixture contained 0.2 ml of 0.2 M potassium phosphate buffer, pH 7.2, 0.05 unit of catalase (crystalline catalase was obtained from the Worthington Biochemical Corporation and freed from ammonium sulfate by dialysis against 0.002 M phosphate buffer, pH 7.2) and enzyme in a volume of 2.4 ml. After equilibration, 6 micromoles (0.1 ml) of spermine hydrochloride (adjusted to approximately pH 7.0) were added from the side arm. The initial rate of oxygen consumption was measured and expressed as microliters of oxygen consumed per 10 min. The unit of enzyme activity was defined as the amount of enzyme catalyzing the consumption of 1 μl of oxygen per 10 minutes. Usually the assays were carried out with 20 to 30 units of enzyme per milligram of protein.

Spectrophotometric Assay. This method was found by Tabor et al. (141) to be more convenient than the manometric one. It is based on the difference between the molar extinction coefficients of benzylamine and benzaldehyde at 250 nm.

The incubation mixtures contained 200 micromoles of potassium phosphate buffer, pH 7.2, 10 micromoles of benzylamine, and enzyme in a total volume of 3 ml. One spectrophotometric unit was defined as the amount of enzyme catalyzing an increase of 0.001 per minute in the optical density reading at 250 nm. One manometric unit was approximately equivalent to 10 spectrophotometric units.

Both the manometric and the spectrophotometric assays gave essentially the same results, when they were used to follow the purification steps of the plasma enzyme.

Protein estimations were carried out by measuring the absorption at 280 nm (1045).

Saturated ammonium sulfate solutions contained approximately 530 g of ammonium sulfate per 1000 ml of solution (about 4 M) at 25 to 30°.

Conductivity Determinations. A Barnstead purity meter (PM–2) was used for the conductivity determinations, which were made to check the salt concentration in all ammonium sulfate fractions. The solutions were diluted 1 : 50,000 in water and the conductivity readings were compared with those of similar dilutions of a standard ammonium sulfate solution (1046).

Spermine was not only very actively degraded but also had a very high affinity for the enzyme. The substrates most rapidly attacked besides spermine and spermidine were benzylamine, homosulfanilamide, furfurylamine, and various aliphatic monoamines (e.g., butylamine). The purified plasma enzyme shows a relatively weak activity against tyramine and especially no activity against tryptamine, ephinephrine, and histamine. Of the diamines only decamethylene diamine is attacked by the enzyme. The plasma amine oxidase is completely inhibited by 10^{-3} M isoniazid as well as by 10^{-3} M iproniazid. Furthermore, this enzyme is fully inhibited by 10^{-3} M KCN, 4×10^{-4} M hydroxylamine, and 10^{-3} M semicarbazide (see Chapter IV.2.A.c,d).

b. PURIFICATION AND CRYSTALLIZATION OF SPERMINE OXIDASE FROM BEEF
   PLASMA, ACCORDING TO YAMADA AND YASUNOBU (143)

The method of purification developed by Yamada and Yasunobu (143) is based mainly on chromatography, using DEAE-cellulose and hydroxylapatite columns. This technique yielded preparations that were purified about 300 to 400 times.

Procedure. All operations were carried out at about 5° and all centrifugations were performed in the Servall model SS3 centrifuge.

Step 1. Beef blood was vigorously mixed according to Tabor, Tabor, and Rosenthal (141; see Section 4.A.a) with ⅙ volume of citrate solution that contained 8 g of citric acid and 26.7 g of sodium citrate. 5·5 H₂O per liter. The citrated blood was centrifuged at 25,000 g for 20 min, and the plasma was collected.

Step 2. First Ammonium Sulfate Fractionation. To 20,700 ml of the plasma, solid ammonium sulfate was added to 0.35 saturation. After standing overnight, the precipitate was removed by centrifugation at 20,000 g for 20 min and discarded. The ammonium sulfate concentration was then increased to 0.55 saturation by the addition of solid ammonium sulfate. After standing overnight, the precipitate was collected by centrifugation at 20,000 g for 20 min and dissolved in 0.01 M sodium potassium phosphate buffer (Na₂HPO₄–KH₂PO₄), pH 7.0. The solution was dialyzed overnight against three changes of 10 liters of 0.003 M phosphate buffer, pH 7.0. An inert protein precipitate was removed by centrifugation. The supernatant usually contained more than 90% of the initial plasma activity.

Step 3. First DEAE-Cellulose Column Chromatography. The dialyzed enzyme solution (2435 ml) was subjected to chromatography on DEAE-cellulose (1047). The resin was packed into four columns (4 × 55 cm) and equilibrated with 0.003 M phosphate buffer, pH 7.0. The enzyme solution was divided into four portions, each of which was passed through a separate column. Each column was then treated with 0.03 M phosphate buffer, pH 7.0, and much of the inert protein was removed. The enzyme was subsequently eluted with 0.1 M phosphate buffer, pH 7.0, at a flow rate of 0.5 ml/min in fractions of 15 ml.

The elution of the protein was followed by the measurement of absorbancy at 280 nm (1045), and the enzyme activity was determined by the spectrophotometric method of Tabor, Tabor, and Rosenthal (141; see this Section 4.A.a) except that the incubation of the enzymic mixture was carried out at 25° rather than 30°. One enzyme unit was defined as the amount of enzyme catalyzing a change of 0.001 absorbancy per minute at 25° under the standard conditions.

The fractions containing enzyme of specific activity greater than 25 were pooled to give a solution of 1257 ml. This solution was concentrated by the addition of solid ammonium sulfate to a saturation of 0.55. The precipitate obtained by centrifugation was dissolved in 0.01 M phosphate buffer, pH 7.0, and dialyzed overnight against 10 liters of 0.003 M phosphate buffer, pH 7.0.

Step 4. Second DEAE-Cellulose Column Chromatography. The dialyzed enzyme solution was subjected again to DEAE-cellulose chromatography. The enzyme solution was placed on a DEAE-cellulose column (3.5 $\times$ 50 cm) previously equilibrated with 0.003 M phosphate buffer, pH 7.0. The enzyme was eluted with 0.05 M phosphate buffer, pH 7.0, in fractions of 15 ml, collected at a flow rate of 0.5 ml/min. Eluates with a specific activity greater than 130 represented the main enzyme fraction and were combined (400 ml).

After elution of the main fraction with 0.05 M phosphate buffer, pH 7.0, a second protein band containing activity was eluted with 0.07 M phosphate buffer, pH 7.0. This fraction showed a specific activity between 65 and 165 and represented 14.6% of the total activity from the preceding step.

Step 5. Second Ammonium Sulfate Fractionation. Solid ammonium sulfate was added to 0.4 saturation to the main eluate (400 ml) from the second DEAE-cellulose chromatography, and the pH adjusted to 7.0 by the addition of ammonia. After standing overnight, the precipitate was removed by centrifugation at 20,000 g for 20 min. The ammonium sulfate concentration was then increased to 0.55 saturation by the addition of solid ammonium sulfate and the pH was adjusted to 7.0. After standing overnight, the precipitate was collected by centrifugation at 20,000 g for 20 min and dissolved in a small amount of 0.01 M phosphate buffer, pH 6.8. The solution was dialyzed overnight against three changes of 2 liters of 0.006 M phosphate buffer, pH 6.8.

Step 6. Hydroxylapatite Chromatography. The dialyzed enzyme solution was subjected to hydroxylapatite chromatography. The hydroxylapatite, prepared according to the method of Tiselius, Hjerten, and Levin

(1048) was packed into a column (4 $\times$ 20 cm) and equilibrated with 0.006 M phosphate buffer, pH 6.8. The enzyme solution was placed on the column, from which the enzyme was eluted stepwise with increasing concentrations of phosphate buffer, pH 6.8. The flow rate was 2 ml/hr, and 3-ml fractions were collected. The behavior of the plasma amine oxidase on the hydroxylapatite column was similar to that observed on DEAE-cellulose chromatography. The enzyme was again resolved into a number of enzymically active fractions. Of the various fractions eluted, the first was desorbed from the column with 0.006 M phosphate buffer, pH 6.8, and showed a specific activity of 330 to 380, the second was eluted with 0.06 M phosphate buffer, pH 6.8, and showed a specific activity of 260 to 310, and the third was eluted with 0.2 M phosphate buffer, pH 6.8, and had a specific activity of 130 to 185. The first and second fractions were pooled and subjected to crystallization. The eluates comprising the third fraction were also combined and stored.

Crystallization of Beef-Plasma Amine Oxidase. The purified enzyme preparation showing a specific activity of 353, obtained from the first fraction of hydroxylapatite chromatography, was first used for crystallization of the enzyme. The enzyme solution was first concentrated by the addition of solid ammonium sulfate to 0.55 saturation. After standing overnight, the precipitate was collected by centrifugation at 15,000 g and dissolved in a minimal amount of 0.03 M phosphate buffer, pH 7.0, to give a solution containing 20 to 40 mg of protein per milliliter. Solid, finely powdered ammonium sulfate was slowly added to the above solution until a faint, permanent turbidity appeared. This precipitate was removed by centrifugation and discarded. Saturated ammonium sulfate solution was then added slowly to the supernatant with stirring until turbidity appeared, after which the mixture was placed in a refrigerator. Crystallization usually began after about 12 hr and was virtually complete within a week.

With the use of a seeding technique Yamada and Yasunobu (143) have been able to obtain also from fraction 2 of the hydroxylapatite chromatography a crystalline enzyme of a specific activity of about 300.

For recrystallization the crystals were dissolved in 0.03 M phosphate buffer, pH 7.0, and the solution was treated as just described. The specific activity of the enzyme increased on crystallization from 300 up to 450 to 500, and the yield of crystals in a typical experiment was 250 to 300 mg from 40 liters of beef blood.

Beef-plasma amine oxidase crystallized in the form of rosettes or as separate needles of faint pink color which easily dissolved in phosphate buffer, pH 7.0, to give a pink-colored solution.

Crude and partially purified preparations of plasma amine oxidase could be preserved at 5° in 0.03 M phosphate buffer, pH 7.0, with little loss of activity. Variation in the pH over the range of 6.0 to 8.0 had little effect on the enzyme stability. Outside this range, however, the rate of enzyme inactivation increased rapidly. Solutions of the highly purified enzyme with a specific activity greater than 350 were less stable than those of only partially purified enzyme preparations; for example, a sample of enzyme with a specific activity of 444 lost 6.1% of its initial activity after standing 2 days at 2° in 0.03 M phosphate buffer, pH 7.0. Both the purified and partially purified preparations, when dialyzed and stored in the absence of electrolytes, became unstable, losing their initial activity and forming an amorphous precipitate.

The crystalline enzyme, however, could be preserved at −5 to +2° as a suspension in 0.03 M phosphate buffer, pH 7.0, containing 55% saturated ammonium sulfate. Since both the specific and the total activity of the enzyme remained constant for periods of over a month, this type of preservation seems to be the most suitable one for the crystalline enzyme.

Lyophilization of the enzyme preparations first caused a loss of 15 to 20% of activity. The remaining activity was retained for at least 1 month.

As to homogeneity of the highly purified recrystallized enzyme, preparations with a specific activity greater than 380 yielded electrophoretic patterns displaying a single boundary. The electrophoretic runs were carried out at 4° in the Perkin-Elmer electrophoresis instrument using phosphate-sodium chloride buffer of 0.1 ionic strength and a pH of 6.0 and 7.0, prepared according to Miller and Golder (1049).

Analysis of the recrystallized enzyme showed in the ultracentrifuge at 24° (Spinco model E analytical ultracentrifuge) a single symmetrical peak. The only impurity found was a small amount of a slower sedimenting material.

The sedimentation coefficient in water at 20° extrapolated to zero protein concentration ($S^0_{20, w}$) was found to be 8.98 Svedberg units, when the protein concentration was varied from 2.8 to 9.3 mg/ml in 0.06 M phosphate buffer, pH 7.0. Assuming values found for most proteins of V, the partial specific volume of the enzyme, to equal 0.74 ± 0.02, a molecular weight of 255,400 ± 20,000 was obtained by the Archibald method (1050).

When the concentration of the phosphate buffer of the assay system was made less than 0.03 M, the enzyme activity was found to be strongly dependent on the ionic strength of the buffer.

With regard to substrate specificity, similar results were achieved with the crystalline preparation to those obtained by Tabor et al. (141) with the partially purified enzyme. However, the activity of the crystalline preparation toward short-chain aliphatic amines, such as butylamine and amylamine, was lower than that reported by Tabor et al. Furthermore, Yamada and Yasunobu (143) found kynuramine to be also a suitable substrate for beef-plasma amine oxidase.

Absorption spectra of the highly purified enzyme were taken with a Beckman model DK–2 spectrophotometer. The enzyme exhibited maxima at about 280 and 480 to 500 nm. A shoulder at 410 nm was also noted in the spectrum. The pink color of the enzyme was related to the maximum at about 480 nm. This absorption was partially abolished by the addition of excess sodium dithionite (which bleached the color) and to almost the same extent by benzylamine under essentially anaerobic conditions. The band and the pink color were subsequently restored by oxygenation. These findings suggest that the pink compound may be involved in the catalytic activity of the enzyme and that it may function as a hydrogen carrier during such an activity.

The difference spectrum (oxidized minus reduced spectrum) shows a maximum at about 450 to 480 nm, which corresponds to the maximum of flavin nucleotides. No flavins were detected, however, in the crystalline plasma amine oxidase (579) after heat denaturation, acid treatment, or digestion with proteases (1051).

In subsequent work Yamada and his colleagues (224, 579, 580, 846) presented evidence suggesting that the crystalline beef-plasma amine oxidase is a copper enzyme, containing 4 g-atoms of cupric copper per mole of protein (224, 579, 580). These workers were able to show that there was a direct proportionality between specific activity and copper content during the isolation of the crystalline enzyme (579). Moreover, they found that dialysis of the enzyme against sodium diethyldithiocarbamate at pH 7.0, removed the copper from the enzyme with a concomitant loss of enzyme activity and that the dialyzed enzyme was specifically reactivated by cupric ions (579). The four times recrystallized enzyme, which was demonstrated to be homogeneous by several physicochemical criteria of purity, was analyzed for its cupric copper content by the method of Peterson and Bollier (931) and for its content of cuprous copper by the technique of Griffiths and Wharton (1052). The latter method was also applied for the analysis of the valency state of copper in plasma MAO.

Analysis of the Valency State of Copper in Plasma MAO (580). To 0.5-ml aliquots of the recrystallized enzyme solution containing 5 mg of protein, were added 30 micromoles of phosphate buffer (KH_2PO_4–Na_2HPO_4), pH 7.2, and the appropriate amounts of benzylamine, sodium dithionite, or hydroxylamine hydrochloride to give a final volume of 0.8 ml. The copper was determined by first adding 0.2 ml of 0.2 M

EDTA and 0.001 M p-chloromercuriphenylsulfonic acid, followed by 1.4 ml of 0.1% neocuproine in glacial acetic acid. After the addition of 1.4 ml of ethanol, the mixture was centrifuged for 10 min at 15,000 g. The optical density at 454 nm was determined by the Beckman model DU spectrophotometer. The results obtained indicated that copper of plasma MAO is present in the cupric form. Aerobic or anaerobic addition of substrate, followed by aerobic or anaerobic addition of neocuproine, showed that no valency change of copper occurred during the catalytic activity. This conclusion was based on the facts observed that direct determination of copper in the enzyme with cuprizone accounted for all the copper, furthermore, that the enzyme was inhibited by cuprizone, which is specific for cupric copper, but not by neocuproine, which is specific for cuprous copper and, finally, that cupric copper was very effective in reactivating the copper-free enzyme. This conclusion has been confirmed by ESR studies (224).

Electron-Spin-Resonance Studies on Plasma Amine Oxidase (224). These investigations were carried out on 30 nanomoles of plasma amine oxidase in 0.2 ml of 0.06 M potassium phosphate buffer, pH 7.2, in a Varian V–4,500 ESR spectrometer using 100-kHz field modulation at −165°. The value of g_{max} was 2.053, and the hyperfine separation was about 155 gauss. This type of spectrum has been previously associated with cupric chelates of amino acids and proteins of a noncatalytic type (573), the hyperfine interval in blue cupric enzymes such as laccase and ceruloplasmin having been found to be about half as great. The content of cupric copper in plasma amine oxidase was determined by comparing the double-integral spectrum of the enzyme with that of cupric EDTA of known concentration as a standard, all other conditions such as temperature and sample geometry being identical. Using this method, 2.7 g-atoms of cupric copper was found per g-molecule of enzyme as compared with 3.7 g-atoms per g-molecule estimated chemically. As pointed out in Chapter IV.1.D, the reason for this discrepancy is unknown. When reducing agents such as sodium dithionite were added to the anaerobic enzyme, the cupric signal disappeared, indicating that the copper was entirely reduced. However, when substrate was added in large excess under the same conditions, the signal magnitude remained unaffected, which suggests that the copper did not undergo a valency change.

From the studies of the spectra of plasma amine oxidase and the fact that the native enzyme reacts with carbonyl reagents Yamada and Yasunobu (579, 846) concluded that pyridoxal phosphate was yet another prosthetic group of this enzyme. Thus, treatment of the enzyme with hydroxylamine resulted in a considerable loss of enzyme activity which was partially restored by the addition of both copper and pyridoxal phosphate (579). Upon the removal of copper by the sodium diethyl-dithiocarbamate method, the enzyme exhibited an absorption maximum at 380 nm, which was reported by Peterson and Sober (1053) as the maximum for pyridoxal phosphate.

Absorption Spectrum of Plasma MAO at Various pH Values. The spectrum of the crystalline plasma amine oxidase was studied by Yamada and Yasunobu (846) at various pH values. In acidic solution, the enzyme exhibited an absorption maximum at 480 nm, but in alkaline solution, an absorption maximum at 410 nm. An isosbestic point was recorded at 466 nm. The peak at 410 nm falls in the wavelength region expected for a Schiff base involving pyridoxal phosphate and an amino group in the enzyme (1054).

The spectrum of the copper-free plasma amine oxidase was also investigated at various pH values. In acidic solutions, the enzyme showed an absorption maximum at 375 nm, at pH 7.0, it exhibited a peak at 380 nm; and in alkaline solution, a maximum at 330 nm. Incidentally, an absorption maximum at 330 nm was obtained by Kent, Krebs, and Fischer (1054) with phosphorylase in neutral solutions and was considered by these authors to be due to the presence of a complicated Schiff base.

Carbonyl reagents exhibited an inhibitory effect not only on crude and partially purified plasma MAO (140, 141) but also on crystalline preparations of this enzyme (846). Marked inhibition was obtained with hydroxylamine, phenylhydrazine, and hydrazine. When any of these reagents was added to the native enzyme, the color of the enzyme solution changed from pink to yellow. Furthermore, the absorption band at 480 nm was replaced by peaks at 447, 370, or 410 nm, depending on whether hydroxylamine, phenylhydrazine or hydrazine, respectively, was added. The oxime of pyridoxal phosphate has been reported to show an absorption maximum at 330 nm (1055), and the phenylhydrazone of pyridoxal phosphate has been found to exhibit a maximum similar to that of the copper-free enzyme (1056).

The pyridoxal phosphate was found to be very tightly bound to the plasma MAO. Methods known to resolve other pyridoxal phosphate–dependent enzymes were unsuccessful with plasma amine oxidase. Thus, treatment of the latter enzyme with cysteine (1057), alkaline hydroxyl-amine (1058), perchloric acid (1059), urea (1060), incubation at an acid or alkaline pH (1061), or irradiation with ultraviolet wavelengths (1062), did not resolve the enzyme (846). Yamada and Yasunobu (846) used in these experiments 10 to 50 mg of crystalline enzyme so that, if pyridoxal phosphate had been liberated by any of the procedures, mentioned above, sufficient amounts of the latter compound would have been available for its spectrophotometric identification. Moreover, attempts to activate apotryptophanase (1063) by deproteinised extracts of plasma MAO were equally unsuccessful (846). From their experimental data Yamada and Yasunobu conclude that two types of pyridoxal phosphate–dependent enzymes seem to exist: one of them exhibiting a readily resolvable pyridoxal phosphate group, and the other one containing pyridoxal phosphate in a tightly bound form for which no method suitable for the resolution of the enzyme has been so far devised (846).

Sodium Borohydride Treatment of Plasma MAO. Since Kent, Krebs, and Fischer (1054) have demonstrated that several pyridoxal phosphate–dependent enzymes can be reduced by treatment with sodium borohydride with the formation of the pyridoxal enzyme derivative, Yamada and Yasunobu (846) allowed sodium borohydride to act at pH 13 on crystalline plasma amine oxidase. The crystalline enzyme (146.3 mg) was dissolved in 7 ml of 0.06 M phosphate buffer, pH 7.0, and the pH of the solution was adjusted to about 13 by the addition of 0.35 ml of 2 N NaOH. Then 5 mg of sodium borohydride were added, and the mixture was allowed to stand for 12 hr. This treatment resulted in the disappearance of the absorption peak at 480 nm and the formation of a shoulder in the region between 310 nm and 330 nm. At the end of the 12-hr period the enzymic activity was found to have completely disappeared. Finally, in another effort to demonstrate the presence of pyridoxal phosphate in plasma MAO Yamada and Yasunobu (846) submitted the crystalline enzyme to digestion with Pronase, and purified the liberated peptide, containing the chromophore, by chromatography on Dowex 50 and Dowex 1 resins. The spectral and fluorescence properties of the isolated chromophore were very similar to, but not identical with those of authentic pyridoxal phosphate.

Isolation of Chromophore from Plasma Amine Oxidase (846). Degradation of the enzyme with Pronase resulted in the liberation of the chromophore in the boiled, centrifuged supernatant. The greenish-yellow chromophore obtained was further purified by ion-exchange chromatography. For the experiment, 165 mg of crystalline enzyme were dissolved in 5 ml of 0.06 M phosphate buffer, pH 7.0, and boiled at 100° for 10 min. The solution was cooled and digested with 8.3 mg of Pronase (*Streptomyces griseus,* protease, B grade, California Corporation for Biochemical Research) for 6 hr, at 35°. The digestion was stopped by boiling, and the insoluble residue was removed by centrifugation. The supernatant was adjusted to pH 2.2 and was applied to a column of Dowex 50–X8, Na+ cycle, 2.0 × 25 cm, that had been equilibrated with 0.2 N acetate buffer, pH 3.1. Gradient elution was carried out by the addition of 2 N acetate buffer, pH 5.1, to 400 ml of 0.2 N acetate buffer, pH 3.1. Fractions of 7 ml were collected at a flow rate of 1 ml/min.

The elution pattern was followed by measuring the absorbancy at 295 nm. From the results of all these studies Yamada and Yasunobu tentatively concluded that beef-plasma MAO could be represented as a protein–$(Cu^{2+})_4$–(pyridoxal phosphate)$_2$ (see Chapter IV.2.A.b). The pyridoxal phosphate and the copper appear to be present as a chelate, since the native enzyme is pink in color and exhibits an absorption maximum at 480 nm, whereas the copper-free enzyme shows an absorption maximum at 380 nm. However, on addition of cupric copper to the copper-free enzyme this maximum tends to shift to the absorption maximum at 480 nm. The absorption maximum at 480 nm is neither characteristic of a pyridoxal-enzyme nor of any copper enzyme reported to date (579).

Gorkin (578) could not derive from his spectral studies on purified beef-plasma amine oxidase any evidence for the presence of pyridoxal phosphate in this enzyme and has claimed that zinc is a prosthetic group of the beef-plasma amine oxidase.

As pointed out in Chapter IV.2.A.b, the question of the prosthetic group(s) of beef-plasma amine oxidase requires further intensive study.

B. Isolation, Purification, and Crystallization of Plasma (Serum) Benzylamine Oxidase (Nonruminant-Plasma Amine Oxidase)

As mentioned in Chapter IV.2.B.a, in 1957 Bergeret, Blaschko. and Hawes (154) discovered that horse serum contained an enzyme that showed most of the properties characteristic of spermine oxidase with the one important exception that it did not possess the rapid action of the latter enzyme on spermine and spermidine. Since benzylamine was found to be rapidly oxidized, the provisional name "benzylamine oxidase" was proposed by the authors for this enzyme (154). Plasma enzymes of the benzylamine oxidase type are widely distributed in mammals (129, 145, 153).

a. PURIFICATION AND CRYSTALLIZATION OF BENZYLAMINE OXIDASE FROM HOG PLASMA ACCORDING TO BUFFONI AND BLASCHKO (146)

Fresh hog blood (20 to 22 liters) was obtained at the slaughterhouse and immediately mixed with 166 ml of the citrate solution used by Tabor et al. (141) and Yamada and Yasunobu (143). The mixture was centrifuged at 2° and the plasma collected.

Methods.

Spectrophotometric Enzyme Assay. A method first described by Tabor, Tabor, and Rosenthal (141), based on the measurement of the increase in absorption at 250 nm occurring on the enzymic formation of benzaldehyde from benzylamine, was used by Buffoni and Blaschko (146) at various stages of purification of the hog-plasma amine oxidase.

For the assay, the incubation mixtures contained in a total volume of 3.0 ml, 170 micromoles of sodium phosphate buffer, pH 7.4, 10 micromoles of benzylamine hydrochloride, and 1 unit of catalase (Worthington). The mixtures were incubated for 5 to 15 min, in air, while the vessel was agitated. The incubation temperature was 37.5° for samples obtained in the fractionation procedure, but pooled fractions were also tested at 25° to make it possible to express the activities in international units. One spectrophotometric unit was defined as the amount of enzyme which causes an increase in the optical density of 0.001/min. One spectrophotometric unit equals 11,300 international units.

Manometric Enzyme Assay. Substrates other than benzylamine were tested by comparing their rates of oxidation with that of benzylamine. In the manometric vessels the reaction volumes were either 2.0 or 0.7 ml, according to the amount of enzyme available. Substrate concentrations were $10^{-2} M$, and sodium phosphate buffer was used. One unit of catalase was added to prevent accumulation of hydrogen peroxide. The gas phase was oxygen, and the incubation temperature was either 25 or 37.5°.

Protein Determination. The protein content of the samples was determined either by measuring the ultraviolet absorption at 260 nm and 280 nm (1045), or by biuret method of Lowry et al. (997) against albumin standards.

Copper Determination. The copper content of the samples was measured by a radioactivation analysis as described by Bowen and Gibbons (1064).

Electron-Spin-Resonance Spectra (570). The ESR spectra of purified hog-plasma benzylamine oxidase were studied at −165°, on a Varian model V 4502X band spectrometer. The instrument was fitted with a dual sample cavity. Hyperfine splittings and G (gauss) values were determined by reference to a sample of 0.1% MnO in MgO that had been calibrated against a proton-resonance probe. Quantitative determination of the copper content was carried out according to the methods of Wyard (1065) and Beinert and Kok (1066) with 2 mM Cu^{2+} in 25 mM disodium EDTA as standard.

Crystalline benzylamine oxidase (570), which had been stored as a suspension in ammonium sulfate solution, was centrifuged and dialyzed at 4° against 0.01 M sodium–potassium phosphate buffer, pH 7.0. The protein concentration was determined by the biuret method of Lowry et al. (997) and was 16.6 mg/ml.

Samples of this enzyme solution were treated at pH 7.0 in anaerobic ESR tubes with either a twofold excess of sodium dithionite or a 60-fold excess of benzylamine hydrochloride. A reaction time of 15 min at 25°

was allowed before freezing in liquid nitrogen and study of the ESR spectra. Further addition of a fourfold excess of dithionite was made to the dithionite-reduced sample, and the ESR spectra were again measured.

In other experiments, duplicate samples (0.2 ml) of the enzyme solution were transferred to ESR tubes and their spectra recorded. Then 72% (w/v) perchloric acid (0.2 ml) was added to each tube, according to the procedure devised by Broman et al. (584), and the ESR spectra were again studied. Portions (0.1 ml) of these perchloric acid–treated samples were further digested with a perchloric acid–sulfuric acid mixture and analyzed for their copper content by the neocuproine method of Smith and McCurdy (1067).

The reaction catalyzed by hog-plasma benzylamine oxidase can be separated into two steps: an anaerobic one in which the addition of substrate bleaches the visible absorption spectrum of the enzyme (146) and an aerobic phase that occurs on oxygenation. Evidence has been obtained that during the anaerobic phase, the substrate is linked to the enzyme by an imine bond (582).

The results from the ESR study (570) clearly indicated that, irrespective of whether low (17.5 mW) or high (175 mW) microwave power was used to observe the resonance condition (1068), the copper in the enzyme remained bivalent after the anaerobic phase with benzylamine as substrate. However, reduction with sodium dithionite did result in a greatly decreased cupric ESR signal. For evaluation of the ESR spectrum of hog-plasma benzlamine oxidase see Chapter IV.1.D.

Electrophoresis. Horizontal starch-gel electrophoresis as described by Smithies (1069) was used. The voltage gradient was 6.1 V/cm.

Ultracentrifugation. The Spinco analytical ultracentrifuge, model E, was employed.

Purification Procedure. Blaschko et al. (145) have demonstrated that benzylamine oxidase activity can be precipitated with ammonium sulfate at saturations between 0.35 and 0.55.

Ammonium Sulfate Purification. Solid ammonium sulfate was added to plasma to 0.35 saturation. The precipitate that formed overnight was removed by centrifugation and discarded. More ammonium sulfate was added to 0.55 saturation. The sample was left standing overnight and was then centrifuged. The sediment was collected, dissolved in 0.01 M phosphate buffer, pH 7.0, and the solution was dialyzed for 18 hr against five changes of 5 liters each of the same buffer. Sörensen buffers (i.e., mixtures of KH_2PO_4 and Na_2HPO_4) were used throughout the entire purification procedure.

Column Chromatography on DEAE-Cellulose. Columns (4 $\times$ 55 cm) were set up with DEAE-cellulose (Whatman powder DE–50; capacity 1 mEq/g; pKa = 8.65), equilibrated with 0.01 M phosphate buffer, pH 7.0. The protein applied to each column was that obtained from about 7 liters of blood, corresponding to 30.4 g of protein in a final volume of 380 ml. The protein solution was applied to the column at a flow rate of 0.8 ml/min and was eluted from the column with two different concentrations of phosphate buffer, pH 7.0—that is, 0.03 M followed by 0.1 M. The flow rate was about 0.24 ml/min. The major portion of the enzyme protein (i.e., about 60%) emerged early in the elution. A second protein peak appeared after the phosphate concentration was increased to 0.1 M. This peak represented about 30% of the protein. This second peak probably contained ceruloplasmin, since the eluted material was blue.

The assay of enzymic activity showed the presence of two peaks with activity, coinciding with the two protein peaks. The first peak contained 38% of the enzymic activity applied to the column (60% protein), and the second peak 61% of the enzyme activity before chromatography (30% protein).

For further purification, the eluted fractions containing the highest specific activity were pooled and concentrated by precipitation with ammonium sulfate at 0.55 saturation. The precipitate obtained was dialyzed against 0.005 M phosphate buffer, pH 7.0.

Column Chromatography on DEAE-Sephadex. A column (4 × 50 cm) was set up with DEAE-Sephadex (A–50, medium; 3.5 mEq/g), pH 7.0. The enzyme was placed on the column and was desorbed from the resin by gradient elution (0–0.3 M NaCl in 0.005 M phosphate buffer, pH 7.0). The enzymic activity was eluted in a single peak, the protein with the blue color (probably ceruloplasmin) having emerged ahead of the peak containing the enzymic activity. The enzymically active samples were collected and pooled; solid ammonium sulfate was carefully added, the pH of 7.0 being maintained by the addition of a dilute aqueous solution of ammonia. The material precipitating between 0.35 and 0.55 saturation was collected by centrifugation at 20,000 g for 20 min. The sediment was dissolved in a small volume of 0.01 M phosphate buffer, pH 6.8, and the solution was dialyzed against 0.006 M phosphate buffer, pH 6.8. It was then subjected to chromatography on hydroxylapatite columns.

Chromatography on Hydroxylapatite Columns. The adsorbent hydroxylapatite was prepared according to Tiselius, Hjerten, and Levin (1048). A hydroxylapatite column (4 × 20 cm) was equilibrated with 0.006 M phosphate buffer, pH 6.8, and the enzyme solution from the preceding step was adsorbed onto it. Elution of the enzyme was achieved by the successive application of three different concentrations of the same buffer, 0.006 M, 0.06 M, and 0.2 M. The flow rate was 11 ml/hr during the loading of the column and 4 ml/hr during the elution of the enzyme. The enzymic activity was eluted with one main peak of high specific activity (90.66), followed by a secondary peak with about one-third of the specific activity of the main peak (29.05).

Crystallization. Crystallization was achieved with the material eluted in the main peak from the hydroxylapatite column; the enzyme eluted in the second peak was rechromatographed on a hydroxylapatite column before further treatment.

The eluates from the main peak of the hydroxylapatite column were concentrated by adding solid ammonium sulfate to 0.55 saturation; the pH was adjusted to 7.0 with ammonia. After 3 days the precipitate was collected by centrifugation at 15,000 g for 20 min, and dissolved in 4.8 ml of 0.05 M phosphate buffer, pH 7.0. To the pink solution, which contained 35.6 mg of protein per milliliter, solid, finely powdered ammonium sulfate was added until a slight turbidity persisted. The latter was immediately removed by centrifugation and discarded. To the supernatant a saturated ammonium sulfate solution was carefully added; when four drops of ammonium sulfate were added, a turbidity appeared, which disappeared again on stirring. The sample was then stored in the coldroom. Crystallization began after 12 hr and was not yet complete when after 16 days a batch of 85.8 mg of crystals was obtained by centrifugation. The main peak yielded 105 mg of crystalline material and from the second peak an additional 11.5 mg of crystals were obtained. Crystals obtained from the second peak, which had a slightly lower specific activity, showed a very slight contamination on electrophoresis. On recrystallization the impurity was removed. An overall 760-fold purification of the hog-plasma amine oxidase was achieved by Buffoni and Blaschko (146) with this technique. Purified hog-plasma amine oxidase has been found to be stable for months, when kept at 0 to 4° in the dark under ammonium sulfate (146).

Electrophoresis. Starch-gel electrophoresis of solutions of the crystals was carried out in various buffer systems: in phosphate buffer of ionic strength 0.1 and pH values of 7.0, 7.4, and 7.8, and also in the tris-citric acid–boric acid buffer system at pH 8.9, described by Poulik (1070). The crystalline enzyme gave one single band upon electrophoresis.

Ultracentrifugation. Ultracentrifugation was carried out in a 0.05 M phosphate buffer, pH 7.0, at 20°, and a sedimentation coefficient $S_{20,W}$, of 8.05×10^{-13} (±0.20 S.E.) was calculated. In the synthetic boundary cell the diffusion constant was determined as 4.01×10^{-7} cm²s⁻¹ (±0.003 S.E.). By assuming a partial specific volume of 0.75, the mixed average molecular weight was found to be about 195,000. In the Archibald approach to equilibrium method virtually the same figure was obtained, and the homogeneity of the enzymic sample proved to be higher than 95%.

Spectrum. Concentrated solutions of plasma amine oxidase are pink; this property is shared by spermine oxidase (143, 580) as well as pea-seedling amine oxidase (872). The absorption spectrum of a solution of crystalline hog-plasma amine oxidase, recorded on the Cary recording spectrophotometer showed a maximum in the ultraviolet at 278 nm and a maximum in the visible range at 470 nm.

For the study of the changes in the enzymic spectrum on the addition of benzylamine as substrate, an experiment was carried out by Buffoni and Blaschko (146). To a solution of the crystalline enzyme in 0.05 M phosphate buffer, pH 7.0, placed in a silica cell with a head similar to that of a Thunberg tube (872), benzylamine was added *in vacuo.* Before tipping the substrate into the enzyme solution, the spectrum was found to be essentially the same as that under aerobic conditions. Upon adding benzylamine *in vacuo,* the pink color and the peak at 470 nm disappeared immediately. The contents of the cell were then removed and dialyzed overnight against 0.05 M phosphate buffer, pH 7.0. The peak at 470 nm reappeared and the enzymic activity was found to be unimpaired. Upon the addition of benzylamine in an atmosphere of nitrogen the peak at 470 nm disappeared again.

Copper Analysis. The crystalline material was used for the determination of copper. A solution of 4.2 mg of protein was found to contain 4.17 μg of copper (i.e., 0.1% of copper). Assuming a molecular weight of 195,000, this would correspond to 3.04 g-atoms of copper per mole of protein.

Substrate Specificity. For the determination of the relative rates of oxidation of different amines the manometric method was used. Most of the experiments with the more impure fractions were carried out at a temperature of 37.5°, but the better purified and more active fractions were tested at 25°. The higher incubation temperature was chosen for the less active fractions, because at a lower temperature the rates of oxygen uptake were unsatisfactorily low, thus causing a great error in the reading. The temperature of 25° was chosen by the authors for the incubation of the more highly purified preparations to enable them to express the enzymic activity in international units.

The substrates chosen were benzylamine, histamine, mescaline, and 4-picolylamine. Benzylamine was chosen, because it has proved to be the most rapidly attacked substrate of hog-plasma amine oxidase and also for the sake of comparison, since this substance was also employed as a substrate in the spectrophotometric assay. Histamine was known to be oxidized by the partially purified hog-serum preparation (145). Mescaline was found by Blaschko, Ferro-Luzzi, and Hawes (865) also to be rapidly oxidized by hog-serum amine oxidase, and finally, the oxidation of 4-picolylamine by hog serum has already been described by Blaschko and Chrusciel (1071). The results obtained by Buffoni and Blaschko (146) indicate that throughout the various stages of purification the relative rates remained essentially unchanged; for example, all fractions with enzymic activity, including the crystalline enzyme, showed histaminase activity. The relative rate of oxidation of histamine was of the same order of magnitude as of the crude hog plasma. The affinity of the hog-plasma amine oxidase for histamine was determined manometrically at a temperature of 37.5° in oxygen using an enzyme preparation with a specific activity of 142.4×10^{-3} (25°, O_2). The Michaelis constant, K_m, was 1.1×10^{-3}. To determine the affinity of the enzyme for benzylamine, the spectrophotometric method was used as 37.5°, in air, and with an enzyme preparation showing a specific activity of 85.8×10^{-3} (25°, O_2). The value of K_m was 9×10^{-5} M, which indicates that the affinity of the hog-plasma enzyme for benzylamine was more than 100 times that for histamine.

The rates of oxidation of mescaline and 4-picolylamine by the crystalline hog-plasma amine oxidase were almost identical. Both mescaline and 4-picolylamine were less readily oxidized by the crystalline enzyme than benzylamine but more rapidly than histamine.

Although histamine appears to be readily oxidized by hog-plasma amine oxidase, Buffoni and Blaschko (146) are reluctant to accept histamine as the main natural substrate of the plasma amine oxidases. Thus, in some species histamine is attacked only very slowly. Furthermore, the intracellular histaminase has a much higher affinity for histamine (145) than the extracellular hog-plasma amine oxidase, and in the work just described, Buffoni and Blaschko have shown that the affinity of the hog-plasma enzyme for benzylamine is much higher than that for histamine. This seems to suggest that *in vivo* the main substrate of the plasma amine oxidase may resemble benzylamine rather than histamine.

As mentioned in Section 4.A.b, Yamada and Yasunobu (846) have recently suggested the presence of pyridoxal-5-phosphate in spermine oxidase, thus giving support to an idea first proposed by Werle and Pechmann (516) for the plant amine oxidase and later by Kapeller-Adler and MacFarlane (553) for the purified hog-kidney histaminase.

In 1965 Blaschko and Buffoni (215) obtained evidence that hog-plasma benzylamine oxidase contains a stably bound pyridoxal phosphate in the proportion of 4 moles of pyridoxal phosphate per mole of enzyme. To judge from the sensitivity of hog-plasma amine oxidase to various inhibitors both copper and pyridoxal phosphate appear to be essential for its activity (see Chapter IV.2.B.b and IV.2.B.d).

b. PURIFICATION AND IDENTIFICATION OF BENZYLAMINE OXIDASE FROM HUMAN PLASMA, ACCORDING TO MC EWEN, JR. (226)

In 1963 McEwen and Cohen (859) demonstrated that human plasma and serum are able to convert benzylamine to benzaldehyde. These authors were able to demonstrate that the enzyme responsible was different from other human-plasma enzymes that catalyze the oxidation of amines, such as DAO or ceruloplasmin.

In the work to be discussed (226), the purification of a soluble amine oxidase from human plasma is described. Although this enzyme has certain characteristics in common with plasma amine oxidases from lower mammals (128, 141, 145, 849), it could be distinguished from these other plasma amine oxidases by its substrate specificity. In his studies on the substrate specificity of the human-plasma enzyme McEwen noted (859) that the oxidation of aliphatic amines displayed a number of distinctive kinetic features. These chiefly concerned enzyme inhibition at high substrate concentration, strict dependence of apparent Michaelis and inhibitor constants on pH, and contributions of nonpolar residues of substrates to their affinities for the enzyme. These findings of McEwen appear to indicate certain characteristics related to the active center of the human-plasma enzyme.

Experimental Procedure. Fresh, heparinized whole blood was obtained from normal human donors or from the pump oxygenation apparatus used in surgical cardiopulmonary bypass. The plasma amine oxidases from both sources behaved identically during purification.

Methods

Standard Assay of Amine Oxidase Preparations. Amine oxidase activity was routinely measured with benzylamine as substrate by the spectrophotometric assay described by Tabor, Tabor, and Rosenthal (141), except that the assay was carried out at 25 rather than 30°. In this standard assay, 1 unit of enzymic activity was defined as the amount of enzyme catalyzing a change in absorbancy at 250 nm of 0.001/min in 3.0-ml reaction mixtures. The latter contained 0.2 M phosphate buffer, pH 7.2 (a mixture of Na_2HPO_4 and KH_2PO_4 was used throughout the purification procedure) and 3.3 mM benzylamine. Under these conditions, rates of aldehyde production were constant for at least 30 min and were proportional to the amount of enzyme added. Specific activities were expressed in units per milligram of protein, and protein was determined by the method of Lowry et al. (997), with crystalline bovine serum albumin as standard.

Modified Standard Assay for the Determination of Benzylamine Oxidase Activity in Crude Human Plasma or Serum (859). The sensitivity of this assay depends on the high molar extinction coefficient of benzaldehyde, the reaction product (141). The procedure was as follows:

Step 1. Assay and control reaction mixtures were placed in open test tubes (0.6 × 10 cm) and incubated in a Dubnoff metabolic shaker bath at 37° for 3 hr. The assay tube contained, in a total volume of 1.5 ml, 0.6 ml serum (or plasma), 0.75 ml of 0.2 M phosphate buffer, pH 7.2, and 0.15 ml of 0.008 M (1.2 micromoles) benzylamine in the same buffer. The control tube was identical except that the substrate benzylamine was added only after incubation was completed.

Incubation of human serum or plasma with benzylamine under standard conditions yields a product that is extractable with cyclohexane from acid solution. This product was identified as benzaldehyde by difference spectra of the cyclohexane extract obtained from the control tube. Difference spectra from the incubation of serum or plasma from over 50 different individuals revealed only the spectrum of benzaldehyde. In particular, the absorption maximum at 232 nm attributable to benzoic acid in cyclohexane was not encountered.

Step 2. After incubation was completed and after the addition of substrate to the control tube, 0.15 ml of 60% perchloric acid and then 1.5 ml of cyclohexane were added to each tube. The contents of all tubes were emulsified with the aid of glass stirring rods, allowed to stand at room temperature for 15 min, and then, after a second emulsification, centrifuged for 10 min at 2000 rpm.

Step 3. The absorbancy of the cyclohexane extract from the assay tube against that of the control extract was measured at 242 nm with a Beckman model DU spectrophotometer and cuvettes of 1-cm light path. The difference in optical density was multiplied by 100 to convert to enzyme units. The spectral characteristics of the cyclohexane extracts were stable at room temperature for 24 hr, if capped before storage. The optical density of the cyclohexane extract of the control tube, as measured against cyclohexane at 242 nm, was less than 0.060 for all sera.

The relation of the modified assay to the standard amine oxidase assay was investigated by McEwen (226), who found that 1 unit of enzymic activity, measured by the standard assay, corresponds to a change in absorbancy of 0.0042/min in the modified procedure.

For the purpose of measuring rates of benzylamine oxidation as well as determining benzaldehyde concentrations, the molar absorptivity of benzaldehyde at 250 nm was assumed to be 1.2×10^4 mole^{-1} cm^{-1} (1072).

Determinations of Hydrogen Peroxide. Hydrogen peroxide was measured by a modification of the peroxidase assay (1030) used for the colorimetric estimation of glucose (Glucostat). Since even purified preparations of the plasma amine oxidase contained catalase activity, the benzylamine oxidase reaction was coupled to the peroxidase reaction. In such an assay, 50 units of plasma amine oxidase, 40 to 160 nanomoles of benzylamine, 300 μg of horseradish peroxidase, 60 μg of o-dianisidine in 25 μl of absolute methanol, and 300 micromoles of sodium pyrophosphate buffer, pH 9.0, were incubated in a total volume of 3.0 ml at 25° until hydrogen peroxide production, as indicated by the change in optical density at 400 nm, was complete. Controls without substrate or without amine oxidase gave no change in optical density at 440 nm. Optical density was converted to nanomoles of hydrogen peroxide. Under these conditions the chromophore resulting from the enzymically produced hydrogen peroxide was stable for 1 hr and was not affected by the amines used as substrates for the plasma amine oxidase. Its absorbancy at 440 nm was proportional to the concentration of hydrogen peroxide in the range of 5 to 75 μM. The concentrations of added hydrogen peroxide solutions, used as standards for the coupled assay, were determined by the change in optical density at 230 nm after the addition of catalase, and the molar absorptivity of hydrogen peroxide at 230 nm was assumed to be 61 mole^{-1}cm^{-1} (1073). With benzylamine as the substrate of the plasma amine oxidase, rates of hydrogen peroxide production agreed well ($\pm$ <5%) with rates of benzaldehyde production.

Assay of Catalase Activity.   This was performed spectrophotometrically by a modification (1074) of the method of Beers and Sizer (1075). One unit corresponds to the decomposition of 1 micromole of hydrogen peroxide per minute at 25° and pH 7.0.

Manometric Measurements.   Oxygen consumption was followed in the Warburg constant volume respirometer with air as the gas phase (1076).

Ammonia Determinations.   Ammonia was determined colorimetrically after separation from the sample by microdiffusion, according to the method of Brown et al. (1077).

Measurement of Kynuramine Oxidation.   The oxidation of kynuramine by mitochondrial MAO results in the production of 4-hydroxyquinoline, which may be measured spectrophotometrically (938; see Section 2.D.e). Difference spectra obtained after incubation of kynuramine with purified human-plasma amine oxidase were identical with the spectrum of authentic 4-hydroxyquinoline. For the spectrophotometric measurement of initial rates of kynuramine oxidation, the molar absorptivity of 4-hydroxyquinoline at 316 nm was taken to be 1.2×10^4 mole^{-1} cm^{-1} at pH 8.9 (1078).

Spectrophotometric Measurements. Initial rates of amine oxidation, as determined spectrophotometrically by the direct and coupled assays, were obtained with a Cary recording spectrophotometer (model 11 M) fitted with a thermostatically controlled cuvette chamber and cuvettes with a light path of 1 cm. For other spectrophotometric measurements a Beckman model DU spectrophotometer was used.

Purification of Human-Plasma Amine Oxidase. Except where otherwise stated, all operations were carried out at 4°. All centrifugations were done at 1300 g for 45 min. The procedure was as follows:

Step 1. Collection of plasma. Plasma was obtained by centrifugation from human whole blood, containing 4500 USP units of heparin per liter. The plasma was used immediately or stored for not more than 2 weeks at —20° before use.

Step 2. First ammonium sulfate and ethanol fractionations. To each liter of plasma was added 400 mg of disodium EDTA and then, with stirring, 300 g of solid ammonium sulfate. The precipitate was collected by overnight filtration and extracted with 200 ml of distilled water. After removal of insoluble material by centrifugation, the extract was diluted to a volume of 400 ml with distilled water and cooled to the point of freezing. The extract was treated with 40 ml of a 0.1 M aqueous solution of manganese chloride and then, with stirring, with 500 ml of 95% ethanol that had been cooled to —10°. After immediate centrifugation at —10°, the precipitate was discarded, and an additional 200 ml of precooled 95% ethanol was added. The mixture was centrifuged and the supernatant solution was discarded. The precipitate was extracted into 200 ml of 0.2 M dipotassium hydrogen phosphate solution containing 1 mM EDTA. Insoluble material was removed by centrifugation, and the extract was dialyzed against three changes of two liters of 0.02 M dipotassium hydrogen phosphate solution containing 0.1 mM EDTA.

Step 3. Second ammonium sulfate fractionation. To each liter of the dialyzed solution were added 250 g of solid ammonium sulfate. The precipitate was removed by centrifugation, and an additional 100 g of solid ammonium sulfate were added to the supernatant solution. The resultant precipitate was collected by centrifugation, and dissolved in 10 ml of 0.2 M phosphate buffer, pH 7.2 (Sörensen phosphate buffer, a mixture of Na_2HPO_4 and KH_2PO_4, was used throughout the entire purification procedure). If the specific activity of this solution was greater than 2.0, it was used directly in step 4. If it was not, the solution was refractionated with solid ammonium sulfate before the next step. On refractionation, the enzyme precipitated from solutions containing 1.3 to 1.5 M ammonium sulfate.

Step 4. Adsorption with Cγ-alumina. Cγ-alumina gel was prepared according to Willstätter and Kraut (925) and aged over 6 months before use. The product of the above ammonium sulfate step was then treated with Cγ-alumina gel (suspended in distilled water) to adsorb inert protein. This "negative adsorption" step gave results that varied from preparation to preparation and necessitated the use of pilot experiments. In general, a procedure was used as follows: 2.5 ml of a 50 g/100 ml (dry weight) suspension of Cγ-alumina were added per 100 mg of protein and the mixture was centrifuged; 0.5 ml aliquots of the suspension per 100 mg of protein were then added until the intensely blue color of the supernatant solution (due to contaminating ceruloplasmin) was completely gone; at this point further additions of C γ-alumina resulted in marked increases in specific activity but at the expense of losses in total activity. For this reason, addition of C γ-alumina was usually not continued after 30% of the enzymic activity had been adsorbed. Specific activities after adsorption varied from 10 to 120. The enzymic activity in the final supernatant solution was precipitated by the addition of solid ammonium sulfate to a final concentration of 2 M and the precipitate was dissolved in a minimal volume of 0.2 M phosphate buffer, pH 7.2. Before use, this concentrated enzyme solution was freed of detectable ammonium ions by dialysis against the same buffer.

By this procedure McEwen (226) achieved an approximately 5000-fold concentration of the plasma amine oxidase. The yield was, however, a very poor one, amounting only to 9%. McEwen found, however, that 90% of the amine oxidase activity, discarded in steps 3 and 4, could be "reworked" in the same manner by adsorbing that activity on C γ-alumina and eluting it with 0.2 M phosphate buffer, pH 7.2.

Properties of Plasma Amine Oxidase. After steps 3 and 4, enzyme preparations in 0.2 M phosphate buffer, pH 7.2, could be stored at —20° for 6 months or more without detectable loss of activity. There was no loss of activity on storing the enzyme at 4° for 48 hr, or keeping it at 25° for 4 hr. On dialysis against repeated changes of 0.2 M phosphate buffer, pH 7.2, for 12 hr approximately 10% of enzyme activity was lost. The specific activity of the purest benzylamine oxidase preparation from human plasma amounts, according to Buffoni (131), to only one-thirtieth of the specific activity of the crystalline benzylamine oxidase of hog plasma. For the discussion of chemical properties, substrate specificity, and effects of inhibitors relating to human-plasma benzylamine oxidase see Chapter IV.2.B.b–d.

C. PURIFICATION AND CHARACTERIZATION OF RABBIT-SERUM MAO, ACCORDING TO MC EWEN, JR., CULLEN, AND SOBER (863)

The authors identified an MAO in rabbit serum and purified it to 450 times the original specific activity by a procedure involving fractionation with ammonium sulfate and adsorption on calcium phosphate gel. The purified enzyme shows a spectral absorption maximum at 470 nm, which is decreased in the presence of substrate. The enzyme oxidizes primary amines with the stoichiometric formation of the corresponding aldehydes, hydrogen peroxide, and ammonia. Substrates most actively oxidized by this enzyme are benzylamine, dopamine, tyramine, and mescaline, as well as long-chain aliphatic amines. Less actively oxidized substrates include tryptamine, serotonin, histamine, and simple aliphatic amines. The ability of rabbit serum to oxidize benzylamine was found to be 10 times that of human serum (863). The authors studied with the purified enzyme the effect of pH on apparent Michaelis constants and maximal velocities of benzylamine oxidation, and found that the protonated form of the amine interacts with the free enzyme and that this interaction is affected by an ionization (pK_a 6.2) of the enzyme. An ionization (pK_a 6.4–6.5) of the enzyme–substrate complex is a determinant of maximal velocities.

In its general substrate specificity the rabbit-serum amine oxidase resembles the plasma amine oxidases of other mammals (e.g., in attacking only primary amines but not secondary and tertiary ones nor amines with α-substitution). It differs, however, from the beef-plasma amine oxidase (141, 143) in not degrading spermine and spermidine, and from the human-plasma amine oxidase in not attacking kynuramine (226), but on the other hand in deaminating mescaline as actively as long-chain aliphatic amines (863). Thus, although similar in many respects to that of the human-plasma amine oxidase, the substrate specificity of the rabbit-serum enzyme resembles more closely that of the hog-plasma benzylamine oxidase (128, 145, 146; see also Chapter IV.4).

Experimental Procedure.   Frozen serum from domestic rabbits was purchased from Pel-Freeze Biologicals, Inc. Recrystallized bovine liver catalase and lyophilized horseradish peroxidase (RZ > 1) were purchased from Worthington, and were freed from ammonium salts before use by dialysis against 0.02 M phosphate buffer (Na_2HPO_4–KH_2PO_4), pH 7.4. Calcium phosphate gel was prepared according to Keilin and Hartree (1044), with the exception that distilled water was used throughout the preparation; before use, the gel was centrifuged at 1300 g for 1 hr and then suspended in an equal volume of distilled water. Benzylamine was purchased as the free base from Distillation Products Industries and purified by distillation under reduced pressure in a helium atmosphere (859). All other amines were obtained from commercial sources.

Methods. *Standard Assay for Measurement of Amine Oxidase Preparations.* This activity was routinely measured by a standard assay with benzylamine as substrate by the spectrophotometric method of Tabor, Tabor, and Rosenthal (141), except that the assay was carried out at 25 instead of at 30°. In this standard assay, 1 unit of enzymic activity was defined as the amount of enzyme catalyzing a change in absorbancy at 250 nm of 0.001/min (0.25 nanomoles of benzaldehyde per minute) in 3.0 ml of reaction mixture 3.3 mM with respect to benzylamine and 0.2 M with respect to phosphate buffer, pH 7.2. Under these conditions, rates of benzaldehyde production were constant for at least 30 min and were proportional to the amount of enzyme added. Addition of 800 units of catalase (with or without 500 micromoles of ethanol) to this standard assay did not affect reaction rates. Specific activities were expressed in units per milligram of protein. Since serum does not contain significant amounts of nucleic acids, protein was estimated by the absorbancy at 280 nm with the use of the $E \begin{smallmatrix} 1\% \\ 1\text{ cm} \end{smallmatrix}$ value of 9.8, applied during the purification of beef-plasma spermine oxidase (143). For the purpose of measuring rates of benzylamine oxidation, the molar absorptivity of benzaldehyde at 250 nm was assumed to be 12 $\times$ 10³ mole⁻¹ cm⁻¹ (1072).

Manometric Method. Oxygen consumption was followed in the Warburg constant volume respirometer (1076) with air as gas phase.

Hydrogen Peroxide Determination. Initial rates of hydrogen peroxide production were measured by a modification (226) of the o-dianisidine peroxidase assay used for the colorimetric estimation of glucose (1030). As previously reported (226; see Section 4.B.b), standard solutions of hydrogen peroxide were used to convert changes in optical density at 440 nm into nanomoles of hydrogen peroxide.

Assay of Catalase Activity. Catalase activity was measured spectrophotometrically by a modification (1074) of the method by Beers and Sizer (1075). One unit corresponds to the decomposition of 1 micromole of hydrogen peroxide per minute at 25° and pH 7.0. With the standard amine oxidase assay, catalase did not oxidize benzylamine.

Spectrophotometric Methods. Initial rates of amine oxidation as well as absorption spectra were measured with a Cary model 11 M recording spectrophotometer, fitted with a thermostatically controlled cuvette chamber, and cuvettes with a light path of 1 cm. A Beckman model DU spectrophotometer was used for other spectrophotometric measurements.

Purification of Plasma Amine Oxidase. All operations were carried out at 4°. Except where otherwise noted, all centrifugations were at 1300 g for 1 hr. The procedure was as follows:

Step 1. First ammonium sulfate fractionation. Each liter of serum was stirred with 200 g of solid ammonium sulfate and then filtered overnight through filter paper. To each liter of filtrate were added 120 g of solid ammonium sulfate. The resultant precipitate, which contained all the enzymic activity, was collected by centrifugation and dissolved in sufficient distilled water to produce a final volume equal to 22.5% of the original serum volume.

Step 2. Second ammonium sulphate fractionation. To each 900 ml of solution from step 1 were added 100 ml of 2 M K_2HPO_4 followed by 100 g of ammonium sulfate. The small precipitate formed was separated by centrifugation and discarded. The supernatant fluid was treated with 20 g of ammonium sulfate over a 10-min period. The precipitate was separated by centrifugation and dissolved in the minimal amount of distilled water. The addition of 20-g amounts of ammonium sulfate was repeated until all the enzymic activity was precipitated. Fractions with specific activity of 2.0, or greater, were pooled. If significant activity was present in other fractions, the latter were combined, diluted with distilled water to a protein concentration of approximately 50 mg/ml, and re-fractionated by the same procedure.

Step 3. Adsorption on calcium phosphate gel. The intensely blue product of step 2 was dialyzed for 24 hr against four changes of distilled water. Each change of distilled water was 20 times the volume of the enzyme solution. The dialyzed solution was diluted with distilled water to a protein concentration of 30 mg/ml and then treated, with stirring, with a volume of a suspension of calcium phosphate gel equal to one-half the volume of the diluted enzyme solution. The gel was then separated by centrifugation at 1300 g for 3 hr and discarded. The supernatant solution was repeatedly treated with volumes of gel equal to one-fourth the gel used in the first addition. After each addition of calcium phosphate, the gel was separated by centrifugation at 1300 g for 3 hr, and the adsorbed protein was eluted by resuspension and centrifugation with increasing concentrations (0.01 to 0.1 M) of phosphate buffer (Na_2HPO_4–KH_2PO_4), pH 7.0. Volumes of buffer used for the elution of the gels were four times the volume of packed gel. Enzyme solutions with specific activity of 40, or greater, were pooled and concentrated by the addition of 500 g of ammonium sulfate per liter of enzyme solution. The precipitate collected by centrifugation was dissolved in 0.2 M phosphate buffer, pH 7.2.

Ceruloplasmin, which accompanies the rabbit-serum amine oxidase through the initial purification steps, could be used as a guide during the gel adsorption. The adsorption of ceruloplasmin preceded that of the amine oxidase. It should be also mentioned that elution of the blue gels in the same fashion as that used for the amine oxidase yielded purified preparations of ceruloplasmin that were free of amine oxidase activity.

By the procedure just described, the rabbit-serum MAO was purified to 450 times its original activity. Its specific activity with benzylamine as substrate was one-sixth that of the crystalline hog-plasma benzylamine oxidase (146), one-fifth that of the crystalline beef-plasma spermine oxidase (143), and approximately equal to the most highly purified preparations of the human-plasma amine oxidase (859).

The spectral properties of rabbit-serum amine oxidase are similar to those obtained under comparable conditions with beef-plasma spermine oxidase (143) and are indistinguishable from those reported for the hog-plasma benzylamine oxidase (146).

Like all purified plasma amine oxidases (131, 143, 146, 226), the rabbit-serum amine oxidase is stable in the frozen state (863).

Reaction Mechanisms. The possibility that the initial steps of enzymic deamination may be identical with those of enzymic transamination has led to the proposal that pyridoxal catalysis is basic to reactions of certain amine oxidases (1079, 1080). Such proposals have been supported by non-enzymic oxidative deaminations catalyzed by pyridoxal in the presence of copper ions and by oxidative deamination of pyridoxamine phosphate itself in the presence of copper ions and oxygen (1079, 1081). They are also consistent with the findings that plasma amine oxidases are sensitive to carbonyl reagents (129, 141) and deaminate primary but not secondary or tertiary amines (129).

Recent evidence indicates that pyridoxal phosphate as well as copper is a tightly bound constituent of the amine oxidases of beef plasma (580, 846) and of hog plasma (146, 215). It has been suggested that pyridoxal phosphate is bound to the apoenzyme by a type of Schiff-base linkage encountered in other pyridoxal enzymes (846, 856). McEwen et al. (863) believe that despite the lack of direct evidence that the rabbit-serum amine oxidase is a pyridoxal protein, the marked similarity of the rabbit-serum enzyme to the hog-plasma enzyme with regard to substrate specificity as well as visible absorption spectrum makes this likely.

C. Methods of Isolation, Purification, and Estimation of Plant Amine Oxidase, Pea-Seedling DAO

a. PURIFICATION, IDENTIFICATION, AND ESTIMATION OF PEA-SEEDLING DAO, ACCORDING TO MANN (592, 872) AND HILL AND MANN (576, 873)

The amine oxidase of pea seedlings belongs to the group of amine oxidases that is inhibited by carbonyl reagents and semicarbazide (872). This finding led to the suggestion that the enzyme contains a functional aldehyde or ketone group. Although evidence has been reported by Werle and Pechmann (516) and Goryachenkova (560) that the prosthetic group of partially purified plant amine oxidase preparations contains pyridoxal phosphate, this claim has not been substantiated with highly purified pea-seedling enzyme preparations (517, 874). The highly purified preparations of pea-seedling DAO give pink solutions, and the absorption spectra show a band in the visible region with a maximum at about 500 nm (872). Spectrophotometric investigations of the reactions of the purified pea-seedling preparations with 1,4-diaminobutane, sodium dithionite, and hydrazine sulfate, supported the suggestion that the pink color is a property of the enzyme. The pure enzyme preparation contained 0.08 to 0.09% of copper, which could be removed only by dialysis against sodium diethyldithiocarbamate; the copper-free preparations were catalytically inactive but most of the original activity was restored by adding Cu^{2+} ions to the dialyzed solution (873).

The copper is firmly bound to the enzyme protein, and the plant DAO may, according to Hill and Mann (873), therefore be classified as a metalloenzyme (1082) containing cupric copper. No cuprous copper was encountered by Hill and Mann in pea-seedling DAO by means of the 2,2'-biquinolyl estimation method for cuprous copper in proteins, as modified by Felsenfeld and Printz (1083) and Felsenfeld (1084). Furthermore, neocuproine, a reagent specific for cuprous copper, did not inhibit the enzyme. These findings suggest that the pea-seedling DAO reaction mechanism does not involve valency change in the copper (see Chapter IV.1.D).

Mann (872) suggested that the copper is present in the enzyme as a complex with a carbonyl compound and that this complex forms the prosthetic group of the enzyme. This theory seems to have been confirmed by recent results of Hill and Mann (576). Werle and his colleagues (517), who also obtained from pea seedlings highly purified enzyme preparations, that were rose red in concentrated solutions and contained 0.12% copper, also suggested that the rose color is a property of the enzyme (see Chapter IV.5.B).

Materials and Methods.   Hydroxylapatite was prepared by the method of Tiselius, Hjerten, and Levin (1048). Sodium phosphate buffers were prepared from Na_2HPO_4 and H_3PO_4. Potassium phosphate buffers were prepared from KH_2PO_4 and KOH.

Estimation of Amine Oxidase Activity. The enzyme activity was estimated manometrically, in air, in the Warburg apparatus at 25° (576, 873). The volume of the reaction mixture was 3 ml, and 0.2 ml of 5 N KOH was present in the center well. A unit of amine oxidase is defined as the amount that at 25° catalyzes the oxidation of 1 micromole of 1,4-diaminobutane per minute, giving an initial rate of uptake of oxygen of 11.2 μl/min in the presence of 10mM 1,4-diaminobutane, 25 μg of catalase, and 67 mM phosphate buffer, pH 7.0, in a total volume of 3 ml. The specific activity is defined as amine oxidase units per milligram of protein.

Estimation of Total Nitrogen (872). The total nitrogen of the amine oxidase preparations was estimated by a micro Kjeldahl method.

Estimation of Copper (872). The copper content of the amine oxidase preparations was estimated colorimetrically. For each estimation, a sample (2–4 ml) was digested with 2 ml of a mixture (1 : 1, v/v) of concentrated sulfuric acid and redistilled nitric acid. The copper content was estimated with sodium diethyldithiocarbamate by the procedure of Forster (1085), except that the extinction of the final solutions was measured at 436 nm.

Spectrophotometry. Extinctions were measured with an Optica **CF4 DR** recording spectrophotometer fitted with silica cells of 1-cm light path (576).

Preparations and Purification of the Enzyme (592, 872). The amine oxidase preparations were obtained from 9-day-old pea seedlings, grown as described by Kenten and Mann (868) except that the roots of the seedlings were discarded. Pea seedlings were washed free from soil, and any obviously diseased material was removed. The washed seedlings were minced in a chilled domestic meat mincer and squeezed through strong cotton cloth. The residual pulp was mixed with 500 ml of 0.067 M phosphate buffer, pH 7.0, macerated for 2 min in a Townson and Mercer (Croydon) macerator and again squeezed through cloth. The combined extracts were cooled in an ice-salt freezing mixture to 0 to 5°, and a mixture of ethanol (200 ml) and chloroform (100 ml), previously cooled to −10°, was added slowly during vigorous mechanical stirring. The stirring was continued for 30 min, and the mixture was then centrifuged. The almost clear yellow supernatant was poured off from the bulky white precipitate and from the bottom layer of chloroform and again cooled. Forty-five grams of ammonium sulfate per 100 ml of supernatant were then added. This precipitated the amine oxidase leaving the bulk of the pea-seedling peroxidase in solution. On centrifuging, the precipitate formed a hard cake on the surface of the liquid. A broad spatula was inserted under the precipitate to support it while the liquid was poured off. The precipitate was ground in a mortar to a smooth paste and 500 ml of 0.02 M phosphate buffer, pH 7.0, was slowly added to give a smooth suspension that was stirred mechanically for 2 hr at room temperature and stored overnight at 0 to 2°. It was then centrifuged and the bulky precipitate was washed on the centrifuge with a further small amount of 0.02 M phosphate buffer, pH 7.0. The extracts were pooled and cooled in ice. The small precipitate obtained by the addition of 18 g $(NH_4)_2SO_4$ per 100 ml of extract was separated by centrifugation and discarded. To each 100 ml of the cooled supernatant a further 18 g $(NH_4)_2SO_4$ was added and the precipitate, which contained the amine oxidase, was collected by centrifugation. It was resuspended in 20 ml of 0.2 M phosphate buffer, pH 7.0, and the suspension was dialyzed for several hours against running tap water and then overnight, at 0 to 2°, against 2 liters of 0.005 M phosphate buffer, pH 7.0. The dialyzed suspension was centrifuged and the cooled supernatant was brought to pH 5.0, by dropwise addition of 0.05 N acetic acid. The resultant suspension was kept at 0 to 2° for several hours until the precipitate flocculated. The precipitate was then collected by centrifugation, triturated with 20 ml of water, and brought into solution by adjusting to pH 7.0 with 0.05 N KOH. The precipitation at pH 5.0 was repeated twice. The final solution at pH 7.0 was centrifuged clear and stored at −10°. Preparations stored in this way showed little loss of activity over several months. Little or no loss of activity occurred during dialysis against distilled water or 0.005 M phosphate buffer, pH 7.0, even when this was prolonged over several days. During prolonged dialysis against distilled water the enzyme tended to precipitate, but the precipitate redissolved on addition of phosphate buffer. By this procedure the amine oxidase of pea-seedling extracts was purified up to 300 times its original activity (872), and this preparation was used as starting material for further enzyme purification by chromatography (576). The weight of the plant material used for each preparation was 1 to 3 kg. The volume of the enzyme preparations in 10 mM potassium phosphate buffer, pH 7.0, was such that 1 ml was equivalent to 100 g of the plant tissue.

Chromatography on Hydroxylapatite. All manipulations during the chromatography of the enzyme on hydroxylapatite and in the subsequent stages of purification were carried out at 0 to 5°. Five to eight of the initial oxidase preparations were combined and centrifuged. A small sample of the supernatant solution was retained for initial enzyme estimations, and the rest was applied to a hydroxylapatite column (4 × 12 cm), equilibrated with 10 mM potassium phosphate buffer, pH 7.0. Elution started with 40 mM potassium phosphate buffer, pH 7.0, and continued by stepwise increments (40 mM) in the concentration of buffer up to a final concentration of 200 or 240 mM. The volume of buffer at each step was 100 ml. The rate of flow varied from 50 to 80 ml/hr in different experiments and the eluate was collected in 5-ml fractions. The fractions of the main amine oxidase peak showing the highest specific activity (i.e., amine oxidase activity in relation to $E_{280\ nm}$) were combined and dialyzed against two changes of 3 liters each of distilled water. The ratio of amine oxidase units/$E_{280\ nm}$ in the pooled fractions generally exceeded 10^4. Fractions in which the ratio was less than 5×10^3 were rarely included.

Adsorption of Inert Material on Aminoethyl- or Diethylaminoethyl-Cellulose. The dialyzed amine oxidase solution was then passed through a column (2×10 cm) of AE-cellulose or DEAE-cellulose which was then washed with 25 ml of 10 mM potassium phosphate buffer, pH 7.0. Whatman AE–50 and DE–50 powders were used for the columns, which were equilibrated with 10 mM potassium phosphate buffer, pH 7.0, before use.

Concentration on Hydroxylapatite (872). The enzyme solutions and washings from AE- or DEAE-cellulose columns were pooled and applied to a column of hydroxylapatite (4×4 cm). The enzyme was eluted with 200 mM potassium phosphate buffer, pH 7.0, and the active fractions of the eluate were combined and stored at $-10°$.

The chromatographic purification procedure together with the preliminary fractionation with ammonium sulfate gave a purification to 800 times the original activity. The final preparations were pink solutions with maximum absorption at about 500 nm; this was the only band detected in the visible part of the spectrum. The pink color of the enzyme solutions was discharged by sodium dithionite and was restored by oxygenation. The enzyme solution changed from pink to yellow when 1,4-diaminobutane was added under anaerobic conditions. The pink color was restored by oxygenation of the mixture. Furthermore, the color of the enzyme solution changed from pink to yellow, when hydrazine was added. On standing in air, the yellow slowly faded but the mixture did not regain its pink color. The copper content of the enzyme preparations increased during purification to 0.08 to 0.09%. The copper was removed by precipitation with sodium diethyldithiocarbamate. The supernatant after dialysis, carried out to remove sodium diethyldithiocarbamate, retained little of the activity of the original preparation. Most of the lost activity was, however, restored by the addition to the dialyzed solution of Cu^{2+} ions (576, 872).

Diamine oxidase solution (20 ml containing 500 units/ml) was mixed with 1 ml of 0.1 M sodium diethyldithiocarbamate and kept at 0 to 5°

for 15 hr. The precipitated copper–diethyldithiocarbamate complex was removed by centrifuging for 30 min at 18,000 g, at 2°. The supernatant solution sometimes remained slightly turbid and was clarified by filtration through a layer of acid-washed kieselguhr. The clear orange-pink solution was dialyzed for 48 hr at 0 to 5° against four changes of 250 ml of 10 mM orthophosphate buffer, pH 7.0. The nondiffusible material was centrifuged for 30 min at 18,000 g and the supernatant solution was stored at −10°.

Reactivation of the Copper-Free Protein by Cu²⁺ Ions (576). The copper-free protein (10 μg) and catalase (25 μg) were added to orthophosphate buffer, 67 mM, pH 7.0, and incubated at 25°, in the presence and absence of added Cu^{2+} ions, for 0 to 3 hr. EDTA (1 mM) was then added to stop reactivation and to prevent inactivation by Cu^{2+} ions during the subsequent assay. Incubation in the absence of Cu^{2+} ions produced a slow partial activation. The rate of reactivation was increased by adding 0.1 μM–Cu^{2+} ion and was then almost complete in 3 hr. Since the enzyme contains 0.087% of copper, complete reactivation of 10 μg of copper-free protein requires, according to Hill and Mann (576), combination with 8.7 ng of Cu^{2+} ions or about 1.4 ml of 0.1 μM Cu^{2+} ion. With 10 μM–Cu^{2+} ion reactivation was complete within 5 min.

b. PURIFICATION AND CHARACTERIZATION OF PEA-SEEDLING DAO, ACCORDING TO WERLE, TRAUTSCHOLD, AND AURES (517)

Methods. Diamine oxidase activity was estimated in conventional Warburg manometers in the presence of catalase using cadaverine as substrate. Enzyme solution (0.05–1.0 ml) and 0.1 ml of a catalase solution [0.3 ml of a suspension of crystalline catalase (Boehringer), made up to 10 ml with 0.067 M phosphate buffer, pH 7.0], placed in the main compartment, were made up with 0.067 M phosphate buffer, pH 7.0, to 2.5 ml. The side arm contained 0.5 ml of a 30 mM cadaverine solution. Substrate concentration in the final volume of 3 ml was 0.01 M. The center well contained 0.25 ml of 10% KOH. Temperature was 37° and the atmosphere oxygen.

Protein was determined in crude extracts by the biuret method of Weichselbaum (1086) and in purified preparations by measuring the absorption at 260 and 280 nm, according to the method of Warburg and Christian (1045). An enzyme unit was defined as the amount of enzyme that under standard assay conditions takes up 1 μl of oxygen in 10 min, and the specific activity as the amount of enzyme units per milligram of protein.

Concentration of Pea-Seedling DAO. Three kilograms of 6- to 10-day old pea seedlings were washed, and then homogenized using an equal amount of cold 0.067 M phosphate buffer, pH 7.0. The mixture was squeezed through a nylon cloth, and the pulp was reextracted with half the amount of 0.067 M phosphate buffer, the mixture being again squeezed through a nylon cloth. To the combined crude extracts 0.5 N acetic acid was then added to reach pH 5 to 6, and the precipitate of inert protein was separated by centrifugation and discarded. The supernatant that contained the DAO was precipitated with solid ammonium sulfate, added to produce 65% saturation. After centrifugation, the precipitate was treated twice with 200 ml each time of 0.02 M phosphate buffer, pH 7.0, per kilogram of pea seedlings initially used; the extracts obtained were combined and precipitated with solid ammonium sulfate to 30 to 35% saturation. The precipitate was separated by centrifugation and discarded. The supernatant was then further treated with solid ammonium sulfate to 60 to 65% saturation. The precipitate thus formed contained the bulk of the DAO. After centrifugation, the precipitate was taken up in 20 ml of 0.2 M phosphate buffer per kilogram of pea seedlings initially used, and the solution was dialyzed against water for 48 hr. After dialysis, this solution was subjected to chromatography on kieselguhr. In a column containing 170 g kieselguhr previously equilibrated with 0.01 M phosphate buffer, pH 6.0, were placed 50 ml of a partly purified pea-seedling DAO solution with a specific activity of 100. The column was washed first with 270 ml of 0.01 M phosphate buffer, pH 6.0, and then with 220 ml of 0.2 M phosphate buffer, pH 6.0 The enzyme was eluted from the column with 0.2 M phosphate buffer, pH 8.0. (Gradient elution with NaCl or buffer solutions gave poor results.) Fractions of 16 ml were collected. The main DAO activity was found in fractions 16 to 22 after the commencement of elution, of which at least four showed a pink color and a specific activity ranging from 14,500 to 21,200.

Whereas partly purified pea-seedling DAO solutions with a specific activity of 3000 to 5000 are stable for months at $-10°$, purified solutions with a specific activity over 10,000 rapidly became inactive. On lyophilization 10% of the activity is lost, but the lyophilized enzyme is stable. At 37° even highly purified DAO appears to be stable for 2 hr between pH 3.0 and 10.0. On incubation at pH 7.0 of 0.5 mg DAO with 400 μg of crystalline trypsin or with 600 μg of chymotrypsin, no loss of DAO activity occurred. With cadaverine and histamine as substrates the pH optimum was found to be pH 7.0. Pea-seedling DAO could not be separated into apoenzyme and coenzyme. Metal chelating agents such as 8-hydroxyquinoline, $\alpha\text{-}\alpha'$ bipyridyl, dithizone, and o-phenanthroline are very active inhibitors of pea-seedling DAO which may indicate that the latter is a metalloenzyme. In some cases reactivation was achieved with Zn^{2+} ions.

D. Methods of Isolation and Purification of Bacterial Amine Oxidases

a. ISOLATION AND ESTIMATION OF AN ADAPTIVE AMINE OXIDASE FROM *Neisseria perflava,* ACCORDING TO WEAVER AND HERBST (156)

An amine oxidase from *Neisseria perflava* degrades spermine and spermidine with the stoichiometric formation of 1,3-propanediamine, an aldehyde, and hydrogen peroxide (see Chapter IV.6). The enzyme is adaptive in nature and differs from plasma spermine oxidase in its substrate specificity and in the products of reaction (see Chapter IV.2.A.b,c).

Experimental Procedure. *Enzyme Assay.* Enzyme activity was measured manometrically with a conventional Warburg apparatus at 37.1°, in air. In a total volume of 3 ml the incubation mixtures usually contained 1 ml of a dialyzed *N. perflava* sonic extract (equivalent to 20 mg of cells), 60 micromoles of potassium phosphate buffer, pH 7.0, and 4 micromoles of spermine hydrochloride, tipped into the main compartment from a side arm after equilibration. The rate of oxygen uptake was measured, and the activity was expressed in the form of microliters of oxygen consumption per 20 mg of cells in the initial 30 min.

Preparation of Cells and Cell Extracts. Neisseria perflava, obtained from the Department of Bacteriology. University of Maryland, used as the enzyme source. A synthetic growth medium (1087) was employed, since cells grown in crude medium (peptone-yeast) yielded a much lower enzyme activity. Since the enzyme is adaptive in this organism, its production was induced by adding spermine to the growth medium. An inoculum was prepared by growing cells in a 30 ml volume for 24 hr at 34°. This volume was then transferred to 3 liters of medium in a 6-liter Florence flask containing 0.5 micromoles of spermine per milliliter. After incubation for 18 hr at 34° with shaking, the cells were harvested in a refrigerated Sharples centrifuge and washed once with 0.033 M phosphate buffer, pH 7.3. Cold water was then added to form a thick suspension, which was immediately and rapidly frozen and lyophilized to dryness. The cells were stored in a desiccator at −10°. Under these conditions the enzyme is stable for months. Although lyophilized cell suspensions can be used in enzyme experiments without further treatment, Weaver and Herbst (156) employed for their experimental work a cell-free extract prepared by sonication. Suspensions containing 20 mg of lyophilized cells per milliliter in either distilled water or 0.02 M phosphate buffer, pH 7.0, were treated for 10 min in a 10-KHz Raytheon oscillator while being cooled with ice water. The extracts were dialyzed for 18 hr at 5° against 100 volumes of either water or phosphate buffer, pH 7.0, and then centrifuged at 15,000 g for 30 min at 0°. The supernatant fluid containing the enzyme was usually used immediately, but it can be frozen and stored at −10° with little loss of activity, at least in the first few days.

The polyamine oxidase of *N. perflava* differs from other known amine oxidases. Whereas it catalyzes an apparently typical amine oxidase reaction, it is differentiated by its substrate specificity. Since the enzyme preparation was inactive on all the monoamines and diamines tested, it seems to lack MAO and DAO activity. In this respect it differs from the beef-plasma spermine oxidase studied by Tabor et al. (141). As already mentioned (see Section 4.A.a), the latter enzyme is active not only on spermine and spermidine but also on a number of monoamines such as benzylamine, homosulfanilamide, and various aliphatic monoamines (141). In addition, the stoichiometry of the degradation of spermine and spermidine by the *N. perflava* enzyme differs markedly from that of plasma spermine oxidase. Whereas in the oxidation of spermine by beef-plasma spermine oxidase 2 moles of oxygen are consumed with the formation of an aminoaldehyde, hydrogen peroxide, and putrescine (141, 843), the amine oxidase of *N. perflava* oxidizes spermine with the consumption of 1 mole of oxygen and the production of 1 mole each of 1,3-propanediamine, hydrogen peroxide, and of an aldehyde (156; see Chapter IV.6). Hence, the two types of enzymic degradation of spermine obviously involve a different attack on the spermine molecule.

The inhibition of amine oxidase from *N. perflava* by carbonyl reagents is suggestive of the presence of an active carbonyl group in the enzyme molecule (156).

b. ISOLATION, PURIFICATION, AND CHARACTERIZATION OF SPERMIDINE
OXIDASE FROM *Serratia marcescens*, ACCORDING TO BACHRACH (513)

Bachrach (513) prepared a partially purified spermidine oxidase from spermidine-adapted cells of *Serratia marcescens*. This enzyme did not oxidize spermine, propane-1,3-diamine, butane-1,4-diamine, pentane-1,5-diamine, or benzylamine. In its attack on spermidine, Δ^1-pyrroline and propane-1,3-diamine were produced in stoichiometric amounts (513; see Chapter IV.6). The formation of the latter compounds was confirmed in experiments using spermidine, labeled with ^{14}C in different parts of the molecule (513).

Materials. ^{14}C-Spermidine, labeled in the three-carbon group, was prepared biologically with a methionine-requiring mutant of *Aspergillus nidulans*. This mutant was grown in a synthetic medium, containing glucose and nitrate (1088), supplemented with 400 mg of Tween 80 (Atlas Powder Company), 15 mg pyridoxine HCl, and 10 micromoles of 2-^{14}C-DL-methionine (Tracerlab), 3.8×10^6 cpm per 100 ml of medium. The culture was shaken at 37° for 70 hr, and the mycelia were collected by filtration through a sintered glass filter, suspended in 20 ml of 0.5% NaCl solution and extracted with 80 ml of hot 0.4 M trichloroacetic acid. The radioactive material extracted was separated on a Dowex 50–H+ column (6 $\times$ 60 mm) with a hydrochloric acid gradient (797). The 180-230-ml fraction was evaporated; an aliquot migrated on paper (879) with the same R_f as did the authentic sample of spermidine.

^{14}C-Spermidine, labeled in the four-carbon group, was synthesized by a modification of the method of Tabor et al. (797) with a putrescine-requiring mutant of *A. nidulans* anol 1,4-^{14}C-putrescine dihydrochloride (New England Nuclear Corporation). The radioactive material was extracted and identified, as described above.

Bacteria. The organism used throughout this work was a strain of *S. marcescens* (879) grown on a medium as follows: yeast extract (Difco), 0.5 g; $K_2HPO_4 \cdot 3H_2O$, 2.0 g; KH_2PO_4, 1.0 g; glucose, 1.0 g; $MgSO_4 \cdot 7H_2O$, 0.2 g; spermidine · 3HCl, 0.1 g; each per 1000 ml of water. The pH of the medium was adjusted with NaOH to 7.0. The medium was sterilized by autoclaving.

Analytical Procedure. Amines, such as propane-1,3-diamine, butane-1,4-diamine, and pentane-1,5-diamine, were separated on a Dowex 50–H+ column (6 $\times$ 70 mm), 2% cross-linked, 200 to 400 mesh, with an HCl gradient (797); fractions of 8 to 10 ml were collected. The amines were quantitatively determined by the 2,4-dinitro-1-fluorobenzene method (795). They were also identified and assayed by paper chromatography with *n*-butanol–acetic acid–water (2 : 1 : 1) (879) and methanol–pyridine–hydrochloric acid–water (80 : 10 : 2.5 : 17.5) as solvents. The papers were sprayed with 0.5% ninhydrin in *n*-butanol. A good separation between propane-1,3-diamine and butane-1,4-diamine was obtained with the methanol-pyridine solvent.

Δ^1-*Pyrroline.* This substance was assayed by a colorimetric method similar to that devised by Holmstedt, Larsson, and Tham (602). The incubation mixture consisted of 0.5 ml of sodium phosphate buffer, pH 7.0, 0.1 ml of enzyme, 0.1 ml of spermidine, 0.1 ml of 0.1% *o*-aminobenzaldehyde solution, and 0.3 ml of 0.85% NaCl solution. This mixture was incubated at 25°, and the intensity of the yellow color was estimated directly at 435 nm in a cuvette of 1-cm light path with a Beckman model DU spectrophotometer. The molar absorbancy index was 1.86 $\times$ 10^3 liter mole–1 cm–1 (602).

Spermidine Oxidase Activity. Activity was determined by measuring the formation of Δ^1-pyrroline. A unit of enzyme activity is defined as the amount of enzyme that causes the formation of 1.0 micromole of Δ^1-pyrroline per hour. Specific activity is expressed in units of activity per milligram of protein. The absorption spectra were measured in a

Cary model 11 M recording spectraphotometer.

Ammonia. This was determined by nesslerization after microdiffusion according to the method of Bessman and Bessman (1089).

Proteins. These were estimated by a spectrofluorometric method (1090).

Manometric Measurements. These were carried out in a Warburg apparatus at 30° with air as the gas phase.

Radioactivity. Measurements of ^{14}C were made at infinite thinness in a Nuclear-Chicago thin-window gas-flow counter.

Enzyme Fractionation. The best preparations of spermidine oxidase were obtained from *S. marcescens,* grown at 30°, with shaking in a Gyrotory shaker (New Brunswick Scientific Co.). Each 2000-ml flask contained 1000 ml of medium. When the bacteria were grown at 37°, or without shaking, the activity of the preparation obtained was very poor. The formation of the pink pigment, prodigiosin, was also optimal, when bacteria were grown at 30° with shaking.

After incubation for 20 hr, the cells were harvested, washed twice with 0.85% NaCl solution, and suspended in 0.2 M sodium phosphate buffer, pH 7.0, containing 0.01% mercaptoethanol. A ratio of 1 g of wet cells per 4 ml of buffer was used. This pink-colored suspension was subjected for 15 min to the action of a Raytheon 10-KHz sonic oscillator. The intact cells and debris were removed by centrifugation at approximately 15,000 g for 15 min, at 4°. The supernatant solution represented fraction 1. All subsequent procedures were carried out with stirring at ice bath temperature. Centrifugations were performed in the cold room.

To 100 ml of fraction 1 were added 12 ml of a solution containing 120 mg of protamine sulfate. The precipitate was immediately separated by centrifugation and discarded. To the supernatant (fraction 2) 31 g of ammonium sulfate was added, and the precipitate obtained after centrifugation was dissolved in approximately 50 ml of 0.2 M sodium phosphate buffer, pH 7.0, containing 0.01% mercaptoethanol. The pH of the solution (fraction 3) was adjusted, when necessary, to 7.0. To each 50-ml aliquot of fraction 3 were added 6 g, wet weight, of hydroxylapatite (1048), and the suspension was shaken gently in a beaker for approximately 5 min. The suspended particles were removed by centrifugation, and the supernatant solution was discarded. To the precipitate were added 50 ml of 0.05 M sodium phosphate buffer, pH 6.7, containing 0.01% mercaptoethanol, and the suspension was again shaken gently for approximately 5 min. The insoluble particles were then collected by centrifugation and treated with 50 ml of 0.1 M sodium phosphate buffer, pH 6.7. The pink supernatant solution (fraction 4), obtained on the removal of hydroxylapatite, was applied to a DEAE-cellulose column in aliquots of 10 ml, as follows: DEAE-cellulose was washed three times with 150 volumes of water and four times in succession with 50 volumes each time of 0.03 M Tris-HCl, at pH 7.0. A DEAE-cellulose column (14 $\times$ 50 mm) was then prepared in a cold room. To 10 ml of fraction 4, 40 ml of water were added and this solution was passed through the prepared column which was subsequently washed with 10 ml of 0.025 M Tris, pH 6.7. The enzyme was eluted from the column by a linear gradient (0.025 M Tris, pH 6.7, and 0.2 M sodium phosphate buffer, pH 7.0). A concentration of approximately 0.1 M phosphate buffer, pH 7.0, was required for complete desorption of the enzyme (fraction 5) from the column. The spermidine oxidase activity in these preparations was determined by the standard assay procedure.

Standard Assay Conditions. Place in a tube 50 micromoles of sodium phosphate buffer, pH 7.0, and add 2.0 micromoles of spermidine, and 0.1 ml of 0.1% *o*-aminobenzaldehyde to a total volume of 1.1 ml. Incubate at 25° for 30 min and read the absorbancy change per min at 435 nm.

Properties of the Partly Purified Spermidine Oxidase. Optimal activity was obtained at pH 6.7 by the standard assay procedure. The enzyme (fraction 3) was 65% inactivated on dialysis against distilled water at 4° for 8 hr. The enzyme was also markedly inactivated (70%) by dialysis for 10 hr against 0.03 M Tris-HCl buffer, pH 7.3. Chilled acetone (—20°) completely inactivated the enzyme, when 0.5 to 1.5 volumes of acetone per volume of enzyme solution were used. Ethanol 25%, 50%, and absolute), previously cooled to —10°, inactivated the enzyme (fraction 3) in the presence or absence of 0.01 M MnCl$_2$. Two volumes of alcohol per volume of enzyme were used. A gradual decrease in enzymic activity was noticed on storage of fraction 3, at —10°, in 0 2 M phosphate buffer, pH 7.0, containing mercaptoethanol (an inactivation of 20% occurred after 1 day and one of 50% after 4 days). Spermidine oxidase was completely inactivated by heating at 55° for 1 min. However, when the enzyme (fraction 3) was heated at 55° for 3 min in the presence of 2 micromoles of spermidine per ml of enzyme, 80% of the activity was retained.

To study the substrate specificity of spermidine oxidase, various amines were tested as substrates using the manometric and the spectraphotometric methods in parallel (602). Significant oxygen uptake was observed only with spermidine and 3,3'-diaminodipropylamine as substrates. On the other hand, only traces of spermine, diacetylspermidine, propane-1,3-diamine, butane-1,4-diamine, pentane-1,5-diamine were metabolized, as was demonstrated by both the manometric and the spectrophotometric methods. Spermine (5 micromoles) or butane-1,4-diamine (5 micromoles) did not inhibit the oxidation of spermidine (10 micromoles) by spermidine oxidase.

Campello, Tabor, and Tabor (891; see Chapter IV.6) have recently further purified spermidine oxidase from *S. marcescens* and reported that the oxidation of spermidine by that enzyme required FAD as well as an additional electron carrier, such as phenazine methosulfate. These authors believe that spermidine oxidase from *S. marcescens* is linked to an electron-transport system and that this enzyme should, therefore, be regarded as a spermidine dehydrogenase with FAD serving as its cofactor.

Finally, according to Bachrach (513) the formation of the pink pigment, prodigiosin, may be associated with the activity of spermidine oxidase. Cells grown at 37° do not produce pigment and show low spermidine oxidase activity. Furthermore, the pigment was found by Bachrach to be present in all active fractions obtained during the purification of spermidine oxidase. In this context, Bachrach recalls findings of Purkayastha and Williams (1091) who reported that prodigiosin is not an easily soluble pigment, but seems to be firmly attached to some protein constituents of the bacterial cell.

References

1. Hare, M. L. C., *Biochem. J.*, *22*, 968 (1928).
2. Best, C. H., *J. Physiol.*, *67*, 256 (1929).
3. Sumner, J. B., *J. Biol Chem.*, *69*, 435 (1926).
4. Northrop, J. H., and M. Kunitz, *J. Gen. Physiol.*, *16*, 267, 295 (1932).
5. Northrop, J. H., M. Kunitz, and R. M. Herriott, in *Crystalline Enzymes*, 2nd Ed., Columbia University Press, New York, 1953.
6. Dixon, M., and E. C. Webb, in *Enzymes*, Longmans, Green, London, 1958, pp. 39, 492.
7. Bernal, J. D., and D. Crowfoot, *Nature*, *133*, 794 (1934).
8. Green, D. W., V. M. Ingram, and M. F. Perutz, *Proc. Roy. Soc.* (London), Ser. A, *235*, 287 (1954).
9. Bokhoven, C., J. C. Schoone, and J. M. Bijvoet, *Acta Cryst.*, *4*, 275 (1951).
10. Blow, D. M., *Proc. Roy. Soc.* (London), Ser. A, *247*, 302 (1958).
11. Srinivasan, R., *Proc. Indian Acad. Sci.*, *A53*, 252 (1961).
12. Blake, C. C. F., in *Advances Protein Chemistry*, *23*, 59 (1968).
13. Crick, F. H. C., and B. S. Magdoff, *Acta Cryst.*, *9*, 901 (1956).
14. Phillips, D. C., in *Advances in Structure Research by Diffraction Methods*, R. Brill and R. Mason, eds., Vol. 2, Vieweg, Braunschweig, 1966, p. 75.
15. Holmes, K. C., and D. M. Blow, in *Methods of Biochemical Analysis*, D. Glick, ed., Vol. 13, Interscience, New York, 1965, p. 113.
16. Sigler, P. B., B. A. Jefferey, B. M. Matthews, and D. M. Blow, *J. Mol. Biol.*, *15*, 175 (1966).
17. Blake, C. C. F., D. F. Koenig, G. A. Mair, A. C. T. North, D. C. Phillips, and V. R. Sarma, *Nature*, *206*, 757 (1965).
18. Blake, C. C. F., L. N. Johnson, G. A. Mair, A. C. T. North, D. C. Phillips, and V. R. Sarma, *Proc. Roy. Soc.* (London), Ser. B, *167*, 378 (1967).
19. Phillips, D. C., *Sci. Am.*, *215*, 78 (1966).
20. Phillips, D. C., *Proc. Natl. Acad. Sci. U.S.*, *57*, 484 (1967).
21. Canfield, R. E., *J. Biol. Chem.*, *238*, 2698 (1963).
22. Jolles, J., J. Tauregui-Adell, and P. Jolles, *Biochim. Biophys. Acta*, *78*, 68 (1963).
23. Goldberger, R. F., and C. J. Epstein, *J. Biol. Chem.*, *238*, 2988 (1963).
24. Kartha, G., J. Bello, and D. Harker, *Nature*, *213*, 862 (1967).
25. Wyckoff, H. W., K. D. Hardman, N. M. Allewell, T. Inagami, T. Tsernoglou, L. N. Johnson, and F. M. Richards, *J. Biol. Chem.*, *242*, 3749 (1967).
26. Wyckoff, H. W., K. D. Hardman, N. M. Allewell, T. Inagami, L. N. Johnson, and F. M. Richards, *J. Biol. Chem.*, *242*, 3984 (1967).
27. Crestfield, A. M., W. H. Stein, and S. Moore, *J. Biol. Chem.*, *238*, 2413, 2421 (1953).
28. Saroff, H. A., *J. Theor. Biol.*, *9*, 234 (1965).
29. Hammes, G., and H. A. Sheraga, *Biochemistry*, *5*, 3690 (1966).
30. Wyckoff, H. W., F. M. Richards, and D. Tsernoglou, in Abstracts of the Gatlinburg Meeting of the *American Crystallography Association*, *38*, 1965.
31. Matthews, B. W., P. B. Sigler, R. Henderson, and D. M. Blow, *Nature*, *214*, 652 (1967).
32. Kraut, J., H. R. Wright, M. Kellerman, and S. T. Freer, *Proc. Natl. Acad. Sci. U.S.*, *58*, 304 (1967).

33. Kraut, J., D. F. High, and L. C. Sieker, *Proc. Natl. Acad. Sci. U.S., 51,* 839 (1964).
34. Dryer, W. J., and H. Neurath, *J. Am. Chem. Soc., 77,* 814 (1955).
35. Desnuelle, P., in *The Enzymes,* Vol. 4, H. Lardy and K. Myrbäck, eds., Academic, New York, 1960, p. 97.
36. Corey, R. B., O. Battfay, D. A. Brueckner, and F. G. Mark, *Biochim. Biophys. Acta, 94,* 535 (1965).
37. Bernal, J. D., I. Fankuchen, and M. F. Perutz, *Nature, 141,* 523 (1938).
38. Hartley, B. S., and D. L. Kauffman, *Biochem. J., 101,* 229 (1966).
39. Keil, B., F. Sorm, and Z. Prusik, *Biochim. Biophys. Acta, 78,* 559 (1966).
40. Bender, M. L., and F. J. Kezdy, *Ann. Rev. Biochem., 34,* 49 (1965).
41. Jencks, W. P., *Ann. Rev. Biochem., 32,* 639 (1963).
42. Mahler, H. R., and E. Cordes, in *Biological Chemistry,* Harper and Row, New York, 1966.
43. Neurath, H., *Federation Proc., 23,* 1 (1964).
44. Oppenheimer, H. L., B. Labouesse, and G. P. Hess, *J. Biol. Chem., 241,* 2720 (1966).
45. Neurath, H., J. A. Rupley, and W. J. Dreyer, *Arch. Biochem. Biophys., 65,* 243 (1956).
46. Oosterbaan, R. A., P. Kunst, J. Van Rotterdam, and J. A. Cohen, *Biochim. Biophys. Acta, 27,* 556 (1958).
47. Schoellman, G., and E. Shaw, *Biochemistry, 2,* 262 (1963).
48. Smillie, L. B., and B. S. Hartley, in *Abstracts, First Meeting Federation European Biochemical Society,* London, 1964, p. A–30.
49. Lawson, W. B., and H. J. Schramm, *Biochemistry, 4,* 317 (1965).
50. Stryer, L., *Ann. Rev. Biochem., 37,* 25 (1968).
51. Ludwig, M. L., J. A. Hartsuck, T. A. Steitz, H. Muirhead, J. C. Coppola, G. N. Reeke, and W. N. Lipscomb, *Proc. Natl. Acad. Sci. U.S., 57,* 511 (1967).
52. Steitz, T. A., M. L. Ludwig, F. A. Quiocho, and W. N. Lipscomb, *J. Biol. Chem., 242,* 4662 (1967).
53. Reeke, G. N., J. A. Hartsuck, M. L. Ludwig, F. A. Quiocho, T. A. Steitz, and W. N. Lipscomb, *Proc. Natl. Acad. Sci. U.S., 58,* 2220 (1967).
54. Riordan, J. F., M. Sokolovsky, and B. L. Vallee, *Biochemistry, 6,* 358 (1967).
55. Roholt, O. A., and D. Pressman, *Proc. Natl. Acad. Sci. U.S., 58,* 280 (1967).
56. Koshland, D. E., Jr., and K. E. Neet, *Ann. Rev. Biochem., 38,* 359 (1968).
57. Crick, F. H. C., and J. C. Kendrew, *Advances Protein Chemistry, 12,* 133 (1957).
58. Dickerson, R. E., in *The Proteins,* Vol. 2, H. Neurath, ed., Academic, New York, 1964, p. 603.
59. Kraut, J., *Ann. Rev. Biochem., 34,* 247 (1965).
60. Atkinson, D. E., *Ann. Rev. Biochem., 35,* 85 (1966).
61. Adelberg, E. A., and H. E. Umbarger, *J. Biol. Chem., 205,* 475 (1953).
62. Brooke, M. S., D. Ushida, and B. Magasanik, *J. Bacteriol., 68,* 534 (1954).
63. Novick, A., and L. Szilard, in *Dynamics of Growth Processes,* E. J. Boell, ed., Princeton University Press, Princeton, N. J., 1954, p. 21.
64. Gots, J. S., *Federation Proc., 9,* 178 (1950).
65. Robertson, R. B., P. H. Abelson, D. B. Cowie, E. T. Bolton, and R. J. Britton, in *Studies of Biosynthesis in Escherichia Coli,* Publ. 607, Carnegie Institute, Washington, D.C., 1955.
66. Abelson, P. H., *J. Biol. Chem., 206,* 335 (1954).
67. Umbarger, H. E., *J. Bacteriol., 65,* 203 (1953).
68. Umbarger, H. E., *Science, 123,* 848 (1956).

69. Yates, R. A., and A. B. Pardee, *J. Biol. Chem., 221,* 757 (1956).
70. Umbarger, H. E., and B. Brown, *J. Biol. Chem., 233,* 415 (1958).
71. Gerhart, J. C., and A. B. Pardee, *Federation Proc., 20,* 224 (1961).
72. Changeux, J. P., *Cold Spring Harbour Symp. Quant. Biol., 26,* 313 (1961).
73. Martin, R. G., B. N. Ames, and B. J. Garry, *Federation Proc., 20,* 225 (1961).
74. Monod, J., and F. Jacob, *Cold Spring Harbour Symp. Quant. Biol., 26,* 389 (1961).
75. Gerhart, J. C., and A. B. Pardee, *J. Biol. Chem., 237,* 891 (1962).
76. Changeux, J. P., *J. Mol. Biol., 4,* 220 (1962).
77. Monod, J., J. P. Changeux, and F. Jacob, *J. Mol. Biol., 6,* 306 (1963).
78. Koshland, D. E., Jr., J. A. Yankeelov, Jr., and J. A. Thoma, *Federation Proc., 21,* 1031 (1962).
79. Koshland, D. E., Jr., *Advances Enzymol., 22,* 45 (1960).
80. Bender, M. L., F. J. Kezdy, and C. R. Gunter, *J. Am. Chem. Soc., 86,* 3714 (1964).
81. Monod, J., J. Wyman, and J. P. Changeux, *J. Mol. Biol., 12,* 88 (1965).
82. Koshland, D. E., Jr., *J. Cellular Comp. Physiol., 47,* Suppl. 1, 217 (1956).
83. Koshland, D. E., Jr., *J. Theor. Biol., 2,* 75 (1962).
84. Wilson, I. B., and F. Bergmann, *J. Biol. Chem., 186,* 683 (1950).
85. Alberty, R. A., in *The Enzymes,* Vol. 5, P. D. Boyer, H. Lardy, and K. Myrbäck, eds., Academic, New York, 1961, p. 531.
86. Jencks, W. P., in *Current Aspects of Biochemical Energetics,* N. O. Kaplan and E. P. Kennedy, eds., Academic, New York, 1966, p. 273.
87. Swain, C. G., and J. F. Brown, *J. Am. Chem. Soc., 74, 2538* (1952).
88. Pocker, Y., *Chem. Ind.* (London), 968 (1960).
89. Fischer, E., *Chem. Ber., 27,* 2985 (1894).
90. Koshland, D. E., Jr., in *The Enzymes,* Vol. 1, 2nd Ed., P. D. Boyer, H. Lardy, and K. Myrbäck, eds., Academic, New York, 1959, p. 305.
91. Koshland, D. E., Jr., *Proc. Natl. Acad. Sci. U.S., 44,* 98 (1958).
92. Koshland, D. E., Jr., *J. Cell. Comp. Physiol., 54,* 235 (1959).
93. Koshland, D. E., Jr., in *Horizons in Biochemistry,* B. Pullman and M. Kasha, eds., Academic, New York, 1962, p. 265.
94. Koshland, D. E., Jr., *Proc. 1st Intern. Pharmacological Meeting* (Stockholm, 1961), 1963, p. 161.
95. O'Sullivan, W. J., and M. Cohn, *J. Biol. Chem., 241,* 3116 (1966).
96. Yankeelov, J. A., Jr., and D. E. Koshland, Jr., *J. Biol. Chem., 240,* 1593 (1965); Neet, K. E., and D. E Koshland, Jr, Proc. Natl. Acad. Sci. U.S., *56,* 1606 (1966).
97. Freundlich, M., and H. E. Umbarger, *Cold Spring Harbour Symp. Quant. Biol., 28, 505* (1963).
98. Srere, P., *J. Biol. Chem., 241,* 2157 (1966).
99. Suelter, C. H., R. Singleton, Jr., F. J. Kayne, S. Arrington, J. Glass, and A. S. Mildvan, *Biochemistry, 5,* 131 (1966).
100. Suelter, C. H., *Biochemistry, 6,* 418 (1967).
101. Matsubara, H., K. Ando, and K. O'Kunuki, *Proc. Japan. Acad., 41,* 408 (1965).
102. Ando, K., H. Matsubara, and K. O'Kunuki, *Biochim. Biophys. Acta, 118,* 256 (1966).
103. Inagami, T., and T. Murachi, *J. Biol. Chem., 239,* 1395 (1964).
104. Theorell, M., B. Chance, and T. Yonetani, *J. Mol. Biol., 17,* 513 (1966).
105. Chance, B., and A. Ravilly, *J. Mol. Biol., 21,* 195 (1966).
106. Chance, B., in *Hemes and Hemoproteins,* B. Chance, R. W. Estabrook, and T. Yonetani, eds., Academic, New York, 1966, p. 213.
107. Doscher, M. S., and F. M. Richards, *J. Biol. Chem., 238,* 2399 (1963).

108. Quiocho, F. A., and F. M. Richards, *Biochemistry, 5,* 4062 (1966).
109. Haurowitz, F., *Z. Physiol. Chem., 254,* 266 (1938).
110. Muirhead, H., and M. Perutz, *Nature, 199,* 633 (1963).
111. Bohr, C., K. Hasselbach, and A. Krogh, *Skand. Arch. Physiol., 16,* 402 (1904).
112. Pauling, L., *Proc. Natl. Acad. Sci. U.S., 21,* 186 (1935).
113. Riggs, A., *J. Biol. Chem., 236,* 1948 (1961).
114. Guidotti, G., *J. Biol. Chem., 242,* 3673 (1967).
115. Zito, R., E. Antonini, and J. Wyman, *J. Biol. Chem., 239,* 1804 (1964).
116. Koshland, D. E., Jr., G. Nemethy, and D. Filmer, *Biochemistry, 5,* 365 (1966).
117. Kirtley, M. E., and D. E. Koshland, Jr., *J. Biol. Chem., 242,* 4192 (1967).
118. Haber, J. E., and D. E. Koshland, Jr., *Proc. Natl. Acad. Sci. U.S., 58,* 2087 (1967).
119. Kingdon, H. S., B. M. Shapiro, and E. R. Stadtman, *Proc. Natl. Acad. Sci. U.S., 58,* 1703 (1967).
120. Caskey, C. T., D. M. Ashton, and J. B. Wyngaarden, *J. Biol. Chem., 239,* 2570 (1964).
121. Kendrew, J. C., and H. C. Watson, in *Principles of Biomolecular Organisation,* G. E. W. Wolstenholme, ed., Churchill, London, 1966, p. 86.
122. Zeller, E. A., in *The Enzymes,* Vol. 2, part 1, 1st Ed., J. B. Sumner and K. Myrbäck, eds., Academic, New York, 1951, p. 536.
123. Blaschko, H., *Pharmacol. Rev., 4,* 415 (1952).
124. Tabor, H., *Pharmacol. Rev., 6,* 299 (1954).
125. Davison, A. N., *Physiol. Rev., 38,* 729 (1958).
126. Zeller, E. A., in *The Enzymes,* Vol. 8, P. D. Boyer, H. Lardy and K. Myrbäck, eds., Academic, New York, 1963, p. 313.
127. Zeller, E. A., and G. Ramachander, *Biochem. J., 92,* 60 P (1964).
128. Blaschko, H., in *The Enzymes,* Vol. 8, P. D. Boyer, H. Lardy and K. Myrbäck, eds., Academic, New York, 1963, p. 337.
129. Blaschko, H., *Advanc. Compar. Physiol. Biochem., 1,* 67 (1962).
130. Werle, E., in *Hoppe Seyler Thierfelder Handbuch der physiologischen und pathologisch-chemischen Analyse,* Vol. VI/A, 10th Ed., Springer, Berlin, 1964, p. 653.
131. Buffoni, F., *Pharmacol. Rev., 18,* 1163 (1966).
132. Mason, H. S., *Advances Enzymol., 19,* 79 (1957).
133. Kapeller-Adler, R., and M. Fletcher, *Biochim. Biophys. Acta, 33,* 1 (1959).
134. Zeller, E. A., R. Stern, and M. Wenk, *Helv. Chim. Acta, 23,* 3 (1940).
135. Zeller, E. A., in *Histamine,* Ciba Foundation Symposium, G. E. W. Wolstenholme and C. M. O'Connor, eds., Churchill, London, 1956, pp. 258, 339.
136. Blaschko, H., and R. Duthie, *Biochem. J., 39,* 478 (1945).
137. Steensholt, G., *Acta Phys. Scand., 14,* 356 (1947).
138. Zeller, E. A., J. Barsky, E. R. Berman, M. S. Cherkas, and J. R. Fouts, *J. Pharmacol. Exp. Therap., 124,* 282 (1958).
139. Werle, E., and E. von Pechmann, *Z. Vit. Horm. Fermentforschg., 2,* 433 (1948/49).
140. Hirsch, J. G., *J. Exp. Med., 97,* 345 (1953).
141. Tabor, C. W., H. Tabor, and S. M. Rosenthal, *J. Biol. Chem., 208,* 645 (1954).
142. Zeller, E. A., J. Barsky, J. R. Fouts, F. A. Kirchheimer, and L. S. Van Orden, *Experientia, 8,* 349 (1952).
143. Yamada, H., and K. T. Yasunobu, *J. Biol. Chem., 237,* 1511 (1962).
144. Kolb, E., *Zentralbl. Veterinärmediz., 3,* 570 (1956).
145. Blaschko, H., P. J. Friedman, R. Hawes, and N. Nilsson, *J. Physiol., 145,* 384 (1959).
146. Buffoni, F., and H. Blaschko, *Proc. Roy. Soc.* (London), Ser. B, *161,* 153 (1964).

147. Hill, J. M., and P. J. G. Mann, *Biochem. J.*, *91*, 171 (1964).
148. Kobayashi, Y., *Arch. Biochem. Biophys.*, *71*, 352 (1957).
149. McHenry, E. W., and G. Gavin, *Biochem. J.*, *26*, 1365 (1932).
150. Zeller, E. A., *Advances Enzymol.*, *2*, 93 (1942).
151. Cotzias, G. C., and V. P. Dole, *J. Biol. Chem.*, 196, 235 (1952).
152. Blaschko, H., and R. Hawes, *J. Physiol.*, *145*, 124 (1959).
153. Blaschko, H., and R. Bonney, *Proc. Roy. Soc.* (London), Ser. B., *156*, 268 (1962).
154. Bergeret, B., H. Blaschko, and R. Hawes, *Nature*, *180*, 1127 (1957).
155. Bernheim, F., and M. L. C. Bernheim, *J. Biol. Chem.*, *123*, 317 (1938).
156. Weaver, R. H., and E. J. Herbst, *J. Biol. Chem.*, *231*, 637, 647 (1959).
157. Zeller, E. A., L. A. Blanksma, W. P. Burkard, W. L. Pacha, and J. C. Lazanas, *Ann. N. Y. Acad. Sci.*, *80*, Art. 3, 583 (1959).
158. Zeller, E. A., *Pharmacol. Rev.*, *11*, 387 (1959).
159. Zeller, E. A., J. Barsky, L. A. Blanksma, and J. C. Lazanas, *Federation Proc.*, *16*, 276 (1957).
160. Zeller, E. A., J. Barsky, and E. R. Berman, *J. Biol. Chem.*, *214*, 267 (1955).
161. Blaschko, H., D. Richter, and H. Schlossmann, *Biochem. J.*, *31*, 2187 (1937).
162. Blaschko, H., and J. Hawkins, *Brit. J. Pharmacol.*, *5*, 625 (1950).
163. Strömblad, B. C. R., *J. Physiol.* (London), *147*, 639 (1959).
164. Bhagvat, K., H. Blaschko, and D. Richter, *Biochem. J.*, *33*, 1338 (1939); Alles, G. A., and E. V. Heegard, *J. Biol. Chem.*, *147*, 487 (1943).
165. Langemann, H., *Helv. Physiol. Acta*, *2*, 367 (1944).
166. Puch, C. E. M., and J. H. Quastel, *Biochem. J.*, *31*, 286 (1937).
167. Langemann, H., F. Roulet, and E. A. Zeller, *Klin. Wochschr.*, *22*, 42 (1943).
168. Birkhäuser, H., *Helv. Chim. Acta*, *23*, 1071 (1940).
169. Arioka, J., and H. Tanimukai, *J. Neurochem.*, *1*, 311 (1957).
170. Bogdanski, D. F., M. Weissbach, and S. Udenfriend, *J. Neurochem.*, *1*, 272 (1957).
171. Weiner, N., *J. Neurochem.*, *6*, 79 (1960).
172. Aprizon, M. H., R. Takahashi, and T. L. Folkerth, *J. Neurochem.*, *11*, 341 (1964).
173. Arnaiz, G. R. De L. and E. D. P. de Robertis, *J. Neurochem.*, *9*, 503 (1962).
174. Marchbanks, R. M., *J. Neurochem.*, *13*, 1481 (1966).
175. Zeller, E. A., and C. A. Joel, *Helv. Chim. Acta*, *24*, 968 (1941).
176. Luschinsky, M. L., and H. O. Singher, *Arch. Biochem.*, *19*, 95 (1948).
177. Thompson, R. H. S., and A. Tickner, *Biochem. J.*, *45*, 125 (1949).
178. Waalkes, T. P., and H. Coburn, *Proc. Soc. Exptl. Biol. Med.*, *99*, 742 (1958).
179. Barsky, J., Ph.D. Thesis, Northwestern University Chicago, 1958.
180. Paasonen, M., *Biochem. Pharmacol.*, *8*, 241 (1961).
181. Robinson, I., *Brit. J. Pharmacol.*, *7*, 99 (1952).
182. Epps, H. M. R., *Biochem. J.*, *39*, 37 (1945).
183. Werle, E., and E. Henning, *Z. Vit. Horm. Fermentforschg.*, *11*, 159 (1960).
184. Blaschko, H., *J. Physiol.*, *99*, 364 (1941).
185. Blaschko, H., and J. Hawkins, *J. Physiol.*, *118*, 88 (1952).
186. Blaschko, H., and D. B. Hope, *Arch. Biochem.*, *69*, 10 (1957).
187. Welsh, J. H., *Ann. N. Y. Acad. Sci.*, *66*, 618 (1957).
188. Werle, E., and B. Z. Roewer, *Biochem. Z.*, *320*, 198 (1950).
189. Werle, E., and B. Z. Roewer, *Biochem. Z.*, *322*, 320 (1952).
190. Korzenovsky, M., C. P. Walters, and M. S. Hughes, *Federation Proc.*, *18*, 1045 (1959).
191. Cotzias, G. C., and V. P. Dole, *Proc. Soc. Exptl. Biol. Med.* (New York), *78*, 157 (1951).

192. Hawkins, J., *Biochem. J., 50,* 577 (1952).
193. Bandhuin, P., H. Beaufay, Y. Rahman-Li, O. Z. Sellinger, R. Wattiaux, P. Jacques, and C. de Duve, *Biochem. J., 92,* 179 (1964).
194. Oswald, E. O., and C. F. Strittmatter, *Proc. Soc. Exptl. Biol.* (New York), *114,* 668 (1963).
195. Aebi, H., F. Stocker, and M. Eberhardt, *Biochem. Z., 336,* 526 (1963).
196. Lehninger, A. L., *Physiol. Rev., 42,* 467 (1962).
197. Gorkin, V. Z., R. S. Krivchenkova, and I. S. Severina, in *Mechanism and Kinetics of Enzymatic Catalysis,* A. E. Braunstein and V. A. Yakovlev, eds., Nauka Press, Moscow, 1964, p. 150.
198. Severina, I. S., and V. Z. Gorkin, *Biokhimiya, 28,* 893, 896 (1963).
199. Gorkin, V. Z., and J. V. Veryovkina, *Vop. Med. Khim., 9,* 315 (1963).
200. Veryovkina, J. V., V. Z. Gorkin, V. M. Mityushin, and J. E. Elpiner, *Biophysics* (Moscow) *9,* 503 (1964).
201. Seiden, L. S., and J. Westley, *Biochim. Biophys. Acta, 58,* 363 (1962).
202. Blaschko, H., *Pharmacol. Rev., 18(1),* 39 (1966).
203. Wright, C. I., and J. C. Sabine, *J. Biol. Chem., 155,* 315 (1944).
204. Lagnado, J. R., and T. L. Sourkes, *Can. J. Biochem. Physiol., 34,* 1095 (1956).
205. Wiseman-Distler, M. H., and T. L. Sourkes, *Can. J. Biochem. Physiol., 41,* 57 (1963).
206. Hawkins, J., *Biochem. J., 51,* 399 (1952).
207. Bachrach, U., and J. S. Oser, *J. Biol. Chem., 238,* 2098 (1963).
208. Tabor, C. W., A. de Campello, and H. Tabor, *Federation Proc., 24,* 232 (1965).
209. Nara, S., I. Igaue, B. Gomes, and K. T. Yasunobu, *Biochem. Biophys. Res. Commun., 23,* 324 (1966).
210. Erwin, V. G., and L. Hellerman, *Federation Proc., 26,* 843 (1967).
211. Erwin, V. G., and L. Hellerman, *J. Biol. Chem., 242,* 4230 (1967).
212. Tipton, K. F., *Biochem. J., 104,* 36P (1967).
213. Tipton, K. F., *European J. Biochem., 4,* 103 (1968).
214. Tipton, K. F., *Biochim. Biophys. Acta, 159,* 451 (1968).
215. Blaschko, H., and F. Buffoni, *Proc. Roy. Soc.* (London), Ser. B., *163,* 45 (1965).
216. Sloane-Stanley, G. H., *Biochem. J., 44,* 567 (1949).
217. Swoboda, B. E. P., in Abstracts, 4th Meeting Federation European Biochemical Society (Oslo, 1967) Universitets Forlaget, Oslo, 1967, p. 82.
218. Gabay, S., and S. R. Harris, *Biochem. Pharmacol., 16,* 803 (1967).
219. Igaue, I., B. Gomes, and K. T. Yasunobu, *Biochem. Biophys. Res. Commun., 29,* 562 (1967).
220. Nara, S., B. Gomes, and K. T. Yasunobu, *J. Biol. Chem., 241,* 2774 (1966).
221. Nara, S., and K. T. Yasunobu, in *The Biochemistry of Copper,* J. Peisach, P. Aisen, and W. E. Blumberg, eds., Academic, New York, 1966, p. 423.
222. Gorkin, V. Z., N. V. Komisarova, H. I. Lerman, and I. V. Veryovkina, *Biochem. Biophys. Res. Commun., 15,* 383 (1964).
223. Mondovi, B., and H. Beinert, Discussion remarks in *The Biochemistry of Copper,* J. Peisach, P. Aisen, and W. E. Blumberg, eds., Academic, New York, 1966, p. 437.
224. Yamada, H., K. T. Yasunobu, T. Yamano, and H. S. Mason, *Nature, 198,* 1092 (1963).
225. Brill, A. S., R. B. Martin, and R. J. P. Williams, in *Electronic Aspects of Biochemistry,* B. Pullman, ed., Academic, New York, 1964, p. 519.
226. McEwen, C. M., Jr., *J. Biol. Chem., 240,* 2003, 2011 (1965).
227. Gabay, S., and A. J. Valcourt, *Biochim. Biophys. Acta, 159,* 440 (1968).

228. Frieden, E., in *Horizons in Biochemistry*, M. Kasha and B. Pullman, eds., Academic, New York, 1962, p. 461.

229. Stark, G. R., and C. R. Dawson, in *The Enzymes*, Vol. 8, P. D. Boyer, H. Lardy, and K. Myrbäck, eds., Academic, New York, 1963, p. 297.

230. Coq, H., and C. Baron, *Experientia, 23*, 797 (1967).

231. Friedenwald, J. S., and H. Herrmann, *J. Biol. Chem., 146*, 411 (1942).

232. Barron, E. S. G., and T. P. Singer, *J. Biol. Chem., 157*, 221 (1945).

233. Lagnado, J. R., and T. L. Sourkes, *Can. J. Biochem. Physiol., 34*, 1185 (1956).

234. Barbato, L. M., and L. G. Abood, *Biochim. Biophys. Acta, 67*, 531 (1963).

235. Gorkin, V. Z., *Vop. Med. Khim., 9*, 616 (1963).

236. Gorkin, V. Z., *Nature, 200*, 77 (1963).

237. Gorkin, V. Z., *Pharmacol. Rev., 18(1)*, 115 (1966).

238. Gorkin, V. Z., and R. S. Krivchenkova, *Biokhimiya, 29*, 992 (1964).

239. Green, A. L., *Biochem. Pharmacol., 13*, 249 (1964).

240. Gorkin, V. Z., *Biokhimiya, 24*, 826 (1959).

241. Youdim, M. B. H., and T. L. Sourkes, *Can. J. Biochem. Physiol., 44*, 1397 (1966).

242. Blaschko, H., *J. Physiol., 153*, 17P (1960).

243. Randall, L. O., *J. Pharmacol., 88*, 216 (1946).

244. Pletscher, A., K. F. Gey, and P. Zeller, in *Progress in Drug Research*, Vol. 2, E. Jucker, ed., Birkhäuser, Basel, 1960, p. 417.

245. Leyden Webb, J., in *Enzyme and Metabolic Inhibitors*, Vol. 1, Academic, New York and London, 1963, p. 194.

246. Augenstine, L. G., in *Proc. 1st Natl. Biophys. Conf.*, (Columbus, Ohio, 1959) p. 154.

247. Hill, R. L., and E. L. Smith, *Biochim. Biophys. Acta, 19*, 376 (1956).

248. Weil, L., and A. R. Buchert, *Federation Proc., 11*, 307 (1952).

249. Neely, W. B., *Arch. Biochem. Biophys., 79*, 297 (1959).

250. Smith, E. L., *J. Biol. Chem., 233*, 1392 (1958).

251. Sarkar, S., R. Banerjee, M. S. Ise, and E. A. Zeller, *Helv. Chim. Acta, 43*, 439 (1960).

252. Zeller, E. A., W. L. Pacha, and S. Sarkar, *Biochem. J., 76*, 45P (1960).

253. Zeller, E. A., *Experientia, 16*, 399 (1960).

254. Hope, D. B., and A. D. Smith, *Biochem. J., 74*, 101 (1960).

255. Zeller, E. A., L. A. Blanksma, and J. A. Carbon, *Helv. Chim. Acta, 40*, 257 (1957).

256. Pratesi, P., and H. Blaschko, *Brit. J. Pharmacol., 14*, 256 (1959).

257. Pletscher, A., and K. F. Gey, *Science, 128*, 900 (1958).

258. Yamada, H., H. Kumagai, H. Kawasaki, H. Matsui, and K. Ogata, *Biochem. Biophys. Res. Commun., 29*, 723 (1967).

259. Carlsson, A., in *Progress in Brain Research*, Vol. 8, H. E. Himwich and W. A. Himwich, eds., Elsevier, Amsterdam, 1964, p. 9.

260. Salmoiraghi, G. C., E. Costa, and F. E. Bloom, *Ann. Rev. Pharmacol., 5*, 213 (1965).

261. Von Brücke, F. Th., and O. Hornykiewicz, *Pharmakologie der Psychopharmaka*, Springer, Berlin, 1966, p. 1.

262. Carlsson, A., in *Handbuch Exp. Pharmakol., 19*, 529 (1965).

263. Glowinski, J., and R. J. Baldessarini, *Pharmacol. Rev., 18*, 1201 (1966).

264. Iversen, L. L., *Nature, 214*, 8 (1967).

265. Iversen, L. L., *The Uptake and Storage of Noradrenaline in Sympathetic Nerves*, Cambridge University Press, London, 1967, p. 432.

266. Kopin, I. J., *Pharmacol. Rev., 16*, 179 (1964).

267. Schildkraut, J. J., and S. S. Kety, *Science*, *156*, 21 (1967).
268. Vogt, M., *J. Physiol.* (London), *123*, 451 (1954).
269. Montagu, K. A., *Nature*, *180*, 244 (1957).
270. Weil-Malherbe, H., and A. D. Bone, *Nature*, *180*, 1050 (1957).
271. Twarog, B. M., and J. H. Page, *Am. J. Physiol.*, *175*, 157 (1953).
272. Amin, A. H., T. B. B. Crawford, and J. H. Gaddum, *J. Physiol.*, *126*, 596 (1954).
273. Carlsson, A., M. Lindqvist, and T. Magnusson, in *Adrenergic Mechanisms*, J. R. Vane, G. E. W. Wolstenholme, and C. M. O'Connor, eds., Little, Brown, Boston, 1960.
274. Falck, B., *Acta. Physiol. Scand.*, *56* (Suppl.), 197 (1962).
275. Falck, B., N. A. Hillarp, G. Thieme, and A. Thorp, *J. Histochem. Cytochem.*, *10*, 348 (1962).
276. Corrodi, H., and G., Jonsson, *J. Histochem. Cytochem.*, *15*, 65 (1967).
277. Falck, B., in *Progress in Brain Research*, Vol. 8, H. E. Himwich and W. A. Himwich, eds., Elsevier, Amsterdam, 1964, p. 28.
278. Hillarp, N. A., K. Fuxe, and A. Dahlstrom, *Pharmacol. Rev.*, *18*, 727 (1966).
279. Von Euler, U. S., *Noradrenaline*, Thomas, Springfield, Ill., 1956.
280. Hökfelt, T., *Z. Zellforsch.*, *79*, 110 (1967).
281. Bloom, F. E., and N. J. Giarman, *Ann. Rev. Pharmacol.*, *8*, 229 (1968).
282. Iversen, L. L., and M. A. Simmonds, in *Metabolism of Amines in the Brain*, G. Hooper, ed., Macmillan, London, 1969, p. 48.
283. Ahlquist, R. P., *Am. J. Physiol.*, *154*, 586 (1948).
284. Ahlquist, R. P., *Pharmacol. Rev.*, *11*, 441 (1959).
285. Ahlquist, R. P., *Arch. Intern. Pharmacodyn.*, *134*, 38 (1962).
286. Ahlquist, R. P., *Ann. Rev. Pharmacol.*, *8*, 259 (1968).
287. Dale, H. H., *J. Physiol.* (London), *34*, 163 (1906).
288. Barger, G., and H. H. Dale, *J. Physiol.* (London) *41*, 19 (1910).
289. Levy, B., and R. P. Ahlquist, *J. Pharmacol. Exptl. Therap.*, *133*, 202 (1961).
290. Moran, N. C., and M. E. Perkins, *J. Pharmacol. Exptl. Therap.*, *124*, 223 (1958).
291. Moran, N. C., and M. E. Perkins, *J. Pharmacol. Exptl. Therap.*, *133*, 192 (1961).
292. De Robertis, E. D. P., in *Progress in Brain Research*, Vol. 8, H. E. Himwich and W. A. Himwich, eds., Elsevier, Amsterdam, 1964, p. 145.
293. Hillarp, N. A., and T. Malmfors, *Life Sci.*, *3*, 703 (1964).
294. Holtz, P., *Pharmacol. Rev.*, *11*, 317 (1959).
295. Holtz, P., and E. Westermann, *Arch. Exptl. Pathol. Pharmakol.*, *237*, 538 (1956).
296. Bertler, A., and E. Rosengren, *Acta Physiol. Scand.*, *47*, 350 (1959).
297. Hornykiewicz, O., *Pharmacol. Rev.*, *18 (2)*, 925 (1966).
298. Masuoka, D. T., H. F. Schott, and L. Petriello, *J. Pharmacol.*, *139*, 73 (1963).
299. MacGeer, E. G., G. M. Ling, and P. L. MacGeer, *Biochem. Biophys. Res. Commun.*, *13*, 291 (1963).
300. Glowinski, J., L. L. Iversen, and J. Axelrod, *Pharmacologist*, *7*, 156 (1965).
301. Blaschko, H., *Experientia*, *13*, 9 (1957).
302. Carlsson, A., M. Lindqvist, and T. Magnusson, *Nature*, *180*, 1200 (1957).
303. Bernheimer, H., W. Birkmayer, and O. Hornykiewicz, *Klin. Wochschr.*, *39*, 1056 (1961).
304. Ehringer, H., and O. Hornykiewicz, *Klin. Wochschr.*, *38*, 1236 (1960).
305. Barbeau, A., G. F. Murphy, and T. L. Sourkes, *Science*, *133*, 1706 (1961).
306. Barolin, G. S., H. Bernheimer, and O. Hornykiewicz, *Schweiz. Arch. Neurol. Psychiat.*, *94*, 241 (1964).
307. Wurtman, R. J., and J. Axelrod, *Science*, *150*, 1464 (1965).

308. Weil-Malherbe, H., J. Axelrod, and R. Tomchick, *Science, 129,* 1226 (1959).
309. Dengler, H. J., I. A. Michaelson, I. A. Spiegel, and E. Titus, *J. Neuropharmacol., 1,* 23 (1962).
310. Iversen, L. L., *Brit. J. Pharmacol. Chemotherap., 17,* 62 (1963).
311. Weil-Malherbe, H., L. G. Whitby, and J. Axelrod, in *Regional Neurochemistry,* S. S. Kety and J. Elkes, eds., Pergamon, Oxford, 1961, p. 284.
312. Whitby, L. G., J. Axelrod, and H. Weil-Malherbe, *J. Pharmacol., 132,* 193 (1961).
313. Mannarino, E., N. Kirshner, and B. S. Nashold, *J. Neurochem., 10,* 373 (1963).
314. Milhaud, G., and J. Glowinski, *Compt. Rend., 255,* 203 (1962).
315. Milhaud, G., and J. Glowinski, *Compt. Rend., 256,* 1033 (1963).
316. Glowinski, J., and L. L. Iversen, *J. Neurochem., 13,* 655 (1966).
317. Glowinski, J., and J. Axelrod, *Pharmacol. Rev., 18,* 775 (1966).
318. Glowinski, J., I. J. Kopin, and J. Axelrod, *J. Neurochem., 12,* 25 (1965).
319. Feldberg, W., and R. D. Myers, *Nature, 200,* 1325 (1963).
320. Feldberg, W., and V. J. Lotti, *Brit. J. Pharmacol. Chemotherap., 31,* 152 (1967).
321. Feldberg, W., in *"Recent Advances in Pharmacology,"* 4th Ed., J. M. Robson and R. S. Stacey, eds., Churchill, London, 1968, p. 349.
322. Simmonds, M. A., and L. L. Iversen, *Science, 163,* 473 (1969).
323. Glowinski, J., J. Axelrod, and L. L. Iversen, *J. Pharmacol., 153,* 30 (1966).
324. Fuxe, K., T. Hökfelt, and U. Ungerstedt, in *Metabolism of Amines in the Brain,* G. Hooper, ed., Macmillan, London, 1969, p. 10.
325. Levitt, M., S. Spector, A. Sjoerdsma, and S. Udenfriend, *J. Pharmacol., 148,* 1 (1965).
326. Spector, S., A. Sjoerdsma, and S. Udenfriend, *J. Pharmacol., 147,* 86 (1965).
327. Nagatsu, T., M. Levitt, and S. Udenfriend, *J. Biol. Chem., 239,* 2910 (1964).
328. Costa, E., and N. H. Neff, in *Biochemistry and Pharmacology of the Basal Ganglia,* E. Costa, L. J. Coté, and M. D. Yahr, eds., Raven, New York, 1966.
329. Holzbauer, M., and M. Vogt, *J. Neurochem., 1,* 8 (1956).
330. Pletscher, A., P. A. Shore, and B. B. Brodie, *J. Pharmacol. Exptl. Therap., 116,* 84 (1956).
331. Carlsson, A., M. Lindqvist, T. Magnusson, and B. Waldeck, *Science, 127,* 471 (1958).
332. Shore, P. A., *Pharmacol. Rev., 14,* 531 (1962).
333. Iversen, L. L., J. Glowinski, and J. Axelrod, *J. Pharmacol. Exptl. Therap., 150,* 173 (1965).
334. Pirch, J. H., R. H. Reech, and K. E. Moore, *Intern. J. Neuropharmacol., 6,* 375 (1967).
335. Brodie, B. B., *The Scientific Basis of Drug Therapy in Psychiatry,* Pergamon, Oxford, 1965.
336. Brodie, B. B., and P. A. Shore, *Ann. N. Y. Acad. Sci., 66,* 631 (1957).
337. Carlsson, A., *Neuropsychopharmacology, 2,* 417 (1961).
338. Shore, P. A., in *Mechanisms of Release of Biogenic Amines* (Proceedings of an International Symposium, Stockholm, 1965), 1st Ed., U. S. von Euler et al., eds., Pergamon, Oxford, 1966.
339. Kety, S. S., *Res. Publ. Ass. Res. Nervous Mental Disease, 40,* 311 (1962).
340. Spector, S., C. W. Hirsch, and B. B. Brodie, *Intern. J. Neuropharmacol., 2,* 81 (1963).
341. Spector, S., P. A. Shore, and B. B. Brodie, *J. Pharmacol. Exptl. Therap., 128,* 15 (1960).
342. Zeller, E. A., and J. Barsky, *Proc. Soc. Exptl. Biol. Med., 81,* 459 (1952).

343. Zeller, E. A., *Ann. N. Y. Acad. Sci., 107,* 811 (1963).

344. Zeller, E. A., and J. R. Fouts, *Ann. Rev. Pharmacol., 3,* 9 (1963).

345. Blaschko, H., in *Progress in Brain Research,* Vol. 8, H. E. Himwich and W. A. Himwich, eds., Elsevier, Amsterdam, 1964, p. 1.

346. Pletscher, A., *Pharmacol. Rev., 18(1),* 121 (1966).

347. Sjoerdsma, A., *Pharmacol. Rev., 18(1),* 673 (1966).

348. Pscheidt, G. R., *Intern. Rev. Neurobiol., 7,* 191 (1964).

349. Ozaki, M., H. Weissbach, A. Ozaki, B. Witkop, and S. Udenfriend, *J. Med. Pharmaceut. Chem., 2,* 591 (1960).

350. Pletscher, A., and K. F. Gey, *Helv. Physiol. Acta, 17,* 635 (1959).

351. Horita, A., *J. Pharmacol. Exptl. Therap., 122,* 176 (1958).

352. Horita, A., *Ann. N. Y. Acad. Sci., 80(3),* 590 (1959).

353. Zeller, E. A., *Neuropsychiatry, 2,* Suppl. 1, 125 (1961).

354. Roewer, F., and E. Werle, *Arch. Exptl. Pathol. Pharmakol., 230,* 552 (1957).

355. Zeller, E. A., J. Bernsohn, W. M. Inskip, and J. W. Lerner, *Naturwiss., 44,* 427 (1957).

356. Lauer, J. W., W. M. Inskip, J. Bernsohn, and E. A. Zeller, *Arch. Neurol. Psych., 80,* 122 (1958).

357. Erspamer, V., *Acta Pharmacologica, 4,* 224 (1948).

358. Kakimoto, Y., and M. D. Armstrong, *J. Biol. Chem., 237,* 422 (1962).

359. Masuoka, D. T., A. Alcaraz, and E. Mansson, *Biochim. Biophys. Acta, 86,* 260 (1964).

360. Sjoerdsma, A., *Circulation Res., 9,* 734 (1961).

361. Randall, L. O., and R. E. Bagdon, *Ann. N. Y. Acad. Sci., 80(3),* 626 (1959).

362. Kopin, I. J., J. E. Fisher, I. M. Musacchio, W. D. Horst, and V. Weise, *J. Pharmacol., 147,* 186 (1965).

363. Zeller, E. A., *Pharmacol. Rev., 18(1),* 141 (1966).

364. Gey, K. F., and A. Pletscher, *J. Pharmacol. Exptl. Therap., 133,* 18 (1961).

365. Gey, K. F., and A. Pletscher, *J. Pharmacol., 145,* 337 (1964).

366. Pletscher, A., K. F. Gey, and E. Kunz, in *Progress in Brain Research,* Vol. 8, H. E. Himwich and W. A. Himwich, eds., Elsevier, Amsterdam, 1965, p. 45.

367. Andén, N. E., B. Roos, and B. Werdinius, *Life Sci., 3,* 149 (1964).

368. Carlsson, A., and M. Lindqvist, *Acta Pharmac. Toxicol. Kbh., 20,* 140 (1963).

369. Camanni, F., G. M. Molinatti, and M. Olivetti, *Nature, 184,* 65 (1959).

370. Ehringer, H., O. Hornykiewicz, and K. Lechner, *Arch. Exptl. Pathol. Pharmakol., 239,* 507 (1960).

371. Costa, E., G. L. Gessa, and B. B. Brodie, *Life Sci., 7,* 315 (1962).

372. Morpurgo, C., *Biochem. Pharmacol., 11,* 967 (1962).

373. Hertting, G., J. Axelrod, and L. G. Whitby, *J. Pharmacol. Exptl. Therap., 134,* 146 (1961).

374. Dengler, H. J., H. E. Spiegel, and E. O. Titus, *Nature, 191,* 816 (1961).

375. Carlsson, A., N. A. Hillarp, and B. Waldeck, *Acta Physiol. Scand.,* Suppl. 215 (1963).

376. Weil-Malherbe, H., and H. S. Posner, *Biochem. Pharmacol., 13,* 685 (1964).

377. Birkmayer, W., and O. Hornykiewicz, *Wien. Klin. Wochschr. 73,* 787 (1961).

378. Cotzias, G. C., P. S. Papavasilion, and R. Gellene, *New England J. Med., 280,* 337 (1969).

379. Calne, D. B., G. M. Stern, D. R. Laurence, J. Sharkey, and P. Armitage, *Lancet, 1,* 744 (1969).

380. Blaschko, H., and B. C. P. Stromblad, *Arzneimittel-Forschg., 10,* 327 (1960).

381. Goodman, L. S., and A. Gilman, eds., *The Pharmacological Basis of Therapeutics,* Macmillan, New York, 1955, p. 2.
382. Glowinski, J., and J. Axelrod, *J. Pharmacol. Exptl. Therap., 149,* 43 (1965).
383. Iversen, L. L., *J. Pharm. Pharmacol., 16,* 435 (1964).
384. Schekel, C. L., and E. Boff, *Psychopharmacologia, 5,* 198 (1964).
385. Moore, K. E., and E. W. Lariviere, *Biochem. Pharmacol., 12,* 1283 (1963).
386. Smith, C. B., *J. Pharmacol. Exptl. Therap., 147,* 96 (1965).
387. Laverty, R., and D. F. Sharman, *Brit. J. Pharmacol. Chemotherap., 24,* 759 (1965).
388. Moore, K. E., and E. W. Lariviere, *Biochem. Pharmacol., 13,* 1098 (1964).
389. Smith, C. B., *J. Pharmacol. Exptl. Therap., 142,* 343 (1963).
390. Van Rossum, J. B., J. A. Van Der Schoot, and T. M. Horkmans, *Experientia, 18,* 229 (1962).
391. Corne, S. I., and I. P. P. Graham, *J. Physiol., 135,* 339 (1957).
392. Udenfriend, S., B. Witkop, B. G. Redfield, and H. Weissbach, *Biochem. Pharmacol., 1,* 160 (1958).
393. Pletscher, A., and H. Besendorf, *Experientia, 15,* 24 (1959).
394. Blaschko, H., and J. M. Himms, *Brit. J. Pharmacol., 10,* 451 (1955).
395. Fastier, F., and J. Hawkins, *Brit. J. Pharmacol., 6,* 256 (1951).
396. Ehringer, H., O. Hornykiewicz, and K. Lechner, *Arch. Exptl. Pathol. Pharmakol., 241,* 568 (1961).
397. Davison, A. N., *Biochem. J., 67,* 316 (1957).
398. Blaschko, H., *Nature, 145,* 26 (1940).
399. Mann, P. J. G., and J. H. Quastel, *Biochem. J., 34,* 414 (1940).
400. Kobayashi, Y., and R. W. Schayer, *Arch. Biochem. Biophys., 58,* 181 (1955).
401. Greig, M. E., R. A. Walk, and A. J. Gibbons, *J. Pharmacol. Exptl. Therap., 127,* 110 (1959).
402. Davison, A. N., *Biochem. J., 64,* 546 (1956).
403. Klerman, G. L., and J. O. Cole, *Pharmacol. Rev., 17,* 101 (1965).
404. Hordern, A., *New England J. Med., 272,* 1159 (1965).
405. Sigg, E. B., L. Soffer, and L. Gyermek, *J. Pharmacol. Exptl. Therap., 142,* 13 (1963).
406. Thoenen, H., A. Huerliman, and W. Haefely, *J. Pharmacol. Exptl. Therap., 144,* 405 (1964).
407. Sigg, E. B., *Can. Psychiat. Assoc. J., 4,* Suppl. 1, 75 (1959).
408. Glowinski, J., and J. Axelrod, *Nature, 204,* 1318 (1964).
409. Schanberg, S. M., J. J. Schildkraut, and I. J. Kopin, *Biochem. Pharmacol., 16,* 393 (1967).
410. Kopin, I. J., *Pharmacol. Rev., 18,* 513 (1966).
411. Eisenfeld, A. J., L. Krakoff, L. L. Iversen, and J. Axelrod, *Nature, 213,* 297 (1967).
412. Sulser, F., M. H. Bickel, and B. B. Brodie, *J. Pharmacol. Exptl. Therap., 144,* 321 (1964).
413. Schildkraut, J. J., *Am. J. Psychiat., 122,* 509 (1965).
414. Malmfors, T., *Acta Physiol. Scand., 64* (Suppl.), 248 (1965).
415. Lindmar, R., and E. Muscholl, *Arch. Exptl. Pathol. Pharmakol., 242,* 214 (1961).
416. Whitby, L. G., G. Hertting, and J. Axelrod, *Nature, 187,* 604 (1960).
417. Ganrot, P. O., E. Rosengren, and C. G. Gottfries, *Experientia, 18,* 260 (1962).
418. Blackwell, B., and E. Marley, *Brit. J. Pharmacol., 26,* 120, 142 (1966).
419. Rand, M. J., and F. R. Trinker, *Brit. J. Pharmacol. Chemother., 33,* 287 (1968).
420. Goldberg, L. I., *J. Am. Med. Assoc., 190,* 456 (1964).
421. Sjöqvist, F., *Proc. Roy. Soc. Med., 58,* 967 (1965).

422. Smith, C. B., *J. Pharmacol. Exptl. Therap.*, *151*, 207 (1966).
423. Axelrod, J., *J. Pharmacol. Exptl. Therap.*, *110*, 315 (1954).
424. Axelrod, J., *J. Biol. Chem.*, *214*, 753 (1955).
425. Horita, A., *Toxicol. Appl. Pharmacol.*, *3*, 474 (1961).
426. Koechlin, B., and V. Iliev, *Ann. N. Y. Acad. Sci.*, *80(3)*, 864 (1959).
427. Weikel, H. Jr., and J. E. Salmon, *Biochem. Pharmacol.*, *11*, 93 (1962).
428. Gorkin, V. Z., in Abstracts, 7th International Congress of Biochemistry, Tokyo, August 19–25, 1967, p. 1023.
429. Axelrod, J., *Physiol. Rev.*, *39*, 751 (1959).
430. Matsuoka, M., *Jap. J. Pharmacol.*, *14*, 181 (1964).
431. Blaschko, H., *Pharmacol. Rev.*, *6*, 23 (1954).
432. Axelrod, J., W. Albers, and C. D. Clemente, *J. Neurochem.*, *5*, 68 (1959).
433. Baldessarini, R. J., and I. J. Kopin, *J. Neurochem.*, *13*, 769 (1966).
434. Nukada, T., T. Sakurai, and R. Imaizumi, *Jap. J. Pharmacol.*, *13*, 124 (1963).
435. Rodriguez De Lores Arnaiz, G., and E. De Robertis, *J. Neurochem.*, *9*, 503 (1962).
436. Alberici, M., G. Rodriguez De Lores Arnaiz, and E. D. P. De Robertis, *Life Sci.*, *4*, 195 (1965).
437. Spector, S., *Ann. N. Y. Acad. Sci.*, *107*, 856 (1963).
438. Shore, P. A., *Pharmacol. Rev.*, *18(1)*, 561 (1966).
439. Bertler, A., *Acta Physiol. Scand.*, *51*, 75 (1961).
440. Crout, J. R., C. R. Creveling, and S. Udenfriend, *J. Pharmacol.*, *132*, 269 (1961).
441. Green, H., and B. W. Erickson, *J. Pharmacol.*, *129*, 237 (1960).
442. Spector, S., D. Prockop, P. A. Shore, and B. B. Brodie, *Science*, *127*, 704 (1958).
443. Costa, E., A. M. Revzin, R. Kuntzman, S. Spector, and B. B. Brodie, *Science*, *133*, 1822 (1961).
444. Pletscher, A., and G. Pellmont, *Clin. Exptl. Psychopathol.*, *19* (Suppl.), 163 (1958).
445. Hamberger, B., T. Malmfors, K. A. Norberg, and C. Sachs, *Biochem. Pharmacol.*, *13*, 841 (1964).
446. Fuxe, K., and N. A. Hillarp, *Life Sci.*, *3*, 1403 (1964).
447. Kopin, I. J., and E. K. Gordon, *J. Pharmacol.*, *140*, 207 (1963).
448. Glowinski, J., L. L. Iversen, and J. Axelrod, *J. Pharmacol.*, *151*, 385 (1966).
449. Shaw, K. N. F., A. McMillan, and M. D. Armstrong, *J. Biol. Chem.*, *226*, 255 (1957).
450. Udenfriend, S., in *5-Hydroxytryptamine* (Proceedings of a Symposium, London, 1957), G. P. Lewis, ed., Pergamon, Oxford, 1958, p. 43.
451. Armstrong, M. D., and A. McMillan, *Pharmacol. Rev.*, *11*, 394 (1959).
452. Neff, N. H., T. N. Tozer, and B. B. Brodie, *Pharmacologist*, *6*, 194 (1964).
453. Sharman, D. F., *Brit. J. Pharmacol. Chemotherap.*, *28*, 153 (1966).
454. Werdinius, B., *J. Pharm. Pharmacol.*, *18*, 546 (1966).
455. Werdinius, B., *Acta Pharmacol. Toxicol.*, *25*, 18 (1967).
456. Neff, N. H., T. N. Tozer, and B. B. Brodie, *J. Pharmacol. Exptl. Therap.*, *158*, 214 (1967).
457. Sharman, D. F., *Brit. J. Pharmacol. Chemotherap.*, *30*, 620 (1967).
458. Sharman, D. F., in *Metabolism of Amines in the Brain*, G. Hooper, ed., Macmillan, London, 1969, p. 34.
459. Pappenheimer, J. R., S. R. Heisey, and E. F. Jordan, *Am. J. Physiol.*, *200*, 1 (1961).
460. Davson, H., C. R. Kleeman, and E. Levin, *J. Physiol.*, *161*, 126 (1962).
461. Pollay, M., and H. Davson, Brain, *86*, 137 (1963).
462. Guldberg, H. C., G. W. Ashcroft, and T. B. B. Crawford, *Life Sci.*, *5*, 1571 (1966).

463. Moir, A. T. B., Ph.D. Thesis, University of Edinburgh, 1967.
464. Ashcroft, G. W., T. B. B. Crawford, R. C. Dow, and H. C. Guldberg, *Brit. J. Pharmacol. Chemotherap., 33,* 441 (1968).
465. Guldberg, H. C., in *Metabolism of Amines in Brain,* G. Hooper, ed., Macmillan, London, 1969, p. 55.
466. Stein, L., *Federation Proc., 23,* 836 (1964).
467. Burgen, A. S. V. and L. L. Iversen, *Brit. J. Pharmacol. Chemotherap., 25,* 34 (1965).
468. Carlsson, A., M. Lindqvist, H. Dahlström, K. Fuxe, and D. Masuoka, *J. Pharm. Pharmacol., 17,* 521 (1965).
469. Ross, S. B., and A. L. Renyi, *Acta Pharmacol. Toxicol. Kbh., 21,* 226 (1964).
470. Goldstein, M., and J. F. Contrera, *Biochem. Pharmacol., 7,* 77 (1961).
471. Axelrod, J., *Science, 127,* 754 (1958).
472. Glowinski, J., S. H. Snyder, and J. Axelrod, *J. Pharmacol. Exptl. Therap., 152,* 282 (1966).
473. Snyder, S. H., J. Glowinski, and J. Axelrod, *Life Sci., 4,* 797 (1965).
474. Weil-Malherbe, H., H. S. Posner, and G. R. Bowles, *J. Pharmacol., 132,* 278 (1961).
475. Green, H., and J. L. Sawyer, *J. Pharmacol., 129,* 243 (1960).
476. Baldessarini, R. J., *Biochem. Pharmacol., 15,* 741 (1966).
477. Axelrod, J., and R. Tomchick, *J. Biol. Chem., 233,* 702 (1958).
478. Belleau, B., and J. Burra, *J. Med. Chem., 6,* 755 (1963).
479. Carlsson, A., H. Corrodi, and B. Waldeck, *Helv. Chim. Acta, 46,* 2371 (1963).
480. Ross, S. B., and O. Haljasmaa, *Acta Pharmacol. Toxicol. Kbh., 21,* 205 (1964).
481. Ross, S. B., and O. Haljasmaa, *Acta Pharmacol. Toxicol. Kbh., 21,* 215 (1964).
482. Matsuoka, M., H. Yoshida, and R. Imaisumi, *Biochem. Pharmacol., 11,* 1109 (1962).
483. Axelrod, J., *Pharmacol. Rev., 11,* 402 (1959).
484. Costa, E. E., and B. B. Brodie, in *Progress in Brain Research,* Vol. 8, H. E. Himwich and W. A. Himwich, eds., Elsevier, Amsterdam, 1964, p. 168.
485. Axelrod, J., *Progress in Brain Research,* Vol. 8, H. E. Himwich and W. A. Himwich, eds., Elsevier, Amsterdam, 1964, p. 81.
486. Spector, S., R. Kuntzman, P. A. Shore, and B. B. Brodie, *J. Pharmacol., 130,* 256 (1960).
487. Holtz, P., *Pharmacol. Rev., 18(1),* 85 (1966).
488. Cuthill, J. M., A. B. Griffiths, and D. E. B. Powell, *Lancet, 1,* 1076 (1964).
489. Cooper, A. J., and K. M. G. Keddie, *Lancet, 1,* 1133 (1964).
490. Glazener, F. S., W. A. Morgan, J. M. Simpson, and P. K. Johnson, *J. Am. Med. Assoc., 188,* 754 (1964).
491. Häggendal, J., *Acta Physiol. Scand., 59,* 261 (1963).
492. Brodie, B. B., and E. Costa, in *Monoamines et système nerveux central,* (Proceedings of a Symposium, Geneva), 1961, J. De Ajuriaguerra, ed., Georg, Geneva, 1961, p. 13.
493. MacGeer, P. L., E. G. MacGeer, and J. A. Wada, *Arch. Neurol., 9,* 81 (1963).
494. Pletscher, A., *Schweiz. Med. Wochschr., 87,* 1532 (1957).
495. Kopin, I. J., G. Hertting, and E. K. Gordon, *J. Pharmacol., 138,* 34 (1962).
496. Lindmar, R., and E. Muscholl, *Arch. Exptl. Pathol. Pharmakol., 247,* 462 (1964).
497. Hamberger, B., and D. Masuoka, *Acta Pharmacol. Toxicol. Kbh., 22,* 363 (1965).
498. Carlsson, A., A. Dahlström, K. Fuxe, and N. A. Hillarp, *Acta Pharmacol. Toxicol. Kbh., 22,* 270 (1965).
499. Barondes, S. H., *J. Biol. Chem., 237,* 204 (1962).
500. Pastan, I., and J. B. Field, *Endocrinology, 70,* 656 (1962).

501. Dale, H. H., and R. P. Laidlaw, *J. Physiol., 43*, 182 (1911/1912).
502. Oehme, C., *Arch. Exptl. Pathol. Pharmakol., 72*, 76 (1913).
503. Guggenheim, M., and W. Loeffler, *Biochem. Z., 72*, 303, 325 (1916).
504. Best, C. H., and E. W. McHenry, *J. Physiol., 70*, 349 (1930).
505. Zeller, E. A., *Helv. Chim. Acta, 21*, 880 (1938).
506. Blaschko, H., P. Friedman, and K. Nilsson, *J. Physiol., 142*, 33P (1958).
507. Blaschko, H., *J. Physiol., 148*, 570 (1959).
508. Buffoni, F., *J. Physiol., 169*, 121P (1963).
509. Kapeller-adler, R., in *Histamine,* Ciba Foundation Symposium, G. E. W. Wolstenholme and C. M. O'Connor, eds., Churchill, London, 1956, p. 356.
510. Werle, E., *Biochem. Z., 309*, 61 (1941).
511. Gale, R. F., *Biochem. J., 36*, 64 (1942).
512. Satake, K., and H. Fujita, *J. Biochem.* (Tokyo), *40*, 547 (1953).
513. Bachrach, U., *J. Biol. Chem., 237*, 3443 (1962).
514. Roulet, F., and E. A. Zeller, *Helv. Chim. Acta, 28*, 1326 (1945).
515. Cromwell, B. T., *Biochem. J., 37*, 722 (1943).
516. Werle, E., and E. von Pechmann, *Liebig's Ann. Chem., 562*, 44 (1949).
517. Werle, E., I. Trautschold, and D. Aures, *Z. Physiol. Chem., 326*, 200 (1961).
518. Lindahl, K. M., S. E. Lindell, H. Westling, and T. White, *Acta Physiol. Scand., 38*, 280 (1957).
519. Kobayashi, Y., and A. C. Ivy, *Am. J. Physiol., 195*, 525 (1958).
520. Kobayashi, Y., and A. C. Ivy, *Am. J. Physiol., 196*, 835 (1959).
521. Shore, P. A., and V. H. Cohn, Jr., *Biochem. Pharmacol., 5*, 91 (1960).
522. Waton, N. J., *Brit. J. Pharmacol., 11*, 119 (1956).
523. Carlsten, A., *Acta Physiol. Scand., 20*, Suppl. 70, 5 (1950).
524. Gesler, R. H., M. Matsuba, and C. A. Dragstedt, *J. Pharmacol. Exptl. Therap., 116*, 356 (1956).
525. Holtz, P., R. Heise, and W. Spreyer, *Arch. Exptl. Pathol. Pharmakol., 188*, 580 (1936).
526. Anrep, G. V., G. S. Barsoum, and M. Talaat, *J. Physiol., 86*, 55 (1936).
527. Burkard, W. P., K. F. Gey, and A. Pletscher, *Helv. Physiol. Pharmacol. Acta, 19*, C61 (1961).
528. White, T., *J. Physiol.* (London), *149*, 34 (1959).
529. Brown, D. D., R. Tomchick, and J. Axelrod, *J. Biol. Chem., 234*, 2948 (1959).
530. Danforth, D. N., and F. Gorham, *Am. J. Physiol., 119*, 294 (1937).
531. Marcou, J., D. Chiriceanu, G. Cosma, N. Gingold, and C. C. Parhon, *Presse Méd., 46*, 371 (1938).
532. Zeller, E. A., and H. Birkhäuser, *Schweiz. Med. Wochschr., 70*, 975 (1940).
533. Werle, E., and G. Effkemann, *Klin. Wochschr., 19*, 717 (1940).
534. Ahlmark, A., *Acta Physiol. Scand., 9*, Suppl. 28, 1 (1944).
535. Kapeller-Adler, R., *Biochem. J., 38*, 270 (1944).
536. Kapeller-Adler, R., *Biochem. J., 51*, 610 (1952).
537. Kapeller-Adler, R., *Arch. Exptl. Pathol. Pharmakol., 219*, 491 (1953).
538. Kapeller-Adler, R., *Clin. Chim. Acta, 11*, 191 (1965).
539. Anrep, G. V., G. S. Barsoum, and A. Ibrahim, *J. Physiol., 106*, 379 (1947).
540. Wicksell, F., *Acta Physiol. Scand., 17*, 359 (1949).
541. Wicksell, F., *Acta Physiol. Scand., 17*, 395 (1949).
542. Swanberg, H., *Acta Physiol. Scand., 23*, Suppl. 79, 7 (1950).
543. Valette, G., B. Maupin, and H. Huidobro, *Compt. Rend. Soc. Biol.* (Paris), *145*, 313 (1951).
544. McEwen, C. H., Jr., *J. Lab. Clin. Med., 64*, 540 (1964).

545. Southren, A. L., Y. Kobayashi, D. H. Sherman, L. Levine, G. Gordon, and A. B. Weingold, *Am. J. Obstr. Gynecol., 89,* 199 (1964).

546. Kobayashi, Y., *Nature 203,* 146 (1964).

547. Okuyama, T., and Y. Kobayashi, *Arch. Biochem. Biophys., 95,* 242 (1961).

548. Kapeller-Adler, R., in Abstracts, 3rd Intern. Pharmacol. Congress, Sao Paulo, July 1966, p. 168.

549. Southren, A. L., Y. Kobayashi, P. Brenner, and A. B. Weingold, *J. Appl. Physiol., 20,* 1048 (1965).

550. Kapeller-Adler, R., and R. Renwick, *Clin. Chim. Acta, 1,* 197 (1956).

551. Tabor, H., *J. Biol. Chem., 188,* 125 (1951).

552. Uspenskaia, V. D., and E. V. Goryachenkova, *Biokhimiya* (English Translation), *23,* 199 (1958).

553. Kapeller-Adler, R., and H. MacFarlane, *Biochim. Biophys. Acta, 67,* 542 (1963).

554. Mondovi, B., G. Rotilio, and M. T. Costa, in *Chemical and Biological Aspects of Pyridoxal Catalysis,* E. E. Snell, P. M. Fasella, A. E. Braunstein, and A. Rossi-Fanelli, eds., Pergamon, Oxford, 1963, p. 415.

555. Mondovi, B., G. Rotilio, A. Finazzi, and A. Scioscia-Santoro, *Biochem. J., 91,* 408 (1964).

556. Mondovi, B., G. Rotilio, M. T. Costa, A. Finazzi-Agro, E. Chiancone, R. E. Hansen, and H. Beinert, *J. Biol. Chem., 242,* 1160 (1967).

557. Hennessey, M. A., A. M. Waltersdorph, F. M. Huennekens, and B. W. Gabrio, *J. Clin. Invest., 41,* 1257 (1962).

558. Smith, J. K., *Biochem. J., 103,* 110 (1967).

559. Sinclair, H. M., *Biochem. J., 51,* X (1952).

560. Goryachenkova, E. V., *Biokhimiya, 21,* 247 (1956).

561. Braunstein, A. E., in *The Enzymes,* Vol. 2, P. D. Boyer, H. Lardy and K. Myrbäck, eds., Academic, New York, 1960, p. 113.

562. Kapeller-Adler, R., *Federation Proc., 24,* 1757 (1965).

563. Mondovi, B., A. Scioscia-Santoro, G. Rotilio, and M. T. Costa, *Enzymologia, 28,* 228 (1964).

564. Mondovi, B., A. Finazzi-Agro, and G. Rotilio, in Abstracts, 7th International Congress of Biochemistry, Tokyo, August 19–25, 1967, p. 826.

565. Kobayashi, Y., *Federation Proc., 24,* 779 (1964).

566. Goryachenkova, E. V., and E. A. Ershova, in *Chemical and Biological Aspects of Pyridoxal Catalysis,* E. E. Snell, P. M. Fasella, A. E. Braunstein, and A. Rossi-Fanelli, eds., Pergamon, Oxford, 1963, p. 447.

567. Goryachenkova, E. V., and E. A. Ershova, *Biokhimiya,* 30, 141 (1965); Goryachenkova, E. V., L. J. Stcherbatiuk, and E. A. Voronina, *Biokhimiya,* 32, 330 (1967); Goryachenkova, E. V., L. J. Stcherbatiuk, and C. J. Zamaraev, in *Proceedings of the 2nd International Symposium on Chemical and Biological Aspects of Pyridoxal Catalysis* (International Union of Biochemistry, Moscow, September 1966), E. E. Snell, A. E. Braunstein, E. S. Severin, and Yu. M. Torchinsky, eds., Interscience, New York, 1968, p. 391.

568. Curti, B., and A. Gosti, *Boll. Soc. Ital. Biol. Sper., 41,* 370 (1965).

569. Buffoni, F., and L. Della Corte, *Boll. Soc. Ital. Biol. Sper., 43,* 1395 (1967).

570. Buffoni, F., L. Della Corte, and P. F. Knowles, *Biochem. J., 106,* 575 (1968).

571. Hamilton, G. A. in *Proceedings of the 2nd International Symposium on Chemical and Biological Aspects of Pyridoxal Catalysis* (International Union of Biochemistry, Moscow, September 1966), E. E. Snell, A. E. Braunstein, E. S. Severin, and Yu. M. Torchinsky, eds., Interscience, New York, 1968, p. 375.

572. Hamilton, G. A., *Advances Enzymol., 32,* 55 (1969).

294 REFERENCES

573. Malmström, B. G., and T. Vänngard, *J. Mol. Biol.*, *2*, 118 (1960).
574. Mason, H. S., Discussion remark in *The Biochemistry of Copper*, J. Peisach, P. Aisen, and W. E. Blumberg, eds., Academic, New York, 1966, p. 437.
575. Mahler, H. R., G. Hübscher, and H. Baum, *J. Biol. Chem.*, *216*, 625 (1955).
576. Hill, J. M., and P. J. G. Mann, *Biochem. J.*, *91*, 171 (1964).
577. Yamada, H., O. Adachi, and K. Ogata, *Agr. Biol. Chem. Tokyo*, *29*, 117, 912 (1965).
578. Gorkin, V. Z., *Vop. Med. Khim.*, *7*, 632 (1961).
579. Yamada, H., and K. T. Yasunobu, *Biochem. Biophys. Res. Commun.*, *8*, 387 (1962).
580. Yamada, H., and K. T. Yasunobu, *J. Biol. Chem.*, *237*, 3077 (1962).
581. Hamilton, G. A., and A. Revesz, *J. Am. Chem. Soc.*, *88*, 2069 (1966).
582. Buffoni, F., in *Proceedings of the 2nd International Symposium on Chemical and Biological Aspects of Pyridoxal Catalysis* (International Union of Biochemistry, Moscow, September 1966), E. E. Snell, A. E. Braunstein, E. S. Severin, and Yu. M. Torchinsky, eds., Interscience, New York, 1968, p. 363.
583. Fasella, P. M., and J. Manning, Abstracts, 7th International Congress of Biochemistry, Tokyo, August, 19–25, 1967, p. 1074.
584. Broman, L., B. G. Malmström, R. Aasa, and T. Vänngard, *J. Mol. Biol.*, *5*, 301 (1962).
585. Knowles, P. F., J. F. Gibson, F. M. Pick, and R. C. Bray, *Biochem. J.*, *111*, 53 (1969).
586. Bray, R. C., *Biochem. J.*, *81*, 189 (1961); Bray, R. C., P. F. Knowles, and L. S. Meriwether, in *Magnetic Resonance in Biological Systems*, B. G. Malmström, A. Ehrenberg, and T. Vänngard, eds., Pergamon, Oxford, 1967, p. 249.
587. Palmer, G., R. C. Bray, and H. Beinert, *J. Biol. Chem.*, *239*, 2657 (1964).
588. Bray, R. C., and P. F. Knowles, *Proc. Roy. Soc.* (London) Ser. A., *302*, 351 (1968).
589. Czapski, G., and L. M. Dorfman, *J. Phys. Chem.*, *68*, 1169 (1964).
590. Zeller, E. A., J. R. Fouts, J. A. Carbon, J. C. Lazanas, and W. Voegtli, *Helv. Chim. Acta*, *39*, 1632 (1956).
591. Fouts, J. R., L. A. Blanksma, J. A. Carbon, and E. A. Zeller, *J. Biol. Chem.*, *225*, 1025 (1957).
592. Mann, P. J. G., *Biochem. J.*, *59*, 609 (1955).
593. Lindell, S. E., and H. Westling, *Acta Physiol. Scand.*, *39*, 370 (1957).
594. Arunlakshana, O., J. L. Morgan, and H. O. Schild, *J. Physiol.*, *123*, 32 (1954).
595. Alles, G. A., B. B. Wisegarver, and M. A. Shull, *J. Pharmacol. Exptl. Therap.*, *77*, 54 (1943).
596. Kapeller-Adler, R., and B. Iggo, *Biochim. Biophys. Acta*, *25*, 394 (1957).
597. Dunathan, H. C., *Proc. Natl. Acad. Sci. U. S.*, *55*, 712 (1966).
598. Jenkins, W. T., *J. Biol. Chem.*, *236*, 1121 (1961).
599. Schuler, W., *Experientia*, *8*, 230 (1952).
600. Lindahl, K. M., *Arkiv Kemi*, *16*, 1, 15 (1960).
601. Holmstedt, B., and R. Tham, *Acta Physiol. Scand.*, *45*, 152 (1959).
602. Holmstedt, B., L. Larsson, and R. Tham, *Biochim. Biophys. Acta*, *48*, 182 (1961).
603. Gey, K. F., W. P. Burkard, and A. Pletscher, *Helv. Physiol. Pharmacol. Acta*, *18*, C27 (1960).
604. De Balbian Verster, F., Ph.D. Thesis, Tulane University, 1960.
605. Fouts, J. R., Ph.D. Thesis, Northwestern University, Evanston, Ill., 1954.
606. Blaschko, H., F. N. Fastier, and J. Wajda, *Biochem. J.*, *49*, 250 (1949).

607. Zeller, E. A., *Helv. Chim. Acta, 23,* 1418 (1940).
608. Blaschko, H., and R. Duthie, *Nature, 156,* 113 (1945).
609. Chapman, J. E., and E. J. Walaszek, *Biochem. Pharmacol., 11,* 205 (1962).
610. Kobayashi, Y., and T. Okuyama, *Biochem. Pharmacol., 11,* 949 (1962).
611. De Marco, C., B. Mondovi, and D. Cavallini, *Biochem. Pharmacol., 11,* 509 (1962).
612. Werle, E., and G. Hartung, *Biochem. Z., 328,* 1228 (1956).
613. Burkard, W. P., K. F. Gey, and A. Pletscher, *Biochem. Pharmacol., 3,* 249 (1960).
614. MacIntosh, F. C., and W. D. M. Paton, *J. Physiol., 109,* 190 (1949).
615. Mongar, J. L., and H. O. Schild, *Nature, 167,* 232 (1951).
616. Lindell, S. E., and H. Westling, *Acta Physiol. Scand., 37,* 307 (1956).
617. Lindell, S. E., *Acta Physiol. Scand., 41,* 168 (1957).
618. Blaschko, H., and S. Kurzepa, *Brit. J. Pharmacol., 19,* 544 (1962).
619. Carlini, E. A., M. Santos, and M. R. P. Sampaio, *Experientia, 21,* 72 (1965).
620. Carlini, E. A., M. R. P. Sampaio, M. Santos, and G. R. S. Carlini, *Biochem. Pharmacol., 14,* 1657 (1965).
621. Ivy. A. C., T. M. Lin, E. K. Ivy, and E. Karvinen, *Am. J. Physiol., 186,* 239 (1956).
622. Tabachnick, I. I. A., and F. E. Roth, *J. Pharmacol., 121,* 191 (1957).
623. Winbury, M. M., J. K. Wolf, and I. I. A. Tabachnick, *J. Pharmacol., 122,* 207 (1958).
624. Urbach, K. F., *Proc. Soc. Exptl. Biol.* (New York), *70,* 146 (1949).
625. Millican, R. C., S. M. Rosenthal, and H. Tabor, *J. Pharmacol. Exptl. Therap., 97,* 4 (1949).
626. Tabor, H., A. H. Mehler, and E. R. Stadtman, *J. Biol. Chem., 204,* 127 (1953).
627. Schayer, R. W., *Brit. J. Pharmacol., 11,* 472 (1956).
628. Livingston, R. H., and C. F. Code, *Am. J. Physiol., 181,* 428 (1955).
629. Sjaastad, V., and O. Sjaastad, *Acta Med. Scand., 172,* 535 (1962).
630. Karjala, S. A., and B. W. Turnquest, *J. Am. Chem. Soc., 77,* 6358 (1955).
631. Karjala, S. A., B. Turnquest, and R. W. Schayer, *J. Biol. Chem., 219,* 9 (1956).
632. Schayer, R. W., and S. A. Karjala, *J. Biol. Chem., 221,* 307 (1956).
633. Schayer, R. W., J. Kennedy, and R. L. Smiley, *J. Biol. Chem., 205,* 1739 (1953).
634. Brown, D. D., J. Axelrod, and R. Tomchick, *Nature, 183,* 680 (1959).
635. Lindahl, K. M., *Acta Chem. Scand., 12,* 2050 (1958).
636. Lindahl, K. M., *Acta Chem. Scand., 13,* 1476 (1959).
637. Lindahl, K. M., *Acta Physiol. Scand., 49,* 114 (1960).
638. Schayer, R. W., *Physiol. Rev., 39,* 116 (1959).
639. Bjuro, T., S. Lindberg, and H. Westling. *Acta Obstr. Gynaec. Scand., 40,* 152 (1961).
640. Bjuro, T., S. Lindberg, and H. Westling, *Acta Obstr. Gynaec. Scand., 43,* 206 (1964).
641. Rothschild, Z., and R. W. Schayer, *Biochim. Biophys. Acta, 30,* 23 (1958).
642. Snyder, S. H., and J. Axelrod, *Biochem. Pharmacol., 13,* 536 (1964).
643. Bjuro, T., *Acta Physiol. Scand, 60,* Suppl. 220, 1 (1963).
644. Kobayashi, Y., *Arch. Biochem. Biophys., 77,* 275 (1958).
645. Schayer, R. W., K. Y. T. Wu, R. L. Smiley, and Y. Kobayashi, *J. Biol. Chem., 210,* 259 (1954).
646. Lindberg, S., S. E. Lindell, and H. Westling, *Acta Obstr. Gynaec. Scand., 42,* Suppl. 1, 35, 49 (1963).
647. Schayer, R. W., and J. A. D. Cooper, *J. Appl. Physiol., 9,* 481 (1956).
648. Tabor, H., and O. Hayaishi, *J. Am. Chem. Soc., 77,* 505 (1955).

649. Crowley, G. M., *Federation Proc.*, *19*, 309 (1960).
650. Crowley, G. M., *J. Biol. Chem.*, *239*, 2593 (1964).
651. Fernandes, J. F., O. Y. Castellani, and M. Plese, *Biochem. Biophys. Res. Commun.*, *3*, 679 (1960).
652. Snyder, S. H., J. Axelrod, and H. Bauer, *J. Pharmacol. Exptl. Therap.*, *144*, 373 (1964).
653. Fox, C. L., and S. E. Lasker, *Am. J. Physiol.*, *202*, 111 (1962).
654. Abdel-Latif, A. A., and S. G. A. Alivisatos, *Biochim. Biophys. Acta*, *51*, 398 (1961).
655. Parrot, J. L., and C. Laborde, in *Histamine*, Ciba Foundation Symposium, G. E. W. Wolstenholme and C. M. O'Connor, eds., Churchill Ltd., London, 1956, p. 52.
656. Zeller, E. A., in *Metabolic Inhibitors*, Vol. 2, J. H. Quastel and R. M. Hoechster, eds., Academic, New York, 1963, Chapter 18.
657. Lilja, B., S. E. Lindell, and T. Saldeen, *J. Allergy*, *31*, 492 (1960).
658. Demis, J. D., and D. D. Brown, *J. Invest. Derm.*, *36*, 253 (1961).
659. Schayer, R. W., *J. Biol. Chem.*, *199*, 245 (1952); *203*, 787 (1953).
660. Kahlson, G., and E. Rosengren, *J. Physiol.*, *149*, 66P (1959).
661. Kahlson, G., and R. Rosengren, *Nature*, *184*, 1238 (1959).
662. Burkhalter, A., *Biochem. Pharmacol.*, *11*, 315 (1962).
663. Kameswaran, L., J. N. Pennefather, and G. B. West, *J. Physiol.*, *164*, 138 (1962).
664. Rosengren, E., *Proc. Soc. Exptl. Biol. Med.* (New York), *118*, 884 (1965).
665. Kahlson, G., E. Rosengren, and T. White, *J. Physiol.*, *145*, 30P (1959).
666. Wicksell, F., *Acta Physiol. Scand.*, *17*, 359 (1949).
667. Bjuro, T., and H. Westling, *Brit. J. Pharmacol.*, *22*, 453 (1964).
668. Kahlson, G., E. Rosengren, and H. Westling, *J. Physiol.*, *143*, 91 (1958).
669. Kahlson, G., E. Rosengren, H. Westling, and T. White, *J. Physiol.*, *144*, 337 (1958).
670. Schayer, R. W., in *Proc. XXII International Congress of Physiology*, Leyden, The Netherlands, 1962, p. 852.
671. Lindell, S. E., and H. Westling, in *Handbuch der Experimentellen Pharmakologie*, Vol. 18, Springer, Berlin, 1966, p. 734.
672. Carlsten, A., and D. R. Wood, *J. Physiol.*, *112*, 142 (1951).
673. Karady, S., B. Rose, and J. S. L. Browne, *Am. J. Physiol.*, *130*, 539 (1940).
674. Telford, J. M., *Nature*, *197*, 600 (1963).
675. Haeger, K., and G. Kahlson, *Acta Physiol. Scand.*, *25*, 255 (1952).
676. Kahlson, G., S. E. Lindell, and H. Westling, *Acta Physiol. Scand.*, *30*, Suppl. 111, 192 (1954).
677. Bartley, A. L., and M. P. Lockett, *J. Physiol.*, *147*, 51 (1959).
678. Hicks, R., and G. B. West, *Nature*, *182*, 401 (1958).
679. Marshall, P. B., *J. Physiol.*, *102*, 180 (1943).
680. Bergman, R. K., and J. Munoz, *Nature*, *205*, 910 (1965).
681. Schapiro, S., *Acta Endocr.*, Copenhagen, *48*, 249 (1965).
682. Falkheden, T., S. E. Lindell, and H. Westling, *Acta Endocr.*, *Copenhagen*, *47*, 581 (1964).
683. Spencer, P. S. J., and G. B. West, *Arch. Intern. Pharmacol.*, *148*, 31 (1964).
684. Andreoli, C., and F. Blanda, *Boll. Soc. Ital. Sper.*, *33*, 306 (1957).
685. Rose, B., and J. Leger, *Proc. Soc. Exptl. Biol.* (New York), *79*, 379 (1952).
686. Code, C. F., D. T. Cody, M. Hurn, J. C. Kennedy, and M. J. Strickland, *J. Physiol.*, *156*, 207 (1961).
687. Code, C. F., D. T. Cody, and J. C. Kennedy, *J. Physiol.*, *159*, 61P (1961).

688. Logan, G. B., *Proc. Soc. Exptl. Biol.* (New York), *111*, 171 (1962).
689. Cody, D. T., C. F. Code, and J. C. Kennedy, *J. Allergy*, *34*, 26 (1963).
690. Logan, G. B., *Proc. Soc. Exptl. Biol.* (New York), *107*, 466 (1961).
691. Giertz, H., F. Hahn, H. Hahn, and W. Schmutzler, *Klin. Wochschr.*, *40*, 598 (1962).
692. Hahn, F., W. Schmutzler, G. Seseke, H. Giertz, W. Bernauer, *Biochem. Pharmacol*, *15*, 155 (1966).
693. Hahn, F., in Abstracts 3rd Intern. Pharmacol. Congress, Saõ Paulo, July 1966, p. 170.
694. Ohkura, Y., *Okayama-Igakkai-Zasshi*, *68*, 1045, 1061 (1956).
695. Albus, G., *Klin. Wochschr.*, *18*, 858 (1939).
696. Werle, E., and O. Koch, *Beitr. Klin. Tbk.*, *96*, 584 (1941).
697. Flechner, I., *Med. Exp.*, *5*, 104 (1961).
698. Hansson, R., C. G. Holmberg, G. Tibbling, N. Tryding, H. Westling, and H. Wetterquist, *Acta Med. Scand.*, *180* (5), 533 (1966).
699. Tryding, N., *Scand. J. Clin. Lab. Investg.*, *17*, Suppl. 86, 197 (1965).
700. Dahl, S., *Ann. Allergy*, *22*, 303 (1964).
701. Maslinski, C., S. M. Maslinski, and H. Weinrander, *Experientia*, *19*, 258 (1963).
702. Maslinski, C., and A. Niedzielski, *Experientia*, *20*, 264 (1964).
703. Rose, B., S. Karady, and J. S. I. Browne, *Am. J. Physiol.*, *129*, 219 (1940).
704. Starr, K. W., J. Gibbons, and J. C. Claudatus, *Nature*, *195*, 1010 (1962).
705. Casati, G., M. Carbone, M. Lops, and R. Italia, *G. Ital. Tuberc.*, *17*, 8 (1963).
706. Borglin, N. E., and B. Willert, *Cancer*, *15*, 271 (1962).
707. Bernstein, J., W. P. Mazur, and E. J. Walaszek, *Med. Exp.*, *2*, 239 (1960).
708. Schayer, R. W., in *Handbuch der Experimentellen Pharmakologie*, Vol. 18, Springer, Berlin, 1966, p. 672.
709. Beall, G. N., and P. P. Vanarsdel, *J. Clin. Invest.*, *39*, 676 (1960).
710. Nilsson, K., S. E. Lindell, R. W. Schayer, and H. Westling, *Clin. Sci.*, *18*, 313 (1959).
711. Kobayashi, Y., and H. Freeman, *J. Neuropsych.*, *3*, 112 (1961).
712. Burkard, W. P., K. F. Gey, and A. Pletscher, *J. Neurochem.*, *10*, 183 (1963).
713. White, T., in Histamine in the Nervous System, a Symposium, W. G. Clark and G. Ungar, eds., *Federation Proc.*, *23*, 1103 (1964).
714. Johnson, B., and T. White, *Proc. Soc. Exptl. Biol.* (New York), *115*, 874 (1964).
715. Adam, H. M., and H. K. A. Hye, *J. Physiol.*, *171*, 37P (1964).
716. Adam, H. M., and H. K. A. Hye, *Brit. J. Pharmacol. Chemotherap.*, *28*, 137 (1966).
717. Halpern, B. N., T. Neveu, and C. W. M. Wilson, *J. Physiol.*, *147*, 437 (1959).
718. Robinson, J. D., and J. P. Green, *Federation Proc.*, *24*, 777 (1965).
719. Green, J. P., *Federation Proc.*, *23*, 1095 (1964).
720. Adam, H. M., in Regional Neurochemistry, S. S. Kety and J. Elkes, eds., Pergamon, London, 1961, p. 293.
721. Adam, H. M., D. C. Hardwick, and K. E. V. Spencer, *Brit. J. Pharmacol. 9*, 360 (1954).
722. Adam, H., D. C. Hardwick, and K. E. V. Spencer, *Brit. J. Pharmacol.*, *12*, 397 (1957).
723. Harris, G. W., D. Jacobsohn, and G. Kahlson, in *Histamine*, Ciba Foundation Symposium, G. E. W. Wolstenholme and C. M. O'Connor, eds., Churchill, London 1955, p. 235.
724. Riley, J. F., *The Mast Cells*, Livingstone, Edinburgh, 1959.

725. Michaelson, I. A., and G. Dowe, *Biochem. Pharmacol., 12,* 949 (1963).
726. McGeer, P. L., *Comparative Neurochemistry,* D. Richter, ed., Pergamon, London, 1964, p. 387.
727. Snyder, S. H., J. Glowinski, and J. Axelrod, *J. Pharmacol. Exptl. Therap., 153,* 8 (1966).
728. Snyder, S. H., J. Axelrod, and H. Bauer, *J. Pharmacol., 147,* 373 (1964).
729. Robinson, J. D., and J. P. Green, *Nature, 203,* 1178 (1964).
730. Carlini, E. A., and J. P. Green, *Brit. J. Pharmacol. Chemother., 20,* 264 (1963).
731. Shore, P. A., A. Burkhalter, and V. H. Cohn, Jr., *J. Pharmacol. Exptl. Therap., 127,* 182 (1959).
732. Green, H., and R. W. Erickson, *Intern. J. Neuropharmacol., 3,* 315 (1964).
733. Sjaastad, V., and H. Stormorken, *Nature, 197,* 907 (1963).
734. Bovet, D., and F. Bovet-Nitti, *Structure et activité pharmacodynamique de medicaments du système nerveux végetatif: adrénaline, acétylcholine, histamine et leur antagonists.* Karger, Bale, 1948.
735. Snyder, S. H., and J. Axelrod, *Federation Proc., 24,* 774 (1965).
736. Lindell, S. E., and R. W. Schayer, *Brit. J. Pharmacol., 13,* 44 (1958).
737. Bjuro, T., H. Westling, and H. Wetterquist, *Arch. Intern. Pharmacodyn., 144,* 337 (1963).
738. Code, C. F., *Federation Proc., 24,* 1311 (1965).
739. Blair, E. L., in *Gastrin,* M. I. Grossman, ed., University of California Press, Berkeley, 1966, p. 255.
740. Grossman, M. I., *Gastroenterology, 52,* 882 (1967).
741. Beaven, M. A., Z. Horakova, W. B. Severs, and B. B. Brodie, *J. Pharmacol. Exptl. Therap., 161,* 320 (1968).
742. Haverbach, B. J., M. I. Stubrin, and B. J. Dyce, *Federation Proc., 24,* 1326 (1965).
743. Kahlson, G., E. Rosengren, D. Svahn, and R. Thunberg, *J. Physiol., 174,* 400 (1964).
744. Levine, R. I., *Federation Proc., 24,* 1331 (1965).
745. Shore, P. A., *Federation Proc., 24,* 1322 (1965).
746. Code, C. F., in *Histamine,* Ciba Foundation Symposium, G. E. W. Wolstenholme and C. M. O'Connor, eds., Churchill, London, 1956, p. 189.
747. Navert, H., E. V. Flock, G. M. Tyce, and C. F. Code, submitted to the *American Journal of Physiology,* 1969.
748. Gregory, R. A. *Gastroenterologia, 76,* 82 (1950).
749. Irvine, W. T., H. D. Ritchie, and H. M. Adam, *Gastroenterology, 41,* 258 (1961).
750. Vasseur, B., F. Rancon, A. Saindelle, F. Ruff, and J. L. Parrot, *J. Physiologie, 59,* 409 (1967).
751. Gunther, R. E., and D. Glick, *J. Histochem. Cytochem., 15,* 431 (1967).
752. Green, J. P., D. H. Fram, and N. Kase, *Nature, 204,* 1165 (1964).
753. Szego, C. M., *Federation Proc., 24,* 1343 (1965).
754. Maslinski, C., and A. Niedzielski, *Experientia, 20,* 372 (1964).
755. Lindberg, S., *Acta Obstr. Gynaec. Scand., 42,* Suppl. 1, 1, 26 (1963).
756. Kahlson, G., and E. Rosengren, *Ann. Rev. Pharmacol., 5,* 305 (1965).
757. Lindahl, K. M., S. E. Lindell, K. Nilsson, R. W. Schayer, and H. Westling, *Arkiv Kemi, 13,* 379 (1958).
758. Guggenheim, M., *Die Biogenen Amine,* 4th Ed., Karger, Basel, New York, 1951.
759. Raina, A., *Acta Physiol. Scand., 60,* Suppl. 218, 7 (1963).
760. Tabor, H., and C. W. Tabor, *Pharmacol. Rev., 16*(3), 245 (1964).

761. Mann, T., *The Biochemistry of Semen of the Male Reproductive Tract*, Wiley, New York, 1964, p. 193.

762. Tabor, H., C. W. Tabor, and S. M. Rosenthal, *Ann. Rev. Biochem., 30*, 579 (1961).

763. Jänne, J., A. Raina, and M. Siimes, *Acta Physiol. Scand., 62*, 352 (1964).

764. Fischer, F. G., and H. Bohn, *Z. Physiol. Chem., 308*, 108 (1957).

765. Stevens, L., *Biochem. J., 99*, 11P (1966).

766. Cohen, S. S., and I. Lichtenstein, *J. Biol. Chem., 235*, 2112 (1960).

767. Spahr, P. F., *J. Mol. Biol., 4*, 395 (1962).

768. Zillig, W., W. Krone, and M. Albers, *Z. Physiol. Chem., 317*, 131 (1959).

769. Keller, P. J., E. Cohen, and R. D. Wade, *J. Biol. Chem., 239*, 3292 (1964).

770. Amotz, S., *Israel. J. Med. Sci., 1*, 486 (1965).

771. Tabor, C. W., and P. D. Kellog, *J. Biol. Chem., 242*, 1044 (1967).

772. Siekevitz, P., and G. E. Palade, *J. Cell Biol., 13*, 217 (1962).

773. Raina, A., and T. Telaranta, *Biochim. Biophys. Acta, 138*, 200 (1967).

774. Keister, D. L., *Federation Proc., 17*, 84 (1958).

775. Huang, S. L., and G. Felsenfeld, *Nature, 188*, 301 (1960).

776. Razin, S., and R. Rozansky, *Arch. Biochem. Biophys., 81*, 36 (1959).

777. Mandel, M., *J. Mol. Biol., 5*, 435 (1962).

778. Herbst, E. J., and E. E. Snell, *J. Biol. Chem., 181*, 47 (1949).

779. Ham, R. G., *Biochem. Biophys. Res. Commun., 14*, 34 (1964).

780. Mager, J., *Biochim. Biophys. Acta, 36*, 529 (1959).

781. Raina, A., *Acta Chem. Scand., 16*, 2463 (1962).

782. Caldarera, C. M., B. Barbiroli, and G. Moruzzi, *Biochem. J., 97*, 84 (1965).

783. Moruzzi, G., and C. M. Caldarera, *Arch. Biochem. Biophys., 105*, 209 (1964).

784. Mahler, H. R., B. D. Mehrotra, and C. W. Sharp, *Biochem. Biophys. Res. Commun., 4*, 79 (1961).

785. Mahler, H. R., and B. D. Mehrotra, *Biochim. Biophys. Acta, 68*, 211 (1963).

786. Mahler, H. R., G. Dutton, and B. D. Mehrotra, *Federation Proc., 23*, 217 (1964).

787. Tabor, H., *Biochemistry, 1*, 496 (1962).

788. Stevens, L., *Biochem. J., 103*, 811 (1967).

789. Krakow, J. S., *Biochim. Biophys. Acta, 72*, 566 (1963).

790. Raina, A., J. Jänne, and M. Siimes, *Biochim. Biophys. Acta, 123*, 197 (1966).

791. Raina, A., and S. S. Cohen, *Proc. Natl. Acad. Sci. U. S., 55*, 1587 (1966).

792. Ochoa, M., and I. B. Weinstein, *Biochim. Biophys. Acta, 95*, 176 (1965).

793. Martin, R. G., and B. N. Ames, *Proc. Natl. Acad. Sci. U. S., 48*, 2171 (1962).

794. Goldberg, A., *J. Mol. Biol., 15*, 663 (1966).

795. Rosenthal, S. M., and C. W. Tabor, *J. Pharmacol. Exptl. Therap., 116*, 131 (1956).

796. Tabor, H., S. M. Rosenthal, and C. W. Tabor, *Federation Proc., 15*, 367 (1956).

797. Tabor, H., S. M. Rosenthal, and C. W. Tabor, *J. Biol. Chem., 233*, 907 (1958)

798. Greene, R. C., *J. Am. Chem. Soc., 79*, 3929 (1957).

799. Tabor, C. W., in *Methods in Enzymology*, Vol. 5, S. P. Colowick and N. O. Kaplan, eds., Academic, New York, 1962, pp. 756, 761.

800. Jänne, J., A. Raina, and M. Siimes, *Biochim. Biophys. Acta, 166*, 419 (1968).

801. Widnell, C. C., and J. R. Tata, *Biochem. J., 93*, 2P (1964).

802. Korner, A., *Biochem. Biophys. Res. Commun., 13*, 386 (1963).

803. Kostyo, J. L., *Biochem. Biophys. Res. Commun., 23*, 150 (1966).

804. Jänne, J., *Acta Physiol. Scand.*, Suppl. 300, 1 (1967).

805. Raina, A., M. Jansen, and S. S. Cohen, *J. Bacteriol., 94*, 1684 (1967).

806. Ballard, P. L., and H. G. Williams-Ashman, *J. Biol. Chem., 241,* 1602 (1966).

807. Pegg, A. E., and H. G. Williams-Ashman, *Biochem. Biophys. Res. Commun., 30,* 76 (1968).

808. Morris, D. R., and A. B. Pardee, *J. Biol. Chem., 241,* 3129 (1966).

809. Morris, D. R., and A. B. Pardee, *Biochem. Biophys. Res. Commun., 20,* 697 (1965).

810. Pegg, A. E., and H. G. Williams-Ashman, *Biochem. J., 108,* 533 (1968).

811. Dykstra, W. G., Jr., and E. J. Herbst, *Science, 149,* 428 (1965).

812. Bucher, N. L. R., in *International Reviews in Cytology,* G. H. Bourne and J. F. Danielli, eds., Academic, New York, 1963, p. 245.

813. Fujioka, M., M. Koga, and I. Lieberman, *J. Biol. Chem., 238,* 3401 (1963).

814. Fox, C. F., and S. B. Weiss, *J. Biol. Chem., 239,* 175 (1964).

815. Schwimmer, S., *Biochim. Biophys. Acta, 166,* 251 (1968).

816. Brewer, E. N., and H. P. Rusch, *Biochem. Biophys. Res. Commun., 25,* 579 (1966).

817. O'Brien, R. L., J. G. Olenick, and F. E. Hahn, *Proc. Natl. Acad. Sci. U. S., 55,* 1511 (1966).

818. Freedlander, B. L., and F. A. French, *Cancer Res.,* 18, 360, 1286 (1958); Mihich, E., *Cancer Res., 23,* 1375 (1963).

819. Schwimmer, S., and A. Aronson, *Biochim. Biophys. Acta, 134,* 59 (1967).

820. Schwimmer, S., and J. Bonner, *Biochim. Biophys. Acta, 108,* 67 (1965).

821. Sartorelli, A. C., A. T. Jannoti, B. A. Booth, F. H. Schneider, J. R. Bertino, and D. G. Johns, *Biochim. Biophys. Acta, 103,* 174 (1965).

822. Bonner, J., and J. Widholm, *Proc. Natl. Acad. Sci. U. S., 57,* 1379 (1967).

823. Tabor, C. W., H. Tabor, C. M. McEwen, and P. D. Kellog, *Federation Proc., 23,* 385 (1964).

824. Bachrach, U., C. W. Tabor, and H. Tabor, *Biochim. Biophys. Acta, 78,* 768 (1963).

825. Bachrach, U., and S. Persky, *J. Gen. Microbiol., 37,* 195 (1964).

826. Bachrach, U., S, Rabina, G. Loebenstein, and G. Eilon, *Nature, 208,* 1095 (1965).

827. Bachrach, U., S. Abzug, and A. Bekierkunst, *Biochim. Biophys. Acta, 134,* 174 (1967).

828. Bachrach, U., and G. Eilon, *Biochim, Biophys. Acta, 145,* 418 (1967).

829. Herbst, E. J., and W. G. Dykstra, in Abstracts, 6th International Congress of Biochemistry, New York, Vol. 3, 1964, p. 24.

830. Bertossi, F., N. Bagni, O. Moruzzi, and C. M. Caldarera, *Experientia, 21,* 80 (1965).

831. Moruzzi, G., B. Barbiroli, and C. M. Caldarera, *Biochem. J., 107,* 609 (1968).

832. Keberle, H. J., J. W. Faigle, H. Fritz, F. Knüsel, P. Loustalot, and K. Schmid, in *Biological Council: Symposium on Embryopathic Activity of Drugs,* J. M. Robson, F. M. Sullivan, and R. L. Smith, eds., Churchill, London, 1965, p. 210.

833. Williams, R. T., H. Schuhmacher, S. Fabro, and R. L. Smith, in *Biological Council: Symposium on Embryopathic Activity of Drugs,* J. M. Robson, F. M. Sullivan, and R. L. Smith, eds., Churchill, London, 1965, p. 167; R. L. Smith, Discussion Remark in *Biological Council, Symposium on Embryopathic Activity of Drugs,* J. M. Robson, F. M. Sullivan, and R. L. Smith, eds., Churchill, London, 1965, p. 230.

834. Kapeller-Adler, R., Discussion remark in *Biological Council: Symposium on Embryopathic Activity of Drugs,* J. M. Robson, F. M. Sullivan, and R. L. Smith, eds., Churchill, London, 1965, p. 229.

835. Bignami, G., D. Bovet, F. Bovet-Nitti, and V. Rosnati, *J. Méd. Chem., 2,* 1333 (1962).

836. Champy-Hatem, S., *Bull. Acad. Natl. Méd.* (Paris), *148*, 301 (1964).

837. Kahlson, G., *Perspectives Biol. Med.*, *5*, 179 (1962).

838. Werle, E., and F. Roewer, *Biochem. Z.*, *325*, 550 (1954).

839. Gorkin, V. Z., A. A. Avakyan, I. V. Veryovkina, and N. V. Komisarova, *Vop. Med. Khim.*, *8*, 638 (1962) (translated in Federation Proc., *22*, T619 (1963).

840. Sakamoto, Y., Y. Ogawa, and K. Hayashi, *J. Biochem.* (Tokyo), *54*, 292 (1963).

841. Unemoto, T., *Chem. Pharm. Bull.* (Tokyo), *11*, 1255 (1963).

842. Tabor, C. W., H. Tabor, and U. Bachrach, *J. Biol. Chem.*, *239*, 2194 (1964).

843. Tabor, C. W., and S. M. Rosenthal, *J. Pharmacol. Exptl. Therap.*, *116*, 139 (1956).

844. Bachrach, U., and R. Bar-Or, *Biochim. Biophys. Acta*, *40*, 545 (1960).

845. Yamada, H., P. Gee, M. Ebata, and K. T. Yasunobu, *Biochim. Biophys. Acta*, *81*, 165 (1964).

846. Yamada, H., and K. T. Yasunobu, *J. Biol. Chem.*, *238*, 2669 (1963).

847. Kihara, H., and E. E. Snell, *Proc. Natl. Acad. Sci. U. S.*, *43*, 867 (1957).

848. Hirsch, J. G., and R. J. Dubos, *J. Exptl. Med.*, *95*, 191 (1952).

849. Hirsch, J. G., *J. Exptl. Med.*, *97*, 323, 327 (1953).

850. Hirsch, J. G., in *Experimental Tuberculosis*, Symposium, G. E. W. Wolstenholme and C. Cameron, eds., Little, Brown, Boston, 1955, p. 115.

851. Alarcon, R. A., G. E. Foley, and E. J. Modest, *Arch. Biochem. Biophys.*, *94*, 540 (1961).

852. Halevy, S., Z. Fuchs, and J. Mager, *Bull. Res. Council Israel*, *11A*, 52 (1962).

853. Holtz, P., K. Stock, and E. Westermann, *Arch. Exptl. Pathol. Pharmakol.*, *248*, 387 (1964).

854. Rosenthal, S. M., E. R. Fisher, and E. F. Stolman, *Proc. Soc. Exptl. Biol.* (New York), *80*, 432 (1952).

855. Risetti, U., and G. Mancini, *Acta Neurolog.* (Naples) *9*, 391 (1954).

856. Fisher, E. R., and S. M. Rosenthal, *Arch. Pathol.*, *57*, 244 (1954).

857. Tabor, C. W., L. L. Ashburn, and S. M. Rosenthal, *Federation Proc.*, *15*, 490 (1956).

858. Alarcon, R. A., *Arch. Biochem. Biophys.*, *106*, 240 (1964).

859. McEwen, C. M., Jr., and J. D. Cohen, *J. Lab. Clin. Med.*, *62*, 766 (1963).

860. Fischer, E. H., A. B. Kent, E. R. Snyder, and E. G. Krebs, *J. Am. Chem. Soc.*, *80*, 2906 (1958).

861. Hughes, R. C., W. T. Jenkins, and E. H. Fischer, *Proc. Natl. Acad. Sci. U. S.*, *48*, 1615 (1962).

862. Fischer, E. H., in *Structure and Activity of Enzymes*, Federation of European Biochemical Societies Symposium, No 1, T. W. Goodwin, J. L. Harris, and B. S. Hartley, eds., Academic, London, 1964, p. 111.

863. McEwen, C. M., Jr., K. T. Cullen, and A. J. Sober, *J. Biol. Chem.*, *241*, 4544 (1966).

864. Slotta, K. H., and J. Müller, *Z. Physiol. Chem.*, *238*, 14 (1936).

865. Blaschko, H., G. Ferro-Luzzi, and R. Hawes, *Biochem. Pharmacol.*, *1*, 101 (1958).

866. Werle, E., and A. Raub, *Biochem. Z.*, *318*, 538 (1948).

867. Werle, E., and A. Zabel, *Biochem. Z.*, *318*, 554 (1948).

868. Kenten, R. H., and P. J. G. Mann, *Biochem. J.*, *50*, 360 (1952).

869. Mann, P. J. G., and W. R. Smithies, *Biochem. J.*, *61*, 89, 101 (1955).

870. Clarke, A. J., and P. J. G. Mann, *Biochem. J.*, *65*, 763 (1957).

871. Clarke, A. J., and P. J. G. Mann, *Biochem. J.*, *71*, 596 (1959).

872. Mann, P. J. G., *Biochem. J.*, *79*, 623 (1961).

873. Hill, J. M., and P. J. G. Mann, *Biochem. J., 85,* 198 (1962).

874. Hill, J. M., *Biochem. J., 104,* 1048 (1967).

875. Suzuki, Y., *Naturwiss., 46,* 427 (1959).

876. Werle, E., C. Beaucamp, and V. Schirren, *Planta, 53,* 125 (1959).

877. Jakoby, W. B., and J. Fredericks, *J. Biol. Chem., 234,* 2145 (1959).

878. Owen, C. A., Jr., A. G. Karlson, and E. A. Zeller, *J. Bacter., 62,* 53 (1951).

879. Razin, S., I. Gery, and U. Bachrach, *Biochem. J., 71,* 551 (1959).

880. Yamada, H., O. Adachi, A. Tanaka, and K. Ogata, in Abstracts, 7th International Congress of Biochemistry, Tokyo, August 19–25, 1967, p. 749.

881. Kim, K., and T. T. Tchen, *Biochem. Biophys. Res. Commun., 9,* 99 (1962).

882. Maggio, E., F. Salvatore, and L. Zarrilli, *Boll. Soc. Ital. Biol. Sper., 32,* 841 (1956).

883. Satake, K., S. Ando, and H. Fujita, *J. Biochem. (Tokyo), 40,* 299 (1953).

884. Silverman, M., and E. A. Evans, Jr., *J. Biol. Chem., 154,* 521 (1944).

885. Razin, S., U. Bachrach, and I. Gery, *Nature, 181,* 700 (1958).

886. Bachrach, U., S. Persky, and S. Razin, *Biochem. J., 76,* 306 (1960).

887. Hasse, K., and H. Maisack, *Biochem. Z., 327,* 296 (1955–1956).

888. Friedemann, T. E., and G. E. Haugen, *J. Biol. Chem., 147,* 415 (1943).

889. Bachrach, U., *Nature, 194,* 377 (1962).

890. Schöpf, C., A. Komzak, F. Braun, and E. Jakobi, *Liebig's Ann. Chem., 559,* 1 (1948).

891. Campello, A. P., C. W. Tabor, and H. Tabor, *Biochem. Biophys. Res. Commun., 19,* 6 (1965).

892. Karström H., *Ergebn. Enzymforsch., 7,* 350 (1938).

893. Bachrach, U., and J. Leibovici, *J. Mol. Biol., 19,* 120 (1966).

894. Bachrach, U., and J. Leibovici, *Biochem. Biophys. Res. Commun., 19,* 357 (1965).

895. Brooks, P., and P. D. Lawley, *Brit. Med. Bull., 20,* 91 (1964).

896. Bachrach, U., and S. Persky, *Biochim. Biophys. Res. Commun., 24,* 135 (1966).

897. Oki, T., H. Yamada, I. Tomita, H. Fukami, and K. Ogata, in Abstracts, 7th International Congress of Biochemistry, Tokyo, August 19–25, 1967 p. 670.

898. Martin, A. J. P., and R. R. Porter, *Biochem. J., 49,* 215 (1951).

899. Hirs, C. H. W., W. H. Stein, and S. Moore, *J. Am. Chem. Soc., 73,* 1893 (1951).

900. Hirs, C. H. W., S. Moore, and W. H. Stein, *J. Biol. Chem., 200,* 493 (1953).

901. Aquist, S. E. G., and C. B. Anfinsen, *J. Biol. Chem., 234,* 1112 (1959).

902. Holtz, P., and H. Büchsel, *Z. Physiol. Chem., 272,* 201 (1942).

903. Hagen, P., and N. Weiner, *Federation Proc., 18,* 1005 (1959).

904. Cotzias, G. C., and J. J. Greenough, *Nature, 185,* 384 (1960).

905. Hardegg, W., and E. Heilbronn, *Biochim. Biophys. Acta, 51,* 553 (1961).

906. Aprison, M. H., M. A. Wolf, G. L. Poulos, and T. L. Folkerth, *J. Neurochem., 9,* 575 (1962).

907. Pscheidt, G. R., and H. E. Himwich, *Biochem. Pharmacol., 12,* 65 (1963).

908. Youdim, M. B. H., and T. L. Sourkes, *Can. J. Biochem., 43,* 1305 (1965).

909. Gorkin, V. Z., *J. D. I. Mendeleev All-Union Chemical Soc.* (Moscow), *9,* 405 (1964).

910. Chodera, A., V. Z. Gorkin, and L. I. Gridneva, *Acta Biol. Med. Germ., 13,* 101 (1964).

911. Youdim, M. B. H., and M. Sandler, *Biochem. J., 105,* 43P (1967).

912. Youdim, M. B. H., G. G. S. Collins, and M. Sandler, *Fed. Europ. Biochem. Soc., 5* (1968); *Nature, 223,* 626 (1969).

913. Burkard, W. P., K. F. Gey, and A. Pletscher, *Biochem. Pharmacol., 11,* 177 (1962).

914. Kobayashi, Y., S. Otsuka, A. L. Southren, and H. Kizuka, in Abstracts, 7th International Congress of Biochemistry, Tokyo, August 19–25, 1967, p. 805.

915. Markert, C. L., and F. Möller, *Proc. Natl. Acad. Sci. U. S.*, *45*, 753 (1959).
916. Anfinsen, C. B., in *The Molecular Basis of Evolution*, Wiley, New York, 1959.
917. Bonner, D. M., J. A. De Moss, and S. E. Mills, in *Evolving Genes and Proteins*, a Symposium, V. Bryson and H. J. Vogel, eds., Academic, New York, 1965, p. 305.
918. Aures, D., R. Fleming, and R. Hakanson, *J. Chromatog.*, *33*, 480 (1968).
919. *Report of the Commission on Enzymes of the International Union of Biochemistry*, IUB Symposium Series, Vol. 20, Pergamon, Oxford, London, New York, Paris, 1961.
920. Weissbach, H., B. G. Redfield, and S. Udenfriend, *J. Biol. Chem.*, *229*, 953 (1957).
921. Schneider, W. C., and G. H. Hogeboom, *J. Biol. Chem.*, *183*, 123 (1950).
922. Giordano, C., J. Bloom, and J. P. Merrill, *Experientia*, *16*, 346 (1960).
923. Guha, S. R., and C. R. Krishna Murti, *Biochem. Biophys. Commun.*, *18*, 350 (1965).
924. Dixon, M., *Biochem. J.*, *54*, 457 (1953).
925. Willstätter, R., and H. Kraut, *Ber. Deut. Chem. Ges.*, *56*, 1117 (1923).
926. Nagatsu, T., *J. Biochem.*, *59*, 606 (1966).
927. Brody, T. M., and J. A. Bain, *J. Biol. Chem.*, *195*, 685 (1952).
928. Tipton, K. F., and A. P. Dawson, *Biochem. J.*, *108*, 95 (1968).
929. Guha, S. R., *Biochem. Pharmacol.*, *15*, 161 (1966).
930. Severina, I. S., and T. N. Sheremetevskaya, *Biokhimiya*, *32*, 843 (1967).
931. Peterson, R. E., and M. E. Bollier, *Anal. Chem.*, *27*, 1195 (1955).
932. Johnston, W. C., in *Organic Reagents for Metals*, Vol. 2, Hopkin and Williams, Chadwell Heath, England, 1964, p. 53.
933. Goa, J., *Scand. J. Clin. Invest.*, *5*, 218 (1953).
934. Creasey, N. H., *Biochem. J.*, *64*, 178 (1956).
935. Cotzias, G. C., and V. P. Dole, *J. Biol. Chem.*, *190*, 665 (1951).
936. Cotzias, G. C., and J. J. Greenough, *Arch. Biochem. Biophys.*, *75*, 15 (1958).
937. Sourkes, T. L., E. Townsend, and G. N. Hansen, *Can. J. Biochem. Physiol.*, *33*, 725 (1955).
938. Weissbach, H., T. E. Smith, J. W. Daly, B. Witkop, and S. Udenfriend, *J. Biol. Chem.*, *235*, 1160 (1960).
939. Green, A. L., and T. M. Haughton, *Biochem. J.*, *78*, 172 (1961).
940. Sjoerdsma, A., T. E. Smith, T. D. Stevenson, and S. Udenfriend, *Proc. Soc. Exptl. Biol. Med.* (New York), *89*, 36 (1955).
941. Slater, E. C., in *Methods in Enzymology*, R. W. Estabrook and B. Pullmann, eds., Vol. 10, Academic, New York, 1967, p. 19.
942. Zeller, V., G. Ramachander, and E. A. Zeller, *J. Med. Chem.*, *8*, 440 (1965).
943. Kohn, H. I., *Biochem. J.*, *31*, 1693 (1937).
944. Smith, T. E., H. Weissbach, and S. Udenfriend, *Biochemistry*, *2*, 746 (1963).
945. Petering, G. H., and F. Daniells, *J. Am. Chem. Soc.*, *60*, 2796 (1938).
946. Longmuir, I. S., *Biochem. J.*, *57*, 81 (1954).
947. Harris, H., and W. R. Barclay, *Brit. J. Exptl. Pathol.*, *36*, 592 (1955).
948. Chance, B., and G. R. Williams, *Nature*, *175*, 1120 (1955).
949. Connelly, C. M., *Federation Proc.*, *16*, 681 (1957).
950. Bellamy, D., and W. Bartley, *Biochem. J.*, *76*, 78 (1960).
951. Clark, L. C., *Trans. Am. Soc. Art. Int. Org.*, *2*, 41 (1956).
952. Reeves, R. B., D. W. Rennie, and J. R. Pappenheimer, *Federation Proc.*, *16*, 693 (1957).
953. Stickland, R. G., *Biochem. J.*, *77*, 636 (1960).

954. Peel, J. L., *Biochem. J., 88,* 296 (1963).

955. Dixon, M., and K. Kleppe, *Biochim. Biophys. Acta, 96,* 357 (1965).

956. Udenfriend, S., H. Weissbach, and C. T. Clark, *J. Biol. Chem., 215,* 337 (1955).

957. Braganca, B. M., J. H. Quastel, and R. Schucher, *Arch. Biochem. Biophys., 52,* 18 (1954).

958. Conway, E. J., and A. Byrne, *Biochem. J., 27,* 419 (1933).

959. Makino, K., K. Satoh, and Y. Joh, *J. Biochem.* (Japan), *42,* 555 (1955).

960. Galat, A., and G. Elion, *J. Am. Chem. Soc., 61,* 3585 (1939).

961. Ukai, T., Y. Yamamoto, M. Yotsuzuka, and F. Ichimura, *J. Pharm. Soc. Japan, 76,* 657 (1956); Chem. Abstr., *51,* 280 (1957).

962. Lovenberg, W., R. J. Levine, and A. Sjoerdsma, *J. Pharmacol. Exptl. Therap., 135,* 7 (1962).

963. Udenfriend, S., in *Molecular Biology,* B. L. Horecker, N. O. Kaplan, and H. A. Scheraga, eds., Vol. 3, 4th printing, Academic, New York, 1966, p. 122.

964. Racker, E., *J. Biol. Chem., 177,* 883 (1949).

965. Weissbach, H., W. Lovenberg, B. G. Redfield, and S. Udenfriend, *J. Pharmacol. Exptl. Therap., 131,* 26 (1961).

966. Gertner, S. B., *Nature, 183,* 750 (1959).

967. Goldberg, L. I., and F. M. Da Costa, *Proc. Soc. Exptl. Biol. Med.* (New York), *105,* 223 (1960).

968. Krajl, M., *Biochem. Pharmacol., 14,* 1684 (1965).

969. Wurtman, R. J., and J. Axelrod, *Biochem. Pharmacol., 12,* 1439 (1963).

970. McCaman, R. E., M. W. McCaman, J. M. Hunt, and M. S. Smith, *J. Neurochem., 12,* 15 (1965).

971. Otsuka, S., and Y. Kobayashi, *Biochem. Pharmacol., 13,* 995 (1964).

972. Satake, K., in *Koso Kenkyu-ho* (Methods in Enzyme Studies), S. Akabori, ed., Vol., 2, Asakura Shoten, Tokyo, 1958, p. 534.

973. Hakanson, R., and C. Owman, *J. Neurochem., 12,* 417 (1965).

974. Goldstein, M., A. J. Friedhoff, C. Simons, and N. N. Procheroff, *Experientia, 15,* 254 (1959).

975. Glenner, G. G., H. J. Burtner, and G. W. Brown, Jr., *J. Histochem. Cytochem., 5,* 591 (1957).

976. Meier-Ruge, W., S. McFadden, and K. Zehnder, *Med. Pharmacol. Exp., 14,* 152 (1966).

977. Francis, C. M., *Nature, 171,* 701 (1953).

978. Francis, C. M., *J. Physiol., 124,* 188 (1954).

979. Koelle, G. B., and A. de T. Valk, Jr., *J. Physiol. 126,* 434 (1954).

980. Graham, R. C., and M. J. Karnovsky, *J. Histochem. Cytochem., 13,* 604 (1965).

981. Adamstone, F. B., and A. B. Taylor, *Stain Technol., 23,* 109 (1948).

982. Weissbach, H., B. G. Redfield, G. G. Glenner, and C. Mitoma, *J. Histochem. Cytochem., 5,* 601 (1957).

983. Cunningham, G. J., L. Bitensky, J. Chayen, and A. A. Silcox, *Ann. Histochem., 1,* 433 (1962).

984. Cohen, S., L. Bitensky, and J. Chayen, *Biochem. Pharmacol., 14,* 223 (1965).

985. Levine, R. J., and A. Sjoerdsma, *Clin. Pharmacol. Therap. 4,* 22 (1963).

986. Levine, R. J., and A. Sjoerdsma, *Proc. Soc. Exptl. Biol. Med.* (New York), *109,* 225 (1962).

987. Smith, R. B. W., H. Sprinz, W. H. Crosby, and B. H. Sullivan, *Am. J. Med., 25,* 391 (1958).

988. Kiese, M., *Biochem. Z., 305,* 22 (1940).

989. Stephenson, N. R., *J. Biol. Chem., 149,* 169 (1943).

990. Laskowski, M., J. M. Lemley, and C. K. Keith, *Arch. Biochem., 6,* 105 (1945).

991. Leloir, L. F., and D. E. Green, *Federation Proc., 5,* 144 (1946).

992. Arvidsson, U. B., B. Pernow, and B. Swedin, *Acta Physiol. Scand., 35,* 338 (1955/56).

993. Swedin, B., *Acta Physiol. Scand., 42,* 1 (1958).

994. Rosenthal, S. M., and H. Tabor, *J. Pharmacol. Exptl. Therap., 92,* 425 (1948).

995. Massey, V., G. Palmer, and R. Bennett, *Biochim. Biophys. Acta, 48,* 1 (1961).

996. Gornall, A. G., C. J. Bardavill, and M. M. David, *J. Biol. Chem., 177,* 751 (1949).

997. Lowry, O. M., N. J. Rosenbrough, A. L. Farr, and R. S. Randall, *J. Biol. Chem., 193,* 265 (1951); Miller, G. L., Analyt. Chem. *31,* 964 (1959).

998. Kapeller-Adler, R., *Biochim. Biophys. Acta, 22,* 391 (1956).

999. Albert-Recht, F., *Clin. Chim. Acta, 4,* 627 (1959).

1000. Kohn, J., *Clin. Chim. Acta, 3,* 450 (1958).

1001. Hsiao, S., and F. W. Putnam, *J. Biol. Chem., 236,* 122 (1961).

1002. Uozumi, K., I. Nakahara, T. Higashi, and Y. Sakamoto, *J. Biochem.* (Japan), *56,* 601 (1964).

1003. Rossi-Fanelli, A., and E. Antonini, *Arch. Biochem. Biophys., 80,* 291 (1959).

1004. Porath, J., and S. Hjerten, *Meth. Biochem. Anal., 9,* 193 (1962).

1005. McFarlane, W. D., *Biochem. J., 26,* 1022 (1932).

1006. Smith, J. L., and R. C. Krueger, *J. Biol. Chem. 237,* 1121 (1962).

1007. Yamada, H., A. Tanaka, and K. Ogata, *Mem. Res. Inst. Food Sci. Kyoto Univ., 26,* 1 (1965).

1008. Kirkman, H. N., *Federation Proc., 18,* 261 (1959).

1009. Hatfield, D., and J. M. Wyngaarden, *J. Biol. Chem., 329,* 2580 (1964).

1010. Valette, G., and Y. Cohen, *Compt. Rend. Soc. Bull.* (Paris), *146,* 714 (1952).

1011. Feulgen, R., and H. Rossenbeck, *Hoppe Seyler's Z. Physiol. Chem., 135,* 203 (1924).

1012. Oster, K. A., and M. C. Schlossman, *J. Cell. Comp. Physiol., 20,* 373 (1942).

1013. Paul, J., in *Cell and Tissue Culture,* 2nd Ed., Livingstone, Edinburgh and London, 1961, p. 254.

1014. Aarsen, P. N., and A. Kemp, *Nature, 204,* 1195 (1964).

1015. Glick, D., in *Quantitative Chemical Techniques of Histo- and Cytochemistry,* Vol. 1, Interscience, New York, 1961, p. 22.

1016. Kirsten, E., C. Gerez, and R. Kirsten, *Biochem. Z., 337,* 312 (1963).

1017. Kapeller-Adler, R., *Biochem. J., 44,* 70 (1949).

1018. Pacha, W. L., and E. A. Zeller, in *Abstracts, 136th Meeting Am. Chem. Soc.* 1959, p. 38C.

1019. Spencer, P. S. J., *J. Pharm. Pharmacol., 15,* 225 (1963).

1020. Tyrode, M. V., *Arch. Int. Pharmacodyn., 20,* 205 (1910).

1021. Wajda, I., M. Ginsburg, and H. Waelsh, *Nature, 191,* 1204 (1961).

1022. Kremzner, L. T., and I. B. Wilson, *Biochim. Biophys. Acta, 50,* 364 (1961).

1023. Bamberger, E., and E. Demuth, *Chem. Ber., 34,* 1309 (1901).

1024. Bamberger, E., *Chem. Ber., 60,* 314 (1927).

1025. Glick, D., and S. Hosoda, *Proc. Soc. Exptl. Biol. Med.* (New York), *119,* 52 (1965).

1026. Glick, D., D. von Redlich, and B. Diamant, *Biochem. Pharmacol. 16,* 553 (1967).

1027. McComb, R. B., and W. D. Yushok, *J. Franklin Inst., 265,* 417 (1958).

1028. Saifer, A., and S. Gerstenfeld, *J. Lab. Clin. Med., 51,* 448 (1958).

1029. Messer, M., and A. Dahlquist, *Anal. Biochem., 14,* 376 (1966).

306 REFERENCES

1030. Guidotti, G., J. P. Colombo, and P. P. Toa, *Anal. Chem.*, *33*, 151 (1961).
1031. Möller, K. M., and P. Ottolenghi, *Compt. Rend. Lab. Carlsberg*, *35*, 369 (1966).
1032. Kapeller-Adler, R., *Biochem. J.*, *48*, 99 (1951).
1033. Zeller, E. A., and H. R. Buerki, in Abstracts, 146th Meeting Am. Chem. Soc., Denver, January 1964, p. 15A.
1034. Zeller, E. A., *Federation Proc.* 24, 766 (1965).
1035. Preisler, W. P., E. S. Hill, R. G. Loeffel, and P. A. Shaffer, *J. Am. Chem. Soc.*, *81*, 1991 (1959).
1036. Mosebach, K. O., *Zeits. Physiol. Chemie*, *309*, 206 (1957).
1037. Wagner, H. E., J. Prakt. Chemie (2) *89*, 377 (1914).
1038. Southren, A. L., Y. Kobayashi, N. C. Carmody, and A. B. Weingold, *Am. J. Obst. Gyn.*, *95*, 615 (1966).
1039. Schayer, R. W., R. L. Smiley, and J. Kennedy, *J. Biol. Chem.*, *206*. 461 (1954).
1040. Greenfield, R. E., and V. E. Price, *J. Biol. Chem.*, *209*, 355 (1954).
1041. Kobayashi, Y., *J. Lab. Clin. Med.*, *62*, 699 (1963).
1042. Keiffer, H., *Gynéc. Obst.*, *14*, 1 (1926).
1043. Danforth, D. N., and R. W. Hull, *Am. J. Obst. Gyn.*, *75*, 536 (1958).
1044. Keilin, D., and E. F. Hartree, *Proc. Roy. Soc.* (London), Series *B 124*, 397 (1938).
1045. Warburg, O., and W. Christian, *Biochem. Z.*, *310*, 384 (1941/42).
1046. Horecker, B. L., P. Z. Smyrniotis, and H. Klenow, *J. Biol., Chem.*, *205*, 661 (1953).
1047. Sober, H. A., F. J. Gutter, M. M. Wyckoff, and E. A. Peterson, *J. Am. Chem. Soc.*, *78*, 756 (1956).
1048. Tiselius, A., S. Hjerten, and O. Levin, *Arch. Biochem. Biophys.*, *65*, 132 (1956).
1049. Miller, G. L., and R. H. Golder, *Arch. Biochem.*, *29*, 420 (1950).
1050. Archibald, W. J., *J. Phys. and Colloid. Chem.*, *51*, 1204 (1947).
1051. Kearney, E. B., *J. Biol. Chem.*, *235*, 865 (1960).
1052. Griffiths, D. E., and D. C. Wharton, *J. Biol. Chem.*, *236*, 1857 (1961).
1053. Peterson, E. A., and H. A. Sober, *J. Am. Chem. Soc.*, *76*, 169 (1954).
1054. Kent, A. B., E. G. Krebs, and E. H. Fischer, *J. Biol. Chem.*, *232*, 549 (1958).
1055. Sutton, C. R., and H. K. King, *Arch. Biochem. Biophys.*, *96*, 360 (1962).
1056. Wade, H., and E. E. Snell, *J. Biol. Chem.*, *236*, 2089 (1961).
1057. Schirch, L., and M. Mason, *J. Biol. Chem.*, *237*, 2758 (1962).
1058. Hayaishi, O., Y. Nishizuka, M. Tatibana, M. Takeshita, and S. Kuno, *J. Biol. Chem.*, *236*, 791 (1961).
1059. Cori, C. F., and B. Illingworth, *Proc. Natl. Acad. Sci. U. S.*, *43*, 547 (1957).
1060. Matsuo, Y., and D. M. Greenberg, *J. Biol. Chem*, *230*, 561 (1958).
1061. Jenkins, W. T., and I. W. Sizer, *J. Am. Chem. Soc.*, *79*, 2655 (1957).
1062. Nashimura, J. S., J. M. Manning, and A. Meister, *Biochemistry*, *1*, 442 (1962).
1063. McCormick, D. B., M. E. Gregory, and E. E. Snell, *J. Biol. Chem.*, *236*, 2076 (1961).
1064. Bowen, H. J. M., and D. Gibbons, *Radioactivation Analysis*, Clarendon Press, Oxford, 1963, p. 232.
1065. Wyard, S. J., *J. Sci. Instr.*, *42*, 769 (1965).
1066. Beinert, H., and B. Kok, *Biochim. Biophys. Acta*, *88*, 278 (1964).
1067. Smith, G. F., and W. H. McCurdy, *Anal. Chem.*, *24*, 371 (1952).
1068. Beinert, H., and G. Palmer, *Advances. Enzymol.*, *27*, 151 (1965).
1069. Smithies, O., *Biochem. J.*, *71*, 585 (1959).
1070. Poulik, M. D., *Nature*, *180*, 1477 (1957).
1071. Blaschko, H., and T. L. Chrusciel, *Brit. J. Pharmacol.*, *14*, 364 (1959).
1072. Dearden, J. C., and W. F. Forbes, *Can. J. Chem.*, *36*, 1362 (1958).
1073. Chance, B., *Methods Biochem. Anal.*, *3*, 412 (1954).

1074. Description Manual No. 11, Worthington Biochemical Corporation, Freehold, N. J., 1961, p. 13.

1075. Beers, R. F., Jr., and I. W. Sizer, *J. Biol. Chem., 195*, 133 (1952).

1076. Umbreit, W. W., R. H. Burris, and J. F. Stauffer, *Manometric Techniques,* Burgess, Minneapolis, 1945, p. 1.

1077. Brown, R. H., G. D. Duda, S. Korkes, and P. Handler, *Archiv. Biochem. Biophys., 66,* 301 (1957).

1078. Tucker, G. F., and J. L. Irwin, *J. Am. Chem. Soc., 73,* 1923 (1951).

1079. Braunstein, A. E., in *The Enzymes,* Vol. 2, Ed. 2, P. D. Boyer, H. Lardy, and K. Myrbäck, eds., Academic, New York, 1960, p. 113.

1080. Braunstein, A. E., in *Chemical and Biological Aspects of Pyridoxal Catalysis,* E. E. Snell, P. M. Fasella, A. E. Braunstein, and A. Rossi-Fanelli, eds., Pergamon, Oxford, 1963, p. 579.

1081. Ikawa, M., and E. E. Snell, *J. Am. Chem. Soc. 76,* 653 (1954).

1082. Vallee, B. L., *Advances Protein Chem., 10,* 318 (1955).

1083. Felsenfeld, G., and M. P. Printz, *J. Am. Chem. Soc., 81,* 6259 (1959).

1084. Felsenfeld, G., *Arch. Biochem. Biophys., 87,* 247 (1960).

1085. Forster, W. A., *Analyst, 78,* 614 (1953).

1086. Weichselbaum, T. E., *Am. J. Clin. Pathol.* (Techn. Sect.), *10,* 40 (1946).

1087. Herbst, E. J., E. B. Glinos, L. H. Amundsen, *J. Biol. Chem., 214,* 175 (1955).

1088. Pontecorvo, G., in *Advances in Genetics,* Vol. 5, M. Demerec, ed., Academic, New York, 1953, p. 150.

1089. Bessman, S. P., and A. N. Bessman, *J. Clin. Invest., 34,* 622 (1955).

1090. Layne, E., in *Methods in Enzymology,* Vol. 3, S. P. Colowick and N. O. Kaplan, eds., Academic, New York, 1957, p. 451.

1091. Purkayastha, M., and R. P. Williams, *Nature, 187,* 349 (1960).

1092. Datta, R. K., S. Sen, and J. J. Ghosh, *Biochem. J., 114,* 847 (1969).

1093. Youdim, M. B. H., and G. G. S. Collins, Proceedings of the 500th Meeting of the Biochemical Society, 15th to 17th December 1969, p. 34.

Index

309

S0-AYO-072

© 2009 by Ray Didinger and Glen Macnow

Published in 2015 by Hallmark Gift Books, a division
of Hallmark Cards, Inc., under license with Running
Press. Photography courtesy of Everett Collection.
Visit us on the Web at Hallmark.com.

All rights reserved. No part of this publication may be
reproduced, transmitted, or stored in any form or by
any means without the prior written permission of the
publisher.

Editorial Director: Delia Berrigan
Editor: Kim Schworm Acosta
Art Director: Jan Mastin
Designer: Brian Pilachowski
Production Designer: Dan Horton

ISBN: 978-1-59530-729-3
BOK2191

Printed and bound in China.

THE ULTIMATE BOOK OF SPORTS MOVIES

FEATURING THE 100 GREATEST SPORTS FILMS OF ALL TIME

Hallmark

BY RAY DIDINGER & GLEN MACNOW

FOREWORD BY GENE HACKMAN

★ CONTENTS ★

Movies have the ability to make us believe. If only we'd had that one little break, that indefinable moment of clarity where we saw that opening between tackle and guard and ran over the linebacker . . . then we could have scored, big time. If we'd just had that clear shot at the basket then it would have put us into the finals. We *coulda* been contenders.

Maybe you'll recognize this scene. *On the Waterfront* certainly wasn't a sports movie, but I love the metaphor. The camera focuses on two men arguing in the backseat of a taxi. The older brother had taken the odds to deliver his younger sibling to the bookmakers in the kid's big fight at the Garden. The younger brother had gone along and taken the dive. Some of the dialogue went like this:

"Remember that night in the Garden, Charlie," Terry, the ex-boxer, tells his older brother. "You came down to the locker room . . . 'It ain't your night kid,' you said, 'We're going for the price on Wilson.' Remember that?" Terry implores. "I *coulda* taken Wilson apart. You *shoulda* looked out for me, Charlie. I *coulda* been somebody."

How many times have we heard "shoulda, coulda?" Most men, when asked if they played sports, will answer either with an equivocal "Yeah, I *coulda*, but I had to work after school." Or better yet, "I was really fast, I *woulda* scored, but they never threw me the ball."

In our memories at least, we seem to be just a smidgen away from that dream of glory Irwin Shaw described so well in his short story *The 80-Yard Run*. Way past his youth, Shaw's hero goes back to the practice field where he had that one spectacular run. He relives the moment, going through all the moves, once again breaking clear into the end zone. If he *coulda* just had the opportunity to do as well in a real game, he laments.

It doesn't seem to matter that a lot of people go on to be successful doctors, lawyers, politicians, writers, and salesmen. There is still that lingering moment we have fantasized about over the course of so many years. We think, surely it must be at least partially true; that "80-yard run," that clear shot at the basket. We've relived it so many times, how could it not be?

The movies solve so beautifully many of those fantasies for us. The quality of many lives, unfortunately, are predicated on how well they adjust to not having been selected to that all-important first team. The movies adjust that for us; we are able to lose ourselves in the pure joy of sport, and forget for the moment that destiny had other things in mind for us.

Film sets up the individual or the team with a conflict—the boxer's lack of connections, the skier's stubborn independence, the baseball player's field of dreams.

As viewers we sit watching, seeing clearly what our team or individual must do, praying that the actors show at least a modicum of athleticism, loving it. "Yeah, that one little break, go man go . . . yesss," and for the next two hours we get to live with the hero's manful quest to overcome his problems, learn something about himself, secure the admiration of girlfriends and town folk, and watch him bathe in the camaraderie of his fellow competitors.

Maybe best of all we are able to empathize for just a few moments with fallen heroes like the ex-boxer as played by Marlon Brando in the backseat of that taxi with Rod Steiger.

"So what happens?" Brando says. "Wilson gets a shot at the title in the ballpark. What did I get? A one-way ticket to Palookaville. You *shoulda* looked out for me, Charlie. I *coulda* had class. *I coulda been a contender. I coulda been somebody.*"

Ahhh . . . the movies; we love them.

❶ ROCKY (1976–PG)

SPORT: Boxing ★ **STARS:** Sylvester Stallone, Talia Shire, Burgess Meredith
DIRECTOR: John Avildsen

How do you choose the best sports movie ever made?

A difficult task. It's a question of judgment and taste, to be sure. But to be the champion among dozens of contenders, the top sports movie must meet five tough standards:

1. It needs to have a powerful story. The script is everything, as they say in Hollywood. There must be challenges and surprises, triumphs and setbacks.

2. It needs to have characters—three-dimensional heroes and bums, interesting folks who make you care about their lives.

3. It needs to have topflight sports action. A compelling story about an athlete who just stands there quickly stops being compelling. A great sports movie needs sweat and blood and speed and power. And the actors in the movie have to be better athletes than those middle-aged guys in your YMCA hoops league.

4. It needs to create goose bumps. There must be at least one scene in the film that sends shivers down your back or raises a lump in your throat.

5. It needs to be realistic—but not too much so. Because a powerful sports movie lets us stretch our imagination, allows us to dream. This is cinema's great advantage over real life.

"Apollo Creed vs. the Italian Stallion. Now that sounds like a damn monster movie."
—APOLLO CREED

More than any movie ever made, *Rocky* meets all five criteria. The script, written by Sylvester Stallone, is the touching story of a hardscrabble club fighter who takes his best shot. Stallone may not have invented the lovable-underdog saga, but he sure perfected it. *Rudy* is a spin-off of *Rocky*. So is *Miracle*, even if it is based on a true story. So is *Hoosiers*. Truth be told, dozens of movies listed in this book owe Stallone a nod.

Rocky is chock full of colorful characters. Paulie (Burt Young), the row-home loser who keeps nipping from a flask and trying to earn a buck off his best friend. Mick (Burgess Meredith), the rheumy-eyed octogenarian trainer seeking one last pass at the brass ring. Apollo Creed (former NFL linebacker Carl Weathers) the brash, angry Ali clone. Even the bit parts—like Gazzo the mobster ("What, you think I don't hear things?") and Buddy, Gazzo's wise-cracking driver ("Take her to the zoo. I hear retards like the zoo.")—add layers of grit to the story.

You want Grade A sports action? The brawl between Balboa and Creed that culminates the film is as good as it gets. From the moment late in the first round when the "badly outclassed challenger" shockingly floors the champ with a left, up through final bell, as the Italian Stallion buries one last hook in Creed's ribcage ("Ain't gonna be no rematch." "Don't want one."), Rocky will have you feinting and cheering for 25 minutes.

Need goose bumps? Watch Rocky in the training sequence performing those one-armed pushups as "Gonna Fly Now" kicks into high gear. Cut to him dashing through the streets of Philadelphia, a quick flash of the fighter punching the sides of beef, and then the iconic shot of Rocky in the gray sweat suit, arms aloft at the top of the Art Museum steps. If that doesn't create tingles, you must be unconscious.

Finally, *Rocky* is plausible—to a point. As most folks know, Stallone based his screenplay on the 1975 fight between Muhammad Ali and little-known Jersey brawler Chuck Wepner (aka "The Bayonne Bleeder"). Wepner never actually knocked down Ali (the champ tripped) and he was called out on a TKO with 19 seconds remaining in the 15th round. So Sly took the real life story and made it better.

Recalling the Ali-Wepner fight to *Playboy*, Stallone said, "The crowd is going nuts. Guys' eyes are turning up white. And here comes the last round, and Wepner finally loses on a TKO. I said to myself, 'That's drama. Now the only thing I've got to do is get a character to that point and I've got my story.'"

For all of those reasons, we deem *Rocky* the best of all time. Whenever we're flipping around the channels and discover it on one of the cable stations, we're hooked for the rest of the night.

"Women weaken legs!"
—MICK

You may disagree with our ranking, which is certainly your right. But just don't hold against *Rocky* the five sequels that followed it—and which we find entertaining to various degrees. The follow-ups diluted the franchise, often making people forget how intelligent and nuanced the original is, full of the tender little moments and layered dialogue that make it worthy of its Best Picture Oscar.

While Ali-Wepner may have inspired the screenplay, *Rocky* really is a metaphor for Stallone's own story. Before this, he was a struggling actor who landed a few small roles—a subway mugger in Woody Allen's *Bananas*, a thug in *Death Race 2000*. Frustrated that he couldn't get a starring part—and coping with a pregnant wife and a $106 bank account—he decided to try writing his own vehicle.

He drafted the original version of *Rocky* into a spiral notebook in one coffee-fueled weekend. There were lots of revisions. In various drafts, Creed was Jamaican, Adrian was Jewish, and

Rocky used his loser's purse to buy a pet shop. One version even had Rocky throwing the fight.

When he finally got it right, Stallone pitched the project to United Artists. Producers Irwin Winkler and Robert Chartoff had been searching for a boxing project and liked the script. Just one problem: No one envisioned Stallone as the lead in his own screenplay. They offered him $20,000 for the rights. He declined. Then $100,000. Then $200,000. Stallone, saying he "had gotten used to being poor," told producers they could have the script for nothing—but he had to star in the film.

Finally, a compromise was reached. Stallone could headline his own project, but it would have a bargain basement budget of $1 million. He would get $20,000 for the script and—we're not kidding—would be paid $1,400 for four weeks of acting. He would also get a small piece of the action—if there was any.

Director John Avildsen (*Lean on Me*, *The Karate Kid*) was hired even though, as he later told *Entertainment Weekly*, "Boxing wasn't something that interested me." What sold him on the script? "By the third page, this guy is talking to his pet turtles, Cuff and Link. I got totally seduced."

The cast was hired (Meredith got the biggest check: $25,000) and the movie was shot in under a month during December 1975. Because of the shoestring funding, some grand ideas were scaled back. For instance, the wonderful scene of the first date between Rocky and Adrian (Talia Shire) was supposed to take place at a packed ice rink with hundreds of skaters enjoying a festive holiday season. But there was no money for extras, so it was rewritten as a closed, empty rink. As they say, necessity is the mother of invention—or in this case, great art.

The unlikely film with the puny budget opened in November 1976 to mixed reviews (Vincent Canby of the *New York Times* panned it, saying that Stallone's performance "reminds me of Rodney Dangerfield doing a nightclub monologue."). But audiences loved *Rocky*, rising for standing ovations at the finale as a blind, bloodied Italian Stallion embraces Adrian in the ring ("I love you." "Where's your hat?"). *Rocky* gained a buzz that made it the nation's smash movie that holiday season.

Ultimately, it took in nearly $120 million at the box office. Overall, it has earned more than $1 billion over the years—not a bad return on investment for United Artists. Stallone earned an estimated $6 million on his small piece of the profits.

Rocky's Best Picture Oscar came over some worthy competition, including *All the President's Men*, *Taxi Driver* and *Network*. Years later, *Network* director Sidney Lumet remains a sore loser. "It was so embarrassing that *Rocky* beat us out," he told *Entertainment Weekly* in 2006. Lumet said he was skeptical when *Network* screenwriter Paddy Chayefsky predicted *Rocky's* win on the flight out to the Oscars.

"I said, 'No, it's a dopey little movie.' And Paddy said, 'It's just the sort of sentimental crap they love [in Hollywood].' And he was right."

Well then, you can mark us down as sentimentalists as well. More than 30 years later, *Rocky* still holds up. Sure, the tempo is a little slower than movies made these days, but the story and sweetness still work.

"Before this, boxing in movies generally spoke to something negative," Talia Shire said in the American Film Institute's salute to the top 10 sports movies (which ranked *Rocky* second to *Raging Bull*). "But here it spoke to something spiritual, transcendent. He just wants to go the distance. That's America. That's what people want in this country."

To which we would only add...

***"Absolutely."*—Rocky**

2 HOOSIERS (1986–PG)

SPORT: Basketball ★ **STARS:** Gene Hackman, Dennis Hopper, Barbara Hershey
DIRECTOR: David Anspaugh

Gene Hackman had his doubts when he was first hired to play embattled coach Norman Dale in *Hoosiers.* "No basketball movie had ever made it commercially," he recalled. "And we're going to make a *high school* basketball movie? I thought, 'We should have our heads examined.' "

Likewise, Dennis Hopper, cast as town drunk and hoops savant Shooter Flatch, was dubious during filming. He and Hackman wondered why David Anspaugh kept focusing on "all those damned basketball games."

"[Gene and I] were sitting on those wooden benches saying, 'Here's the big money. Why aren't they shooting us?' " Hopper mused.

It's said that during the filming of *The Godfather*, Francis Ford Coppola fretted that he had a disaster on his hands. If Hackman and Hopper had similar concerns, their fears were equally unfounded. Just as Coppola produced a masterpiece, so too did *Hoosiers* director

David Anspaugh. The little Cinderella story about the Hickory High Huskers is a nearly perfect sports movie.

Hoosiers is a predictable David vs. Goliath tale without being cliché. It is emotionally stirring without being manipulative. It is heart-stirring without being mushy; a feel-good film without being preachy. It boasts brilliant performances—from Hackman and Hopper right down to the flannel-shirt-wearing bit characters in bad haircuts who seem like they just walked out of small-town America, circa 1952.

Add to that some rousing sports action, plus the fact that the film is based on a true story and, well, you've got an enduring classic on your hands.

We surveyed more than 100 prominent people for this book, asking them to name their favorite sports movie. No film was cited more than *Hoosiers*. From legendary golfer Arnold Palmer ("It had a lot of heart. It was a great portrait of America in the 1950s.") to Super Bowl XXXIV MVP Kurt Warner ("They weren't supposed to be there and they won it all. It is really like my story."). From retired Edmonton Oilers left winger Ryan Smyth ("Small-town people rising to win a championship in the big city.") to WNBA star Sue Bird ("Back in CYO we used to watch it before games to get psyched up."). From former Red Sox third baseman Mike Lowell ("I love that Jimmy Chitwood *never* misses a shot.") to University of Wisconsin basketball coach Bo Ryan ("How could anyone involved in basketball *not* love it?").

To, yes, even President Barack Obama, who calls *Hoosiers* "truly inspirational." We can almost imagine the president's next State of the Union Address to Congress: "If you put your effort and concentration into playing to your potential, to be the best that you can be, I don't care what the scoreboard says at the end of the game. In my book we're gonna be winners."

Okay, maybe not.

On the surface, *Hoosiers* is a simple tale about tiny Hickory High (just 64 boys) which sends its team all the way to the state title game against a school 20 times its size. Its template is the true story of the 1954 Milan Indians, who won the Indiana championship in a game that ranks up there with the U.S. Olympic Hockey "Miracle on Ice" in terms of improbable victories. Anspaugh and screenwriter Angelo Pizzo—Indiana U college buddies from the 1960s—embellished history with the subplots that any textured film needs. (By the way, the old frat brothers peaked here in their first collaboration. Their second film, *Rudy*, was a few notches down. Their third, *The Game of Their Lives*, was a bomb.)

In *Hoosiers*, Anspaugh and Pizzo take you to a time and place where high school hoops mean everything to a small village. Sometimes that's positive—as when the caravan of townspeople cheerily follows the team bus to road games. Sometimes it's not—as when suspicious local hicks cross-examine the new coach on everything from his religion to his attitude on zone defense. Regardless, there is an organic sense of a real community throughout the film.

And there's something more, because *Hoosiers* isn't just about basketball or a long-gone America. It is also about redemption. The comeback of the small team sets the plot arc for the comeback of its coach, a puzzling 50-ish man with a shadowy past who seems a little too qualified for this outpost. It's also the chance of a comeback for Hopper's village drunk character. Once, 20-plus years ago, Shooter had his own chance for glory. But he missed the winning shot in a high school title game and his life since has been on a downward spiral. He is a disgrace to his son Everett (David Neidorf), one of the current players.

Coach Dale offers Shooter rebirth—over everyone's objections—in the form of an assistant-coaching job. The catch is that Shooter must remain sober. He manages for a few games and then stumbles into a critical semifinal contest bombed. It's a great scene, as the rowdy drunk humiliates his son, who then injures himself in a fistfight borne out of embarrassment. Hopper asked Anspaugh for a 20-second warning before his entrance into the scene. He then spun in circles so that he'd be able to reel around the hardwood like a true drunk.

Two other characters are worth mentioning here. The fabulously named Myra Fleener (Barbara Hershey) is the flinty teacher (and "flinty" is an extremely kind way of describing her) who has a lifelong grudge against high school hoops and seems intent in taking it out on Coach Dale.

Myra: "A basketball hero around here is treated like a god. . . . I've seen them, the real sad ones. They sit around the rest of their lives talking about the glory days when they were 17 years old."

Coach Dale: "You know, most people would kill to be treated like a god, just for a few moments."

The other intriguing character is Jimmy Chitwood (Maris Valainis), the lights-out shooter who quit playing after the Huskers' previous coach died. Jimmy also apparently quit talking, since he doesn't utter a word in his first few scenes. Anyway, he, too, finally joins the ride, saving the coach's job and bringing peace and harmony to tiny Hickory.

It's at this point that *Hoosiers* really picks up. The first half of the movie focuses on Coach Dale's travails; the second half is the run to the state tournament. The final game, of course, ties it all together.

And the pep talk before the final game? Well, we rank the all-time top five movie locker-room scenes like this:

5. Tony D'Amato (Al Pacino) in *Any Given Sunday*: ***"We fight for that inch. On this team, we tear ourselves and everyone else around us for that inch. . . . I'm still willing to fight and die for that inch!"***

4. *Rudy's* **Dan Devine** (Chelcie Ross, who also appears in *Hoosiers*) telling his Notre Dame seniors before their final home game: ***"Remember no one, and I mean no one, comes into our house and pushes us around."*** This clip has been played *ad nauseum* at every single NHL and NBA game over the past decade.

3. Bluto Blutarsky (John Belushi) psyching the *Animal House* Deltas for their day of revenge: ***"Over? Did you say over? Nothing is over until we decide it is! Was it over when the Germans***

bombed Pearl Harbor? Hell no!"

2. Knute Rockne (Pat O'Brien) channeling George Gipp in *Knute Rockne All-American* with the ***"Win one for the Gipper"*** speech.

And #1:

Coach Dale: ***"We're way past big-speech time. I want to thank you for the last few months. They've been very special to me. Anybody have anything they want to say?"***

Merle: ***"Let's win this one for all the small schools that never had a chance to get here."***

Everett: ***"I want to win for my dad."***

Buddy: ***"Let's win for coach, who got us here."***

Preacher Doty: ***"Blah, blah, blah."*** (This part slows the whole thing down and should have been cut.)

Reverend Purl: ***"And David put his hand in the bag and took out a stone and flung it. And it struck the Philistine in the head, and he fell to the ground. Amen."***

Coach Dale: ***"I love you guys."***

All (clasping hands): ***"Team!"***

Works every time.

There has always been debate about how strong a coach Norman Dale really is. Wrote Mike Vaccaro in the *New York Post*: "What do you get when you mix one part Dean Smith, one part Vince Lombardi, one part Casey Stengel, and one part Red Auerbach? About half the coach that Norman Dale is."

Truth be told, Dale's game strategy is sometimes lacking much imagination beyond "Give the ball to Jimmy." But put us in those old black Chuck Taylor high tops and we couldn't name anyone we would rather play for. And no one, and we mean *no one*, comes into our house and suggests that any movie coach ever was a better motivator.

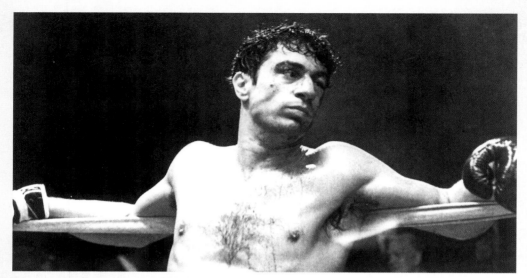

RAGING BULL (1980—R)

SPORT: BOXING ★ STARS: ROBERT DE NIRO, JOE PESCI, CATHY MORIARTY
DIRECTOR: MARTIN SCORSESE

Martin Scorsese never intended for *Raging Bull* to be another *Rocky*. He knew his film would not be embraced or loved because his central figure—we can't call him a hero—was Jake LaMotta, who is neither embraceable nor lovable.

Red Smith, the Pulitzer Prize-winning columnist, called the middleweight LaMotta, "A crumbum, a creep, a highly unappetizing specimen of the human race …a louse through and through, foul of speech, vile of temper, a bullying wife-beater, a pimp, altogether self-centered without a spark of decency."

Okay, Red, we get the picture, and so did Scorsese. He knew LaMotta, the former world boxing champion, was a violent, abusive man with a soul as dark as the night. He knew if he filmed LaMotta's story—no matter how brilliantly—it would not appeal to everyone. He was proven correct.

Raging Bull divided movie critics and audiences like few films before or since. It grossed $23 million at the box office, which was good but not great. It received some enthusiastic reviews—Vincent Canby of the *New York Times* called it "an achievement"—but it was hammered by other critics who were repelled by its characters, principally LaMotta.

Frank Deford, reviewing the film for *Sports Illustrated*, found it "so peopled with despicable human beings and so gratuitously violent that all the good stuff is cancelled out." Kyle Smith of the *New York Post* wrote, "If it leaves us with any feeling at all, it is smug superiority to its punishingly unpleasant, but not meaningful or important, central character."

Still, *Raging Bull* was nominated for eight Academy Awards, including Best Picture, Best Director, Best Actor (Robert De Niro), Best Supporting Actor (Joe Pesci) and Best

Supporting Actress (Cathy Moriarty), De Niro and Thelma Schoonmaker (Best Film Editing) won Oscars. Scorsese lost to Robert Redford, whose film *Ordinary People* also earned Best Picture honors.

Over time, *Raging Bull* has grown in stature. The American Film Institute ranked it among the top five American movies of all time. *Entertainment Weekly* placed it fifth on its list of the 100 Greatest Movies. *Halliwell's Film Guide* put it seventh on its Top 1,000. Yet Andre Soares of the *Alternative Film Guide* calls it "the most overrated American movie of the last 30—or 40 or 50—years.

"I'm not quite sure how far back one must go to find so much praise bestowed upon so much artifice," Soares wrote. "Much of the acting is as out of control as a bull running through the streets of Pamplona."

In boxing, the reviews would be called a split decision, but clearly Scorsese comes out ahead. *Raging Bull* is not the people's champion in the way that *Rocky* is and always will be, but it is widely recognized as the greater artistic achievement. Roger Ebert of the *Chicago Sun-Times* called *Raging Bull* the best film of the 1980s, and one of the four greatest movies ever made.

It is an unusual sports film in that we can't bring ourselves to root for the main character. That means there is no joy and no happy ending. It may leave you feeling as bruised and battered as one of LaMotta's opponents, but that's what Scorsese wanted. He wanted the audience to feel the punches and the rage that defined LaMotta. There are no soft edges, there is only a willful brutality that carries LaMotta to the top and then brings him down.

Author Joyce Carol Oates, who often writes about boxing, once talked about the boxer's "divided self," that is, the duality of a man who lives by one code in the ring and another in his personal life. Inevitably, a boxer suffers when he can no longer separate the two. In the case of LaMotta, there never was a separation: he was the same person inside the ring and out. In *Raging Bull*, De Niro brings that monster to life in a chilling and unforgettable way.

Canby called De Niro's Oscar-winning performance, "the best of his career, a titanic character, a furious original, a mean, inarticulate Bronx-bred fighter whom the movie refuses to explain away in either sociological or psychiatric terms. He is propelled not by his milieu, his unruly id or by his guilt, but by something far more mysterious. Just what that is, I'm not at all sure."

No one understood LaMotta, even those closest to him, including his five wives and his brother/manager, Joey. He was proud, yet self-destructive. He always said his toughest opponent was Sugar Ray Robinson, whom he fought six times (winning only once). But it would be more accurate to say Jake LaMotta's toughest opponent was himself, a point *Raging Bull* makes over and over again.

Scorsese, as he did in his previous films *Mean Streets* and *Taxi Driver*, brings the texture of New York City into the story. Scenes of Jake and Joey (Pesci) walking down the street on a summer day, past kids playing by an open fire hydrant, evoke a powerful sense of time and place. It is The Bronx in the '40s—men in sleeveless T-shirts, suspenders and dress pants sitting at the community pool, swapping dirty jokes and talking business with the Mafia wiseguys. You can almost feel the sweat on Jake's eight-ounce bottle of Coke.

Scorsese shot *Raging Bull* in black-and-white, in part, because he wanted his film to stand apart from the *Rocky* series, which had already cranked out two installments and was working on a third. Scorsese also felt black-and-white would be more effective in capturing the mood of LaMotta's time. As the director, he moved the point of view back and forth between objective and subjective in a way that enhanced, rather than confused, the narrative, which is not easy to do.

When Scorsese wanted to show the world as seen by Jake, he went to slow-motion. The scene in which Jake first sees Vickie (Moriarty), the 15-year-old blonde who becomes his obsession, is especially memorable. Vickie is sitting on the edge of the pool, her legs paddling in the water. As the camera lingers on her seductive figure, we know what's going on in Jake's mind and we can almost feel what's happening in his Fruit of the Looms.

In the fight scenes, Scorsese insisted on staying in the ring. He did not want to show the action from outside the ropes. He wanted the audience to experience the brutality, so he shot it tight, mixing film speeds and occasionally used a series of explosive stills. When LaMotta breaks an opponent's nose, you hear the crunch, and you see the flesh collapse and the blood spray like water from a ruptured hose.

The violence shocked audiences when the film was first released, and it was reason enough for some critics to pan it. But the photography is spectacular (cinematographer Michael Chapman earned an Oscar nomination), and when the punches land, the sound is unlike anything we've heard before. The audio engineers mixed a variety of sounds—the splat of a melon being split, the crack of a gun being fired—and the effect is mesmerizing.

But for all of Scorsese's mastery, the film really belongs to De Niro. Much has been written about the actor gaining 60 pounds to play the bloated, middle-aged LaMotta and, granted, that's taking art to the extreme. But it is in the first half of the film, when he is the young LaMotta, twisted with ambition

and jealousy, raging against the world, that he carries *Raging Bull* to the heights. He commands the screen, and his profane dialogue with Joey is alternately funny and frightening.

The film shows LaMotta as a tough-as-nails middleweight contender who insists on doing things his way. He refuses to play ball with the mob—Mafia boss Tommy Como is played by Nicholas Colasanto, "Coach" on TV's *Cheers*—and as a result, he is denied a title shot.

Como tells Joey, "Your brother can beat all the Sugar Ray Robinsons and Tony Janeros in the world, but he's not getting a title shot. Not without us, he's not."

Reluctantly, LaMotta agrees to tank a fight against Billy Fox so the mob can make a killing betting against him. He is rewarded by getting a chance to fight Marcel Cerdan of France for the middleweight championship. LaMotta defeats Cerdan on a TKO to finally win the championship belt.

Lean and muscular thanks to months of training, De Niro looks like a world-class boxer. He studied tapes of LaMotta for hours, so he has the Bronx Bull's crouching, wade-in style down to perfection. There was nothing pretty or scientific about LaMotta. He was a brawler, plain and simple, and when De Niro lowers his shoulders and moves in on an opponent, you feel like it is 1948 and you're ringside at the old Garden.

Even more compelling is the character De Niro creates outside the ring. In the ring the violence is understood; blood flies, but there are rules. It is the rest of the time that LaMotta is the most frightening. There is a look of malevolence in his eyes, and you never know when it will erupt. He could lash out at any second. It could be something he sees or, worse, something he imagines. De Niro puts a very unsettling tick in that time bomb.

Much of LaMotta's rage is rooted in jealousy. He is convinced Vickie, his beautiful young wife, is unfaithful. He believes she's sleeping with the same Mafia guys he has hated for years. Later, he adds Joey to the list of suspects. When a brooding LaMotta walks into the bedroom and asks Vickie, "Where were you today?" we cringe because we know what's coming. There is no indication she has done anything wrong, but the look on his face says it doesn't matter.

"Jake, don't start, OK?" she says.

She has been through this dozens of times, deflecting Jake's questions about various men, including total strangers, such as Tony Janero whom she off-handedly refers to as "good-looking." When Jake meets Janero in the ring, he smashes his face to a bloody pulp and then smirks at Vickie as Janero falls to the canvas. It is unspeakably cruel, but for the Raging Bull, totally in character.

This time, Jake attacks Vickie, slapping her and throwing her to the floor. Then he storms into his brother's house, pulls him away from the dinner table and, while Joey's wife and kids look on, he knocks him through a glass door and pummels him. He is convinced the two are having an affair. Why? He didn't like the way Joey kissed Vickie earlier in the day. "You kissed her on the mouth," Jake says. That was enough to start him thinking.

Is it any wonder that, by the end of the film, Jake is alone? Vicky has left him and Joey has disowned him and when he finally goes to jail for pimping out an underaged girl in his seedy Miami nightclub, he has no one to call and nowhere to turn.

Raging but no longer bullish, the flabby ex-champion bangs his head against the wall of his cell, crying, "Why, why, why?" He pounds his fists into the concrete, the same fists that brought down the best middleweights in the world, but now it is just sound and fury signifying nothing.

"I'm not an animal," he says, sobbing. "I'm not an animal."

If it were anyone else, we would be sympathetic.

But it is Jake LaMotta. So we're not.

☛ **DON'T FAIL TO NOTICE:** The scene in which LaMotta warms up in the dressing room, then makes the long walk through the tunnel to the ring in Briggs Stadium for the fight with Marcel Cerdan. Scorsese does it in one tracking shot. There are no cuts, no second or third cameras. It is one continuous shot that allows you to take that long walk with Jake and Joey and listen as the roar of the crowd grows louder.

Scorsese went on to use the same tracking technique for the famous nightclub scene in *Goodfellas*.

🎬 **PIVOTAL SCENE:** LaMotta agrees to throw the fight with Billy Fox, but he can't bring himself to hit the canvas. LaMotta never was knocked down, not once in his entire career, and he wasn't about to let the light-hitting Fox put him on the deck. So he leans against the ropes, drops his hands and lets Fox pound on him.

"Hit me, you bum," he mutters through his mouthpiece.

The referee stops the fight and the crowd boos.

LaMotta insists on doing it his way, even when he takes a dive.

❝ **BEST LINE:** "He ain't pretty no more." — Tommy Como after LaMotta, in a jealous rage, brutalizes Tony Janero.

🍎 **BET YOU DIDN'T KNOW:** Robert De Niro broke two of Joe Pesci's ribs in the scene where they spar in the gymnasium.

4 THE NATURAL (1984–PG)

SPORT: Baseball ★ **STARS:** Robert Redford, Glenn Close, Wilford Brimley
DIRECTOR: Barry Levinson

When TriStar Pictures agreed to produce *The Natural*, many in the movie industry felt it was a mistake. The script, adapted from Bernard Malamud's 1952 novel, had been kicking around Hollywood for years. Jon Voight, Michael Douglas, and Nick Nolte all expressed an interest at one time or another, but no studio was willing to invest in it.

Sports movies are a risky proposition. A few succeed at the box office, but more fail. Also, Malamud's original story was rather dark with overtones of Homer and Greek mythology. How were you going to sell that to baseball fans? And if a movie about baseball couldn't even attract baseball fans, well, what was the point?

But in 1983, Robert Redford saw the script. He was looking for an opportunity to work with director Barry Levinson and he knew Levinson had always wanted to make a baseball movie. They took the idea to TriStar, and on the strength of Redford's name, and Levinson's success with the film *Diner*, the studio agreed to do the film.

Still, there were doubts. "Redford goes into *The Natural* with two strikes against him," wrote syndicated columnist Ivor Davis. "Baseball, that traditionally iffy film subject, and middle-aged disillusionment and failure in the character." There was a third potential strike: the 46-year-old Redford trying to pass for a major league ballplayer.

The producers resolved one issue—the disillusionment—by changing the story. In Malamud's novel, Roy Hobbs is a tragic figure who literally strikes out at the end. In the film, the script by Roger Towne (brother of Oscar-winner Robert Towne) has Hobbs hitting a mammoth home run to win the pennant. As endings go, that's a little more uplifting.

"The movie's tone is very different," producer Mark Johnson said. "Ours is a story about a man who is given a second chance and what he does with it. The book is cynical, ours is hopeful."

On the other matter—Redford's credibility as a ballplayer—the producers and writers did not have to do a thing. Redford played a lot of baseball in his youth, and even in his forties he was in good shape, so he was believable as Roy Hobbs. Okay, it was a stretch when he played Hobbs as a teenager, but when he threw the ball, that smooth southpaw motion was a pleasure to watch. And when he picked up a bat as the older Roy, well. . . .

"Robert Redford's swing is maybe the best acting he's ever done," wrote Wilfrid Sheed in the *New York Times*.

"Redford plays so authentically, you want to sign him up," wrote Roger Angell in *The New Yorker*.

The film opens with Hobbs as a boy, playing catch in the fields with his farmer father. One night, lightning splits a tree behind the house and young Roy carves a bat out of the wood. He engraves the name "Wonderboy" on the barrel.

It isn't long before Roy (now played by Redford) is 19 and riding the train to Chicago for a tryout with the Cubs. Along the way, he encounters "the Whammer" (Joe Don Baker), a thinly disguised version of Babe Ruth. The scout who found Roy bets sportswriter Max Mercy (Robert Duvall) that the kid can strike out the Whammer on three pitches. When the train stops, they find an open field and Roy does, indeed, blow three fastballs past the Whammer.

The mysterious Harriet Bird (Barbara Hershey) watches the contest and the next thing you know, she is sitting with Roy on the train asking, "Have you ever read Homer?" Right away, you get a bad feeling. She has a weird look in her eye and she is dressed entirely in black. You keep waiting for the alarm to go off in Roy's head, but he is so naive he tells this creepy stranger about his dreams.

"Someday when I walk down the street, people will say: 'There goes Roy Hobbs, the best there ever was, the best there will ever be in the game,' " he says.

Roy doesn't realize it, but he has just walked into the black widow's web. In Chicago, Harriet invites him to her hotel room where she shoots him with a silver bullet. It seems Harriet is a deranged woman on a mission to kill "the best" in every sport. She planned to take out the Whammer (she was giving him the eye on the train) but when she saw Roy in action, he became the target instead.

It is a bizarre scene—and not very well explained in the movie—but it is based on a true story. In 1949, a woman named Ruth Ann Steinhagen shot Eddie Waitkus, first baseman of the Philadelphia Phillies, in Chicago's Edgewater Beach Hotel. Police reports indicate she was a delusional baseball groupie who was infatuated with Waitkus. We've seen that stalker syndrome repeated many times.

Harriet Bird is a different story. She appears out of nowhere, shoots the hero and leaps to her death from the hotel window. We're left wondering: "What was all that about?" It is the weakest part of the movie, but it sets up the real story which is Roy's comeback.

Like Waitkus, Hobbs survives the shooting. Waitkus returned to baseball the following season and helped the Phillies win the National League pennant. Hobbs isn't as lucky. He drops out of sight for 16 years. He resurfaces in 1939 as a middle-aged rookie with the New York Knights, a last-place team in the National League.

When Hobbs reports, manager Pop Fisher (Wilford Brimley) doesn't want any part of him. "Mister," he says, "you don't start playing ball at your age, you retire." Hobbs will not discuss his past or why he was away from the game for more than a decade. He was playing outfield for a semi-pro team when a Knights scout saw him and signed him for $500. Fisher puts Hobbs on the bench and ignores him.

Finally, Fisher tires of the lackadaisical play of right fielder Bump Bailey (Michael Madsen). He benches Bailey and sends Hobbs up to hit for him. "Come on, Hobbs," Fisher says. "Knock the cover off the ball." Which he does—literally. Hobbs hits the ball so hard, the seams explode and the insides unravel. The outfielders are chasing a tangle of string while the winning runs cross the plate. The legend is born.

Given the chance to perform, Hobbs becomes the best player in baseball. But evil forces are at work. The team owner, the Judge (Robert Prosky), will gain full control of the ball club from Fisher if the Knights fail to win the pennant. It looked like a sure bet until Hobbs came along. Now the team is winning. The Judge offers Hobbs a bribe to tank the rest of the season. Hobbs flatly refuses.

But the Judge and his gambler sidekick Gus (Darren McGavin) have a Plan B. Actually, it's Plan Double B as in "Blonde Bombshell." They introduce Hobbs to Memo Paris (Kim Basinger). Just like with Harriet Bird, Roy ignores the warning signs. Even Pop Fisher, Memo's uncle, tells Roy, "She's a jinx." But it's too late, Roy is smitten. Pretty soon, he is scoring in the bedroom but striking out on the field. He falls into a deep slump and the Knights go into a tailspin.

The team goes to Chicago and his old girlfriend from back home, Iris (Glenn Close), surprises him by coming to the game. The sight of Iris in her shimmering white dress snaps Roy out of his funk. He hits the game-winning home run one day and hits four more homers the next. After the last game, he walks Iris home and tells her the whole story of what happened to him.

"My life didn't turn out the way I expected," he says.

"I believe we have two lives," Iris says. "The life we learn with and the life we live with after that."

Second chances, in other words.

Iris tells Roy she has a 16-year-old son whose father lives in New York. Hello, Roy? This is your son, get it? It takes awhile—in fact, Iris has to put it in the form of a note and send it to the dugout—but finally Roy figures it out.

The Knights' season comes down to the final weekend. They need one win to clinch the pennant. The Judge will do anything to stop them. At a party, Memo puts poison in Roy's drink and it causes the bullet wound to rupture in his stomach. The doctor tells Roy if he plays again, it could cost him his life. Without Roy, the Knights lose three in a row and it looks like the season is down the drain.

But on the final night, Roy shows up and with the blood seeping through his jersey, he hits a ball that clears the roof and crashes into the stadium lights, setting off a series of explosions, allowing Roy to do his final home run trot through a shower of fireworks.

"*The Natural* is a movie that stays with you," said actor Tom Selleck. "You think the movie is over when he hits the home run but then it goes to the shot of him playing catch with his son on the farm, and it adds another layer to it. His life has come full circle. Now he's back where he started as a boy and he's playing catch with his son. That's how most people think of baseball—a game that links the generations."

Chase Utley, the Philadelphia Phillies All-Star second baseman who has been called "The Natural" by some baseball people, loves it as well. "It's old-school in a lot of ways, but I could relate to what [Hobbs] was going through and how much it meant to finally get a chance to play in the big leagues. He knew he might only have that one season, so when he says, 'You'll get the best I've got,' you know he means it.

"Some things are exaggerated, obviously, like the part where he knocks the cover off the ball and they're throwing around a ball of string. But it's such a good movie, you don't mind. There's a part of you that says, 'That could never happen,' but there's another part of you that says, 'That's pretty cool.' If you're really into the story, you go with it."

❝ BEST LINE: "My Dad always wanted me to be a baseball player," Hobbs says before the final game.

"Well, you're the best one I ever had," Pop Fisher says, "and you're the best ------- hitter I ever saw. Go suit up."

⑤ BULL DURHAM (1988–R)

SPORT: Baseball ★ STARS: Kevin Costner, Susan Sarandon, Tim Robbins DIRECTOR: Ron Shelton

The standard rap on *Bull Durham* over the years has been that while it's a sexy and comedic love triangle, it's more of a chick flick than a sports movie. The portrayal of minor league baseball, some critics charge, just doesn't ring true.

So we asked some real ballplayers what they thought.

Brad Lidge, retired Philadelphia Phillies closer: "Of every baseball movie I've ever seen, it's the one that tells it like it is. The bus rides, the sweltering hot days, the knuckleheads you encounter along the way. Trust me, that was exactly what the minors were like."

Joe Morgan, Hall of Fame second baseman: "I played a season in Durham and watching *Bull Durham* made me feel like I was reliving it. The scenes of the players doing kid things, like wetting down the field and sliding around in the mud, is the way it is, basically, because most minor leaguers are kids."

Prince Fielder, Texas Rangers first baseman: "My favorite scene is when the pitcher is called into the office and told he is going up to the majors. It reminds me of when I got the call up from Nashville. You just walk around with this big smile on your face."

Aaron Heilman, former Chicago Cubs reliever: "Kevin Costner looks like the veteran guy who was stuck in the minors, kind of resigned to it, but still with enough pride to do the job. The scenes on the bus where the young guys are listening to him, trying to soak up his knowledge—that's how it is."

Good enough for us.

In the course of researching this book, we interviewed dozens of major leaguers, asking each his favorite sports movie. The one they mentioned the most? *Bull Durham.* In fact, more than a few players say they routinely slip lines from the film into their daily ballpark conversation.

Former Major League catcher Chris Coste— whose 10 years in the minors make him a real-life version of Costner's Crash Davis character—says, "The conversation at the pitcher's mound [from *Bull Durham*] actually became a running joke between [former Phillies teammate] Clay Condrey and me. Every time he came in, I would go out there and say, 'I hear candlesticks make a good wedding gift.' That kept things loose. It's amazing how accurate the dialogue from that movie is."

Not surprising, however, since *Bull Durham* was written and directed by Ron Shelton, who spent five seasons in the Baltimore Orioles minor league system playing with future major league stars Don Baylor and Bobby Grich. When Shelton lovingly shows the quirks and daily patterns of life in the Class-A Carolina League, he does so from the vantage of a guy who's been there.

Bull Durham—the first (and best) of a half-dozen sports movies created by Shelton— is a rarity in the genre. First off, it combines sports, comedy, and romance—a difficult triple play to pull off. It is neither sentimental nor reverential. It is not about superstars, but mostly the fringe players who scruff around in baseball's minor leagues. It doesn't end in tragedy, or in the underdogs rising to their shining moment. In fact, the hero of the film ends up getting cut. At the same time, it shows a love and respect for the game, embodied in Crash's impassioned creed calling for (among other things) a Constitutional ban on Astroturf and the designated hitter.

The set-up for the movie is this: Crash, the proverbial "player to be named later," is a 30-something minor league catcher who finds his Triple-A contract re-assigned to the Single-A Durham Bulls. Seems the organization has a big-time prospect, brilliantly named Ebby Calvin "Nuke" LaLoosh (Tim Robbins), who possesses a "million-dollar arm and a five-cent head." In exchange for tutoring Nuke on and off the field, Crash gets to keep getting paid to play baseball.

His resentment is palpable. Crash is a guy who would cut off his arm to make it, who speaks of his brief stint in the majors as "the 21 greatest days of my life." His student, meanwhile, is a callow kid who defines success by his Porsche 911 with a quadrophonic Blaupunkt.

Still, Crash knows he must point Nuke toward "The Show." So he offers these lessons:

- **Don't think.** You can only hurt the ball club. Never shake off your catcher, or he may just tip off the hitter on what pitch is coming.
- **When you get in a fight with a drunk,**

never throw a punch with your pitching hand.

- **Don't try to strike everybody out.** Strikeouts are boring, in addition to being fascist. Ground balls are more democratic.
- **A player needs to learn his clichés** before talking to the media. Study them, memorize them. They are your friends.
- **You'll never make it to the big leagues with fungus on your shower shoes.** Think classy and you'll be classy. If you win 20 in The Show, you can let the fungus grow back and the press will think you're colorful. Until you win 20, however, it just means you are a slob.
- **Be cocky,** even when you're getting beat. Play the game with fear and arrogance.
- **Respect a streak.** If you think you're suddenly winning because you are having sex, or not having sex, or wearing a woman's garter belt—well then, you are.

Speaking of sex, that's where Nuke's other teacher comes in. Annie Savoy (Susan Sarandon) is the ultimate groupie, a William Blake-quoting community college instructor who, each season, pledges her heart and body to a single player on the Bulls. Nuke—the future superstar—is the obvious candidate except, well, the older and wiser Crash keeps getting in the way. There's a brilliant scene early on, where Annie auditions the two men for the role of lover. Crash walks out on the opportunity for sex (something players *never* do), saying he's "been around too long to try out for anything." If he doesn't win her heart right there, he certainly wins her mind.

"And so this strange kind of ménage thing happens among the three of them," Sarandon said on a 2008 TV show in which the American Film Institute named *Bull Durham* the No. 5 sports movie of all time (hmmm, we agree). "Who does she choose? The young buck with lots of energy—not too much knowledge—but lots of raw talent? Or the guy who maybe doesn't have the talent as much, but has the brains and the love of the sport? He can still hit them out of the park, but isn't going to go any further."

The romantic triangle evolves over the course of the 1987 Carolina League season. While Annie keeps sleeping with Nuke (or not sleeping with him, so as to avoid jinxing a winning streak), she and Crash find themselves irresistibly pulled together. It all ends up in satin bathrobes, bathtub sex, dancing to the Dominoes' "Sixty Minute Man," and sharing a bowl of Wheaties. Very erotic stuff, trust us.

In the end, Nuke (thanks to lessons learned) gets promoted all the way from Single-A to the Big Leagues in one season—a bit of a stretch, to be sure. Crash gets released, plays out the string on another team (supposedly breaking the all-time minor league record for home runs), retires and comes back to Annie. Presumably, they are looking forward to a life together; he as a coach for Visalia in the California League and she, we suppose, as a composition teacher at the College of the Sequoias. You can start a good debate on the topic of whether they'd still be together after all these years.

Bull Durham captures the authenticity, the lingo and the funny side of baseball better than any other film ever made. If Annie Savoy is the high priestess of what she calls "The Church of Baseball," consider us devotees.

🍎 **BET YOU DIDN'T KNOW:** The character of Nuke LaLoosh is loosely based on minor-league legend Steve Dalkowski. In a nine-year career, Dalkowski struck out an astounding 12.6 batters per nine innings—but he also walked 12.3. Unlike Nuke, he never made it to The Show.

⑥ SLAP SHOT (1977–R)

SPORT: Hockey ★ **STARS:** Paul Newman, Michael Ontkean
DIRECTOR: George Roy Hill

When it first came out, the comedy *Slap Shot* was dismissed by critics as vulgar and trivial and a misplaced tribute to on-ice thuggery. Many theatergoers were appalled to hear such blue language from Paul Newman (even Butch Cassidy never used *those* words). Others couldn't relate to a two-hour paean to the garage sport of minor league hockey.

The movie pulled in $28 million at the box office, ranking it below sports-comedy stink-bombs of the era such as *The Main Event* and *Semi-Tough*. Within a matter of weeks, *Slap Shot* skated in and out of the theaters. It appeared destined to be forgotten.

Except...

Except that somewhere along the way, this uproarious movie evolved from being an amusing throwaway into a cult classic—if we define cult classic as one whose fans see it dozens of times and gleefully recite every line.

"It's the best bus movie ever," said former Philadelphia Phillies catcher Chris Coste, referring to the films that teams like to watch on those 300-mile road trips. Coste, who played beer-league hockey as a young man in North Dakota (his missing front tooth serves as proof), added, "I've quoted every line from that movie millions of times—'putting on the foil' or, 'Effing machine stole my quarter' or, 'Old-time hockey, like Eddie Shore and Dit Clapper.'"

And Toe Blake.

So how did *Slap Shot* grow in prestige over the years to the point that most sports fans (including us, obviously) place it on their top-ten lists?

We're not quite sure. But we figure that once people got past the obscenity (admittedly, it's tough to sail past "Hanrahan—Suzanne sucks ----! ...She's a ----!") and accepted the minor-league theme (perhaps the success of *Bull*

Durham opened minds), they discovered that *Slap Shot* might just be the most hilarious sports movie ever made.

Hey, the critics also panned *It's a Wonderful Life* when that first came out.

Granted, Newman's portrayal of Coach Reggie Dunlop isn't as heart-warming as Jimmy Stewart's George Bailey. But let's just say that it would be a tough Christmas Eve call if competing networks aired the bucolic idealism of Bedford Falls against the grimy games at the War Memorial Auditorium.

Slap Shot is set in the fictional town of Charlestown, PA. Newman's Reg Dunlop is the aging and unsuccessful player-coach of the Federal League's Charlestown Chiefs, who play uninspired hockey before sparse and hostile crowds.

The town's largest mill is set to close, which sparks rumors that the Chiefs will be shut down as well. So Dunlop is struggling to hold his team together and, at the same time, trying to ascertain the identity of the Chiefs' owner so that he can lobby for his players.

Meanwhile, the Chiefs' cheapskate general manager, Joe McGrath (Strother Martin), reaches down to the low-level Iron League to sign three siblings—the bespectacled, semi-adolescent Hanson Brothers. The boys bring their toy cars along on road trips and wrap their knuckles with aluminum foil before games, the better to pack a nasty punch.

Dunlop is initially horrified and won't even put the trio on the ice. When he eventually lets them play, the Hansons create chaos, brutalizing the opposition and thrilling Charlestown's long-suffering loyalists. Other Chiefs players follow their goonish lead—except for leading scorer Ned Braden (Michael Ontkean), a Princeton grad repulsed by the Cro-Magnon approach.

Anyway, the Chiefs start winning games, drawing crowds and inciting police actions.

To old heads, this may sound a lot like the Philadelphia Flyers of the mid-1970s. So it shouldn't be surprising that Hockey Hall of Famer Bobby Clarke—the captain of those Broad Street Bullies—cites *Slap Shot* as his favorite film. "I started laughing during the opening scene and I never quit laughing during the whole movie," Clarke said.

And does Clarke see any parallel between the Chiefs and his Flyers? "Well . . . we only had two brothers on our team [Joe and Jimmy Watson]," he said. "They had three."

After watching *Slap Shot* for about the 874th time, we decided to list the reasons why it holds up better than ever three decades after its initial lukewarm reception:

1. For a farce, the screenplay rings extremely true to life. *Slap Shot* was written by Nancy Dowd (yes, a woman!) whose brother Ned played for the Johnstown (PA) Jets, the team on which the Chiefs team is based. Intrigued by Ned's tales, Nancy gave him a tape recorder to document life behind the scenes in minor league hockey. Dowd, by the way, shared an Oscar for best screenplay one year later for writing *Coming Home*, a super-serious film about returning Vietnam veterans.

Nearly every character, from barfing goalie Denis Lemieux ("Who own da Chief?") to toupee-wearing announcer Jim Carr ("Everyone is on their feet yelling, 'Kill! Kill!' This is hockey!") to reporter/stooge Dickie Dunn ("I tried to capture the spirit of the thing.") is based on a real person. Indeed, to this day, hockey players refer to a reporter who writes false rumors as a "Dickie Dunn."

"It's a spoof, but anyone who has spent time around hockey knows all those characters," said veteran broadcaster Gary Thorne. "That *is* the minor league hockey life. That's how those guys live, how they travel, how they talk. You walk through a locker room before a game and you'll hear guys say: 'I'm going to get so-and-so tonight.' Hockey people decried the movie and said it would hurt the sport. I think it actually helped the sport."

A couple of side notes here: First, Dowd originally planned her project to be a documentary. Director George Roy Hill convinced her it would work better as a full-length comedy. Second, the real-life Ned Dowd has a cameo in the movie, playing ultimate goon Ogie Ogilthorpe.

2. Newman's presence gives *Slap Shot* more than just star power. His acting chops (he appears in 80 percent of the scenes) help it graduate from being a series of goofy skits into a taut, plot-driven film. Okay, maybe we just exaggerated. But without Newman, this movie is *The Fish that Saved Charlestown*. With him, it's a film about broken dreams, working-class angst and love's longings—not to mention guys who flash the audience at fashion shows.

Newman said for years that he had more fun filming *Slap Shot* than any movie he ever made. "I always wanted to play for a team," he told Time magazine. "This gave me that chance." It also taught him how to curse like a player. "This is the raunchiest film I have ever done," he said in a promotional interview at the time of *Slap Shot's* release. "The language is heavy beyond blue, into purple almost. I think it will be tastefully vulgar."

The amazing thing to watch is how well Newman moves on the ice. He was 51 years old when *Slap Shot* was filmed (Dunlop is said to be 38), and the guy could flat-out skate. The only injury he sustained during filming came when he was sitting down. Watch the scene where Hanrahan (the goalie with the lesbian wife) dives into the penalty box to pummel Dunlop. Somehow, both actors managed to pull groin muscles at the same moment.

3. The marauding Hanson Brothers are the greatest auxiliary characters in the history of sports comedy, bar none. Better than Tanner Boyle in *The Bad News Bears*. Better than Pedro Cerrano in *Major League*. Better than Bob Barker in *Happy Gilmore*.

The look—stringy hair and black-rimmed coke-bottle glasses—is not contrived. That's who these guys were. The language, too, is genuine. Director Hill told the Hansons (actually brothers Steve and Jeff Carlson and David Hanson) to speak on camera as they spoke in real life. Even the toy racing cars are real.

The three, in fact, did play for the 1974-75 Johnstown Jets (as did a third genuine Carlson brother, Jack, who lost his part in the film when he was called up to the Edmonton Oilers). They finished 1-2-3 on the team in penalty minutes, started more than one brawl at the opening faceoff and, according to Jets coach Dick Roberge, really wrapped their fingers with aluminum foil until a league ruling outlawed taping their hands.

"They were very professional," Newman told *Sports Illustrated* in 2007. "And they were completely crazy. We drank a lot of beer."

The three Carlsons, plus Hanson, all eventually got to the NHL or its 1970s rival, the World Hockey Association. To this day they make a living by, well, being themselves—traveling to conventions and bar mitzvahs in full uniform and kibitzing with an adoring public.

And that's kind of the bottom line. More than 30 years after its release, *Slap Shot* is more popular than ever with the public. In a recent letter to the fan-based website *Slap Shot Nation*, Nancy Dowd wrote, "You wore the Halloween costumes, hosted the *Slap Shot* parties, memorized the lines, and laughed and laughed. That is the real measure of a motion picture, not the opening weekend grosses. When an object is embraced by a popular culture, it takes on a life of its own."

⁊ THE LONGEST YARD (1974–R)

SPORT: Football ★ **STARS:** Burt Reynolds, Eddie Albert, Michael Conrad
DIRECTOR: Robert Aldrich

How is it that we wind up rooting for Paul Crewe (Burt Reynolds)? When *The Longest Yard* opens, he is shown swilling down liquor and slamming his girlfriend against the wall. He makes off with her Maserati and drives it into a lake. He slugs a cop and is sent to prison where we learn he is a former NFL star who was banished for shaving points.

Clearly, this is not a nice guy. He is an abusive, selfish lout with no sense of decency or morals. So why is it that by the end of *The Longest Yard*, we're on our feet rooting for him?

Two reasons. One: the warden and the guards at Citrus State Prison are so evil, they make Crewe and his fellow convicts look almost saintly by comparison. Two: Burt Reynolds plays Crewe with a raffish charm that wins us over. We're willing to believe he has a good heart, even if there is plenty of evidence to the contrary.

Director Robert Aldrich did a similar good-guy/bad-guy flip-flop with *The Dirty Dozen* in 1967. The two films are very similar. You might even call *The Longest Yard* "The Dirty Eleven" because it uses the same thematic playbook: hardened criminals come together, develop a sense of pride and mission, spit in the eye of cynical authority figures, and prove themselves on the field of battle—in this case, the gridiron.

Reynolds earned good reviews two years earlier for his performance in *Deliverance*, but *The Longest Yard* is the film that made him a star. In her *New York Times* review, Nora Sayre wrote, "[Reynolds] manages to make even the prison chains look like fashionable men's wear, as though he were modeling the latest in metal accessories." In other words, the man is very cool.

"I was very proud of the film, extremely proud," Reynolds told Bob Keisser of the *Los Angeles Herald-Examiner*. "It was a real entree for me and the rest of my career. We wanted it to be authentic. We wanted to get as much realism as possible."

Why'd you drive her car into the bay?
—POLICE OFFICER

Couldn't find a car wash. —PAUL CREWE

In the film, upon arriving at Citrus State Prison, Crewe meets Warden Hazen (Eddie Albert) who greets him with a smile and the warm words, "It's nice to have you here." Crewe responds with a grunt and a limp handshake.

Hazen tells Crewe of his passion for football. He proudly shows off the trophies won by the prison guards, a beefy collection of sadistic brutes who amuse themselves by kicking around the prisoners when they're not practicing the power sweep. The guards play in a semi-pro league and while they've won a lot of games, they have never won the national championship. The warden feels Crewe can put them over the top if he agrees to join the coaching staff.

"What do you think of semi-pro football, Mr. Crewe?" Hazen asks.

Crewe smirks. "Semi-pro is a joke," he replies.

The look on the warden's face is priceless. Crewe might as well have kneed him in the groin.

Hazen comes right to the point: Crewe can have easy duty and a (relatively) pleasant stay at Citrus State if he agrees to coach the team.

"I'm not interested in football anymore," Crewe says dismissively. "I just want to do my time and get out of here."

Seething, Hazen tells the guard, "Get him out of here."

Crewe is tossed into the cell block where he is not exactly welcomed with open arms. As one prisoner—a resourceful scrounger known as Caretaker (James Hampton)—tells him, "Most of these old boys don't have nothing. Never had nothing to start with. But you had it all. Then you let your teammates down and got yourself caught with your hand in the cookie jar."

"Oh, I did, did I?" Crewe says.

"I ain't saying you did or you didn't," Caretaker replies. "All I'm saying is you could've robbed banks, sold dope or stole your grandmother's pension checks and none of us would've minded. But shaving points off a football game? Man, that's un-American."

To make Crewe pay for his disrespect, the warden orders him to assemble a football team from the prison population and play a game against the guards. Crewe wants no part of it, but he gives in when Hazen promises him an early parole.

Crewe begins recruiting players. He enlists a giant weightlifter (Richard Kiel), a mass murderer (Robert Tessier), a dim-witted country boy (Wilbur Gillian) and others, all of whom welcome the chance to play against the guards. It is their one shot at payback for all the abuse they suffered.

Aldrich hired dozens of real players and divided them between the teams. Among the prisoners are Pervis Atkins, a halfback for the Los Angeles Rams; Sonny Sixkiller, a quarterback at the University of Washington; and Ernie Wheelwright, a fullback with the New Orleans Saints. The guards include former Minnesota Viking Joe Kapp (Walking Boss) and Rams linebacker Mike Henry (Lt. Rasmussen). Green Bay linebacker Ray Nitschke had no acting experience, but he has a featured role as Bogdanski, the biggest and meanest of the guards.

"Originally, Ray was supposed to be like the other [players] and be in the background," said Dave Robinson, Nitschke's Green Bay teammate who passed up a chance to be in the film to play another season in the NFL. "But when Ray met with the director, he took out his false teeth and put them on the table. Ray was really scary looking with his teeth out.

"The director said, 'This is the guy I want. And no makeup. I don't want any makeup. I want to be able to see all those scars.' So he wound up giving Ray quite a few lines. For a guy who wasn't an actor, Ray did a good job. You talk to anyone who saw the movie and they all remember Ray's character."

But most of all, people remember Reynolds as Paul Crewe. A former running back at Florida State, Reynolds has the fluid stride of a big-time player. He also has the attitude. Whether he is smooth-talking his fellow prisoners, facing off with the warden, or making love to the warden's secretary (Bernadette Peters in a memorable beehive hairdo), he is pitch-perfect as the cocky smart-ass who, even in handcuffs and leg irons, still swaggers like a one-time NFL MVP.

The Longest Yard builds to the big game between the guards and the prison team, which calls itself "The Mean Machine." With Crewe flashing his old moves at quarterback, the prisoners play the guards on even terms for a half. During intermission, the warden orders Crewe to tank the second half. He doesn't want the cons pulling off an upset and shaking the prison's power structure.

"I can't do that," Crewe says.

"You've done it before," Hazen replies, reminding Crewe of his earlier point shaving and threatening to implicate Crewe in the arson death of Caretaker. Crewe had nothing to do with it, but Hazen is ready to frame him for the crime. "You'll be with us until you're old and gray," Hazen says, telling Crewe he wants the guards to win by no less than 21 points.

Reluctantly, Crewe agrees, but he gets a promise from the warden that the guards will ease up once they have the game in hand. Hazen gives his word, but of course he has no intention of honoring it. Crewe throws two interceptions and fumbles a snap allowing the guards to build a three-touchdown lead. But at that point the guards, on Hazen's orders, start playing even more viciously and send several prisoners off the field on stretchers.

Realizing he has been double-crossed, Crewe turns the tables on Hazen. He leads a fourth quarter rally that climaxes with Crewe scoring the winning touchdown on the final play, a one-yard run that takes one minute and 50 seconds of super, super, super slow-motion to complete. It truly is *the* longest yard.

Aldrich does not shortchange the football action. The big game lasts 47 minutes, more than a third of the film, but we aren't complaining. It neatly weaves comedy and drama and ends, fittingly, with Crewe handing the game ball to the ashen-faced warden.

"Stick this in your trophy case," he says with a go-to-hell grin.

Crewe disappears up the tunnel as the credits roll. We can only assume he is sent directly to solitary confinement where the warden sees to it that he spends the next 50 years living on bread and water. But we also get the feeling that Crewe would say it was worth it.

★ **SPORTS-ACTION GRADE:** A. Even Brian Dawkins, a seven-time Pro Bowl safety, feels it looked and sounded like the real thing. "I love the tough football scenes," Dawkins said. "That's the way football ought to be played."

8 THE HUSTLER (1961–NR)

SPORT: Pool ★ STARS: Paul Newman, Jackie Gleason, George C. Scott
DIRECTOR: Robert Rossen

There was a time in this country when pool halls were as plentiful as McDonalds. They were populated not by college kids looking for a Friday night pick-up, but by murky characters of all ages angling to take your money over a game of nine-ball.

The Hustler returns you to those days. It is set in the early sixties, but really more closely resembles the Great Depression, when there were 40,000 billiard parlors across the country. There weren't glossy state lotteries to bet on back then. Working-class folks didn't shuttle to Las Vegas for $99 on the weekends. You wanted to wager a few bucks, you went downtown and picked up a cue stick and some chalk.

Paul Newman stars as Fast Eddie Felson, a young hustler from California. In the opening scene at a dingy watering hole, he sinks a nearly impossible shot, misses it the second time, then feigns drunken bravado and bets the predatory clientele he can make it one more time. Of course he does, departing with a fat wad of crumpled bills.

But Eddie is aiming for bigger game. He is hunting for Minnesota Fats (Jackie Gleason), the whale of a legend known in every seedy pool hall in the nation. The two meet soon enough and lock horns in a $200-a-game showdown.

What ensues is one of cinema's great scenes, backed by a jumpy jazz score, in which two great actors strut their stuff. Director Robert Rossen devotes a full 20 minutes to this one claustrophobic battle, trusting his audience's patience in a way that no director would today.

Initially, Fats controls the table, and Eddie is in awe. "He is great, that Fat Man," he says. "Look at him move, like a dancer. And those chubby fingers, that stroke, like he's playing a violin."

But Eddie finds his game and starts winning. He struts like Deion Sanders, boasting, "Fat Man, I've got a hunch that it's me from here on in...I've dreamed about this game every night... You know, this is my table. I own it."

The marathon match goes back and forth. After eight hours, Eddie is up $1,000. The stakes rise. After 25 hours, he's up $18,000. His manager (Myron McCormick) wants his player to leave ahead, but Eddie isn't here to take Fats' money—he's here to take his crown.

"The game ends when Fats says it does," he says.

Of course Minnesota Fats didn't become a legend by losing. More than a full day into the marathon, he sends out for booze, knowing that Eddie will follow suit. The veteran can handle his liquor, the kid cannot. Fats never sweats, never loosens his tie, never loses composure. Eddie, meanwhile, succumbs to drunkenness and exhaustion. After 40 hours, Fats has all the money and Eddie is broken—his skill undercut by his own demons and weaknesses.

The rest of the movie leads up to the inevitable rematch, traveling down a road of smoky flophouses and seedy bars. Eddie shacks up with the wrong dame, Sarah Packard (Piper Laurie), a broken woman who goes to college two days a week and drinks on the others. He gets his thumbs cracked by dock workers who spot him as a hustler.

He dumps one manager and finds another, the evil Bert Gordon, brilliantly played by George C. Scott as Satan in sunglasses. Gordon berates his new charge by calling him a loser and saying he lacks "character." The insults will either crush Eddie or prompt him to play better in anger. Gordon doesn't care either way; he's just as happy to bet against his man.

There's a pivotal seduction-rape scene between Gordon and Packard (off-camera, of course—this was 1961) that ends with Sarah's suicide at a Louisville hotel. That cements Eddie's transition into a hardened cynic as he returns to New York for the rematch with Fats.

The second showdown with Newman and Gleason is even better than the first. Both men glide around the table, shooting fast, as world-class players really do. Gleason's Minnesota Fats doesn't have a lot of dialogue, but his grace and sad, expressive face are brilliant. Newman's Eddie, meanwhile, is in a zone borne of rage.

"Fat Man, you shoot a great game of pool."
—EDDIE

"So do you, Fast Eddie."
—FATS

"How can I lose?" he shouts, at Gordon more than Fats. "It's not enough that you just have talent. You have to have character, too. Right? I sure got character now. I picked it up in a hotel room in Louisville."

You can watch for yourself to see how it all ends, but the final exchange between the two main characters provides perfect punctuation:

"Fat Man, you shoot a great game of pool."

"So do you, Fast Eddie."

The Hustler was a minor box office success when it came out, spurring a brief resurgence in pool around the nation. It was more so a critical hit, winning Academy Awards for both Art Decoration Black-and-White and Cinematography Black-and-White. It was nominated for seven others, including Best Picture (losing out to *West Side Story*). All four of the major characters (Newman, Gleason, Scott, and Laurie) were nominated for acting Oscars, although none won.

We would argue that Newman's performance in *The Hustler* makes his personal Mount Rushmore of excellence, along with those from *Hud*, *Cool Hand Luke*, and *The Verdict*.

It's interesting to note that although he was nominated for nine acting Oscars over his career, the only one he won was for playing Fast Eddie Felson the second time around, in 1986's *The Color of Money*. Consider that an honor handed out 25 years too late.

In *The Color of Money*, Newman plays a wiser and wizened version of Eddie. A young Tom Cruise, fresh off of *Top Gun*, plays his protégé. It's not bad but, like most sequels, it doesn't match up to the original.

🔊 **CASTING CALL:** Director Rossen said over the years he had hoped Frank Sinatra would play the role of Fast Eddie Felson—reviving his strung-out performance from *The Man with the Golden Arm*—but Old Blue Eyes was unavailable. Jack Lemmon turned down the gig, and Cliff Robertson auditioned but lost out to Newman. Lemmon, Robertson and Newman were all 35 years old at the time of shooting.

CADDYSHACK (1980–R)

SPORT: Golf ★ STARS: Bill Murray, Chevy Chase, Rodney Dangerfield, Ted Knight DIRECTOR: Harold Ramis

9

You probably couldn't get rivals Tiger Woods and Phil Mickelson to agree on much. But the top two golfers in the world see one thing the same: Both regard *Caddyshack* as the best movie ever made.

"It's my favorite by far," says Tiger. "It's just a classic." Woods so adores the film that he channeled greenskeeper Carl Spackler in a 2007 American Express commercial paying homage to many of *Caddyshack's* touchstone scenes.

Says Mickelson, "Every other movie is a couple of notches down." What most impresses Lefty is how much of *Caddyshack*—including the Dalai Lama scene, his favorite—was ad-libbed. "That was pretty amazing. It makes me more enchanted with the movie when I find out how it was made and the talent that was required."

The two Masters champs are not alone. Every weekend hacker adores *Caddyshack* and has memorized dozens of classic lines from the film. In fact, we can't think of a single movie this side of *The Godfather* that has so influenced the vernacular of the American man.

Who doesn't quote from Bill Murray's "Cinderella story" monologue ("It's in the hole!") at least once a round? How often have you insulted a partner's garish clothing with Rodney Dangerfield's great putdown, "When you buy a hat like that, I bet you get a free bowl of soup, huh? It looks good on *you* though." Heck, you can even go to mini-golf and find some guy steadying his nerves over a putt while his taunting buddies yell, "N-n-n-n-ooooonan!"

knock you dead."

Okay, count us in.

This is the part of the chapter where we usually lay out the plot synopsis. But we're going to make a leap of faith here and assume you've already watched *Caddyshack* countless times. So, rather than lay out the ABCs of something you already know ("Danny Noonan wanted a golf scholarship . . ."), we're just going to tell you 10 things we love about it:

1. Most of *Caddyshack,* as Mickelson notes, was improvised. We're not just talking about a few lines here and there. Entire scenes were added on the fly following daily brainstorming sessions among first-time director Harold Ramis, head writer Brian Doyle-Murray (who also plays Lou, the caddy master) and several of the lead actors. The best addition, perhaps, came after

"I don't play golf for money." —TY
"What are you, religious or something?" —AL

Nearly 30 years after it was made, *Caddyshack* is that rare movie you can watch repeatedly and still glean new nuggets—hmmm, how come we never noticed how *funny* Doctor Beeper is? It boasts three great comics—Murray, Dangerfield, and Chevy Chase—at the top of their game (and Ted Knight is no slouch himself). It satirizes class warfare, but also offers poop and booger jokes. All that plus pointless nudity.

And somehow, amazingly, critics panned it upon release. "Immediately forgettable," sniffed Vincent Canby of the *New York Times.*

Time has proven Canby wrong, don't you think?

"If you're still at the age when farting and nose-picking seem funny," offered the *Time Out Film Guide,* "then *Caddyshack* should

Ramis realized there was no joint screen time between his two biggest stars—Chase and Murray. He hurriedly wrote a premise in which Ty Webb (Chase) sends an errant shot into Spackler's (Murray) ramshackle home/garage and wanders in to retrieve it. Everything after that—the inane conversation about cutting Judge Smails' hamstring, the joint smoking ("Cannonball!"), Spackler spitting on his own living room floor—was ad libbed. The scene has nothing to do with the overall plot, but is completely hilarious.

2. "So we finish the 18th and he's gonna stiff me. And I said, 'Hey, Lama, how about a little something, you know, for the effort?' And he says . . . 'There won't be any money, but when you die, on your deathbed, you

will receive total consciousness.' So I got that going for me, which is nice."

3. The film was inspired by Brian Doyle-Murray's memories of working as a caddy from age 11, as well as writer/producer Doug Kenney's recollections of growing up as the son of a country club tennis pro. Both Murray and Ramis also worked as caddies in their youth. Nearly all of the characters, from the despotic Judge Smails (Knight) to the sexy Lacey Underall (Cindy Morgan) to the fossilized Havercamps ("That's a peach, Hon.") were based on real people they encountered. So, too, were most of the setups, including the infamous Baby Ruth scene.

4. Judge Smails: "You know, you should play with Dr. Beeper and myself. I mean, he's been club champion for three years running, and I'm no slouch myself."

Ty Webb: "Don't sell yourself short, Judge. You're a tremendous slouch."

5. There is a lot of creative tension at work here. First off, Chase and Morgan (a former model for Irish Spring soap, by the way) did not get along at all—which made their sex scene a challenge. "The bottle of massage oil he spilled on my back certainly wasn't in the script," Morgan said. "You can look in my eyes and almost hear me say, 'You --- -- - -----, now the gloves are off.'"

A few others in the cast didn't approve of the atmosphere on the set, which Clark Collis of the *London Daily Telegraph* described as "an 11-week party, legendary even by Hollywood standards." Knight, particularly, was troubled by what executive producer Jon Peters called "nightly debauchery."

6. Al Czervik (Dangerfield) owns the greatest golf bag in history. In addition to holding about 20 clubs, it boasts a beer dispenser, a stereo system that plays Journey, a phone (in the days before cell phones), a mini television, an ejector button that prompts all of his clubs to fly into the air, and a putter with sonar tracking. We want that bag.

7. Judge Smails christening his yacht: "It's easy to grin / When your ship comes in / And you've got the stock market beat. / But the man worthwhile / Is the man who can smile / When his shorts are too tight in the seat."

8. *Caddyshack* is the final work of co-producer Doug Kenney, a comic genius. Kenney co-founded the 1970s satire magazine *National Lampoon* and wrote *National Lampoon's Animal House*, which became the top-grossing comedy in history upon its release in 1978.

Kenney was a hugely important influence in moving humor from its tame form in the early 1970s, to the subversive, sophomoric, and edgy form that emerged with the *Lampoon*, the early years of *Saturday Night Live,* and the sometimes vulgar, often hysterical films that followed *Animal House*. One month after *Caddyshack* opened, his body was found at the bottom of a cliff in Hawaii. The death was ruled accidental. He was 32 years old.

9. Al Czervik to Judge Smails' wife: "Oh, this your wife, huh? A lovely lady. Hey baby, you must've been something before electricity. . . . You're a lot of woman, you know that? Yeah, wanna make 14 dollars the hard way?"

10. Call it a multiple tie. The lead song, *I'm Alright*, by Kenny Loggins. . .Spackler overly enjoying the golf ball washer. . . Spaulding Smails drinking the leftover liquor and barfing into Dr. Beeper's

moon roof. . . "Elihu, will you loofah my stretch marks?". . . Spaulding picking his nose. . . "Hey everybody, we're all gonna get laid!". . . The water ballet scene during the lifeguard's swim session. . .

The Napoleon Dynamite-looking caddy who flashes eight fingers when Bishop Pickering says, "Put me down for a five.". . . "You'll get nothing and like it!"

DRY OFF ON RUG

10 NORTH DALLAS FORTY (1979–R)

SPORT: Football ★ STARS: Nick Nolte, Mac Davis, Bo Svenson
DIRECTOR: Ted Kotcheff

Football fans see only what happens on Sunday. They see the game, the cheering crowd, the adulation. They don't see Monday when the real price is paid by the players who must put their battered bodies back together. They don't see the politics of the front office, the mind games played by the coaches, and the unspoken fear that grips each player who knows that any game could be his last.

Wisely, that's where *North Dallas Forty* begins. It doesn't start with Phil Elliott (Nick Nolte) scoring the winning touchdown for the North Dallas Bulls. It doesn't show him

signing autographs for his adoring fans. Rather, the film opens with Elliott, the morning after the game, waking up on a blood-stained pillow, wincing as he feels the effects of every hit he took the day before.

Elliott gulps some pain pills, washes them down with beer, and slowly drags himself down the hall to the bathroom. Every few steps he reaches for an aching body part and flashes back to a moment in the game: a hit that flipped him in the air, an elbow that hit him in the face, a tackle that twisted his head around. He remembers each blow. He will live with them all week, if not longer.

Finally, Elliott lowers his scarred and swollen body into a warm bath, and just when you are wondering why anyone would put

The Cowboys, on whom Gent based the story, blasted the film. Team president Tex Schramm called it total fiction. Pete Rozelle, then commissioner of the NFL, dismissed it as

"Alcohol and fear are a potent combination."
—PHIL

"What are they afraid of?" —CHARLOTTE
"Falling on their asses in Chicago Monday night."
—PHIL

himself through this, he begins to smile. Then he flashes back to the one-handed catch he made to win the game. Suddenly, the pain is gone. Elliott is back in that moment and that moment makes it all worthwhile.

It is a powerful scene, and it lets you know immediately that *North Dallas Forty* is not going to be your typical rah-rah football film. Based on the best-selling novel by former Dallas Cowboys receiver Pete Gent, *North Dallas Forty* strips away the sugar-coating on the NFL to reveal a world that is brutal and dehumanizing. There are excesses in the film, some of the characters are exaggerated stereotypes, but there is a lot of truth here as well.

"Seeing Nick Nolte trying to get out of bed the morning after a game, I said, 'That's me,'" said Cris Collinsworth, who played eight seasons as a wide receiver with the Cincinnati Bengals. "I've tried to explain it to people, what it's like to go through that, and most of them have no clue. But that scene brings it home.

"I played when the fields were all Astroturf and you had all the skin scraped off your arms and elbows. The morning after a game, I was stuck to the sheets because of all the turf burns. I was a skinny guy, and I'd be so beat up and sore I couldn't stand up most Mondays. I'd literally roll out of bed onto my knees and crawl to the bathroom."

"just a movie." But many players said otherwise. Jeff Severson, who played six years in the league, said the film had "the authenticity of a newsreel." Carl Weathers, who played two seasons in Oakland before pursuing an acting career, said, "I think it's the film that comes closest to capturing the drama of what goes on in a locker room."

"Pete got the alienation part right, how the players hate the system and the crap that goes on," said Fred Dryer, who played in the league for 13 seasons before becoming an actor. "I never sat at my locker and talked with [teammate] Jack Youngblood about how to defend the 44 trap. I talked about kidnapping the general manager, taking him into the desert and burying him. That's what we talked about on the bus and walking to practice.

"One day we were taking a break in the shade. The coach sounded the horn and said, 'OK, let's go. Back to work.' I said to Jack, 'Don't you hate yourself for doing this? Here we are, trapped inside this world. Isn't it a shame?' Jack said, 'Yeah, you're right.' But then we got up and went back to practice. We hated it and we hated ourselves for doing it, but we did it. People would say, 'Why? The money, right?' No.

"Pete touched on it. There are those once-in-a-lifetime moments, those explosions of talent when it all comes together and you do something truly beautiful. For Pete, it was a catch. For me [a defensive end], it might be a tipped-ball interception. But those are the moments you live for. That's why you put up with the other stuff. It's the price you pay for the chance at one of those moments."

North Dallas Forty focuses on Elliott, a veteran wide receiver (Gent based the character on himself) who still enjoys playing the game, but hates the rules put in place by owners and coaches. Elliott is a rebel who smokes pot, mutters wisecracks under his breath in team meetings, and sleeps with the fiancee of the owner's brother.

Head coach B.A. Strothers (a stern Tom Landry figure played by G.D. Spradlin) has benched Elliott because he doesn't like his attitude, but he won't cut him because Elliott can still make a big play when called upon.

When Strothers accuses him of immaturity, Elliott replies, "I've scored five touchdowns coming off the bench. That's pretty mature." The coach leaps from his chair. "You scored five touchdowns?" he says. "We let you score those touchdowns. It's a team game, it's not about the individual. When are you going to learn that?"

Elliott's best friend is quarterback Seth Maxwell (Mac Davis playing a character loosely—or not-so-loosely—based on Don Meredith). Maxwell is a good old boy who shares Elliott's distain for management, but unlike Elliott, he knows how to hide his feelings behind a wink and a smile.

"You better learn how to play the game," Maxwell tells Elliott, "and I don't mean the game of football. Give 'em what they want. Hell, I've been fooling them bastards for years."

"I can't do it, Seth," Elliott says. "You start acting like somebody else and pretty soon that's what you become, somebody else."

Elliott meets Charlotte Caulder (Dayle Haddon) at a team party. She is standing off to the side, appalled by the raunchy goings-on. Elliott, who is something of an outsider himself, is drawn to her. At one point, the hulking Jo Bob Priddy (Bo Svenson) hoists a terrified woman over his head while the other players and groupies cheer him on.

"Jo Bob is here to remind us that the biggest and the baddest get to make all the rules," Elliott says.

"I don't agree with that," Charlotte replies.

"Agreeing doesn't play into it," Elliott says.

Disgusted, Charlotte heads for the door, only to be intercepted by Jo Bob, who announces in graphic terms what he would like to do to her. Elliott intercedes, but Jo Bob, fueled by drugs and booze, turns on Elliott and begins choking him. Maxwell has to rush in and save him.

"What do you expect?" Maxwell says. "These girls know what goes on at these parties."

"She didn't seem like that to me," Elliott says.

He follows Charlotte home. Reluctantly, she lets him in. She explains she went to the party with a friend. "I thought it would be fun," she said. "Obviously, I was wrong." She is unlike other women Elliott has known. She cares nothing about football. To her, it is an alien world, violent and barbaric. What she saw at the party only confirmed that. But in Elliott she sees a decent guy—complicated, conflicted, and maybe a little messed up, but basically decent. Soon, they are in bed together.

The film plays out over one week. That's one of the weaknesses of the film compared to the novel. Gent's book spanned an entire season so all the violence, treachery, and debauchery was spread out over four months. In the film, it is compressed into seven days which makes it seem unrealistic. We don't doubt that Gent saw all the things he wrote about at various points in his five-year NFL career, but when he tries to cram a lifetime's worth of baggage into a two-hour screenplay, sometimes things get messy.

There are a few gratuitous shots at players who read the Bible and the team chaplain is portrayed as a hapless jock-sniffer, all of which is unnecessary. But when Gent sticks to the meat of the story, it is prime stuff.

The big game—the Bulls meet Chicago on Monday night television—is a savage affair with a Rambo-like body count. Jo Bob and O.W. Shaddock (ex-Raider John Matuszak) intentionally break the leg of Chicago's best lineman, Alcie Weeks (ex-Rams tackle Doug France). Wide receiver Delma Huddle (ex-Chiefs halfback Tommy Reamon) tears his hamstring and takes a vicious shot that shatters his cheekbone.

On the final play, Elliott catches a touchdown pass to pull the Bulls within one point of a tie, but the holder drops the snap on the extra point attempt and the Bulls lose to fall from championship contention. In the locker room Shaddock explodes when an assistant coach (Charles Durning) tells the players they lost because they didn't study the computer tendencies.

"That's all you give us, your computer tendencies and other ----------," Shaddock says. "You've got no feeling for the game."

"You're professionals," the coach says. "This is a business."

"Every time I call it a game, you say it's a business," Shaddock roars. "And every time I say it's a business, you call it a game."

Matuszak is no Olivier, but he delivers those lines with a passion that is totally convincing. You have to believe those are sentiments he felt many times in losing locker rooms and this scene was a cathartic experience, not just for Matuszak but for hundreds of NFL players watching and thinking, "Right on."

"Most of us [players] saw it as a movie about a sport that, when you look at it, is kind of absurd," Collinsworth said. "We're a bunch of adult men out there trying to kill each other. *North Dallas Forty* raises that question, 'Are these guys crazy?' It's a question I think we all asked ourselves at one time or another."

🍎 **BET YOU DIDN'T KNOW:** Tom Fears, a Hall of Fame receiver with the Los Angeles Rams and a head coach with the New Orleans Saints, was the technical adviser on the film.

☞ **DON'T FAIL TO NOTICE:** Stallings, the lineman who is cut for missing a block, is Jim Boeke, who played guard for the Cowboys. Boeke had a similar experience, jumping offside at a critical moment in the 1966 NFL championship game. He was traded to New Orleans one year later.

🎬 **PIVOTAL SCENE:** When the coach lectures Elliott about giving something back to the game, Elliott lists his many injuries.

"There's pieces of me scattered from here to Pittsburgh on these football fields," he says. "Isn't that giving something back to the game?"

Strong stuff and effectively delivered by Nolte.

⭐ **SPORTS-ACTION GRADE:** B. A dozen former NFL players appear in the cast, so the stances and moves look authentic. The offensive line copies the Cowboys with linemen straightening up before dropping into a three-point stance. The one problem: Mac Davis throwing the football. If he's supposed to be Don Meredith, Dandy Don has grounds to sue.

11 FIELD OF DREAMS (1989–PG)

SPORT: Baseball ★ **STARS:** Kevin Costner, James Earl Jones, Ray Liotta
DIRECTOR: Phil Alden Robinson

There are many who regard *Field of Dreams* as the greatest sports movie ever made—a wondrous allegory about life, fathers and sons, lost dreams, and a bygone America, all staged in an Iowa cornfield. It is, to reverential devotees, the modern version of *It's a Wonderful Life*.

"That movie had a mystical feel to it," says Pete Carroll, head coach of the Seattle Seahawks. "It was very inspiring."

And then there are those who regard it as so much pap. *Premiere* magazine listed it among the 20 most overrated films of all time. "A gooey fable," declared Peter Travers of *Rolling Stone*.

Bill Simmons of ESPN wrote, "I think the world is separated into two kinds of people—people who loved *Field of Dreams*, and people who don't have a heart. If I were dating a woman and she said she didn't like *Field of*

Dreams, I'd immediately dump her."

We agree. Certainly, Simmons offers a better method of judging a potential mate than the unlock-the-car-door test in *A Bronx Tale*. Anyone—man, woman, child—who isn't touched by the innocence and fantasy of *Field of Dreams* is too much a grinch for us.

Writer and broadcaster Peter King, who has witnessed thousands of inspiring moments over his career, called *Field of Dreams* the most emotional sports movie he has ever seen.

"Every kid who grew up playing baseball at some point played catch with his father," says King. "You don't think of it as an emotional experience. But when your dad is gone, you do think of it as an emotional experience. Every time I see that movie, I bawl. I can't help it."

So why doesn't it crack our Top Ten? Because, great as it is, that sweetness sometimes seeps into sappiness. Occasionally, the movie becomes just a tad too reverential for our taste—worshiping that deity called baseball. Moses saw a burning bush, Ray Kinsella (Kevin Costner) sees a vision of a baseball diamond in his cornfield. With a chorus of angels singing in the background, just in case you missed the subtlety.

Plus this: The entire movie leads up to the final seven-minute scene, the climactic reconciliation between Ray and his long-estranged dad, John Kinsella (Dwier Brown). The two had parted ways—emotionally and physically—when Ray was a teen, and never squared matters before Dad's early death. Years later, thanks to Shoeless Joe Jackson (Ray Liotta), they get that second chance out in the magical Iowa ball field.

We are told that Dad was a damned good catcher in his day. Now, as father and son meet again, there's some talk about Iowa and heaven and how easily the two are confused. As we watch, we're feeling the lump growing in our throat.

"Hey Dad," says Ray. "You wanna have a catch?"

We're welling up at this point.

"I'd like that," says John, as the music nears crescendo.

Ray picks up his glove, and we're totally verklempt. Searching for Kleenex in the pocket.

So Ray throws, in that convincing Costner-as-Crash-Davis overhand motion. Dad catches the ball, and then . . . well, he executes the feeblest toss this side of Biddy Ball. Pushes it like a nine-year-old girl.

Almost ruined the movie right there.

Okay, call us sticklers. We apologize. We don't want to get bogged down in the negative. As we said, we're willing to buy into the overall vision of *Field of Dreams*. Once upon a time,

Hollywood regularly produced this kind of movie—usually starring Jimmy Stewart. Nowadays, such flights of fancy are too rare.

Field of Dreams has been parodied so many times that it's easy to forget the original premise: An Iowa farmer, still wracked by guilt over that suspended relationship with his dad, is told by voices to plow under his field and build a baseball diamond. This leads to the appearance of his father's hero, the disgraced Shoeless Joe.

But it does much more than that. The voices then tell Ray to enlist reclusive writer Terence Mann (James Earl Jones, playing a character based on author J.D. Salinger), and track down Minnesota doctor Archibald "Moonlight" Graham (Burt Lancaster), who never got his full shot in the majors. And it concludes with our hero gaining peace of mind, closure with his father, and financial success far beyond his imagination.

Director Phil Alden Robinson never really tries to explain the far-fetched plot twists, which is smart. Instead, he relies on the audience leading with its heart. "This story teaches you that you've got to have dreams," he said in a *Newsweek* interview soon after its release, "but you should keep yourself open to the possibility that a fork in the road may lead you to something just as good."

It all works. Well, almost all. We just wish they had taught the dad how to throw.

BET YOU DIDN'T KNOW: The movie was shot in 1988, as Iowa was going through its worst drought since the Dust Bowl of the 1930s. Producers spent $25,000 watering the cornfield to keep it from dying. "We had the only corn for 100 miles in any direction," noted director Robinson.

12 MILLION DOLLAR BABY (2004–PG-13)

SPORT: Boxing ★ STARS: Clint Eastwood, Hilary Swank, Morgan Freeman
DIRECTOR: Clint Eastwood

Million Dollar Baby is the most honored sports movie of all time. It was nominated for seven Academy Awards and swept four of the top prizes—Best Picture, Best Director (Clint Eastwood), Best Actress (Hilary Swank), and Best Supporting Actor (Morgan Freeman). It dominated the Oscars in a year when most predicted Martin Scorsese's lavish film *The Aviator* would clean up.

There must have been a lot of awkward foot-shuffling in the audience when the awards show turned into a *Baby* parade. Most of the big-name producers had the script on their desk at one time or another but rejected it. It was passed around to all the major studios. The executives would read a few pages and groan, "What, another boxing movie?" The project was tossed in every trash can in town.

Al Ruddy, who produced *The Godfather*, loved the story, but for four years he tried to find backers without success. "I couldn't get anybody interested," Ruddy told *Sports Illustrated*, "and I'm talking about people who are friends of mine, people I've done business with for years. They'd tell me, 'Who wants to see a movie about two old grizzled guys and a girl fighter?' "

Finally, Ruddy found a believer in Clint Eastwood. "It's a downer," Eastwood said after reading the script, "but, God, it's gorgeous."

A year earlier, Eastwood proved he could take a painful subject and turn it into gold. His *Mystic River*, a film that dealt with death and child abuse, was nominated for six Academy Awards. Eastwood convinced long-time backers Warner Brothers and Lakeshore

Entertainment that he could have similar success with *Baby*.

On Eastwood's word, the two studios agreed to split the $30 million production cost. Eastwood, one of Hollywood's most efficient directors, completed the shoot in 37 days and brought the film in under budget. It wound up earning more than $200 million at the box office, in addition to winning two Golden Globes, two Screen Actors Guild awards and a New York Film Critics Award, as well as the four Oscars.

Kenneth Turan of the *Los Angeles Times* called *Baby*, "The director's most touching, most elegiac work yet . . . a film that does both the expected and the unexpected [with] the emotional daring of the great melodramas of Hollywood's golden age when films considered it a badge of honor to wear their hearts on their sleeves.

"What can be said about Clint Eastwood that hasn't been said before?" Turan wrote. "That he is American film's last and best classicist, a 74-year-old director who's aged better than a *Sideways* Pinot Noir? That his increasingly fearless and idiosyncratic choice of material has made him more of an independent filmmaker than half the people at Sundance? That he continues to find ways to surprise audiences, yet remain inescapably himself? It's all true, and never more so than in Eastwood's latest, *Million Dollar Baby*."

The screenplay was written by Paul Haggis, who combined two short stories authored by F.X. Toole, the pen name of veteran boxing cornerman Jerry Boyd. They were originally published as part of a collection entitled *Rope Burns*. They have since been re-published borrowing the title *Million Dollar Baby*.

The focus of the story is Maggie Fitzgerald, a 31-year-old waitress and amateur boxer who is determined to turn pro. She wants Frankie Dunn (Eastwood), a crafty old trainer, to teach her the ropes. She introduces herself one night at the arena and Frankie does his best to brush her off.

"I don't train girls," he says, never breaking stride. He keeps walking and Maggie keeps walking with him, telling him she won her bout earlier that night. Frankie hates the whole notion of women boxing. He calls it "the latest freak show out there." He leaves Maggie at the arena, drives off, and assumes that's the last he will ever see of her.

The next day, Frankie is walking through the Hit Pit, the tattered old gym he owns in downtown Los Angeles, when he sees Maggie punching a heavy bag. "What are you doing?" he asks. Again, she tells Frankie she wants him to train her. Again, he refuses, but Maggie grins and we can see where this is headed. She is going to wear the old coot down.

If that was the whole of the story, it wouldn't be anything special. The reluctant mentor-irrepressible pupil story has been done a million times. The poor boxer-makes-good story has been done more than a million times. Just slapping one on top of the other like a slice of salami on a ham sandwich wouldn't make it a gourmet meal. But there are many layers to *Million Dollar Baby*, all beautifully rendered, and they combine to make the film truly memorable.

While Frankie is portrayed as a gruff and distant loner, there is a deep hurt in his life—his estrangement from his daughter. He writes to her every week and each letter returns to Frankie unopened. The look on his face as he picks up his mail and sees the latest letter in the stack, the quick flicker of pain you see in his eyes, is so sad. He never says a word but he doesn't have to. It is an eloquent silence.

We learn Maggie comes from a white trash background. She lost her father and did not have much of a relationship with her mother and siblings, so she took off for L.A. She is alone and looking for a father figure and Frankie has a void in his life due to the rift with his daughter—so it is clear they are destined to come together.

It starts with left jabs and right hooks, but becomes much more than that. Frankie and his sidekick Eddie "Scrap Iron" Dupris (Freeman) become Maggie's family, and Maggie becomes Frankie's surrogate daughter. The same material in the hands of less skillful performers could have become trite—especially with Freeman's character serving as an off-camera narrator, always a tricky device. But Eastwood, Swank, and Freeman are so good, we want to see them through every crisis.

Swank is superb as Maggie—earnest, gutsy, yet so emotionally needy and vulnerable. When Frankie finally caves in and agrees to train her, she flashes a smile that seems to fill her entire face. When she says, "Boxing is the only thing I ever felt good doing," we believe her. As good as Swank was in her other Oscar-winning role in *Boys Don't Cry*, she is even better here.

Freeman's character is a former heavyweight contender whose career was ended by a ring injury that left him blind in one eye. Frankie was his manager and he still blames himself for letting it happen. Now Scrap Iron works at the Hit Pit, sweeping the floors and emptying the spit buckets, but his real job is serving as a common-sense dispenser for the irascible Frankie.

When a delusional but harmless kid wanders into the gym and announces he wants to be the next welterweight champ, Scrap Iron indulges him, allowing him to hang around and hit the bags even though it is obvious it is all a fantasy. The kid—whom Scrap Iron calls "Danger"—has no money, and he shows up every day in the same shirt and sweatpants.

Frankie asks Scrap Iron if Danger is paying gym dues. Scrap Iron leans on his mop and gives Frankie his "Are-you-serious?" look. "The boy can't even afford pants, Frankie," he says. "How's he gonna pay dues?" Frankie wants to kick him out, but Scrap Iron says, "Aw, he ain't hurting nobody. Let him be."

Frankie grumbles, but Danger stays.

Eastwood was nominated for Best Actor and did not win, but he is the equal of Swank and Freeman, which is saying a lot. Turan calls Eastwood's performance "the most nakedly emotional of his 50-year career." Eastwood and Freeman established a comfortable on-screen chemistry in *Unforgiven*, another Oscar winner, in 1992 and they find the same smooth old-buddy groove here.

But the real heart of *Baby* is the relationship between Frankie and Maggie and what happens once Frankie begins to manage her career. He takes her burning desire, adds his years of boxing knowledge and turns the thirtysomething diner waitress into a contender. Maggie wants to move the process along quickly, but Frankie is the cautious type. (Early in the film, we see a heavyweight prospect named Big Willie Little leave Frankie because he wants to step up in class. Frankie doesn't think he's ready.)

Frankie, who has been denied an emotional connection to his own daughter, develops one with Maggie. He still is her trainer and 24/7 taskmaster, but he smiles more and he listens when she talks about her life and her family. He has a robe made for her with the words "Mo Cuishle" on the back. It is Gaelic for "My darling, my blood" although he doesn't tell her that until the very end of the film.

Maggie gets her title shot in Las Vegas against the fearsome Billie the Blue Bear, the WBA women's champion, played by Lucia Rijker, a real-life kick-boxer whose scowl and brooding eyes would have intimidated Mike Tyson in his prime.

We won't reveal what happens next because if you haven't seen the film there is no way to address it without spoiling everything. It is as emotionally wrenching as any film you are likely to see. The ending caused a considerable uproar when the film was released and there was speculation it might hurt the movie's Oscar chances. Obviously that proved not to be true, but the final act is widely debated among movie critics and fans alike.

Andrew Sarris of the *New York Observer* said, "No movie in my memory has depressed me more than *Million Dollar Baby*." Meanwhile, Roger Ebert of the *Chicago Sun-Times* called it "a masterpiece, pure and simple."

"A movie is not good or bad because of its content, but because of how it handles its content," Ebert wrote in calling it the best film of 2004. "*Million Dollar Baby* is classical in the clean, clear, strong lines of its story and characters and it had an enormous emotional impact."

CHEERS: Hilary Swank trained with a boxing coach for three months and packed on almost 20 pounds of muscle to play Maggie. She is simply great, balancing the poor waif Maggie (scraping up the pennies and nickels that pass for tips in the diner) with the ferocious Maggie we see in the ring.

BET YOU DIDN'T KNOW: In addition to producing, directing, and starring in the film, Eastwood also wrote the musical score.

CASTING CALL: Sandra Bullock was originally cast as Maggie Fitzgerald, but dropped out before the filming began.

GOOFS: When Scrap Iron tells Maggie the story about how he lost the sight in his eye, he talks about "going 15 rounds." It would have only been 15 rounds if it was a title fight and we already know Scrap Iron never fought for the title.

REALITY CHECK: We can't be sure because we weren't at the weigh-in, but we find it hard to believe that Maggie and Billie The Blue Bear are in the same weight class. Billie looks like a linebacker. Maggie is buff, granted, but she still looks at least 20 pounds lighter.

PIVOTAL SCENE: Maggie wins a bout out of town and on the way back to L.A., she asks Frankie if they can stop and visit her family. Maggie has been sending her mother and siblings a portion of her winnings ever since she started boxing.

Maggie finds her family hasn't changed at all. They are still redneck lowlifes—and ungrateful ones at that. Instead of thanking Maggie, her mother looks at the fresh bruise on her face and mocks her for boxing.

"People hear what you're doing and they laugh at you," the mother says.

Maggie leaves more determined than ever to win the championship, if only to prove something to her family.

SPORTS-ACTION GRADE: B. The boxing scenes are well-staged and Swank throws a nifty combination, but some of the bouts would have been stopped by a referee or ringside physician. Swank is allowed to continue fighting with a busted, bloody nose. That would not happen in any responsible arena.

13 ▶ THE PRIDE OF THE YANKEES (1942–NR)

SPORT: Baseball ★ **STARS:** Gary Cooper, Teresa Wright, Walter Brennan
DIRECTOR: Sam Wood

Most movies don't age well. The plots and the dialogue go stale. The clothes and the music become dated. The camerawork and sound isn't as good. This is especially true for sports movies. Today, the games are more sophisticated and the action is more visceral so it makes for better filmmaking.

But the truly great films retain their power and that is the case with *The Pride of the Yankees*. It is the only film in our top 15 that was made before 1960 and, really, it was an easy call. *The Pride of the Yankees* is a work of art and, as such, it is timeless. Black and white or color, it doesn't matter. The story of baseball legend Lou Gehrig, beautifully played by Gary Cooper, still touches your heart.

It is interesting to watch *The Pride of the Yankees* now and consider the forces that shaped it. The film was made in 1942 when the United States was reeling from the attack on Pearl Harbor. The war was not going well for the Allies. President Roosevelt was calling for Americans to demonstrate their courage and strength. *The Pride of the Yankees* offered Gehrig as a symbol of those qualities.

To drive home the point, the film opens with a prologue written by Damon Runyon describing Gehrig as a man who "faced death with the same valor and fortitude that has been displayed by thousands of young Americans on the far-flung fields of battle."

RKO Studios also factored the war into how Gehrig's story was told. Typically, a baseball film appeals to a male audience, but since most American men were overseas, producer Sam Goldwyn and director Sam Wood shifted the focus. They made it more of a love story so it would appeal to women who were buying most of the movie tickets in 1942. In that sense, it was a forerunner of *Bull Durham* and *Jerry Maguire*.

The original movie poster shows Eleanor Gehrig (Teresa Wright) melting in the arms of her husband (Cooper). She is in a dress, he in a suit. There is no sign of baseball in the illustration or the tag line ("It's the Great American Love Story."). When the film was re-issued after the war, the poster was changed to show Lou in a Yankees uniform swinging for the fences.

One can only imagine the impact the film had when it was first released less than a year after Gehrig's death on June 2, 1941. He was when he said his goodbye at Yankee Stadium and walked away, I couldn't believe it. Sad as it was, (the ending) made it very real and it made Gehrig very real to me.

"Years later when I played in Yankee Stadium (1969-71 with the football Giants), I walked around the field and thought about that. I thought about Gehrig, about the speech, about the emotion that must have been in the stadium that day. It was like daydreaming, but in black and white because the images in my mind were the images from the movie."

The Pride of the Yankees follows the standard rags to riches path. Gehrig is born in New York and attends Columbia, where he follows the wishes of his German immigrant parents and studies to be an engineer. He

"How's Tanglefoot? Has he come to yet or can't you tell?"—ELEANOR

"What does it mean when a girl says you remind her of a Newfoundland puppy?" —LOU

just 37 when he succumbed to amyotrophic lateral sclerosis (ALS), the crippling illness now known as Lou Gehrig's Disease.

Fred Dryer, who played 13 seasons in the NFL and later became an actor and producer in Hollywood, considers *The Pride of the Yankees* the finest sports film ever made.

"I saw it when I was 10 or 11 and it had a real effect on me," Dryer said. "I was living in California and I'd never seen major league baseball. The Dodgers didn't come west until 1958, so *The Pride of the Yankees* was my introduction to the mythology of baseball. It made the whole ballpark thing come alive for me.

"It hit me like a sledgehammer when Gehrig fell ill (in the film). I wasn't prepared for that. Back then, movies had happy endings. I thought, 'He's sick, but he'll come back.' When he didn't, also excels in sports. The Yankees offer him a contract which he signs to help pay the bills incurred by his mother's illness. She is furious when she is told.

"You gave up everything we planned and for what? For you to play ball?" she says. She wants Lou to be an engineer like his Uncle Otto back in the old country.

"People have to live their own lives," Lou says. "Nobody can live it for you. Nobody could have made a baseball player out of Uncle Otto and nobody can make anything but a baseball player out of me."

After a brief stop in the minors, Lou is called up to the Yankees. He takes over one day when the regular first baseman Wally Pipp complains of a headache. The rookie goes in the lineup, and that's the end of Wally Pipp. Lou plays the next 2,130 consecutive games to set a major league record and he teams with Babe Ruth to give the Yankees the greatest one-two punch in baseball history.

The Pride of the Yankees contrasts the personalities of the swaggering, larger-than-life Babe (played effectively by The Babe himself) and shy, salt-of-the-earth Lou. A cynical sportswriter covering the team mocks Lou. "He's a chump, a rube with a batting eye," he says. "He plays the game, goes back to the hotel, reads the funny papers, gargles and goes to bed." The Babe, obviously, cut a wider and more colorful swath.

Gehrig and Ruth combine to hit more than 700 home runs and they help the Yankees win four world championships. After The Babe is dealt away in 1935, Gehrig plays another four seasons, wins one home run title and grows into the role of team captain.

In 1938, Gehrig notices he is slowing down. He thinks at 34 he is just feeling the effects of middle age. But the following season, he goes 4-for-28 in the first eight games and takes himself out of the lineup. He goes to the hospital for tests and learns he has a nerve disease for which there is no cure.

"Is it three strikes, Doc?" Lou asks.

"You want it straight?" the doctor replies.

"Yeah," Lou says.

"It's three strikes."

The Yankees honor him at the stadium. That's where Lou gives the speech in which he refers to himself as "the luckiest man on the face of the earth." In the film, they change the speech slightly but it does not lessen the impact. Cooper delivers it with just the right feeling. He is emotional but not maudlin. Standing at the microphone in that baggy white uniform, Cooper actually seems to become Lou Gehrig.

The Pride of the Yankees was nominated for 11 Academy Awards, including Best Picture, Best Actor (Cooper), and Best Actress (Wright). It won one Oscar: Daniel Mandell for film editing. Cooper lost out to James Cagney (*Yankee Doodle Dandy*). The American Film Institute ranks *The Pride of the Yankees* #22 on its list of the top 100 most inspiring movies in American cinema. The "luckiest man" speech ranks #38 among greatest movie quotes.

There is some baseball in the film, but not a lot. In part, that is because the studio wanted the love story of Lou and Eleanor in the foreground. But there was another reason for limiting the baseball action: Gary Cooper wasn't much of an athlete. The 6-foot-3 Cooper was actually bigger than Gehrig. The Iron Horse was 6-0 and 200 pounds in his prime. Cooper also projected the stoic strength that was so much a part of Gehrig's character. But the similarity ended there. On the field, Gehrig was a natural and Cooper was the unnatural. Cooper did not play sports and he had a lot to learn taking on the role of Gehrig.

Lefty O'Doul, who played 11 years in the big leagues, worked with Cooper, teaching him the basics. Cooper only felt comfortable swinging the bat right-handed. Gehrig, of course, was left-handed. O'Doul tried to turn Cooper around without success. Director Wood came up with a solution. He filmed Cooper batting righty and flipped the negative so on the screen it appears he is a lefty. To make it work, the lettering on Cooper's cap and uniform had to be reversed and when he hit the ball, he ran to third base, not first. It sounds comical, but it worked.

"I saw the movie as a teenager and I was so impressed," said Robin Roberts, the Hall of Fame pitcher who broke in with the Philadelphia Phillies in 1948. "I had three sports idols: Whizzer White, the football player who became a Supreme Court justice; Otto

Graham, who was a basketball and football star in college; and Lou Gehrig.

"When I got to the big leagues, I met a lot of people who knew Gehrig. I asked them about Gehrig because I was curious. They all said the same thing: 'What you saw in the movie, that was Lou. He was a great ballplayer, but he was also a great man.'"

☞ **DON'T FAIL TO NOTICE:** The bracelet Teresa Wright wears in the film is the bracelet Lou gave Eleanor on their fourth wedding anniversary. It has 17 medallions, each one representing a Yankee world championship or an All-Star appearance. The bracelet is now on display at the Baseball Hall of Fame in Cooperstown, N.Y.

14 HOOP DREAMS 1994–PG-13)

SPORT: Basketball ★ STARS: William Gates, Arthur Agee
DIRECTOR: Steve James

"People always say to me, 'When you get to the NBA, don't forget about me.' Well, I should have said back, 'If I don't make it to the NBA, don't you forget about me.'"—William Gates, in the final scene of *Hoop Dreams*

Hoop Dreams is a movie about the 99.99 percent of those kids who grow up *not* to become Kobe Bryant. It is about two adolescent boys who work and sweat and imagine themselves as the next NBA superstar. It is about the impossible dream of escaping a grueling inner-city destiny by devoting yourself to basketball.

It may also be the most powerful sports movie ever made. Because it is all true.

This 1994 documentary follows the lives of

William Gates and Arthur Agee, whom we meet as two 14-year-olds living in Chicago's Cabrini Green projects. From the jump, basketball is everything to them. They watch the NBA All-Star Game on TV and then try to imitate Michael Jordan's tongue-wagging moves on a neighborhood court that has weeds growing through the asphalt. Each boy is certain that stardom awaits him. "I'm going to get my dad a Cadillac," pronounces Arthur, "so he can cruise to my games."

Both teens, in fact, are talented enough to be recruited to St. Joseph's High School, a private academy in the Chicago suburbs most famous as the alma mater of Isiah Thomas. When Arthur gets to play a little one-on-one with the Detroit Pistons superstar at a summer camp at St. Joseph's, you see him sizing himself up for a bright future.

But it is William, more mature emotionally and physically, who makes the better early impression. As a freshman, he wins a spot on the St. Joseph's varsity squad and even starts at point guard, the first time that has happened since . . . well, you know who.

"I think I may have seen the next Isiah Thomas," Chicago sports talker Bill Gleason proclaims on the show *The Sportswriters on TV*. Young William, he tells the audience, has the right stuff.

If you think this story is just about teens playing ball, you are wrong. From the start the boys are encircled by various characters and hangers-on—some aiming to be helpful, some no better than parasites. There are talent scouts, coaches, journalists and recruiters, all of whom want their piece of the action. Before he is even 15, William is shown opening a stack of brochures from top colleges—Marquette, Rice, Illinois. He glances over glossy pictures of well-scrubbed coeds and grassy quads that must look enticing to a youngster seeking to escape

the concrete claustrophobia of the projects.

There is also family to deal with. William's older brother, Curtis, was once a talented athlete whose career peaked playing junior college ball. Bad breaks, he says. Uncoachable, others say. Now, Curtis sees his own failed possibilities being resurrected through his younger sibling. He is desperate for William to succeed. Admirable goal, except that it comes with the burden of unwanted pressure. "I always felt that Curtis should not be living his dream through me," William wisely notes.

Arthur must contend with tougher family issues. His father, Bo Agee, drifts between jobs and lapses into drug use. In perhaps the film's most wrenching scene, father and son play a good-natured game of one-on-one. They hug goodbye at the end and, as Arthur stays to practice his shooting, the camera follows Bo as he wanders to the shadows beyond the court and negotiates a crack deal. The camera pans back to Arthur and you see the pain in the 15-year-old's eyes.

Hoop Dreams pulls no punches. You watch the breakup of the 20-year marriage between Bo and Sheila Agee and learn later that he comes back to assault her. You peek inside the Agees' darkened apartment after the power is cut off by the electric company and they are living by the light of a lantern. But you also see hope: Sheila Agee enrolls in nursing school, aiming to pull herself from welfare. She graduates with the highest grade in the class, and you'll likely cry along with her at the graduation.

The power of *Hoop Dreams* is that it is at times disturbing, at times uplifting, and always compelling. Director Steve James and his crew shot more than 250 hours of film and edited it down to a 170-minute gut punch.

By sophomore year, the boys are going in separate directions. William is making honor roll and emerging as a player with college—and perhaps pro—potential. Because his family cannot afford St. Joseph's hefty tuition, a well-heeled alum appears as an angel to take care of his financial needs.

But not so Arthur. His family doesn't have the tuition money either. He sits home for two months, waiting—but, in his case, no fairy godmother shows up. And, he has fallen into disfavor with St. Joseph's coach Gene Pingatore. "Coach keeps asking me, 'When are you gonna grow?' " he notes.

Finally, he is forced to leave the posh academy and enroll at his local high school, located in a ramshackle neighborhood of broken houses and broken dreams. In a heart-rending moment, a wiser-than-his years Arthur suggests that Pingatore quit on him once his prospects as a star player faded. In the final irony, his family is forced to pay $1,300 in back tuition for St. Joseph's to release his transcript. The wealthy school that plucked him from the pond has thrown him back—with a debt to boot.

The coaches and faculty members at St. Joseph's were upset with their portrayal in *Hoop Dreams*, although every character gets his say on camera. School administrators later filed a lawsuit, saying they were promised the film was going to be a non-profit venture shown on PBS, rather than in theaters. We're not really sure why that would make a difference. Regardless, it grossed about $8 million, which ranks among the top 20 documentaries ever. Director James split the profits—about $200,000—with the Agee and Gates families.

By junior year, William is a nationally ranked prospect, getting recruiting mail from the likes of Duke, Georgetown and Michigan State. Everything seems perfect until he injures his knee in practice. The trip to the doctor will remind you of Boobie Miles' tragic moment in *Friday Night Lights*. The youngster, who has never been hurt before, believes he will spring right back. The doctor must tell him otherwise, and sitting for nearly a full season is a hard nut for an idealistic teen to swallow.

Still, William works his way back. The following summer, he wins an invitation to the prestigious Nike Camp in Princeton, N.J., where the top college coaches in the nation—Bob Knight, Mike Krzyzewski, Bobby Cremins—peer from the bleachers at the future of their programs. It all goes well, until William again injures his knee. As he writhes on the floor, you feel his pain.

Many of the coaches in *Hoop Dreams* are shown as manipulators, making promises to the young athletes while ignoring their parents' concerns about academic achievement. One coach who comes across better is Marquette's Kevin O'Neill. The camera is there as O'Neill visits the Gates household, and again as William makes his recruiting visit to the Wisconsin campus. O'Neill offers a four-year scholarship, saying it remains valid even if William's tender knee gives out.

William takes the offer, and celebrates at his 18th birthday party. "He's a great kid and some kids don't even live to this age," says his mother, Emma Gates. "He's 18 and he's lived this long and I'm very proud of him."

Arthur takes longer to find himself, but eventually gets there. As a senior, his high school finishes third in the state and he is a leading scorer. He wins a scholarship to Mineral Area Junior College in southern Missouri, which—truth be told—resembles a prison. All seven of the school's black students play on the basketball team and live together in one house.

Ultimately, of course, neither William Gates nor Arthur Agee became the next Isiah Thomas. Both played college ball (William at

Marquette and Arthur, eventually, at Arkansas State). William got a degree in communications and Arthur later formed a foundation to help other kids get into college.

The journey we watch over five years is dramatic and revealing. It will both outrage and inspire you.

"I KNOW THAT GUY": Look closely at the aspiring college players during the scene at Nike camp. Two of them are teenaged Jalen Rose and Chris Webber, both of whom realized their own hoop dreams.

15 BRIAN'S SONG (1971–NR)

SPORT: Football ★ **STARS:** Billy Dee Williams, James Caan
DIRECTOR: Buzz Kulik

In 2008, the web site eHarmony compiled a list of "20 Movies that Make Men Cry." *Brian's Song* finished No. 1.

Larry Csonka, the Hall of Fame fullback, admits he choked up watching *Brian's Song*. It's hard to imagine Csonka with that battle-scarred face and outta-my-way scowl getting all misty over a movie, but he swears it really did happen.

"I was breaking into the league around the time [Gale] Sayers and [Brian] Piccolo were playing in Chicago," Csonka said. "Like everyone else, I was shocked when Piccolo died. I heard they were doing this movie and I thought it would be one of those sappy TV deals.

"I was surprised at how good it was. I didn't know the depth of their friendship.

I was watching it and thinking, 'Wow, this is really something.' The scene where Gale visits (Piccolo) in the hospital for the last time, the love those two guys had for each other, it really hits you."

It is a tribute to the power of *Brian's Song* that a TV movie made almost 40 years ago still resonates with athletes and fans. When we surveyed people for their all-time favorite sports movies, *Brian's Song* was on most short lists. Csonka had it at No. 1, as did Rich Gannon, the former Oakland quarterback, and Brian Baldinger, the former NFL lineman. He is currently an analyst with the NFL network

"I was 10 when I saw it," said Baldinger, "and I decided that night that sports would be my life. That's how much of an impression it made on me."

"I saw it at my grammar school, St. Cecilia's in Philadelphia," said Gannon, the NFL's Most Valuable Player in 2002. "I was a kid who dreamed of playing in the NFL and that movie was my first behind-the-scenes look at pro football. It showed the training camp, the film study, guys rehabbing injuries, stuff I never thought about before. And, yeah, I cried."

The film stars Billy Dee Williams as Sayers and James Caan as Piccolo. They were unknowns at the time, although Williams would soon emerge as a romantic leading man, and Caan would create an unforgettable role as Sonny Corleone in *The Godfather*.

Brian's Song is the poignant story of two Chicago Bears who go from rookie rivals to best friends over the course of five seasons. The film ends tragically as Piccolo succumbs to cancer, but his courage inspires Sayers on and off the field.

It begins in 1965 as rookies Sayers and Piccolo join the Bears. Sayers is a first-round draft pick, an All-American from Kansas, a dazzling runner who is destined for greatness.

Piccolo is a free agent from Wake Forest, short on talent but long on desire.

Director Buzz Kulik and writer William Blinn draw contrasting portraits of the two players. Sayers, for all of his acclaim, is painfully shy. Piccolo is an extrovert with a wry sense of humor. When they first meet, Sayers is on his way to see Bears coach George Halas (Jack Warden). Piccolo tells Sayers that Halas is deaf in one ear so he should keep that in mind.

Piccolo's story is nonsense, but Sayers believes him. When he meets with Halas, the coach is moving around his office, filing papers. Sayers keeps running circles around the coach, trying to stay on his "good" side. Perplexed, Halas says, "I know you have good moves, Sayers, but what are you doing?" Sayers realizes he's been duped.

Over the course of the summer, Sayers and Piccolo become friends. Gradually, Piccolo brings Sayers out of his shell. As the team prepares for the regular season, Halas asks the two men how they would feel about being roommates on the road. He explains it would be the first time in Bears history that a black player and a white player shared a room.

"It will be news," Halas says.

Sayers and Piccolo shrug. They're friends, their wives are friends, they go out together. Roommates? Sure, why not? So a racial barrier comes down and hardly anyone bats an eye. Given the context of America in the turbulent '60s, that's no small accomplishment.

Sayers has one of the greatest rookie seasons in NFL history, scoring 22 touchdowns, including six in one game against San Francisco (still tied for an NFL record). Over the next two years, he becomes the most celebrated player in the game, winning the rushing title in 1966. Piccolo plays sparingly. Mostly, he cheers on his roommate.

Sayers is off to another blazing start in 1968 when he suffers a devastating knee injury. Piccolo steps into Sayers' spot and plays well, including a 112-yard rushing performance against New Orleans. When Sayers recovers the following year, he is paired with Piccolo in the backfield. This is what they wanted, the chance to play together, but it is cut short when a malignant tumor is discovered in Piccolo's chest.

Piccolo's illness is not widely known until after the season, when Sayers accepts the Most Courageous Athlete award for coming back from his knee injury. He tells the audience that Piccolo is more deserving of the honor for his courage in battling cancer.

"I love Brian Piccolo," Sayers says, "and I'd like all of you to love him, too. Tonight, when you hit your knees, please ask God to love him."

It is an emotional scene, and Williams plays it very well. Overall, it is the acting of Williams and Caan that keeps *Brian's Song* from becoming maudlin. It is a pleasure to watch Williams as Sayers opens up. When he stands in the locker room and tells the team about Piccolo's illness, he is not the quiet kid we saw earlier. He is strong and confident, and we know it is largely due to Piccolo's influence.

Caan does his part by maintaining Piccolo's irreverence. Lying in the hospital, barely able to speak, he tells Sayers: "When you dedicate a game to someone, you're supposed to win it." (The Bears lost after Sayers' locker room speech). After Sayers gives him a blood transfusion, Piccolo says, "I'm developing this tremendous craving for chitlins."

The laugh the two men share is a touching reminder of happier times. And if it moves you to tears, well, you are not alone.

"That was the first sports movie I can remember where the people, not the games, were the most important thing," Baldinger said.

"I watched it and thought, 'That's what sports should be about—friendship and teamwork.' When you get right down to it, that's what life should be about."

PIVOTAL SCENE: When Sayers returns home following his knee surgery, he finds Piccolo in the basement, assembling workout equipment. Sayers is irritated, and tells Piccolo he doesn't want to use it. Piccolo gives Sayers a verbal kick in the butt.

Piccolo tells Sayers how he went to Wake Forest, led the nation in rushing, and still was passed over in the draft because the scouts thought he was too slow. He came to the Bears as a free agent, fought for a spot on the roster, and then sat on the bench and waited for a chance to play. Sayers' injury gave Piccolo that chance.

"So next season if I beat you out, I want to know I beat you at your best," Piccolo says.

Piccolo pushes Sayers through his rehabilitation and Sayers comes back to win his second NFL rushing title.

BET YOU DIDN'T KNOW: This isn't in the movie, but NFL Commissioner Pete Rozelle spent Thanksgiving Day 1969 at the hospital with Piccolo, watching the Lions-Vikings game on television.

IF YOU LIKED THIS, YOU'LL LIKE: *Something for Joey*, a 1977 TV movie about Penn State's John Cappelletti and his kid brother, Joey, who was stricken with leukemia. Cappelletti dedicated his 1973 Heisman Trophy to his brother.

16

MIRACLE (2004–PG)

SPORT: Hockey ★ **STARS:** Kurt Russell, Patricia Clarkson, Noah Emmerich
DIRECTOR: Gavin O'Connor

There are no surprises in *Miracle*. There is no M. Night Shyamalan twist at the end. Even those who weren't born in 1980 when the United States hockey team upset the seemingly invincible Soviets know the story. They have heard Al Michaels' famous call—"Do you believe in miracles? Yessssss!"—countless times.

It is familiar territory, we know exactly where we're going the whole time, but thanks to the uncanny performance of Kurt Russell as coach Herb Brooks and a convincing cast of young athletes/actors, we thoroughly enjoy the ride. Credit director Gavin O'Connor and screenwriter Eric Guggenheim for developing an insightful portrait of the team and what it overcame on its way to winning the Olympic gold medal.

Miracle establishes the context of the times during the opening credits. Newsreel footage of the Iran hostage crisis, the meltdown at Three Mile Island, and the long lines at the gas pumps paint a picture of an America experiencing what President Carter called "a crisis of confidence." The Cold War is still going on and the Soviet invasion of Afghanistan is casting a dark shadow across the globe.

Against this backdrop, Brooks is given the task of assembling a U.S. hockey team to take on the world at the Winter Olympics.

The American team will consist of college kids while other countries send their battle-tested veterans. The Soviets and Czechs still are operating behind the Iron Curtain so their top players have not yet found their way to the National Hockey League. They play for their national teams.

The Soviets have won the hockey gold medal in every Olympics since 1960, when the U.S. won. They routed the NHL teams in a series of exhibition matches in the '70s and defeated a team of NHL All-Stars in a best-of-three series in 1979. The Soviets are the best hockey team in the world, a finely tuned machine that crushes everything in its path. If the North American pros are no match for them, what chance do a bunch of amateurs have?

"Put your street clothes on. I've got no time for quitters."
—HERB BROOKS

"You want me to play on one leg? Huh? I'll play on one leg."
—ROB McCLANAHAN

"That'll get 'em going."
—HERB BROOKS

But Brooks has a plan: He will beat the Soviets at their own game. For years, the North American teams have played their plodding dump-and-chase game against the Soviets. They shoot the puck into the corner and chase after it, hoping to outmuscle the Soviets along the boards. But that strategy rarely worked because the Soviets were much faster. On the larger international ice surface, they would get to the puck first and with one or two crisp passes be back on the attack.

"You can't beat those guys playing that way," Brooks says. "I'm going to take their game and throw it right back in their face."

Brooks' blueprint calls for a team of hungry kids, not fat-headed all-stars, who are willing to learn a new style of play. They will have to endure a grueling training program. Brooks will make them skate until they're on their knees with their tongues hanging out.

"I can't promise you we'll be the best team at the Olympics, but we'll be the best conditioned," Brooks says. He adds, "You don't have enough talent to win on talent alone."

Seventy of the top collegiate players in the country are invited to a tryout camp. On the first day, Brooks hands a sheet of paper to assistant coach Craig Patrick (Noah Emmerich).

"What's this?" Patrick asks.

"Twenty-six names," Brooks says.

He has already decided which players to keep and which ones to send home.

"You're missing some of the best players," Patrick says.

"I'm not looking for the best players, Craig," Brooks says. "I'm looking for the right ones."

Miracle does several things very well, which is why the film holds our interest even though we know the whole story before we even settle into our chairs. First, it takes us deep into the scheming mind of Brooks, who pushes every button and pulls every string to get the most out of his players. He gives written psychological tests and most of the time he seems as cold as the ice he stands on.

"I'll be your coach, I won't be your friend," he tells the players.

Deep down, of course, he cares deeply for them. He sees in them the same qualities that drove him when he was trying out for the 1960 U.S. Olympic team. He was the last man cut from that squad and was forced to watch on television as the team won the gold medal.

Brooks has been trying to recapture that moment ever since.

The film also shows the rift that existed on the team and for awhile threatened to tear it apart. There was a split between the Eastern players (Boston University, Boston College, and Harvard) and the Midwestern players (Minnesota, Wisconsin, and North Dakota). They played against each other in junior tournaments and then in college so there was a lot of baggage that needed checking before they could come together as a team.

One reason Brooks was so hard on the players was he wanted them to bond and he felt the quickest way to do that was to have them unite against him. As the team doctor (Kenneth Welsh) says, "If they all hate him, they won't have time to hate each other."

So there was a method to Brooks' madness. As he told *Sports Illustrated*, "It was a lonely year for me. This team had everything I wanted to be close to, everything I admired: the talent, the psychological makeup, the personality. But I had to stay away. There wasn't going to be any favoritism."

Everyone agrees Brooks was fair. Tough, but fair. When he was named coach, most people assumed he would favor the players who helped him win an NCAA championship at the University of Minnesota. He did not. He played Jim Craig (Boston University) in goal ahead of Steve Janaszak (Minnesota) and he picked Mike Eruzione (Boston U.) to be the team captain. He cut Minnesota kids to keep players from other schools.

"I gave our guys every opportunity to call me an honest SOB," Brooks said. "Hockey players are going to call you an SOB at times anyway, in emotion. But they could call me an honest one because everything was up front."

Well, not everything. Brooks was always playing mind games. In the film, less than a month before the Olympics, Brooks brings in a new player. The other players, who have been together for seven months, are resentful. Several players, including Eruzione, fear losing their spot on the team to the newcomer.

A group of players, led by Eruzione, confront Brooks.

"This is crazy, bringing him in this late," Eruzione (Patrick O'Brien Demsey) says.

"The kid can flat-out play," Brooks says.

"And we can't?" says defenseman Jack O'Callahan (Michael Mantenuto).

"He's got the attitude I want," Brooks says. "Somebody here better tell me why I shouldn't be giving him a hell of a look."

"Because we're a family," says center Mark Johnson (Eric Peter-Kaiser).

"What?" Brooks says.

"We're a family," replies winger Rob McClanahan (Nathan West).

It is the validation Brooks has been waiting for—Eruzione and O'Callahan (Boston), Johnson (Wisconsin), and McClanahan (Minnesota), all standing as one. Brooks has his team. The new guy has served his purpose. He is sent packing and the U.S. squad heads to Lake Placid ready to shock the world.

"It's a wonderful film," wrote E.M. Swift of *Sports Illustrated* who covered the 1980 Winter Olympics. "My 11-year-old son also loved it, leading me to believe that another generation will fall under the team's magical spell. Who could have known that two superlative weeks of play by that group of fresh-faced kids would keep its hold on the American imagination for so long?

"[Brooks] brought out in them qualities they didn't know they had. And together as a team, they did the same for us. For Americans. I've always believed that was the miracle—that a hockey team could do such a thing."

17

CHARIOTS OF FIRE (1981–PG)

SPORT: Track ★ STARS: Ben Cross, Ian Charleson, Ian Holm
DIRECTOR: Hugh Hudson

When *Chariots of Fire* won the Oscar for Best Picture of 1981, it was a huge upset. The film was up against a strong field—*Raiders of the Lost Ark, Reds, On Golden Pond,* and *Atlantic City* were the other nominees—and the tale of two British track stars seemed to have about as much chance as a sprinter with a pulled hamstring.

When Loretta Young opened the envelope to announce the winner, she could not hide her surprise. Her eyes widened and she paused for a second as if to make sure there wasn't a mistake. When she announced "*Chariots of Fire,*" it was hard to tell if the screams that followed were screams of delight or disbelief.

Even now, you can still start a lively debate asking people what they think of the film. Frank Deford of *Sports Illustrated* considers it the best sports film ever made. *Premiere*

Magazine rated it one of the 20 Most Overrated Movies of All-Time. Many critics adored it—as did many athletes. Dara Torres, the only American swimmer to compete in five Olympics, says it is her favorite movie ever.

It is easy to find fault with *Chariots.* The first half is downright boring at times. Certain plot points are repeated until they are beaten into the ground. Harold Abrahams feels persecuted because he is Jewish. Eric Liddell is devoted to his Christian faith. OK, we get it. Can we move the story along now?

For a movie about men who run fast, *Chariots* often seems stuck in molasses. And when Abrahams and Liddell finally do run, they are shown in slow-motion, which annoys even Deford. ("A ghastly cliche," he calls it.)

"It's a shame that once again, even in a movie as true and sophisticated as this one, the sporting scenes are marred by slow-motion sequences," Deford wrote in his review. "Can't people in movies run and jump and bat and throw at the same speed with which they do everything else?"

The reason the film works—and it really does work—is the actors (Ian Charleson as Liddell, Ben Cross as Abrahams) are so good they make you care about their characters. The fact that it is a true story, and the script—which won an Oscar for screenwriter Colin Welland—is pretty faithful to the real events, gives *Chariots* almost a documentary feel.

Abrahams, the son of a Lithuanian immigrant, enrolls in Cambridge. As one of the few non-Anglo-Saxons on the hallowed grounds, he feels like an outsider. He says he feels "a cold reluctance in every handshake." But it only makes him more determined to succeed. "I'm going to take them on, one by one, and run them off their feet," he vows.

Abrahams makes good on his promise. In one of the film's best scenes, he becomes the first student to run a complete lap around the Great Court in less time than it takes the steeple clock to strike 12 bells at midday. Looking on, the school master (Sir John Gielgud) says, "I doubt there is a swifter man in the kingdom."

That is the cue to introduce Liddell, the son of Scottish missionaries and a world-class sprinter. He enjoys the competition, but it is secondary to his religion. Liddell is shown winning races and preaching sermons to his adoring fans afterwards.

As the 1924 Olympics approach, Liddell's sister Jennie (Cheryl Campbell) accuses him of putting his training ahead of God's work. She says, "You are so full of running, you have no time for standing still." She is preparing to meet their parents at their mission in China. Eric is expected to join them, but he wants to run in the Olympics first. He tries to make his sister understand that his athletic pursuits are really an extension of his faith.

"I believe God made me for a purpose, but He also made me fast," he says. "And when I run, I feel His pleasure."

Abrahams is running against prejudice. Liddell is running to honor his God. They are very different sorts of heroes, but by the time they climb aboard the steamer in London and set sail for the Olympic Games in Paris, you are rooting for both of them. You are also prepared to root against their jaunty American rivals, Charlie Paddock and Jackson Scholz, who are expected to kick butt in Paris.

Abrahams and Liddell are entered in the 100 meters, but as it turns out the qualifying heats are on Sunday. Liddell refuses to run on the holy day. The head of the British delegation pulls him into a meeting with the Prince of Wales and other politicians who try to pressure him into changing his mind. Staring into the eyes of the future king, Liddell stands his ground.

"God made countries," he says. "God makes kings and the rules by which they govern. And those rules say that the Sabbath is His. And I for one intend to keep it that way."

Another member of the British team offers Liddell his spot in the 400 meters which is scheduled for later in the week. Liddell, who has successfully competed at the longer distance, accepts. But word of his principled stance makes headlines back home and assures his sister that he still has his priorities in order.

When the Games begin, director Hugh Hudson faces a structural dilemma. Filmmakers talk about "delivering the moment"—that is, building the emotion to one climactic scene, one event that pays off everything that has gone before. Hudson is faced with *two* endings: Abrahams' quest for gold in the 100 *and* Liddell's race in the 400.

It is to Hudson's credit that the two scenes play so well. Abrahams' victory, which comes first, is not shortchanged and Liddell's triumph does not feel like an anti-climax. It is a neat bit of storytelling on the part of the director, who was nominated for an Oscar but lost to Warren Beatty (*Reds*).

Hudson makes it work by showing the two races from different perspectives. He lets us see Abrahams' victory through the eyes of his coach Sam Mussabini (Ian Holm), who is not allowed in the Olympic stadium because he is considered a professional. Sam stands at the hotel window, waiting to see which flag is raised for the medal ceremony. It is not until Sam sees the Union Jack go up that he knows his man has won the gold. All alone, Sam celebrates by punching his fist through his straw hat.

Liddell's victory is told in more straightforward fashion with the runner—yes, in slow-motion—hitting the tape, his head thrown back, smiling as his words ("When I run, I feel His pleasure") echo above the roar of the crowd.

66 **BEST LINE:** Harold Abrahams: "In one hour's time, I will raise my eyes and look down that corridor, four feet wide, with 10 lonely seconds to justify my whole existence. But will I? I've known the fear of losing, but now I am almost too frightened to win."

18 BODY AND SOUL (1947–NR)

SPORT: Boxing ★ STARS: John Garfield, Lilli Palmer, William Conrad
DIRECTOR: Robert Rossen

Body and Soul was the first boxing movie to shine a light on the sport's ugly underbelly. Previous boxing films were mostly tender love stories or lighter-than-air comedies, such as the popular *Joe Palooka* series. So when *Body* and *Soul* came out, it hit audiences like an uppercut to the chin.

Body and Soul took its lead from the New York State Senate inquiry into corruption in sports, specifically boxing. The investigation was attracting national headlines at the time and Americans were learning more about the shady dealings between gangsters and many boxing promoters. *Body and Soul* put real flesh and blood on those headlines. The film is more than 60 years old, yet it still is regarded as one of the best boxing movies ever made.

"Here are the gin and tinsel, squalor and sables of the Depression era, less daring than when first revealed in *Dead End* or *Golden Boy*, but more valid and mature because it is shown without sentiment or blur," wrote the *National Board of Review* in praising *Body and Soul* as a triumph for director Robert Rossen and star John Garfield.

"What are you gonna do, kill me? Everybody dies." —CHARLIE

Body and Soul is the fictional story of Charlie Davis (Garfield), a Jewish kid from New York's East Side who piles up an impressive record as an amateur boxer. His father runs a candy store and his mother wants Charlie to forget boxing and get an education. "That is no way to live, hitting people and knocking their teeth out," she says.

But when Charlie's father is killed (a bomb meant for the speakeasy next door takes out the candy store instead), his mother applies for welfare. When a woman from the agency comes to the house, Charlie throws her out. "We don't need your help," he says. He's too proud to take a handout.

"Do you think I did it to buy myself fancy clothes?" his mother asks. "Fool, it's for you. To learn, to get an education, to make something of yourself."

Enraged, Charlie tells his friend Shorty (Joseph Pevney) to get in touch with Quinn (William Conrad), a boxing manager, and tell him he's ready to turn pro.

"Tell Quinn to get me a fight," Charlie says. "I want money, you understand? Money, money."

"I forbid it," his mother says. "Better you should buy a gun and shoot yourself."

"You need money to buy a gun," Charlie shouts back.

What follows is the standard montage of Charlie, now a professional, knocking out one opponent after another. There are the usual shots of spinning newspaper headlines, pages flipping on a calendar, and train wheels rolling down the tracks. This kind of storytelling was hardly new even in 1947.

But *Body and Soul* takes a sharp turn when a dapper mobster named Roberts (Lloyd Gough) shows up to get his piece of the action. Charlie has been piling up the wins, but they've all been in small towns for small purses. If he wants to break into the big time, he'll have to cut a deal with Roberts. Shorty sees through Roberts right away and tells Charlie that the mobster is bad news.

"He doesn't care about you," Shorty says. "He just wants his piece of the pie."

"What's the difference?" Charlie replies. "It'll be a bigger pie, more slices for everybody."

As soon as Roberts moves in, we see Charlie change. He becomes more greedy and ruthless. He dumps his girlfriend Peg Born (Lilli Palmer) for Alice (Hazel Brooks), a nightclub singer who was seeing Quinn. His relationship with Shorty becomes strained. But Charlie knows Roberts can pave his way to the title and that's what he wants.

Roberts gets Charlie his shot at champion Ben Chaplin (Canada Lee). Chaplin has a blood clot on the brain, the result of a previous bout, and he is just looking for one more payday. He takes the fight with Davis knowing he will lose, but with the assurance that Davis will take it easy on him. Roberts tells Chaplin and his manager they have nothing to worry about.

"Nobody gets hurt, you have my word," Roberts says.

But Roberts doesn't tell Charlie anything. Fighting with his usual ferocity, Charlie knocks out Chaplin and sends him to the hospital. Shorty overhears Chaplin's manager and Roberts arguing in the dressing room. Learning what happened, Shorty tells Roberts he's finished with him. "And when Charlie finds out, he'll quit, too," Shorty says.

But Charlie doesn't quit. He feels badly for Chaplin, he knows he got a raw deal, but Charlie also knows Roberts is where the money is. "I'm the champ," he says. "If I walk away now, what will I do?" Disgusted, Shorty leaves Charlie. (He dies when he is struck by a car outside the nightclub.)

As the champion, Charlie succumbs to the pitfalls of wealth and celebrity. We see him dressed in flashy clothes, slurping down drinks with Alice, ripping up losing tickets at the race track and it's all subsidized by Roberts, who keeps handing him fat envelopes full of cash with a smile and a slimy, "Here ya go, champ."

Then one day Charlie comes around for his envelope and Roberts says, "Not so fast." Uh oh, the bill is coming due. Roberts wants Charlie to fight the top contender, a rising star named Jackie Marlowe. And that's not all. Roberts wants Charlie to lose.

"It's time, Charlie," Roberts says. "You don't like fighting anymore, you like living too much. Take the money from this fight and buy yourself a restaurant."

Charlie gets the picture. Roberts has his hooks in Marlowe; that's his new meal ticket. Charlie is yesterday's news. He is being tossed onto the trash heap as Ben Chaplin was before him. He doesn't like the idea of tanking his last fight and he balks until Roberts slaps another fat envelope on the table.

"Money's got no conscience, Charlie," Roberts says. "Here, take it." So he does.

Wrestling with his emotions and with no one else to turn to, Charlie reaches out to his mother and to Peg, his old girlfriend. When he tells them this will be his last fight, they are elated. But when he tells them it's fixed, the mood changes. Peg slaps him across the face and walks out. . . .

Now, we don't want to ruin the big fight, but suffice it to say, it brings *Body and Soul* to a rousing finish.

The film was a career highlight for Garfield, who was nominated for an Academy Award (he lost to Ronald Colman in *A Double Life*). It drew two other nominations—Best Original Screenplay (Abraham Polonsky) and Best Editing (Francis Lyon, Robert Parrish)—with Lyon and Parrish winning the Oscar.

Yet *Body and Soul* has a bitter legacy. Many of its cast and crew—including Garfield, Anne Revere, Gough, Lee, and Polonsky— were called before the House Un-American Activities Commission for alleged affiliations with the Communist Party. All were blacklisted and their careers effectively ended.

 BET YOU DIDN'T KNOW: Canada Lee, who plays Ben Chaplin, and Artie Dorrell, who plays Jackie Marlowe, were professional boxers. Lee fought a 10-round draw with future middleweight champion Vince Dundee in 1928. Dorrell compiled a record of 38-9-6 and fought former welterweight champion Fritzie Zivic in 1944 (he was TKO'd in the 7th round).

 BEST LINE: Shorty (Joseph Pevney): "Charlie's not just a kid who can fight. He's a machine that makes money."

19 REMEMBER THE TITANS (2000–PG)

SPORT: Football ★ **STARS:** Denzel Washington, Will Patton, Ryan Hurst
DIRECTOR: Boaz Yakin

Remember the Titans is based on the true story of T.C. Williams High School in Alexandria, Va. The football team, the Titans, won the state championship in 1971, but accomplished much more than that. President Richard Nixon was quoted as saying, "The team saved the city of Alexandria."

That may be stretching the truth, but there is a reason why people remember the Titans. At a time when racial hatred was tearing the country apart, the coaches and players at T.C. Williams High School rose above it in an inspiring way. As Ben Morris wrote in the *Alexandria Gazette*, "It all came about through the hard work and sacrifice of these young men and their coaches who worked together,

proving to everyone that different ethnic groups can mesh unanimously for a singular ambition."

That is the theme of the film: There is no limit to what people can accomplish when they recognize that their common dreams outweigh their differences. In the case of the Titans, it is the coaches and players, black and white, who are thrown together when their schools are merged. They learn to work together and they learn to trust each other. In the end, they learn to love each other and their success sends a healthy message to the whole town.

The film was produced for Disney Studios by Jerry Bruckheimer, who is better known for his bombastic visuals (*Pearl Harbor*, *Armageddon*) than coherent story-telling. His heavy hand does slam down on *Titans* now and then—usually in the form of cartoonish football violence—but the acting is so good, especially Denzel Washington as head coach Herman Boone, the film stays solidly on course.

Titans is built on two storylines: One is the relationship between Boone and assistant coach Bill Yoast (Will Patton), the other is the relationship between Gerry Bertier (Ryan Hurst) and Julius Campbell (Wood Harris), the two best players on the team—one white, one black—who conquer their mutual distrust and become best friends.

This is familiar ground, rutted with cliches, and if a film hits too many of them, it can wind up in a ditch. *Titans* is able to avoid that by allowing the characters, especially Boone and Yoast, to think, speak, and act like real people and not mere devices.

It would be easy, for example, to make Boone, the African-American coach brought in to take over the team, a man of unshakable conviction, a combination of Vince Lombardi and Martin Luther King. But writer Gregory Allen Howard puts enough rough edges on Boone that we watch him with keener interest because we never know just how he will react to any given situation.

He describes himself as "a mean cuss," and at times he goes out of his way to prove it.

In hiring Boone, the school board passes over Yoast, a long-time coach in the district. The move is designed to appease the African-American portion of the newly merged school district. "We have to give them something," a board member says.

No one spells it out for Boone, but he is smart enough to know what's going on. He knows Yoast is getting the shaft, yet Boone is ambitious enough to take the job anyway.

Yoast ends up as his defensive coordinator.

When he moves in, Boone is welcomed as a hero by the African-American families. In a town where integration is greeted by picketers, racial taunts, and bottle-tossing, the coach is seen as a symbol of hope, an authority figure who can exert an influence beyond the field and into the community. Boone, rather coldly, tells them he's not that kind of leader.

"I'm not the answer to your prayers," he says. "I'm not the savior, Jesus Christ, Martin Luther King, or the Easter Bunny. I'm just a football coach."

Boone is tough and does not play favorites. He is tough on everyone, including Yoast. Boone sees Yoast talking to one of the African-American players after Boone benches him for blowing an assignment. Boone accuses Yoast of coddling the kid, claiming he would not have done it if the player were white.

"I'm the same mean cuss with everybody out there on that football field," Boone says. "The world don't give a damn about how sensitive these kids are, especially the black kids. You ain't doing these kids a favor by patronizing them. You're crippling them. You're crippling them for life."

Yoast is a product of the Old South who isn't thrilled with integration, but he is trying to make the best of it. He talked the white players on team out of quitting (they were ready to walk off to protest the hiring of Boone) and he is juggling football with his duties as a single father. He doesn't like Boone (understandable since Boone took the job that should have been his), but he begins to understand him when some rednecks toss a brick through Boone's window one night when Yoast and his daughter are visiting.

"Maybe your little girl got a taste of what my girls go through," Boone says. "Welcome to my world, Yoast."

The relationship between the two men grows and they learn from each other. The same is true of Bertier and Campbell. Forced to room together at the training camp in Gettysburg, they can't stand each other (Bertier tells Campbell, "You're nothing but a pure waste of God-given talent") but they grow so close, Bertier winds up walking away from his foxy girlfriend (Kate Bosworth) when she tells him to choose between her and "Big Ju." (She won't even shake Campbell's hand).

"You've got your priorities all messed up," she tells Gerry.

"I don't think so," he says.

The bonding of the coaches and players is the real story, more than the touchdowns and the victories. In truth, the Titans were even better than the film suggests. They not only went undefeated (13-0) but they dominated. They outscored their opponents 357-45 and their defense, led by Bertier, posted nine shutouts. The producers built a few close games into the film for the sake of drama.

"That movie is all about unity," San Francisco 49ers receiver Anquan Boldin told us. "I am a huge believer in unity because that is what sports are all about. No matter what your race is, team unity and great teammates make sports something special to be a part of."

What made the film special—and a $115 million hit for Disney—is the emotional center embodied by Boone.

"Tonight we got Hayfield; like all the other schools in this conference they're all white," Boone says in one stirring locker room speech. "They don't have to worry about race, we do. Let me tell you something. You don't let anyone come between us. Nothing tears us apart.

"In Greek mythology, the Titans were greater even than the gods. They ruled their universe with absolute power. Well, that football field out there is our universe. Let's rule it like Titans."

📽 **PIVOTAL SCENE:** During camp, with the white players and black players refusing to work together, Boone takes the team on a 3 a.m. run through the Gettysburg battlefield. When they finish, he delivers a speech that pulls the team together.

"Fifty thousand men died on this field, fighting the same fight that we're still fighting among ourselves," he says. "Listen to their souls, men. You listen and you take a lesson from the dead. If we don't come together right now on this hallowed ground, we, too, will be destroyed."

❝ **BEST LINE:** Herman Boone: "We will be perfect in every aspect of the game. You drop a pass, you run a mile. You miss a blocking assignment, you run a mile. You fumble the football and I will break my foot off in your John Brown hind parts and then you will run a mile."

☺ **GOOFS:** When the Titan player picks up an opponent's fumble and runs for a touchdown, he is running the wrong way.

☞ **DON'T FAIL TO NOTICE:** To ensure this football movie appealed to non-football fans (read: women) Yakin built in a soundtrack that quickly switches to upbeat '70s rock music every time something positive happens on the field for the Titans.

😵 **"I KNOW THAT GUY":** Ryan Gosling has a small role as defensive back Alan Bosley. Gosling has gone on to become a leading man, and was nominated for an Academy Award for Best Actor in the 2006 film *Half Nelson*.

★ **SPORTS-ACTION GRADE:** C-minus. The game action is way over-the-top and the real Titans did not dance onto the field. The cheesy choreography is a Hollywood invention.

20

BREAKING AWAY (1979—PG)

SPORT: Bicycle Racing ★ **STARS:** Dennis Christopher, Dennis Quaid, Daniel Stern, Jackie Earle Haley, Paul Dooley ★ **DIRECTOR:** Peter Yates

Say the words "teen movie" and what comes to mind is a beer-soaked bacchanal that opens with overcharged high schoolers conspiring to lose their virginity while Mom and Dad are out of town, and ends with someone barfing in the pool.

Not that we didn't enjoy semi-classics such as *American Pie* and *Porky's*. But it would be nice, just once, to see a film about adolescents that goes beyond beer bongs, torturing the class nerd, and the predictable jokes about masturbation.

Breaking Away is that movie.

This little film from 1979 explores issues like teen identity, parent-child relationships, self-confidence, and class bias, and does so without bathing them all in alcohol and testosterone. Not once does it insult your intelligence. The result is one of the greatest

coming of age movies ever made. And, oh yes, the sports scenes are terrific.

Breaking Away focuses on four 18-year-olds, a few months out of high school, struggling to decide whether they want to spend their future hanging out together, finding a job or— perish the thought—moving on to college. They live in Bloomington, Indiana, where the tension between the snobby college kids and the working-class "cutters" (so known because their fathers cut the stones to build the university) leaves the locals feeling inadequate and resentful.

Dave Stoller (Dennis Christopher) is the most intriguing of the quartet, a restless kid who's obsessed with bike racing and inspired by the famous Italian Cinzano race team. He lives out his fantasy listening to opera, speaking with a faux-Italian accent, and driving his father

(Paul Dooley) crazy by changing the family cat's name from Jake to Fellini. So immersed is he in this daydream that he begins an unlikely romance with a beautiful Indiana University coed (Robyn Douglass) by convincing her he is an Italian exchange student.

It may be second to none among coming-of-age films. We'll put it right up there with *American Graffiti, Breakfast Club, Boyz n the Hood,* and *Basketball Diaries* in the pantheon of teen flicks that show respect for their characters—and audience—and go above and beyond rampant

"Hey, are you really gonna shave your legs?" —CYRIL
"Certo. All the Italians do it." —DAVE
"Some country. The women don't shave theirs." —MIKE

Mike (Dennis Quaid) is the former high school jock, embittered and afraid of the future. He aims to keep the group together, even if that means dead-end jobs for all of them. Cyril (Daniel Stern) is the wise-guy slacker, and Moocher (Jackie Earle Haley) is the undersized loner.

While the film focuses on the foursome's friendships and fear of the so-called real world, we learn that Dave, to everyone's amazement, has developed himself into a world-class cyclist. We get to see his talent—chasing down a truck on the interstate, and challenging his Italian heroes in a 100-mile race (accompanied by Rossini's *Barber of Seville* overture). It all leads up to the 12-minute climactic finale (predictable but still fist-pumping exciting), in which Dave takes on the frat-boy elitists in the Indiana University "Little 500" cycling race.

Every few years, a Slumdog Millionaire or Juno comes out of nowhere and becomes the little independent movie that could. *Breaking Away* was among the first of that sort. Its author, Steve Tesich, won the 1979 Academy Award for Best Original Screenplay. Overall, it was nominated for five Oscars, including Best Picture (won by *Kramer vs. Kramer*). In 2006, the American Film Institute placed it eighth on the list of America's 100 Most Inspiring Movies—second only to *Rocky* among sports films.

raunchiness and trivial nonsense.

On a side note, it's interesting to examine the long-term careers of the four aspiring actors in this movie. For Quaid, 24 at the time, *Breaking Away* was his breakout movie. He's all over this book. Stern, as well, went onto a long career, often playing characters with the same dry wit and Woody Allen outlook as he showed here with Cyril. Think of his character Shrevie in *Diner*.

The bantam-sized Haley was a successful teen actor (he plays Kelly Leak in *The Bad News Bears*), before disappearing into B-movies for a few decades. He re-emerged in 2006, earning a Best Supporting Actor Oscar nomination portraying a pedophile in *Little Children*.

It's Christopher, among the four, whose career ended with a flat tire. He followed up *Breaking Away* with a fine performance in 1981's *Chariots of Fire*. And then? Well, nothing notable, really, unless you want to include the freakish made-for-TV clown horror mini-series *It*. Which we won't.

So what happened? We figured it out recently by catching Christopher on an episode of HBO's *Deadwood*. At age 50, he still looks 18 years old, still looks like Dave Stoller about to slap on his bike helmet. In this case, staying youthful proved to be a bad career move.

21

WHEN WE WERE KINGS (1996–NR)

SPORT: Boxing ★ STARS: Muhammad Ali, George Foreman
DIRECTOR: Leon Gast

Leon Gast was a young filmmaker who went to Zaire, Africa, in 1974 to chronicle an international Woodstock featuring artists from America, including James Brown, and performers from Africa, such as Miriam Makeba. The week-long festival would conclude with a heavyweight title fight: the undefeated champion George Foreman against former champion Muhammad Ali.

Originally, Gast was more interested in the music. He was known for filming concerts—he produced films on B.B. King and the Grateful Dead—not sports. But when Foreman suffered an eye injury in training, the bout was pushed back five weeks. To fill the time, Gast began trailing after Ali, watching him interact with the people of Zaire. He quickly realized he had a whole different kind of film on his hands.

In *When We Were Kings*, Gast has total access to Ali, and the former champ is at his audacious and boastful best, offering playful predictions such as: "You think the world was surprised when Nixon resigned? Just wait till I whip Foreman's behind." He delights the crowds with riffs that sound like modern rap. A sample: "I wrestled an alligator, I tussled with a whale. I done handcuffed lightning and thrown thunder in jail. I injured a rock, I murdered a brick. I'm so mean, I make medicine sick."

But Ali also has many introspective moments as he walks through the villages and bonds with the people. It is clear the trip to Africa touches an emotion deep inside him. "I live in America," he says, "but Africa is the home of the black man. I'm going home to fight among my brothers."

Desson Howe of the *Washington Post* describes Ali as "a champion with grace, charm, an oversized ego, formidable fighting skills, a cheekily inspired sense of poetry, political savvy, religious conviction and defiantly independent opinions." At 32, Ali is all that. Gast's film shows the people of Zaire reaching out, touching his hands and face as if he were the Messiah.

Foreman, by contrast, alienates the people when he steps off the plane with his German Shepherd. He is unaware that German Shepherds were used against the citizens of Zaire by the Belgian military when they occupied the nation then known as the Belgian Congo. The sight of Foreman's snarling dog sends an ugly message to the populace and it is only reinforced by the champion's sullen manner. While Ali mingles with the people, Foreman stays secluded with his entourage.

This is not the affable George Foreman of today, the smiling grandfather who sells grills and appears in sitcoms. This is Foreman at 25, a dour and inscrutable wrecking machine fresh off devastating knockouts of Joe Frazier and Ken Norton, both of whom had wins over Ali. Foreman is a heavy betting favorite. Even Howard Cosell, an Ali loyalist, picks Foreman to win.

A French reporter asks Foreman what he would do if he lost to Ali. Foreman glares at the man. "I beg your pardon?" he says. The reporter nervously tries to rephrase the question, but Foreman cuts him off. "I beg your pardon?" he says, ending the interview.

Given the personalities of Ali and Foreman, plus the backdrop of a newly independent African nation, Gast has all the elements for a great documentary. He shot more than 300,000 feet of film—which helps to explain why it took him 22 years to edit it down to 90 minutes—so even if the fight had ended the way most people expected, with Foreman crushing Ali, Gast would have had an intimate character study of the two men.

But when the fight turns out to be the now-legendary "Rumble in the Jungle" with Ali shocking the world by knocking out Foreman in the eighth round, well, Gast has a film for the ages. *When We Were Kings* is the toast of the Sundance Film Festival and wins the 1997 Academy Award for Best Documentary Feature.

"*Kings* limns the meaning of the fight," writes David Armstrong of the *San Francisco Examiner*, "and shows why it transcends the parochial world of sports. The filmmakers blend vintage interviews with a romantic young Ali who extols Islam and idealizes Mother Africa. . . . In the hands of director Gast and chief editor and interviewer Taylor Hackford, *Kings* becomes a rich social document."

There are many levels to the story, including the rise of Don King. The bombastic promoter makes the decision to stage the fight in Zaire when that nation's president, Mobutu Sese Seko, agrees to put up the $10 million purse. Mobutu sees the event, with the accompanying music festival, as a way of promoting his country, but some critics find it hard to justify spending $10 million on what amounts to a PR campaign when the nation itself is wracked by poverty.

Gast sets what is surely a bizarre scene. The fight is held in a 70,000-seat soccer stadium rumored to have been the site of mass executions during Mobutu's rise to power. Mobutu does not attend the bout, fearing an assassination attempt. The fight is scheduled for 3 a.m. Zaire time so it will be prime time in the United States. There are native drummers and witch doctors at ringside. As the fighters enter the arena, the crowd chants, *"Ali, boma ye."* Translation: "Ali, kill him."

Gast uses assorted talking heads to reconstruct the round-by-round drama. Authors Norman Mailer and George Plimpton, who were at ringside, eloquently describe the mood, which bordered on dread for those who were rooting for Ali. Mailer says Ali's dressing room was "like a morgue." There was a feeling, even in Ali's camp, that Foreman was invincible. They feared he may demolish Ali and even humiliate him.

Mailer offers a vivid account (which is supported by Gast's footage) of Ali sitting on his stool after the first round, staring across the ring at the hulking Foreman and wondering if he could stand up to that power.

"The nightmare had finally come to visit [Ali]," Mailer says. "It was the only time I've ever seen fear in his eyes. He was up against a man who was younger and bigger and stronger than he was. A man who was not afraid of him. [Ali] had never been in this position before.

"Watching [Ali], you could see him nodding as if he were talking to himself, asking himself, 'Do you have the guts?' You could almost see him summoning the courage and as he nodded, saying, 'The time has come. I will find a way to master this man.' "

Ali mastered Foreman using the technique he called the "rope-a-dope," leaning back against the ropes, going into a defensive shell and inviting Foreman to pound away. At the time, it seemed suicidal. Plimpton recalls thinking, "The fix is in." It appeared Ali was making no attempt to fight back. But Ali let Foreman punch himself into a state of exhaustion. Then he chopped the big man down with a series of counterpunches, finally knocking him out

with a straight right hand to the jaw.

With the stunning victory, Ali regained his heavyweight title, but more than that, he was once again an icon. He had done what no one else thought possible: he felled the mighty George Foreman. It was the high point of his career, and there is an undeniable sadness in seeing the film and comparing the charismatic Ali of Zaire with the sad figure of today, trembling with Parkinson's disease, his eyes glassy and unfocused.

But thanks to Gast, *When We Were Kings* stands as a vivid reminder of what he once was. As Desson Howe writes: "The movie is a metaphor for any of us who lament the passing of time, who remember when we were younger, quicker, and more alert. Ali was magic; he was lightning in gloves. He was the best and the brightest. *Kings* proves it."

CHEERS: The film has no narrator. The story is told entirely through interviews and clips from the original broadcast, and it flows seamlessly.

❝ BEST LINE: George Plimpton describes Ali's rope-a-dope as resembling "a man leaning out his window to see if there is something on his roof." Plimpton also describes Don King as having "a great uprush of hair."

BET YOU DIDN'T KNOW: When *Kings* won the Oscar for Best Documentary Feature, Ali and Foreman joined Leon Gast on stage. They received a standing ovation.

"The sight of Foreman helping Ali onto the stage was so poignant," said Jeff Wald, executive producer of the TV series *The Contender*. "No one even noticed the producer [Gast], they were applauding the two fighters. Afterwards, the press asked George what he thought of the film. George said, 'I've seen it several times. I keep thinking the ending will change.' "

SPORT: Baseball ★ **STARS:** Tom Hanks, Geena Davis, Madonna
DIRECTOR: Penny Marshall

Most baseball seasons don't play out according to plan. Key players are injured and have to be replaced. Lineups are juggled and sometimes it turns out for the better. It can be the same way with baseball movies. Take *A League of Their Own*, for instance.

When Columbia Pictures began production on its film about the All-American Girls Professional Baseball League (AAGPBL), Debra Winger was cast in the role of Dottie Hinson, the star catcher of the Rockford Peaches. But Winger suffered a back injury and dropped out, so producers Elliot Abbott and Robert Greenhut had to find a replacement. They brought in Geena Davis.

It was a tough break for Debra Winger, but it worked out well for Columbia.

Watching the film now, it is hard to imagine anyone other than Davis, the statuesque, red-haired beauty, in the role of Hinson. She was the pinup girl of the AAGPBL, "The Queen of Diamonds," whose combination of baseball ability and good looks helped to sell the girls' league in America during World War II.

Winger may have been able to catch and throw, but she would not have looked as regal on the cover of *Life* magazine, and it is unlikely she could have played off co-star Tom Hanks as effectively as Davis. The on-screen chemistry between the leads—Davis as the star player, Hanks as the drunken slob of a manager—is one of the film's real strengths.

Davis joined the cast only days before filming was due to start. The other actors—Rosie O'Donnell, Madonna, Ann Cusack among them—had already completed the Hollywood equivalent of spring training. They worked with professional baseball coaches, including Joe Russo, the coach at St. John's University, to learn the fundamentals. "It sounds silly," Russo said, "but what I have to do is get them not to throw like a girl."

The cast bonded as a team during the experience. When Davis came in, she was a bit of an outsider. She auditioned for the part by playing catch with director Penny Marshall in Marshall's backyard. Davis proved to be a good enough athlete that she was able to outperform the other actors within a matter of weeks. There never was a question about her acting ability: she won an Oscar for her role in *The Accidental Tourist* in 1988.

The script was written by Lowell Ganz and Babaloo Mandel, who have a way with witty dialogue, as they demonstrated in *City Slickers*, *Splash,* and the underrated *Night Shift*. As a director, Marshall tends to smear sentimental goo over her projects, and the result is like eating a cake with about eight inches of icing. *Big*, *Awakenings,* and *The Preacher's Wife* are so sugary, they should come with a warning for diabetics.

A League of Their Own has its too-sweet moments—especially the bracketing device of opening and closing the film with a reunion of the AAGPBL alumni 50 years later at the Baseball Hall of Fame—but for the most part, Marshall keeps the sappy stuff in check. The result is a very pleasant two hours at the ballpark.

Vincent Canby of the *New York Times* called it "one of the year's most cheerful, most relaxed, most easily enjoyable comedies . . . a serious film that's lighter than air, a very funny movie that manages to score a few points for feminism in passing."

The film is a fictional account of the AAGPBL, a professional league for women which helped fill the void created when most major league stars went off to military service in World War II. There were four original teams—the Rockford Peaches, the Racine Belles, the Kenosha Comets, and the South Bend Blue Sox. *League* focuses on the Rockford team, which is managed (sort of) by Jimmy Dugan (Hanks).

Dugan is a former major league star whose career was cut short by alcoholism. He takes the Rockford job because he needs the money, but he has no intention of actually managing the team. "Girls are what you sleep with after the game," he says, "not what you coach during the game."

Schram) for throwing to the wrong base.

"Start using your head," he says. "That's the lump that's three feet above your ass."

Evelyn's lip quivers, then she begins to cry. Dugan, the hard-boiled baseball lifer, is aghast.

"Are you crying?" he says. "There's no crying in baseball!"

The American Film Institute put the "no crying" line on its list of the 100 greatest film quotes of all time. It placed 54th, but there is no doubt it is repeated more often than half of the quotes ranked ahead of it. It is very funny, and the story Dugan tells about how he didn't cry when his manager Rogers Hornsby called him "a talking pile of pig ----" is a hoot.

There's no crying in baseball.

Thanks to Jimmy Dugan, we'll never forget that cardinal rule.

"I was in the toilet reading my contract and it turns out, I get a bonus when we get to the World Series. So let's go out there and play hard."
—JIMMY DUGAN

So while Dugan gets drunk and dozes in the dugout, Dottie Hinson (Davis) runs the team. Hinson is a no-nonsense Oregon farm girl. A married woman whose husband is at war, she has no patience for the dissolute Dugan. She also has ambivalent feelings about baseball. She was talked into joining the league by her kid sister Kit (Lori Petty), a pitcher who knew she could ride Dottie's coattails into the AAGPBL.

But there is a part of Dottie that enjoys the game, more than she is willing to admit even to herself. Dugan sees it and it helps bring him around. He sobers up (he never does get around to shaving) and he begins to function as a real manager, encouraging his players and, when necessary, chewing them out.

The most famous scene from *League* is Dugan ripping outfielder Evelyn Gardner (Bitty

One of the most interesting aspects of the film is the understated way the relationship between Dugan and Dottie evolves. From the early scenes when Dugan is in a drunken haze and Dottie can barely tolerate the sight of him, they develop a respectful manager-player relationship and, finally, they become friends. They bond over baseball. Dottie keeps insisting the game isn't that important to her, but Dugan has seen her play and he knows better.

"I gave away five years at the end of my career to drink," Dugan says. "Five years, and now there isn't anything I wouldn't give to get back any one day of it."

"Well, we're different," Dottie says.

Dugan doesn't buy it.

"Baseball is what gets inside you," he says. "It's what lights you up, Dottie. You can't deny that."

When Dottie says the game is too hard, Dugan answers like a man who has been to the big leagues and won a home run title or two.

"It's supposed to be hard," he says. "If it wasn't hard, everyone would do it. The *hard* is what makes it great."

League was the No. 1 film in America when it opened in July 1992. It made $107 million domestic, and another $25 million overseas. Nice return for a film that cost just $40 million to make.

🍎 **BET YOU DIDN'T KNOW:** The original story was written by Kelly Candaele, whose mother played in the AAGPBL. Kelly's brother Casey played 9 seasons in the major leagues.

23 FRIDAY NIGHT LIGHTS (2004—PG-13)

SPORT: Football ★ STARS: Billy Bob Thornton, Derek Luke, Lucas Black
DIRECTOR: Peter Berg and Josh Pate

Turning a successful book into a movie can be tricky. Four hundred pages of plot don't always translate neatly onto film; readers may have preconceptions of their favorite characters; a downer of a story ending can spell box-office disaster.

Sometimes the adaptation works (*The Godfather, Jaws*). Sometimes it doesn't (*Bonfire of the Vanities, The Da Vinci Code*).

Friday Night Lights, the laser-sharp look at West Texas high school football, works magnificently. Based on the acclaimed 1990 best-seller by H.G. "Buzz" Bissinger, the movie captures the culture of gridiron as a small town's identity, where 17-year-old boys become gods for a few months, but the humiliation of failure is worse than death.

It helps that one of the film's directors, Peter

Berg, happens to be Bissinger's cousin. Clearly, Berg got the author's underlying unease over how an economically ravaged oil town can spend millions building a state-of-the-art stadium, when the school district itself is close to broke. Priorities. As one caller to a talk-radio show says during the movie, "It's about coaching. And you know what else? They're doing too much learning in that school."

But that's the way it works in Odessa, Texas, where the Permian High Panthers have won four state football titles (plus two more since the 1988 season about which the book was written). Half the town's men seem to wear championship rings from their youth, and most of them show up to watch the high-school two-a-days. "It is a religion," producer Brian Grazer says on the DVD commentary. "And people go to that church every Friday night."

That said, *Friday Night Lights* is not an attack on football. The practice and game scenes are outstanding, especially the final game, which was filmed at the old Astrodome in Houston. And the young men who play the game come off as likeable and sympathetic.

Consider Boobie Miles (Derek Luke), the cocky star running back who sits at his locker gazing through the glossy catalogs of the dozens of colleges aiming to recruit him. Miles destroys his knee early in his senior season, and suddenly yesterday's toast of the town is today's stale crust. His dreams are gone. As Bissinger says, "One millisecond changed his whole life."

We meet quarterback Mike Winchell (Lucas Black), an unsmiling, twitchy Peyton Manning look-alike who copes each day with an overriding sense of doom; tailback Don Billingsley (Garrett Hedlund), whose drunken failure of a dad tries to cure the kid's fumble-itis by duct-taping his arms around a football;

and safety Brian Chavez (Jay Hernandez), the deep thinker of the team, who is the first to recognize that the glare of the bright stadium lights can burn a young man.

In one of the most telling scenes, those three seniors sneak off to the edge of town to relax by skeet shooting—and to consider the overwhelming pressure placed upon their bulked-up shoulders.

"We got to lighten up. We're 17."—BRIAN CHAVEZ "I sure don't feel 17." —MIKE WINCHELL

The young actors are genuine, but the movie belongs to Billy Bob Thornton, playing Coach Gary Gaines. As the son of a high school basketball coach, Thornton was able to nail the role without having to do much research. "I mean he's a coach from Texas," Thornton told ESPN.com. "It's not like a stretch. It's not like they hired a British theatre actor to play a Texas football coach. They hired a guy from Arkansas."

Thornton's Coach Gaines is alternately fuming, compassionate, fearful, and stoic. Much of his time is spent suffering through local boosters—everyone from the bank president to the school's janitor—who tell him how to coach and let him know that anything short of a state title will be considered grounds for dismissal. After one humiliating loss, Gaines and his wife return home to find a dozen for-sale signs propped on their lawn.

The Permian Panthers don't quite bring home the state title in 1988, but their journey will keep you enthralled. The back end of the film is dominated by on-field action, which is nothing short of superb. Plus, by that point, you know the characters and find yourself caring about them. There's a wonderful small moment right near the end between Billingsley and his abusive dad. And there's also one of those almost-cliché "where are they now" finishes, where we learn that the young men who spent their high school lives as supermen and scapegoats grow up to become lawyers, surveyors, and insurance men.

The book works, the movie works. And of course, *Friday Night Lights* was turned into a television series by Berg and Grazer in 2006. The show, although a critical success, spent much of its first two seasons on the verge of being cancelled. While it strays further from Bissinger's original premise, it is well worth watching. Other than *Friday Night Lights* and *M*A*S*H*, we can't think of another worthwhile book-movie-TV series triple-play, can you?

24 THE WRESTLER (2008–R)

**SPORT: Professional Wrestling ★ STARS: Mickey Rourke, Marisa Tomei
DIRECTOR: Darren Aronofsky**

There was a time early in his career when Mickey Rourke was compared to a young Robert De Niro. In his early roles—Boogie in *Diner*, Harry Angel in *Angel Heart*, Charlie in *The Pope of Greenwich Village*—he was both dangerous and delicate, confident but still a little vulnerable.

By all accounts, he was going to be a big star. And then. . . .

And then Rourke took a disastrous career path. He took inane roles in cringe-inducing films. He left acting for four years to become a boxer—winning a few fights, but taking shots that scarred his soulful eyes. He got arrested, more than once. His career—and his looks—died in a haze of temper tantrums, spousal abuse charges, rumors of drug use, and undeniably bad plastic surgery. And then. . .

And then he came back. A few small roles in decent films (Francis Ford Coppola's *The Rainmaker*, the graphic-novel *Sin City*) helped Rourke emerge from the depths. The De Niro comparisons were long gone, but the man could still act. And then. . .

And then, in 2008, Rourke got his star vehicle, playing, well, in many ways, himself. In *The Wrestler*, he stars as Randy "The Ram" Robinson, a one-time pro wrestling icon who packed-in the fans at Madison Square Garden back in the 1980s but is now reduced to grappling third-rate opponents in half-empty VFW halls.

It's a story about faded glory. In his prime, Randy inspired action figures and video games. (There's a poignant scene where the aging wrestler implores a neighbor kid to play against him in a prehistoric Nintendo game that bears The Ram's likeness. "This is sooo old," gripes the kid.)

"The Shiek? Last I heard he was selling used cars." —RANDY

Now he lives in a grungy North Jersey trailer park, lugs boxes part-time at an overly fluorescent supermarket, and takes whatever matches he can get. He wears a hearing aid and reading glasses and limps through painful arthritis. Still, he'll do anything to continue in the business—including shooting steroids into his butt, or engaging in self-mutilation during matches. In one tragic scene The Ram sits at a nearly empty nostalgia show, surrounded by other broken down relics of the ring, hoping to hawk Polaroids of himself for a few dollars. Ahh, show business.

The sad story is made sadder when Randy suffers a heart attack right after a match. Wrestling is all he knows, all he wants to do. When a doctor tells him he must retire, he realizes he has nowhere to go. Somewhere along the way he neglected to come up with a Plan B.

The Wrestler is pumped full of clichés. But it works because Rourke is wonderful at portraying, as The Ram calls himself, "a broken down piece of meat." His appearance is startling—a puffy, reconstructed face, stringy hair made blond through trips to the beauty parlor and a muscled-up body that looks like it would fail any performance-enhancing-drug test. The growling voice is something off a Tom Waits recording. Seriously, if you were not told in advance that is Mickey Rourke on the screen, you would never guess who you are watching. Overall, it was impressive enough to earn Rourke the 2009 Golden Globe Award for Best Actor in a Motion Picture–Drama.

Anyway, Randy tries for awhile to adjust to life without wrestling which, of course, isn't going to work. There's a powerful (albeit obvious) parallel story in which he falls for a stripper named Cassidy (Marisa Tomei) who won't commit to any relationship that lasts longer than a lap dance. Just like him, Cassidy is trying to survive flaunting a body that isn't getting any younger. Just like him, she uses a stage name. Just like him, she is a pro at faking something that the fans want to pretend is genuine.

Tomei is terrific in the role. In fact, our only criticism would be that her body—extremely naked throughout much of the movie—looks too good to play a fading stripper. Trust us, we're not really complaining here.

There's an exchange in *The Wrestler*, when The Ram and The Stripper are sipping beers in a dive bar, as he tries to convince her to care for him. "Round and Round," the heavy-metal standard from the band Rat Attack comes on the jukebox, and both agree that, as Randy says, "The '80s ruled, man, before that Cobain ----- had to come along and ruin it." Really, for both parties here, it's not just the music that peaked decades earlier.

There have been previous efforts to put pro wrestling on the screen, but most were played for laughs. (Does anyone remember *The One and Only*, with Henry Winkler as a Gorgeous George clone?) *The Wrestler*, while eliciting a few chuckles along the way, is definitely not designed to show the fun and happy side of "sports entertainment," or whatever Vince McMahon calls his seedy empire these days. Rather, it's a magnifying glass aimed at the underbelly of the business—pills and shots, gruesome physical punishment, decaying hearts and bodies that give out before their owners hit age 50.

The Wrestler leads up to an ending that can be interpreted several ways—in fact, the two authors of this book debated exactly what it meant. That's a good thing. Nearly two hours in you will deeply care for Randy "The Ram," and worry about what happens to him. Credit writer Robert D. Siegel's layered script, credit director Darren Aronofsky. But mostly, credit Rourke, who creates an unforgettable character.

🔊 **CASTING CALL:** At one point, Rourke was fired from the project and Nicolas Cage was set to replace him, until Aronofsky had a change of heart.

Most of the wrestlers in the movie are played by veterans of the real circuit, including Ernest "The Cat" Miller, Dylan Keith Summers (aka Necro Butcher) and Ron "The Truth" Killings.

25 THE SET-UP (1949–NR)

SPORT: Boxing ★ STARS: Robert Ryan, Audrey Totter, George Tobias
DIRECTOR: Robert Wise

Shot in 20 days and only 72 minutes in length, *The Set-Up* is a small film, but it packs a heavyweight wallop.

It was overshadowed by *Body and Soul*, the John Garfield classic which was released two years earlier and earned four Academy Award nominations, and *Champion* with Kirk Douglas, which came out the same year and picked up six Academy Award nominations. *The Set-Up* wasn't nominated for anything, yet it is arguably the superior film.

Body and Soul and *Champion* were more traditional films. They told an entire life story, following Charlie Davis (Garfield) and Midge Kelly (Douglas) from their dirt-poor youth to

their reign as world champions. We saw the hard times and we saw the high times, the wealth, the flashy clothes, the hot babes, all the trappings of success.

The Set-Up has a more narrow focus. The entire story takes place in one night. In fact, it takes place in just 72 minutes. When we first see the clock, it is 9:10 p.m. The film ends at 10:22 P.M., which means the story unfolds in real time. All the action takes place in a two block stretch of a seedy town where Stoker Thompson (Robert Ryan) is scheduled to fight that night.

We don't see any montages of Stoker's career. We don't see any flashbacks to how he got started or how high he climbed in the heavyweight rankings. We don't see any of the glory, if there was any. All we see is the present and the present isn't very pretty.

Stoker is a 35-year-old pug with scar tissue around his eyes and a brain that's turning to tapioca pudding. There is the suggestion that once upon a time he was a decent fighter, maybe even a borderline contender, but that was many years and many miles ago. Now he is what promoters call "an opponent," a guy who goes from town to town fighting whatever local boy they put in front of him. He lives on chump change and memories.

"I'm just one punch away."
—STOKER

"Don't you see? You'll always be just one punch away." —JULIE

Stoker still has his pride, and he clings to the belief that he is just one win away from getting back in the big time. He hasn't totally lost his skill—he still has knockout power in his right hand—but he is old and slow and his reflexes are gone. He takes too many punches and usually finishes the night flat on his back.

His wife Julie (Audrey Totter) wants him to quit but he refuses. As he packs his gear in his shabby hotel room, preparing to walk across the street to the arena, he says he has a feeling he is going to win this fight. And a win, he says, will earn him a main event spot next time out.

"I'm just one punch away," he says.

Julie has absorbed all the emotional punishment she can stand. On this night, she finally tells Stoker the truth.

"I remember the first time you told me that," she says. "You were just one punch away from a title shot then. Don't you see? You'll always be just one punch away."

"But if I don't do this, what would I do?" he asks.

"I don't care," she says. "It's better than seeing you hurt. It's better than having you dead."

Julie refuses to go with him so he leaves her ticket on the bureau. He walks across the street to the rundown arena. It is such a low budget operation there are only two locker rooms so Stoker shares a dressing space with four other fighters: a nervous teenager who is making his pro debut, a punchy old-timer who still dreams of winning a title, and the two headliners who are ripe with bravado. As Stoker undresses, he studies the others in the room and sees his own career from how it began to where it is headed.

What Stoker doesn't know is his manager Tiny (George Tobias) has sold him out. His opponent Tiger Nelson (Hal Baylor) is an up-and-comer and the mob boss who owns Nelson doesn't want to take any chances on Stoker landing a lucky punch. They slip Tiny $50 for a guarantee that Stoker will throw the fight. But Tiny is so certain that Stoker will lose, he doesn't bother to tell him he is supposed to take a dive.

"There's no percentage in smartening up a chump," Tiny says. He knows Stoker has lost his last four fights, all by KO. He also knows there is no way Stoker would agree to take a dive. Tiny pockets the money and waits for Nelson to send Stoker to dreamland. No one will be any the wiser and Tiny will be $50 richer.

But Stoker, stung by Julie's words and feeling more alone than ever, sees the fight as his last stand. He goes toe-to-toe with Nelson and after three rounds of savage fighting, it is clear Stoker isn't taking a dive. The mob boss (Alan Baxter) is giving Tiny that "so-you-want-to-wind-up-in-the-river?" look. Before the final round, a desperate Tiny tells Stoker about the arrangement. He begs Stoker to go down.

"You gotta do it," Tiny says. "You can't double-cross these guys."

The look on Stoker's bloody face—a mix of disbelief, bewilderment, and finally anger—is unforgettable. We don't want to spoil the ending, but we promise it will bring a lump to your throat.

Director Robert Wise, who later would win Oscars for the musicals *West Side Story* and *The Sound of Music*, makes the dingy arena so real, you can almost feel the peanut shells crunching beneath your feet. He cuts between the violence in the ring to the people in the audience, all screaming with blood lust, and makes the sport seem as barbaric as anything staged in ancient Rome.

🍎 **BET YOU DIDN'T KNOW:** Hal Baylor, who plays Tiger Nelson, was the California state heavyweight boxing champion. He had a professional record of 52-5.

26 MAJOR LEAGUE (1989—R)

SPORT: Baseball ★ STARS: Tom Berenger, Charlie Sheen, Corbin Bernsen DIRECTOR: David S. Ward

We did not have very high hopes for *Major League*. The previews made the film look like a bad slapstick comedy and the plot read like a rip-off of *Bull Durham*. Let's see, we have the wise old catcher, the rookie pitcher with control problems, the first baseman who practices voodoo, the crusty manager. I mean, haven't we seen this before?

When the team reports to spring training, we see the players living in military-style barracks with bunk beds. It's all we can do not to flee the theater. Major league ballplayers don't live like that. Most of them have million-dollar condos on the beach. The rookies might stay in a hotel, but sleeping in a barracks? What utter nonsense.

But if you stay with *Major League* awhile, the strangest thing happens. You get sucked in. Maybe it's the first few wisecracks from play-by-play announcer Harry Doyle (Bob Uecker). Maybe it's Rick "Wild Thing" Vaughn (Charlie Sheen) throwing fastballs that look like real fastballs. It's hard to say why, exactly, but you find yourself settling in for the full nine innings.

Major League isn't as good as *Bull Durham*. It isn't as cleverly written and the romance between catcher Jake Taylor (Tom Berenger) and Lynn Wells (Rene Russo) lacks the sizzle of Crash Davis (Kevin Costner) and Annie Savoy (Susan Sarandon). But there are a lot of fun moments and when the woeful Cleveland Indians turn it around and make their late season run for the pennant, you are on the Wahoo bandwagon.

The movie was written and directed by David S. Ward, who also wrote *The Sting*. Ward is a life-long Indians fan and his screenplay draws much of its humor from the team's bleak history. (Doyle says, "Remember, fans. Tuesday is Die Hard Night. Free admission for anyone who was actually alive the last time the Indians won the pennant.")

The film opens with Rachel Phelps (Margaret Whitton) inheriting the team from her late husband. Rachel is a former showgirl who has no interest in baseball and no love for Cleveland. She wants to move to sunny Miami and finds a clause in the stadium lease which allows her to relocate the franchise if season attendance falls below 800,000.

To ensure that happening, she assembles a team of misfits, including the broken down Taylor, the fresh-from-prison Vaughn and the voodoo man Pedro Cerrano (Dennis Haysbert). She hires manager Lou Brown (James Gammon) who spent 30 years knocking around the bush leagues. When they call to offer Brown the job, he is selling tires. "Can't talk now," he says. "I got a guy on the other line who wants to buy four whitewalls."

The Indians start the season losing games in an empty stadium. For Rachel, everything is going according to plan. The general manager, whose sympathies lie with the players, tells Brown about the owner's scheme. Brown tells the players, saying: "Once she moves the team to Miami, she'll get rid of all of us for better personnel."

"There's only one thing to do," Taylor says. "Win the whole ------- thing."

The inspired Indians go on a tear, climbing from last place to a tie for first in the American League. The season comes down to a one-game playoff against the New York Yankees with Rachel watching glumly from the owner's box. In what hardly qualifies as a surprise, the Indians pull out a victory in the bottom of the ninth.

If it sounds like the re-working of a tired formula, well, it is. But it does so with just enough irreverence to keep you entertained. There is a great scene in which Rachel, in her silk blouse and stiletto heels, walks through the clubhouse, patting players on the butt and snapping their jock straps. With her naughty strut and withering stare, she's the perfect She-Devil owner, a cross between George Steinbrenner and Morgan Fairchild.

"It takes all the little moments and actions of baseball and makes them funny," says Boston Red Sox closer Jonathan Papelbon. "Hey, those stupid things do happen. There are players who act like that. It's a stretch—but not too much."

Uecker, the former big league catcher who now calls games for the Milwaukee Brewers, provides the most laughs with his portrayal of Doyle, the tippling announcer. His call on a Vaughn wild pitch—"Jusssst a bit outside"—is now widely imitated throughout baseball.

Early in the season, when the Indians are in last place and the cavernous ballpark is deserted, Doyle is wrapping up another loss. "For the Indians, no runs, one hit …That's all we got, one ------- hit?" His broadcast partner slaps his hand over the microphone. "Harry, you can't say ------- on the air." Doyle says, "Don't worry, nobody's listening anyway."

Among the players, only Sheen as Vaughn looks like the real thing, but he is so good he makes up for the rest. As a youth, Sheen attended the Mickey Owen Baseball School and in high school, he was scouted by the pros.

One of the few flaws in *Bull Durham* was the casting of Tim Robbins as pitcher Nuke LaLoosh. There was nothing in his mechanics that suggested he could throw a ball hard enough to knock a tin can off a fence. In the film, when they talk about Nuke hitting 100 on the radar gun, we roll our eyes and say, "Yeah, right."

With Sheen as the Wild Thing, we believe it. And when he comes in to get the final out in the playoff game with the Yankees and he throws that high cheese past the cleanup hitter, we're right there with him.

Former Dodgers catcher Steve Yeager was technical advisor for the film and he has a small role as Indians coach Duke Temple. But even Yeager could not make Corbin Bernsen into a believable major league third baseman. Bernsen's Roger Dorn, a prima donna who worries more about his stock portfolio than infield practice, is the weak link in the cast.

KNUTE ROCKNE ALL AMERICAN (1940—NR)

SPORT: Football ★ STARS: Pat O'Brien, Ronald Reagan, Gale Page
DIRECTOR: Lloyd Bacon

The Gipper is fading fast. The strep throat was bad enough, but now pneumonia has set in. The doctors have given way to priests at his bedside. It's only a matter of time, they say.

Rockne is there, of course. They've been in tough spots before, the Gipper and the Rock. There was the Notre Dame-Indiana game in 1920. The Gipper had a separated shoulder that day. The Irish fell behind 10-0, but the Gipper brought 'em back. They pulled it out, they always did.

But this time it's different. The Gipper and the Rock both know that. Gipp asks his coach to lean forward. He has something to say, but it's hard. His voice is almost gone.

"Sometime, Rock," the Gipper whispers, "when the team is up against it, when things are going wrong and the breaks are beating the boys, tell them to go in there with all they've got and win just one for the Gipper. I don't know where I'll be then, Rock, but I'll know about it and I'll be happy."

It is one of the classic movie moments of all time. Pat O'Brien in a fake, flat nose is the Rock—Knute Rockne, head football coach of the Fighting Irish. Ronald Reagan, his pompadour sinfully lavish by Notre Dame standards, is the Gipper—George Gipp, the team's All-American halfback. Together, they spilled enough tears to float the Golden Dome.

The film, *Knute Rockne All American*, might not have launched Notre Dame's national stock, but it surely enhanced it. The movie was a hit with football fans and non-fans, men and women, just about everyone except avowed Notre Dame haters. The scene set in the 1928 season where Rockne gives his "win one for the Gipper" speech at halftime of the Army game has been repeated—and parodied—so often, it is part of our cultural identity.

Who doesn't know "Win one for the Gipper"? We've all heard it, we've all repeated it, and some of us probably voted Reagan into the White House because of it. As David Casstevens wrote in the *Dallas Morning News*, "Ronald Reagan is said to have immortalized George Gipp. A more accurate statement is that the role of George Gipp immortalized Ronald Reagan."

It is true. Throughout his public life, Reagan was referred to as "The Gipper." Political opponents used it mockingly, suggesting Reagan was a shallow wanna-be jock. Allies used it with affection, because they felt Gipp and Reagan shared an All-American, can-do spirit that was instrumental in their success. Even now, years after his death, when people talk about Reagan regardless of the context, they usually slip in a reference to "The Gipper."

Considering how the role defined him, it is worth noting that Reagan almost didn't get the part. The film's producer, Hal Wallis, considered William Holden, John Wayne (a former football player at Southern Cal), Robert Young, and Robert Cummings for the role. He also tested actor Dennis Morgan, who was a bigger name than Reagan at the time, but Reagan made such a passionate pitch for the role, Wallis hired him. Also, Pat O'Brien liked Reagan and lobbied with the producer on his behalf.

"I've always suspected that there might have been many actors in Hollywood who could have played the part better," Reagan said in a 1983 interview. "But no one could have wanted to play it more than I did. And I was given the part largely because Pat O'Brien kindly and generously held out a helping hand to a beginning young actor."

"I'd been trying to write a screenplay about Knute Rockne," Reagan said. "I didn't have many words on paper when I learned the studio that employed me [Warner Brothers] was already preparing a story treatment for the film."

Reagan got remarkable mileage out of very little screen time. *Knute Rockne* runs 98 minutes, and Reagan is on the screen for approximately eight minutes. He has only four scenes, but all are memorable.

Scene One: Gipp (Reagan) is walking across the field in street clothes, picks up a loose ball, and punts it out of the stadium. "What's your name?" Rockne asks. "George Gipp," the freshman replies. "What's yours?"

Scene two: Gipp is stretched out on the practice field, bored, possibly sleeping, when Rockne sends him in to play against the varsity defense. On his first carry, Gipp breaks away on a spectacular touchdown run. When Rockne meets him in the end zone, Gipp flips him the ball and says, "Guess the boys are just tired."

Scene three: Gipp is now the best halfback in the nation, having completed a record-setting senior year which saw him outgain the entire Army team (332 yards) in a 27-17 victory. When the coach's wife (Gale Page) praises him for his great career, the cocky Gipper turns all mushy.

"No, Bonnie, Rock's the rare one, not me," he says. "There will be new fellows coming along year after year, a lot them much better football players than I ever was. There will never be but one Rockne, here at Notre Dame or anywhere else. He gives us something they can't teach in schools, something clean and strong inside.

Not just courage, but a right way of living that none of us will ever forget.

"Don't tell Rock I said that, Bonnie," the Gipper says. "He'll think I was an awful sap."

Gipp coughs, but brushes it off as "just a little sore throat." We know better.

Scene four is the Gipper's dying request to Rockne.

If you're only going to have four scenes in a movie, those are four pretty good ones. Reagan rode them to the Governor's Mansion in California and, finally, to two terms in the White House.

"I have a copy [of the film] and I told President Reagan about it," said Jack Kemp, a former Buffalo Bills quarterback and U.S. Congressman (R-NY). "He asked, 'What's your favorite scene?' I said it was when Rockne sends him in and tells him to run. He [Reagan] asks in a cocky way, 'How far?' Then he runs for a touchdown. To me, it was very much like Reagan. He had the same sort of optimism, that's what made him a great leader."

While historians value the film more for its ties to Reagan, it stands up as a fitting tribute to Rockne, who still holds the record for the highest winning percentage of any Division I football coach (.881). In 13 seasons at Notre Dame, his teams won 105 games and lost 12, with five ties. They were undefeated five times, and in one five-year stretch, Rockne's team lost just once in 40 starts.

The film traces Rockne's story from his arrival in the United States as a boy emigrating from Norway. With little money, Rockne (O'Brien) has to work four years in a Chicago post office to earn enough to afford the tuition at Notre Dame. He is 22 when he finally enrolls and he shows as much promise in chemistry class as he does on the football field.

Father Callahan (Donald Crisp) asks Rockne to stay on as a graduate assistant in the chemistry department, but Rockne says his heart is in coaching. "You think I'm making a mistake, don't you?" Rockne asks. "Anyone who follows the truth in his heart never makes a mistake," the priest replies.

Combining creative brilliance with a dynamic personality, Rockne builds Notre Dame, a small Catholic school in Indiana, into a major force in American sports on a par with the New York Yankees. Millions of people—many of whom never graduated high school or ever set foot in Indiana—became fans of the Irish. That passion still exists among the so-called "Subway Alumni" and it all started with the charismatic Rockne.

Knute Rockne All American was released in 1940, nine years after Rockne's death in a plane crash. In 1997, the film was deemed "culturally, historically or aesthetically significant" by the United States Library of Congress and selected for preservation in their National Film Registry.

🍎 **BET YOU DIDN'T KNOW:** Knute Rockne was only 43 years old when he died.

🎬 **CASTING CALL:** James Cagney tried to convince Warner Brothers to cast him as Rockne. He felt he was typecast as a gangster and saw this role as a chance to break out. However, the studio chose O'Brien.

🎬 **PIVOTAL SCENE:** Rockne attends a Broadway show, and while watching the dancers move about the stage, he is inspired to create the "shift." That was an innovation in which the backfield would line up in one formation and shift to another formation just before the ball was snapped. That was the Notre Dame offense until the 1940s, when coach Frank Leahy went to the T-formation.

28 BANG THE DRUM SLOWLY (1973–PG)

SPORT: Baseball ★ **STARS:** Michael Moriarty, Robert De Niro
DIRECTOR: John D. Hancock

Mark Harris wrote the best-selling novel *Bang the Drum Slowly* in 1956, but it wasn't made into a motion picture until 1973. Why the delay? Well, most people in Hollywood thought the story—the friendship between two major league ballplayers, one of whom is dying—was too depressing. They didn't think audiences would pay to see it.

However, the success of tear-jerkers such as *Love Story* (1970) and *Brian's Song* (1971) made the studios reconsider. *Love Story* was a box office smash and was nominated for seven Academy Awards. *Brian's Song* is one of the most popular made-for-TV movies of all time. So if you have the right story, people will watch, even if it means stocking up on Kleenex.

Bang the Drum Slowly is certainly a compelling story. It is, in essence, *Brian's Song* in baseball, the major difference being *Brian's Song* was based on a true story and *Drum* is fiction. But the themes of friendship, teamwork and—dare we say it?—love are much the same. Audiences and critics alike embraced the film.

Drum features a young Robert De Niro as Bruce Pearson, the catcher who is dying of Hodgkin's Disease. Michael Moriarty is Henry Wiggen, Pearson's roommate. Pearson is a country bumpkin with limited baseball skills. Wiggen is an All-Star pitcher and Renaissance man who writes books and sells insurance when he is not winning games for the New York Mammoths.

They are an odd couple whose paths never would have crossed if it were not for baseball. Pearson chews tobacco and wears a smiley face T-shirt under his sports coat. Wiggen is a bit of a snob, intellectually superior to the other players and rather aloof. His teammates call Wiggen "Author," but Pearson is so dim, he doesn't get the joke. He calls the pitcher "Arthur."

Pearson is an easy target for clubhouse pranks and the other players pass the time by ragging on him. When he learns he has Hodgkin's Disease, Pearson turns to Wiggen, whom he considers a friend. He begs his roommate to keep his illness a secret. He fears the team will get rid of him if it finds out.

Wiggen, who serves as the film's narrator, says in a voiceover, "Suddenly, you're driving along with a man who's been told he's dying. It was bad enough rooming with him when he was well."

Wiggen, who merely tolerated Pearson earlier, becomes his protector. He sympathizes with this simple, good-natured soul who made it to the big leagues only to learn that he has less than a year to live. As Pearson's father (Patrick McVey) tells Wiggen, "I swear, my son's been handed one ------ deal."

"Skip the facts. Just give me the details."
—DUTCH

Wiggin nods in agreement.

When Wiggen negotiates his new contract with the Mammoths, he insists the team include a guarantee that Pearson will not be released or sent to the minors. Wiggen convinces management that he pitches better when Pearson is behind the plate. Since he is the ace of the staff, they give him what he wants.

As the season goes along, Wiggen and Pearson are inseparable. People begin to, shall we say, wonder, what's up with those two guys? Manager Dutch Schnell (Vincent Gardenia) hires a private eye to follow them around. Dutch, an old-school baseball man, won't stand for any hanky-panky on his ball club.

There are some funny scenes with Wiggen and Pearson making visits to the Mayo Clinic and explaining them away as fishing trips. They even invent a story about Pearson having to see a specialist for a case of the clap. It's a tricky thing going for laughs in a film about terminal illness, but it is done to good effect in *Bang the Drum Slowly*. Those moments—which usually include Dutch sucking on a cigarette and arching a suspicious eyebrow—give the audience some needed relief from all the sadness.

Wiggen finally tells a few teammates about Pearson's illness. They feel ashamed for all the times they needled and mocked Pearson. They close ranks around him and—here is where the film gets a little sticky—the Mammoths launch an inspired pennant drive with Pearson playing the best baseball of his career. He is dying, but somehow he turns into a combination of Carlton Fisk and Johnny Bench.

It is a stretch, especially given De Niro's awkward attempts to swing a bat. He never played baseball as a kid, and it shows. When he was cast in the role of Pearson, he went to the Cincinnati Reds training camp and studied the catchers, including Bench. He picked up a few things—he is very good at chewing tobacco and spitting, for example—but he isn't convincing as a player. Pearson is supposed to be a fringe big leaguer, but De Niro isn't even that good.

Moriarty, on the other hand, is very good as Wiggen. His right-handed throwing motion is smooth and his little tics on the mound—the way he fingers the ball, the way he fiddles with his cap and looks in for the catcher's sign—are those of someone who played a lot of baseball.

Moriarty also impressed as a hockey player in *The Deadliest Season* (1977) before joining the original cast of the TV series *Law and Order*.

Bang the Drum Slowly walks a fine line between solid storytelling and sentiment. There are a few slips into the sugary stuff, but for the most part director John D. Hancock keeps things on course. Late in the film, Pearson—his health deteriorating—staggers under a pop fly. It's a scene that could have been maudlin, but it is handled delicately and well. (De Niro, the method actor, ran in circles to make himself dizzy before each take.)

The film ends with Pearson's inevitable passing, and Wiggen, as the narrator, gets the final word. His tribute to his late roommate is understated, but it stays with you for that very reason.

"He wasn't a bad fellow," Wiggen says. "No worse than most and probably better than some. And he wasn't a bad ballplayer when they gave him half a chance.

"From now on," Wiggen says, "I rag nobody."

29) JERRY MAGUIRE (1996—R)

SPORT: Football ★ STARS: Tom Cruise, Cuba Gooding Jr., Renee Zellweger DIRECTOR: Cameron Crowe

Jerry Maguire is actually two movies in one. Men will watch it and see a solid sports film. Women will enjoy a delightful chick flick. The great thing is—it works for both.

So we had He and She sit down together and take notes on this Tom Cruise vehicle about an idealistic sports agent (boy, there's an oxymoron) and the woman and wide receiver who love him for different reasons.

He: "This might be the best movie ever about sports business. It shows the seamy side of football as an industry, without destroying your love for the game. Plus, did you catch all those great cameos? Give a big thumbs up to Drew Bledsoe and Troy Aikman and Warren Moon and—"

She: "What are you talking about? Jerry Maguire is about relationships. It's about a man who is everybody's friend, but can't show intimacy. It's about a woman—and bravo to Renee Zellweger as Dorothy Boyd—who loves him for the man he wants to be or, as she says,

'I love him for who he almost is.' "

He: "Yeah, well, she does a fine job, I guess. But let's be honest—the best character in this movie is Rod Tidwell. Cuba Gooding Jr. won the Best Supporting Actor Oscar for portraying Tidwell as the proud, hot-dogging, sometimes angry wide receiver that we've seen so many times in the NFL. He's Chad Johnson, or Terrell Owens. Gooding nailed the role. I challenge you to name any actor ever who so accurately got into the skin of the modern athlete."

She: "Funny, I though Gooding was great, but more for how he displayed all that love for his family, more than all that preening. And, good as he was, the character who steals *Jerry Maguire* is Jonathan Lipnicki as Dorothy's son, Ray. That little boy with the glasses was adorable. Didn't you notice how Jerry falls in love with him before he falls in love with Dorothy? And how about how Ray walks around saying, 'Do you know the human head weighs eight pounds?' How endearing was that?"

He: "Ah, you give any little kid a couple of funny lines and he'll try to steal the movie. Let's get to the part that really surprised me about *Jerry Maguire*. The football scenes are terrific—I give it an A-minus on sports action. That pivotal scene—where the Cardinals play the Cowboys on *Monday Night Football*—seems lifelike. Yes, some of it actually was clips of Aikman and Emmitt Smith making plays. But the shot of Tidwell catching that touchdown pass in traffic, being dumped on his head, laying there unconscious and—finally—getting up to perform the greatest touchdown dance ever? That was phenomenal action."

She: "You call that action? I'll give you A-plus action—the pivotal scene, which takes place on Dorothy's front porch after her first date with Jerry. He breaks the strap on her dress (maybe accidentally, but I don't think so), and when he goes to fix it, it evolves into one of the great foreplay scenes in movie history."

He: "To be honest with you, I found Jerry's fiancé, Avery Bishop (Kelly Preston), a lot hotter than Dorothy. But that's just how a guy looks at things. We tend to like sexy more than cute. Anyway, let's get to some of these issues we're supposed to debate. Best line? Easy. When Rod says to Jerry, 'Show me the money.' Sure, we all grew tired of the line after it became a cliché. But that's every self-centered pro athlete talking."

She: "I suppose it is, but it's not as good as two lines right near the end of the film. Jerry comes home to win back his wife. He gives her the whole close-the-deal love speech, leading up to 'You complete me.' Ooh, that was romantic. And she stops him right there and says, 'Shut up. Just shut up. You had me at hello.' How great was that moment?"

He: "Not as good as it would have been if they didn't exchange those lines in a room full of bitching divorcees. Actually, that'll be my thumbs down—those repeated scenes of the cackling manhater's club. It seems like the henhouse where Dorothy and her sister (Bonnie Hunt) live is permanently inhabited by those shrews."

She: "You know what? We can agree on that—but I take a little umbrage at your characterization there. Let's just say that those scenes with the angry women could have been cut way down. By the way, the older woman who talks about 'getting in touch with my anger?' That's director Cameron Crowe's mother."

He: "Hey, can I give another compliment here? Jay Mohr was perfect as Bob Sugar, the oily, despicable anti-Jerry agent who reminds me of every Scott Boras/Drew Rosenhaus greed-hound polluting sports these days."

She: "Okay, if despicable is what you're looking for. I'll give my 'cheers' to the soundtrack. Everything from Springsteen to Nirvana to Paul McCartney to Charlie Mingus. Best soundtrack since *Ghost*. Speaking of which, I'd list *Ghost* under the category of, 'If you liked this, you'll like . . .' but I know you want me to name another sports movie here. So I'll say *Cinderella Man*. Another movie about a good-hearted man in a tough business who wins over Renee Zellweger."

He: "Not a bad choice. I'll watch that with you any time. But I have to say that if you like this, you'll like *Any Given Sunday*, which is also a critical look at the business side of sports. As far as 'repeated watching quotient,' I'd say I could see *Jerry Maguire* about once a year."

She: "I'm with you. You bring the popcorn, I'll bring the hankies."

🍎 **BET YOU DIDN'T KNOW:** *Jerry Maguire* grossed more at the box office—$154 million—than any sports drama in history. It marked the fifth consecutive Tom Cruise movie to crack the $100 million mark, following *A Few Good Men*, *The Firm*, *Interview With the Vampire*, and *Mission: Impossible*.

30 DOWNHILL RACER (1969–NR)

SPORT: Skiing ★ **STARS:** Robert Redford, Gene Hackman
DIRECTOR: Michael Ritchie

There is much to admire in *Downhill Racer*, director Michael Ritchie's taut study of the World Cup ski circuit. There is the camerawork, which conveys the speed and danger of the sport. There is the spectacular scenery. There is the acting of Robert Redford as skier David Chappellet and Gene Hackman as Eugene Claire, coach of the U.S. team.

But what is most intriguing about the film is its portrait of the world-class athlete. In this case, the athlete is Chappellet, and behind the blonde hair and perfect smile is what? An All-American hero? Uh, no. Chappellet is a ski bum with little in the way of education, a shallow narcissist incapable of carrying on a conversation or sustaining a relationship. Think Bode Miller, toothy smile and all.

"All you ever had was skis," Claire (Hackman) says, "and that's not enough."

Well, for Chappellet (Redford), it is. His life has been defined by one thing: his ability to ski downhill faster than anyone else. He discovered that talent at an early age and spent years developing it—to the exclusion of everything else. He is a skier, but he could be a swimmer, a gymnast, a skater, or a fencer, and it would be the same thing. The dedication required to reach the elite level is the same across the board.

"This is the point we miss when we persist in describing champions as regular, all-around Joes," wrote Roger Ebert in the *Chicago Sun-Times*. "If they were, they wouldn't be champions."

This is not to say all Olympic athletes are like that, but some are. Chappellet is one of those who operates in a narrow tunnel, but Ritchie and writer James Salter don't condemn him for it. They make it clear that if he were a regular, all-around Joe, he'd still be on the family farm in Idaho Springs, Colorado. He wouldn't be tearing up the slopes in France and signing fat endorsement deals with ski manufacturers.

Chappellet is ego-driven and selfish, which leads to clashes with Claire, who runs the U.S. team with a football coach's "we're-all-in-this-together" mentality. Chappellet is standoffish with the other skiers, and his rivalry with the other top American, Johnny Creech (Jim McMullan), contributes to Creech crashing and wrecking his knee.

"He's not a team man," Claire says angrily.

"Well, this isn't exactly a team sport," another coach points out.

He is not trying to defend Chappellet, exactly, but he makes a valid point. Chappellet is a skier. He's not a point guard or a tight end. When he is careening down the hill at 90 miles per hour, he is all alone. Can you really blame him if he is all about himself?

"His world is that international society of the well-exercised inarticulate, where the good is known as 'Really great' and the bad is signified by silence," wrote Roger Greenspun of the *New York Times*. "In appreciating that world, its pathos, its tensions and its sufficient moments of glory, *Downhill Racer* succeeds with sometimes chilling efficiency."

"He's not for the team. He never will be."
—CREECH

Ritchie and Salter draw the character of Chappellet in subtle strokes, some of which make him sympathetic. He doesn't say much, and when he does speak, his conversations tend to trail off. He seems incapable of explaining himself, or maybe there just isn't that much to explain. Perhaps he is uncomfortable around the other skiers, most of whom are rich kids with college educations. Maybe that's why Chappellet keeps to himself. We're left to wonder.

There is a poignant scene when Chappellet, now a rising star on the ski circuit, goes back to visit Idaho Springs. When he returns to the farm, his father (Walter Stroud) practically ignores him. There are long stretches of silence and, painful as they are, they help explain why Chappellet is so cut-off from those around him. At one point, he tries to make his father understand his competitive desire.

"I'll be a champion," he says.

His father barely looks up. "The world's full of them," he grunts.

The moment is as chilling as a head-first tumble into a snow bank. From then on, you view Chappellet a little differently. He is still a cynic—he'd sneer at having his face on a Wheaties box, but he'd gladly pocket the money—and he has a hollow affair with a French woman who works for a ski company, but what lingers is the sense that he has no life away from the slopes. As glamorous as it is, it is also rather empty and sad.

"Some of the best moments in *Downhill Racer* are moments during which nothing special seems to be happening," Ebert wrote. "They're moments devoted to capturing the angle of a glance, the curve of a smile, an embarrassed silence. Together they form a portrait of a man that is so complete and so tragic that *Downhill Racer* becomes the best movie ever made about sports without really being about sports at all."

🍎 BET YOU DIDN'T KNOW: Robert Redford did most of his own skiing in *Downhill Racer*. Ritchie only used a double (former ski racer Joe Jay Jalbert) for scenes that required Chappellet to take a tumble.

HE GOT GAME (1998–R)

SPORT: Basketball ★ **STARS:** Denzel Washington, Ray Allen
DIRECTOR: Spike Lee

Do you remember LeBron James' senior year of high school in 2003? The national ESPN-fueled mania over an 18-year-old with a regal nickname; the premature parallels to Jordan; the unconfirmed stories of illegal gifts and shady hangers-on; the debate over whether King James should go pro or spend a year at college? Remember?

Well, take that story, turn it into a Spike Lee film (or "joint," as the director calls it), add Denzel Washington for depth, a dynamite soundtrack, and a few extraneous subplots, and you've got *He Got Game*. This story is more about temptation than about basketball, more about greed than about sport. And it

works—in large part because Lee loves hoops enough to both embrace it and shine a light on its dirty underbelly.

"It's my favorite movie," says former Cleveland Cavaliers guard Delonte West, "because it really shows the reality of making the move from high school basketball to the next step. It shows the other side of basketball—pressure—that people don't get to see."

The story opens with Jake Shuttlesworth (Denzel) shooting jumpers on a run-down court at the Attica Correctional Facility in upstate New York. Jake is serving a lengthy sentence for murdering his wife—although we later find out there's more to that story.

Meanwhile, the son he left behind, Jesus (Ray Allen), has grown to become the top-ranked high school player in the country.

Jake is summoned to see the warden (Ned Beatty), who offers a proposition: If Jake can convince his son to play at the governor's alma mater, Big State, The Guv will agree to reduce Jake's interminable sentence. Jake is given a cheap hotel room in the old neighborhood on Coney Island, a week's time, and a letter of intent he must persuade Jesus to sign.

The problem is that Jesus is not happy to see the man who killed his mother. The more Jake tries for reconciliation, the more he is resisted. And, soon enough, Jesus comes to regard his father as one more slimy bloodsucker trying to latch onto him to achieve his own ends.

Which, of course, is true. In fact, the movie is populated by smarmy characters who praise Jesus as their ticket to the main chance. They include—and give us a little space here to roll out this roster—the conniving girlfriend (Rosario Dawson) who's convinced she'll be left behind; the pedophile uncle who quotes Don Fanucci from *The Godfather II* ("All I'm asking is that you let me wet my beak."); the oily sports agent who dangles a platinum Rolex in the kid's face; the viscous high school coach who lays $10,000 on the table; the holy-roller coach of Tech U (John Turturro); the swaggering college star (Rick Fox), and the silicon-enhanced "recruiters." And, of course, the tainted governor of New York. We think his name was Spitzer.

Throughout the betrayal, Jesus tries to stay pure. He accepts no graft (well, very little . . . what man could turn down those nubile twins at Tech?) and repeatedly tells everyone he is weighing his options. The more he demurs, the more pressure is exerted on him.

It's all a little black and white, and we don't just mean that in the way you might expect from a Spike Lee film (yeah, yeah, "joint."). *He Got Game* would be improved if Lee introduced one or two sympathetic characters to his story. And never mind the Coney Island hooker with the heart of gold (Milla Jovovich) who takes a liking to Jake Shuttlesworth. We still can't figure out what she had to do with the story.

But overall, this is one of the better basketball movies ever made—not in a lead-up-to-the-big-game *Hoosiers* kind of way, but in an athletes-as-commodities kind of way. And if you think the movie is a big stretch on the truth, listen to the words of New Orleans Pelicans guard Tyreke Evans, who spent one year at Memphis University after being a high school basketball phenom in Chester, PA.:

"I felt I was watching my own life when I saw that movie. Everything that happened in *He Got Game* really happens. Trust me."

And give Spike Lee credit for taking risks. Casting Allen—a great player, but a novice actor—works, because his inexperience on screen plays well to the character's innocent persona. The soundtrack combines Public Enemy with Aaron Copland in a way some critics hated, but, well, we always loved the "Beef. It's what's for dinner" song, whatever that's called.

🔊 **CASTING CALL:** Spike Lee sure knows how to call on friends. Cameos come from more than a dozen top-shelf NCAA coaches, as well as an NBA Dream Team of Shaquille O'Neal, Scottie Pippen, Reggie Miller, Charles Barkley, and Michael Jordan (the only one in the movie who actually utters, "He got game."). Jesus' high school teammates include Travis Best, Walter McCarty, and John Wallace. No wonder they won the city title. And it's always good to catch the great Jim Brown, here stealing all the good lines as a mean-spirited probation guard.

32 REQUIEM FOR A HEAVYWEIGHT (1962–NR)

SPORT: Boxing ★ **STARS:** Anthony Quinn, Jackie Gleason, Mickey Rooney
DIRECTOR: Ralph Nelson

Before Rod Serling launched his landmark TV series *The Twilight Zone*, he wrote the powerful screenplay *Requiem for a Heavyweight*. Serling, who did some boxing in his younger days, painted a bleak picture of the fight game. *Requiem* is a brilliant piece of work, but, man, is it depressing.

It was first produced for *Playhouse 90*, a live TV show, on October 11, 1956. Jack Palance played the boxer Mountain McClintock, Keenan Wynn was his unscrupulous manager Maish, and Ed Wynn was the faithful trainer, Army. The show won four Emmys, and Serling's script earned a Peabody Award.

One year later, British television produced the show (changing the title to *Blood Money*) with the pre-James Bond Sean Connery in the lead. The story also has been revived a few times on Broadway, most recently in 1985 with John Lithgow as the broken-down fighter and George Segal as the manager. Segal's Maish was so hateful, *New York Times* theatre critic Frank Rich described him as "serpentine."

So *Requiem* has been around for a long time, and some talented actors have taken on the roles, but it is hard to imagine anyone else in those parts once you've seen the 1962 film with Anthony Quinn, Jackie Gleason, and Mickey Rooney. They are so good, they blow you away.

Quinn and Gleason were at their peak in 1962 when the film was made. Quinn was fresh off two hits, *The Guns of Navarone* and *Lawrence of Arabia*, and Gleason had just earned an Academy Award nomination for his performance in *The Hustler*. The *film noir* tone of *Requiem* was right in step with *The Hustler* and Gleason stepped seamlessly into the role of Maish.

The film represented a comeback for Rooney, whose movie career had faded. He does not have many lines—Army is a quiet sort who serves as caretaker as well as trainer of Mountain Rivera (Quinn)—but when he finally tells Maish off, the diminutive Rooney cuts the hulking Gleason off at the knees.

"You fink," he says. "You dirty, stinking fink."

Nothing more needs to be said. It is a killer line and Rooney drives it home with the force of a sledgehammer.

Requiem is the story of a fictional heavyweight who was once ranked fifth in the world, but after 17 years in the ring he is battered and used up. The film opens with Mountain Rivera (Quinn) in the ring against Cassius Clay (Muhammad Ali in his film debut). The viewer sees the fight through Rivera's eyes as the lightning-fast Clay throws a flurry of punches into the camera lens en route to a knockout victory.

In the dressing room, the doctor examines Rivera's bloody eye and tells him it's all over. If he fights again, he likely will go blind. "Tell him to buy a scrapbook," the doctor says with little sympathy. He obviously has delivered this message before.

Maish and Army have been Mountain's manager and trainer from Day One. The three men have been a team, traveling the country together, drinking and playing cards. They were in the big time for awhile when Mountain was a contender, but those days are long gone. Things have been sliding downhill for years, and now they're at an end. So what are the three men to do?

Mountain would seem to face the toughest road. He is 37 and punch drunk, with a face that has been beaten into a mask of pulpy scar tissue. He has no education and no skills other than fighting. But Maish has even greater problems. He has lost a ton of money to local mob boss Ma Greeny (a truly scary Madame Spivy), and he has no way of repaying her now that Mountain, his meal ticket, is out of business.

Army takes Mountain on a series of job interviews, hoping to find something the big guy can do. He almost gets a job as an usher in a movie theater, but they don't have a uniform to fit him. He goes to an employment agency where kindly counselor Grace Miller (Julie Harris) takes an interest in him. She sees that beneath the misshapen nose and cauliflower ears, Mountain is really a gentle soul with a good heart. She suggests a job as a counselor at a summer camp. She arranges for Mountain to meet the camp director for an interview.

Maish, meanwhile, is frantically trying to save his butt. He makes a deal with a sleazy promoter (Stan Adams) for Mountain to become a professional wrestler. He will wear a feathered head dress and wrestle as "Big Chief Mountain Rivera." It's humiliating, but it is a paycheck and once Maish gets his cut, he can pay off Ma Greeny.

Maish tells Mountain about the wrestling offer. Mountain wants no parts of it. He had 111 professional fights and never took a dive. He has too much pride to get involved in a freak show. He tells Maish about the job interview. Maish pretends to be happy for him. He suggests they celebrate. Maish gets Mountain drunk so that when he goes to meet the camp director, he is an incoherent, falling-down mess.

Grace is heartbroken by Mountain's humiliation. A spinster, she feels drawn to him—and he feels the same for her—but it is awkward. Neither one knows how to move beyond the first tentative kiss. As Grace is leaving Mountain's room, she runs into Maish. She slaps him for double-crossing Mountain. He responds with a look that says, "Lady, you really don't get it."

"Do you really want to help him?" Maish asks. "Here's how you can help him. Leave him alone. If you gotta say anything to him, tell him you pity him. Tell him you feel so sorry for him you could cry. But don't con him. Don't tell him he could be a counselor at a boys' camp. He's been chasing ghosts so long, he'll believe anything, any kind of ghost. Championship belt, pretty girl, maybe just 24 hours without an ache in his body. It don't make any difference. It all passed him by."

Like we said, *Requiem* is depressing. But it's also one of the finest boxing films ever made.

66 BEST LINE: Mountain Rivera: "Mountain Rivera was no punk. Mountain Rivera was almost Heavyweight Champion of the world."

PIVOTAL SCENE: Maish admits to Mountain that toward the end of his career, he made money by betting against him. Mountain cannot believe the betrayal.

"Why, Maish?" he asks. "Why did you bet against me?"

"Would it have made any difference?" Maish replies. "Would it have made any difference if I hocked my left leg to bet on you? You're not a winner anymore, Mountain. There's only one thing left and that's make money from the losing."

"In all the crummy 17 years I fought for you, I wasn't ashamed of one single round, not one single minute," Mountain says. "Now you make me ashamed."

33 PHAR LAP (1983–PG)

SPORT: Horse Racing ★ **STARS:** Tom Burlinson, Ron Leibman, Martin Vaughan
DIRECTOR: Simon Wincer

Phar Lap is to Australia what Seabiscuit is to America—a racehorse whose rise to prominence during the Depression lifted the spirits of a nation. Both horses overcame humble beginnings to gallop into legend. Eventually, both horses were immortalized in books and motion pictures.

Seabiscuit, the movie, made more money and earned more praise, including seven Academy Award nominations, but we think *Phar Lap,* a much smaller film made in Australia, is better. Even if you don't have the slightest interest in horse racing, the story is so beautifully told that you will be cheering.

The film is directed by Simon Wincer, who also directed the American TV series *Lonesome Dove.* It is photographed by Russell Boyd, whose other credits include *The Year of Living Dangerously, Tender Mercies,* and *Gallipoli.* David Williamson (*The Club*) provides the screenplay. Their combined talents make *Phar Lap* a film you won't soon forget.

Phar Lap died in 1932, yet the big chestnut colt still is revered in Australia. His enormous heart—which weighed 6.2 kilograms compared to 3.2 kilograms for most horses—is on display in the National Museum of Australia. In fact, it is the object visitors ask to see most often. His body is stuffed and mounted in the Australia Gallery at the Melbourne Museum. Fans still cry and leave flowers at his feet.

"In my opinion, there is nothing maudlin in a nation mourning the loss of a racehorse when that horse is Phar Lap," said Sir Hubert Opperman, Australia's champion cyclist.

In the film, Phar Lap is purchased sight unseen by American sportsman David Davis (Ron Leibman) on the recommendation of Australian trainer Harry Telford (Martin Vaughan). Telford feels the horse has a winner's bloodlines. But when Phar Lap arrives in Australia from his native New Zealand, he looks like a dud. He is gangly and shows little promise on the track.

"He looks like a cross between a sheepdog and a kangaroo," Davis says.

The horse loses its first five races, and Telford's wife, Vi (Celia De Burgh), tells him to sell the worthless nag and be done with it. But the trainer believes he can turn the horse into a winner, and he does so with the help of stable boy Tommy Woodcock (Tom Burlinson).

The horse literally rips the shirts off the other stable boys, but he bonds with the kindly Woodcock. Phar Lap trusts Woodcock, and the stable boy's patience unlocks the horse's potential. When Woodcock tells him to run, he runs like the wind.

Telford and Woodcock disagree on how to train the horse. Telford trains him hard, almost to the point of cruelty. Woodcock, who loves the horse and calls him by the pet name "Bobby," prefers a gentler hand. When Woodcock objects to Telford's methods once too often, the old trainer fires him. However, Phar Lap becomes depressed and stops eating, so Telford is forced to bring Woodcock back. The horse immediately perks up.

"Your trouble is you think horses are human," Telford tells Woodcock as the boy slips Phar Lap another sugar cube. "You can't train them that way. You ease up and they'll take advantage of you."

"Bobby is different," Woodcock says.

Phar Lap begins winning races and setting track records across Australia. In 1930 and '31, he wins 14 consecutive races. The fans love him because, like Seabiscuit, he is an underdog, a horse that came from nowhere to dominate the sport. But the stuffed shirts who run the racing industry aren't nearly as fond of him. Phar Lap is so good, they say, he is taking the competition out of the sport. He is also costing the bookmakers a bundle because the adoring public keeps betting on him and he keeps cashing their tickets.

At one point, a promoter takes Davis aside and says, "Look, Dave, if something is good, that's OK. But if something is too good, it upsets the entire system." In other words, look out.

One morning, Phar Lap is out for his morning romp and a shotgun blast is fired from a passing car. The horse is not injured, but Davis and Telford agree to move him to a secret location and hire guards for protection.

Things aren't much better on the track. Racing officials try to handicap Phar Lap by putting lead weights in his saddle to slow him down. They keep increasing the weight until Woodcock and jockey Jim Pike (James Steele) are almost in tears. "I've never had a horse die under me," Pike says. "I don't want Phar Lap to be the first." Yet the horse—which the press dubs "The Red Terror"—keeps winning.

Williamson's script makes the point that Phar Lap isn't a horse that runs fast. Rather, he is a horse that runs as much with his heart as with his legs. Like any great athlete, he has an innate competitive drive that carries him beyond exhaustion and beyond pain. He runs one race on a cracked hoof and as he comes down the stretch, the camera focuses on his bloody leg wrappings, yet Phar Lap never breaks stride until he crosses the finish line.

"He's a freak," a rival owner says.

In 1932, Davis takes Phar Lap to Mexico to race in the Agua Caliente Handicap for the largest purse in North American track history. The horse wins again in track record time, and Davis pockets a cool $50,000.

OK, here is the spoiler alert. If you haven't seen Phar Lap but you have an interest in the film, you might want to stop reading now, because we're about to discuss the ending.

After his win in Mexico, Phar Lap is sent to a ranch in California, where he will stay while Davis lines up a series of races for him in the United States. One night Woodcock is awakened by the sound of moaning from Phar Lap's stall. The horse is terribly ill and there is nothing the doctors can do. He dies the next morning, just 16 days after his spectacular run at the Agua Caliente Stakes.

While the film does not come out and say it, it certainly implies that Phar Lap was the victim of foul play. It is believed the horse was killed either by professional gamblers or jealous rivals. Tests found Phar Lap's stomach and intestines inflamed, which suggests poisoning, most likely arsenic.

It is interesting to note the film was edited differently for Australian and American audiences. The version that played in Australia opened with Phar Lap's death and then told his story in flashback. The Phar Lap that played in America opened with Davis and Telford purchasing the colt and followed his rise to stardom.

Why the change? Well, the studio knew that in Australia most people were familiar with the story and if the film was structured in the traditional way, audiences would spend the whole night waiting for—and dreading—the sad ending. It was better, the producers felt, to get it out of the way early. The American audience was less likely to know the story, so the ending would come as a surprise.

Regardless of which version you see, be forewarned: The sight of Tommy Woodcock crying "Bobby" as he hugs the dying Phar Lap will rip your heart out.

❝ BEST LINE: Jim Pike: "If anything catches us today, mate, it'll have to have wings."

★ SPORTS-ACTION GRADE: B. Director Simon Wincer, as he demonstrated in *Lonesome Dove* and *The Lighthorsemen*, knows how to film a story on horseback.

34 THE ROOKIE (2002—G)

SPORT: Baseball ★ STARS: Dennis Quaid, Rachel Griffiths, Brian Cox
DIRECTOR: John Lee Hancock

The story behind the making of *The Rookie* is almost as improbable as the story itself. You find it hard to believe that a 35-year-old high school teacher could become one of the oldest rookies in major league history? Well, that is the tale of Jim Morris, who had a brief but inspiring career with the Tampa Bay Devil Rays in 1999-2000. But how that story made it to the screen is just as unlikely.

It began with a young film producer named Mark Ciardi flipping through a magazine at the doctor's office. He saw an article about a Texas school teacher who was signed to a professional baseball contract after throwing 98 MPH fastballs in an open tryout.

"I saw the picture and said, 'That's Jimmy Morris,' " Ciardi said.

Ciardi and Morris were both signed by the

Milwaukee Brewers in 1983. They roomed together in spring training. Ciardi made it to the big leagues (he pitched four games for the Brewers in 1987) but Morris did not. He had a series of arm injuries that ended his career in 1989. Ciardi quit around the same time. Morris went back to Texas; Ciardi went to Hollywood.

A decade later, their lives intersected again, this time with Morris' dramatic resurrection as a ballplayer. When Ciardi saw the article, Morris was pitching for the Devil Rays' top farm team in Durham, N.C.

"My first thought was, 'This is a movie,'" Ciardi said. "My second thought was, 'I want to make it.'"

Ciardi and his business partner Gordon Gray were still trying to establish themselves in the movie industry. They were working out of a garage, searching for just the right project. The Jim Morris story, they felt, was it.

Ciardi called the clubhouse in Durham and spoke with Morris. He was excited about the idea of a movie and referred Ciardi to his agent to work out the details.

"We were up front with him," Ciardi said. "We told him we didn't have much money, but we knew people who would write the check. We took the idea to Disney and they loved it. They agreed to back us 100 percent. We tried to get back to the agent to sign the deal, but he was away. We couldn't contact him."

That weekend, Morris was called up to the Devil Rays. He made his first major league appearance in Texas and struck out the first batter he faced (Royce Clayton). The team went to California a few days later and a writer from the *Los Angeles Times* did a front page profile on Morris. The headline read: "This Should Be a Movie." The piece caught the eye of every studio executive in town. Pretty soon, Morris and his agent were flooded with offers.

"I thought we were sunk," Ciardi said. "When we reached (the agent), he kept cutting out to take other calls. He got over 200 calls in three days. Every major studio wanted in. But Jim and his agent stuck with us. They said we came to them first so it was only right that we get the first crack. Thank goodness for loyalty."

The Rookie was a hit, taking in more than $75 million at the box office. It launched the careers of Ciardi and Gray, who produced two more inspirational sports films, *Miracle* (2004) and *Invincible* (2006) and the father-daughter football comedy *The Game Plan* (2007). Ciardi and Gray have long since moved out of the garage. They are now major players in Hollywood and it all started with the story of Jim Morris.

Set in a dusty West Texas town called Big Lake, Morris (Dennis Quaid) teaches science and coaches baseball at the local high school. He was once a hot pitching prospect (the No. 4 overall pick in the 1983 amateur draft) but four arm surgeries later, he was given his release. Now 35, married with three children, he is doing OK, but that lost opportunity still nags at him. At night, he throws pitch after pitch into the rusty backstop on the high school diamond.

One day while pitching batting practice to his team, Morris is goaded into putting a little "something" on the ball. OK, he says. He tugs on his cap, cranks up the old left arm and throws a fastball that explodes into the catcher's mitt. The kids can't believe it. Neither can Morris, who says, "Gee, I forgot how good that sounds."

The Big Lake kids are headed for another losing season when Morris lectures them on never quitting. "If you don't have dreams," he says, "you don't have anything."

One of the kids asks, "What about your dreams, Coach?" Another says, "Yeah, you're the one who should be wanting more."

They strike a deal: If the team wins the district championship, Morris will take another shot at the big leagues. In what seems like a totally implausible storyline (except it really happened), the Big Lake team catches fire and wins the district title. When they pin the newspaper clipping on the bulletin board, one of the players writes on it, "It's your turn, Coach."

Morris hears about a tryout in nearby San Angelo. His wife (Rachel Griffiths) is going out that day so he tosses the kids in his pickup truck and drives to the field where he is surrounded by a hundred hard-throwing studs half his age. The Tampa Bay scouts take a look at Morris and wave him to the back of the line.

When they finally call his name, Morris is in the middle of changing his baby's diaper. He rushes onto the mound, flexes his arm and uncorks a heater that lights up the radar gun. The great mystery is how Morris, after four arm surgeries, can throw in the high 90s when he topped out in the 80s ten years earlier. Asked to explain it, Morris shrugs. He still doesn't know how it happened, it just did.

The Devil Rays offer Morris a minor league contract which he signs only after his wife first objects ("You can't eat dreams, Jim") and then relents, knowing she can't deny him another shot at pro ball. He starts in Double-A Orlando, earns a promotion to Durham, and finally gets the call to The Show.

Only the worst kind of cynic would point out the Devil Rays were an awful team in 1999 and that Morris was only added when the rosters were expanded in September. That's all true, but it's beside the point. So is the fact he pitched in just 21 games over two seasons and his statistics—no wins, no losses, a 4.80 ERA in 15 innings—are unremarkable. The mere fact that he got there at all is what makes his story worth telling.

🎬 **PIVOTAL SCENE:** Morris calls his wife from a pay phone in Durham and asks her to meet him in Arlington. "I need you to bring my blue blazer," he says. She looks puzzled. He explains, "They have a dress code in the big leagues."

Her reaction when the words sink in is one of the best moments in the film.

❝❝ **BEST LINE:** Morris' minor league teammates enjoy teasing him about his age. Two of the better lines: "So what was it like watching The Babe play?" And "How many fans did you lose when they raised the ticket prices to 50 cents?"

35 ▶ HEAVEN CAN WAIT (1978—PG)

SPORT: Football ★ STARS: Warren Beatty, Julie Christie, Jack Warden
DIRECTOR: Warren Beatty, Buck Henry

Heaven Can Wait is, in a sense, a precursor of many sports movies that followed, including *Bull Durham, Jerry Maguire,* and *Fever Pitch.* Call it a romantic comedy focused around a sports theme. The sports element works well; the screwball romance works even better.

The story centers on Joe Pendleton (Warren Beatty), a veteran quarterback for the Los Angeles Rams—and, yes, kids, there was a real NFL team in Los Angeles once upon a time. Joe is struggling to come back from injury to start for a Rams squad that seems bound for the Super Bowl. From the start, we see that Joe is an earnest guy who treats everyone well, plays an off-key clarinet, and feeds his body liver-and-whey milkshakes, with a little spinach mold tossed in.

Joe also goes on long bike rides, which proves to be a problem when he pedals through a tunnel and is broadsided by a vehicle.

Well, almost. Turns out that the angel (Buck Henry) assigned to oversee Joe's life anticipates the accident and—aiming to spare him the pain—sends him to heaven a second too early. With his quick reflexes, it turns out, Joe would have avoided the accident. He is due another 50 years on Earth. What to do?

The case is appealed to Mr. Jordan (James Mason), the director of heaven's way station. He offers to send Joe back—except that the poor quarterback's remains have already been cremated. So Joe is returned to Earth in the body of wingnut billionaire Leo Farnsworth, whose wife and top aide are scheming to murder him and who owns a canning-plastics-nuclear-energy company apparently responsible for most of the planet's pollution.

Now, Joe's got two aims. One is to win the love of British activist Betty Logan (Julie Christie), which he can do by halting the ecological plunder. The other, more difficult challenge, is to get Farnworth's soft body into football shape so that he can return to the Rams. He hires team trainer Max Corkle,

played perfectly by Jack Warden, and throws a lot of passes to clumsy butlers. Soon enough, his skills return.

Of course, the Rams aren't about to give the job to a wacky industrialist, so Joe—now Farnsworth—buys the team and installs himself as first stringer. Sure he does. Hey, some movies ask you to take a leap of faith. *Heaven Can Wait* asks you to vault the Grand Canyon.

Anyway, everything seems perfect until the heavenly escorts return with more bad news: The murder plot against Farnsworth is about to succeed. Another new body is needed. All that arduous training—for nothing.

Then comes a wacky murder investigation, a few plot twists, the Super Bowl, and they all live happily ever after. Needless to say, it all works out for Joe's third persona. Did you expect Beatty—who produced, co-wrote, and co-directed this movie—to end it any other way?

A critical sports fan may dismiss some of the far-fetched plot twists as naïve pap. Certainly they are. But, you know, sometimes there's nothing wrong with a corny ending.

🍎 **BET YOU DIDN'T KNOW:** Beatty originally wanted to stay true to the boxing theme and tried to cast Muhammad Ali as Joe. But Ali declined, saying he needed to train for his pending title fight against Earnie Shavers.

36 HEART LIKE A WHEEL (1983–PG)

SPORT: Drag Racing ★ STARS: Bonnie Bedelia, Beau Bridges
DIRECTOR: Jonathan Kaplan

Shirley Muldowney was the first woman licensed by the National Hot Rod Association to drive a gas-powered car, and in 1977 she rose to the top of that high-octane macho world by winning an NHRA points championship. She was called a pioneer and was often compared

to Jackie Robinson, who broke the color line in major league baseball.

But to Shirley, it was never about being a pioneer. As she said in a 2003 interview, "It was always about the racing." That was it, pure and simple. She loved the roar of the engines and the competition. Most of all, she loved the winning.

She was a drag racer, all guts and grease, just like Don (Big Daddy) Garlits, Connie (The Bounty Hunter) Kalitta, Don (The Snake) Prudhomme, and the rest of the male hot rodders. She didn't do it for Gloria Steinem or the women's lib movement. She wasn't out to prove any point except that she could drive really fast.

"I didn't have to put up with all that chauvinism," she once said. "I chose to put up with it. This life just suited me. I've always loved everything about this sport."

Heart Like a Wheel, the story of Muldowney's life, makes that clear. The film opens with Shirley as a little girl, sitting in her daddy's lap, clutching the steering wheel as they drive the back roads in upstate New York. She loves driving the car, and when she begs her dad (singer Hoyt Axton) to go faster, we're seeing a glimpse of the future.

At 16, Shirley (Bonnie Bedelia) drops out of school and marries Jack Muldowney (Leo Rossi), a mechanic who competes in local drag races. One night Shirley convinces Jack to let her drive. She wins that race and that whets her appetite. Pretty soon, she is urging Jack to let her try the NHRA circuit. He tells her there is no way that the other drivers will let a woman crack their fraternity.

Shirley will not be deterred, however. She goes to an NHRA event, where she is told she needs the signature of three drivers to get her license. She goes from pit to pit, paper in hand, asking the drivers to sign. Most laugh her off, but Garlits sees it as a way to bring new fans to the sport so he signs. Kalitta also signs—although the way he eyes Shirley, it's clear his motives aren't the best—and he strong-arms another driver into providing the third signature.

Shirley competes in her first NHRA event and in her qualifying run, she sets the track record. She wants to hit the circuit full-time, but Jack resists. He still sees racing as a hobby; he doesn't believe they can make it work as a career. They argue and Shirley takes off.

Shirley hooks up with Kalitta (Beau Bridges), who helps her get a car and a sponsor. They become romantically involved as Kalitta remakes Shirley's image, outfitting her in white boots and hot pants and giving her the nickname "Cha Cha."

But Kalitta is a world-class horn dog who spends most of the film with his arm around various women, telling them, "You're the number one thing in my life." Shirley knows what Kalitta is up to, but he is her business partner as well as her paramour, so she hangs in longer than she should. Finally, she reaches her limit and breaks off the relationship.

Without Kalitta's backing, Shirley's career suffers. She loses her sponsorships and she is forced to work with a young, inexperienced crew. She goes through two lean years on the circuit before rebounding.

Kalitta returns to driving and in 1980, the NHRA top fuel championship final comes down to a race between the former lovers. Shirley beats Kalitta to claim her second world title. As she removes her helmet, she looks at Kalitta and says, "Not bad for an old broad in a used car, eh?" He laughs, she smiles, and it seems like they put their differences behind them.

Well, maybe not entirely.

In a 2003 interview, Shirley was asked to pick a career highlight. She picked the 1980 victory over Kalitta. "I don't think I ever got as much satisfaction as I did in beating Connie after our falling out at the end of the '77 season," she said, still sounding like a woman scorned.

One year after the release of *Heart Like a Wheel*, Shirley was involved in the worst crash of her career. She was in a qualifying run at the Sanair Speedway near Montreal when the front wheels of her dragster locked, sending her car tumbling 600 feet down the track.

She suffered two broken legs, a fractured pelvis, and two broken hands. The doctors needed six hours working with wire brushes to clean the dirt and grease from her skin before they could operate. She was hospitalized for two months and required five more surgical procedures, including a skin graft, over the next two years.

Even racing enthusiasts were surprised when Shirley returned to the track in 1986 and resumed her career. Why risk a comeback at age 46? She simply said, "I missed it. It's what I did best."

She finally retired in 2003 at the age of 62. The following year, she was inducted into the International Motorsports Hall of Fame along with Indianapolis 500 winner Bobby Rahal.

37 JIM THORPE—ALL AMERICAN (1951—NR)

SPORT: Football/Track and Field ★ **STARS:** Burt Lancaster, Charles Bickford, Phyllis Thaxter ★ **DIRECTOR:** Michael Curtiz

John Riggins learned the story of Jim Thorpe by reading a book he checked out of the library at his grammar school in Centralia, Kansas.

"I stumbled upon a true American hero," said Riggins, who read the book as a third-grader and went on to become a star running back with the New York Jets and Washington Redskins. "The book made him larger than life and the film, *Jim Thorpe—All American*, gave him life.

"It's a film I could watch every day just because Jim Thorpe was the real deal," said Riggins, who now occupies a space alongside Thorpe in the Pro Football Hall of Fame. "And Burt Lancaster gave an authentic performance of the Indian superstar."

Lancaster was well coached. That's because the real Jim Thorpe was on the set every day serving as technical advisor. Lancaster said it was the only time in his career that he felt intimidated.

"Imagine how I felt," Lancaster recalled in a 1983 interview, "trying to do all the things Jim Thorpe did with the man himself sitting just a few feet away. He was very gracious. He kept saying, 'You're doing fine, young man.'"

Lancaster was an accomplished athlete. He competed in baseball, track, and gymnastics at New York University. He performed as a trapeze artist with Ringling Brothers. He could handle most of the sports scenes. The one thing he couldn't do was throw the discus.

"My footwork was all wrong," Lancaster said. "One day Jim got up, took off his jacket and said, 'Here, let me show you.' He was in his 60s, I'm sure he hadn't touched a discus in 30 years, yet he threw it flawlessly. We were speechless. Then we began to applaud."

Lancaster and Thorpe became friends during the filming and they stayed in touch until Thorpe's death in 1953. Lancaster had many memorable roles in his acting career, but Thorpe remained one of his favorites. "I'll always remember the look in his eyes," he said. "There was great pride, but also a lot of sadness."

Thorpe did not have a very happy life; the film makes that clear. It shows his triumph at the 1912 Olympics when he won two gold medals, only to have them later taken away. It shows him earning All-America honors in football, but it also shows him hitting rock bottom late in his career. This is a bittersweet portrait of the man chosen as the greatest athlete of the first half of the 20th century.

Like many sports films, *Jim Thorpe* is told in flashback with Thorpe's college coach, the legendary Glenn "Pop" Warner (Charles Bickford), as the narrator. It opens with Thorpe being honored at a dinner in his native Oklahoma. From there, it rolls back in time to the turn of the century and a young Thorpe, an Indian from the Sac and Fox Nation, running across the reservation with the speed and endurance of a stallion.

Jim's father sends him to the Carlisle (Pa.) Indian school to learn the skills of the white man's world. The students, who range in age from six to 25, spend half the day learning to read and write, the other half in the shop learning a trade. Jim is an indifferent student, but he is excited by his introduction to organized sports. He has no training and no grasp of the rules, but he is blessed with natural ability. Soon, he is excelling in both track and football.

The real Thorpe was just over six feet tall and weighed 200 pounds, with the speed of a sprinter. The Carlisle Indians had only 12 players on the football squad—with names like Jesse Young Deer, Little Boy, and Sunny Warcloud—but Thorpe made them a national power. In four seasons, Carlisle won 43 games and lost only five, with two ties. As a senior, Thorpe scored 198 points and 25 touchdowns in 14 games.

In the film, Thorpe (Lancaster) pursues a coaching job after college, but he is passed over, presumably because he is an Indian. Pop Warner suggests Thorpe take a crack at the 1912 Olympics. He enters the decathlon and the pentathlon and wins the gold medal in both. King Gustav V of Sweden presents Thorpe with a laurel wreath and says, "Sir, you are the greatest athlete in the world."

Thorpe returns home a hero, but his joy is short-lived. A sportswriter discovers that he played two summers of semi-pro baseball for which he was paid $2 a game. By the rules of the time that makes him a professional, and therefore ineligible for the Olympics. Even though Thorpe apologizes, explaining he was unaware of the rules, the International Olympic Committee strips him of his medals.

Thorpe suffers an even greater loss a few years later when his son, Jim Jr., whom he adores, dies of infantile paralysis. Devastated, Thorpe begins drinking heavily and his bitterness drives away his wife (Phyllis Thaxter). He continues his athletic career—he helps launch the first professional football league in 1920, starring for the Canton (OH) Bulldogs—and while he still is a big name, his performance tails off due to his boozing and disruptive behavior.

Thorpe ends up a shabby figure, walking into a practice and asking, "Do you need a back?" The coach first brushes him off, then he notices the initials J.T. on his bag. He realizes who this is. "Can you still go, Jim?" he asks. Thorpe says, "Give me a chance." In his first scrimmage, Thorpe breaks into the clear but collapses from exhaustion. He is carried off the field for the last time.

After that, he becomes a sideshow act, calling out the contestants in a marathon dance contest. Barely able to stand and slurring his words, Thorpe is an embarrassment. "This guy is laying the biggest egg since the dinosaurs left town," the promoter says as he prepares to give him the boot.

Thorpe is in the dressing room, pouring himself a drink when Warner walks in. The old coach gives him a tough-love lecture about being a quitter and feeling sorry for himself and Thorpe, looking in the mirror, knows he is right.

Warner invites Thorpe to join him for the opening ceremonies of the 1932 Summer Olympics in Los Angeles. When Thorpe enters the Coliseum, he receives a hero's welcome. Basking in the applause, the gleam returns to his eyes and he appears ready to get his life back on track.

In reality, Thorpe continued to have ups and downs. He earned a living through personal appearances and remained a prickly character. He feuded with Warner Brothers over the handling of the film. He said the studio paid him just $7,000 for the rights to his story. He also was paid $250 a week for his role as technical advisor. "But when they take the taxes out," he said, "I get $211."

Thorpe boycotted the premiere of the film in Muskogee, Ok., because they would not pay him an appearance fee. "Warner Brothers," he said, "are a bunch of damned stinkers."

Despite Thorpe's bitterness, many viewers have found the movie inspiring over the years. Among them was Herman Edwards, a 10-year NFL player and, later, head coach of two NFL teams. "Jim Thorpe made me want to become an athlete," Edwards recalled. "I was about 10 years old, flipping around the channels and I saw this black and white movie. I said, 'You gotta be kidding me. He was the best track man *and* football player?' And look at what he overcame.

"I knew if I was going to go anywhere in life and make something of myself, I'd have to do it through athletics, the way Jim Thorpe did. I loved the scene where he raced his Dad home. Twelve miles, his Dad's driving a truck and Jim's running, but Jim gets there first. I think he's the greatest athlete of all time."

🍎 **BET YOU DIDN'T KNOW:** Director Michael Curtiz won an Academy Award for directing the classic *Casablanca* (1942).

SPORT: Boxing ★ **STARS: Charles Bronson, James Coburn**
DIRECTOR: Walter Hill

Charles Bronson and James Coburn made some classic films together, including big-budget epics *The Magnificent Seven* and *The Great Escape*. *Hard Times* was their third pairing, and while it was a much smaller film—shot in just 38 days—it may have been the most enjoyable.

In *The Magnificent Seven* and *The Great Escape*, Bronson and Coburn were secondary characters in large ensemble casts. In *Hard Times*, they are front and center and playing to their strengths—Bronson as the stoic loner, and Coburn as the fast-talking hustler who always has a smile on his face and a trick up his sleeve.

Roger Ebert of the *Chicago Sun-Times* is right on when he calls this "a definitive Charles Bronson performance." Bronson himself said it was one of his favorites. He was a man of action, and in *Hard Times* his character—a bare-knuckles fighter named Chaney—is all action. He has very little dialogue. Mostly, he speaks with his eyes and his fists.

Bronson said he spoke fewer than 500 words in the entire film. We haven't counted them—you can if you want—but that seems about right. It suits his character, a drifter with one name who rolls into New Orleans on a freight train looking to make a few bucks in the Big Easy's underground boxing scene.

Chaney is a mystery. He has the gray hair and weathered face of a middle-aged man, but he has the muscular torso of a welterweight contender. So how old is he? He doesn't say. Where is he from? He doesn't say. Is he a former boxer? He doesn't say. Is he on the run from the law? From the mob? From an ex-wife? He doesn't say.

Writer-director Walter Hill reveals nothing of Chaney's past and that—combined with Bronson's inscrutable persona—makes him even more interesting. He gives no hint of how long he plans to stick around, or where he may be going next. Each time he finishes a fight and pockets his winnings, there is a chance he'll just hop the next freight and vanish into the night. He keeps everyone, including the viewer, guessing.

Hard Times is set, appropriately, in the 1930s during the Great Depression. When Chaney first meets Spencer "Speed" Weed (Coburn), a New Orleans gambler and wheeler-dealer, he is just off the train with six dollars in his pocket.

"I suppose you've been down the long, hard road," Speed says, eyeing the shabbily dressed stranger.

"Who hasn't?" Chaney replies.

"Every town has somebody who thinks he's tough as a nickel steak," Speed says. "But they all come to old Speed for the dough-re-me."

Speed manages bare-knuckles fighters (they are called "hitters") in illegal matches held on barges and in warehouses. It just so happens the night Chaney arrives, Speed's hitter loses to a hitter managed by another gambler, Chick Gandil (Michael McGuire). Chaney tells Speed he can beat Gandil's guy. Speed is skeptical until Chaney says he will bet his last six dollars on himself.

When Chaney steps in the ring, his burly young opponent sneers. "Hey, Pops," he says, "you're a little old for this, ain't ya?" Chaney, who is outweighed by at least 50 pounds, knocks the big guy stiff with one punch.

Speed realizes he has struck gold with this grizzled hobo. He tells Chaney he will set up the matches and they will split the winnings. Chaney agrees but tells Speed not to count on a long-term partnership. "I'm just looking to fill a few in-betweens," he says.

Chaney derives no pleasure from fighting. He doesn't take pride in being the toughest guy in town. To him, it is just a way to make a buck. He tries to explain it to Lucy (Jill Ireland, his real-life wife), a woman he meets in New Orleans.

muscles in a bar. They are professionals, they fight only for the cash. But we all know it's inevitable that Chaney and Street will square off to answer the question of who's better.

As Ebert writes, "They fight in a steel-mesh bullpen, and there's a certain nobility about them. They may seem to be animals, but they're craftsmen in a way and they respect each other. The real animals are the spectators."

👍 **CHEERS:** Cinematographer Philip Lathrop, who shot the classic gambling film *The*

"You know what they say, Chick, the next best thing to playing and winning is playing and losing."
—SPEED

"What does it feel like to knock somebody down?" she asks.

"It makes me feel a hell of a lot better than it does him," Chaney replies.

"That's a reason?" she says.

"There's no reason about it," Chaney says. "Just money."

Chaney knocks off every hitter in New Orleans and takes out a few more in the Bayou. There is no one left for Chaney to fight, so he is preparing to leave when Chick brings in the best hitter from Chicago, a man named Street (Nick Dimitri). Chaney isn't interested. He has enough money and he is ready to move on. Chick and Street find Chaney in a bar, drinking a beer.

"There's no point in avoiding this thing," Chick says. "It's going to happen."

"I could start it right here and now," Street says.

"Yeah, but you won't," Chaney replies.

"You don't think so?" Street says.

"You're not going to do it for free," Chaney says, finishing his beer and heading for the door.

He is right, of course. He and Street aren't two macho knuckleheads looking to flex their

Cincinnati Kid, uses the same lighting effects to bring just the right atmosphere to the fight scenes.

🍎 **BET YOU DIDN'T KNOW:** Charles Bronson was 54 years old when he filmed *Hard Times*.

☞ **DON'T FAIL TO NOTICE:** Chick Gandil is the name of one of the Chicago Black Sox who threw the 1919 World Series.

66 **BEST LINE:** Speed Weed: "To the best man I know. To the Napoleon of Southern sports. Me."

😎 **"I KNOW THAT GUY":** Robert Tessier, who plays Chick's henchman, was Shokner, the shaved head convict in *The Longest Yard*.

⭐ **SPORTS-ACTION GRADE:** B-plus. This is Walter Hill's directing debut, and he displays a knack for filming a fight scene. The big fight between Chaney and Street took a week to film, but it was worth it. Bronson and Dimitri (a veteran stuntman) are so into it, you almost wonder if they are fighting for real.

THE BINGO LONG TRAVELING ALL-STARS & MOTOR KINGS (1976–PG)

SPORT: Baseball ★ **STARS:** Billy Dee Williams, James Earl Jones, Richard Pryor
DIRECTOR: John Badham

Billy Dee Williams plays a Satchel Paige spin-off and James Earl Jones evokes Josh Gibson in a funny and sometimes poignant look at black baseball before integration of the major leagues.

The two are stars in the Negro National League, a throwback to the 1930s when players on opposing teams joked with each other during games and socialized off the field in a close-knit society. Quickly, they grow tired of their owners' cheapness and mistreatment of players. Led by Jones's character, slugging catcher Leon Carter (who quotes W.E.B. Du Bois about self-determination), they decide to quit working for others and form their own barnstorming team—known as the Bingo Long Traveling All-Stars & Motor Kings. Problem is, the owners of existing black teams won't play them. And white crowds in small towns in the South and Midwest react angrily to seeing their local nine soundly beaten by the visiting "coloreds."

What to do? Bingo (Williams) decides to play it all for laughs, and soon enough the All-Stars and Motor Kings become hardball's version of the Harlem Globetrotters—batting backwards, tossing out exploding balls, and playing with enormous gloves.

There's a fine line between humor and humiliation, and the story sometimes crosses it, such as the scene where Bingo pitches in a gorilla suit. But the point that's being nailed home here (albeit heavy-handedly) is that the black players of the era must do whatever they must to survive.

The screen chemistry between Williams and Jones is terrific. Hell, we'll take it any day over their work in *The Empire Strikes Back*. Bingo is impulsive, creative, and romantic; Carter is methodical and intellectual. But perhaps the best performance in the movie comes from Richard Pryor, who plays a light-skinned black man so intent on getting to the big leagues that he tries posing first as a Spanish-speaking Puerto Rican (taking the name Carlos Nevada) and then as a Mohawk-wearing Native American (calling himself Chief Takahoma).

🍎 **BET YOU DIDN'T KNOW:** In the opening scene, the supremely self-confident hurler Bingo stands on the field alone—his teammates lined up in foul territory—and challenges the opposition's leadoff batter to "hit my invite pitch." Of course, any ball hit into fair territory would almost certainly go for a home run.

The bit was inspired by Satchel Paige, who reportedly started games for more than a decade with this combination of showmanship and arrogance. Paige swore no one ever connected with that first pitch.

🔊 **CASTING CALL:** Well, directing call, anyway. Producers were close to hiring Steven Spielberg for the project in 1975—just as the summer blockbuster *Jaws* was released. Spielberg's huge success with the shark film gave him the opportunity to pursue any movie he wanted. He instead chose to direct *Close Encounters of the Third Kind*.

THE ROCKET: THE LEGEND OF MAURICE RICHARD (2005–PG)

SPORT: Hockey ★ STARS: Roy Dupuis, Stephen McHattie, Julie LeBreton
DIRECTOR: Charles Biname

The year is 1955 and the city of Montreal is engulfed in a riot. Cars are aflame, angry people fill the streets, and the issue is . . . hockey?

Indeed it is, and if you're not a devotee of the sport, the opening of *The Rocket* will have you wondering: Why would otherwise level-headed folks lose their sanity over what we learn is the suspension of their star player by the National Hockey League commissioner?

The so-called "Richard Riot" serves as the opening and closing bookends for this marvelous movie. Maurice Richard, star winger for the Montreal Canadiens, was punished with a season-ending ban for slugging a referee. What *The Rocket* explains is why Richard acted as he did and why the good people of one of North America's great cities followed suit in their rage.

The record books show Richard to be one of the all-time superstars of his sport, perhaps the most dynamic player of the NHL's pre-expansion era. But he was more than that. He was, to French Canadians, an icon—an ethnic hero in an age when Quebecois were regarded as second-class citizens, even in a city where they were the majority. That he emerged from humble beginnings in Montreal only served to make him more of a superman to the local working class.

The prejudice even carried into the locker rooms. In one notable scene, Canadiens coach Dick Irvin (Stephen McHattie) berates his players: "You Frenchies are all alike—nothing but cowards. It's in your blood. You're nothing more than chicken droppings." They are not even allowed to speak French on the bench.

Richard, who is modest and stoic, absorbs the insults through most of his career. He overcomes a double-whammy reputation of being brittle (sort of an Eric Lindros of his era) and afraid to fight (conquered when he kayos a New York Ranger tough guy) to eventually become a star with the blend of skill and determination rarely seen in sport.

And, over time, he gains the courage to speak up off the ice. In one memorable scene, Richard (played magnificently by Quebec native Roy Dupuis) spills his guts to a newspaper columnist, creating a national buzz with his stories of the taunting and slighting of French-Canadian players.

> **"You Frenchies are all alike –nothing but cowards. It's in your blood. You're nothing more than chicken droppings."**
> —DICK IRVIN

The politics in *The Rocket* are certainly interesting. But the blood-and-guts portrayal of the sport of hockey in the era is what makes this movie. "These guys were hurt all the time," director Charles Biname told Matthew Hays of the Canadian Broadcast System. "It was routine. And they were exhausted. We talked a lot about (Ridley Scott's) *Gladiator*. And I really wanted to capture hockey the way (Martin) Scorsese captured boxing with *Raging Bull*."

Biname succeeds. *The Rocket* is not without

flaws—the character of Richards's wife, played by Julie LeBreton, doesn't get to do much more than fret and stare over two hours. In addition, the movie was filmed with some actors speaking English and others speaking French, so half of it is clumsily dubbed for either audience.

Overall, however, it terrifically evokes a time when sport and society were much different. Dupuis, unknown to American audiences, seems to have been born to play this role.

Any sports fan should enjoy *The Rocket*. Hockey fans will marvel at it. Fans of the *bleu, blanc et rouge* will be overcome with joy.

PIVOTAL SCENE: Once labeled as soft and injury prone, Richard is taken down by a cheap shot in Game 7 of the 1952 Semi-Finals against the Boston Bruins. Concussed and unconscious, he is carried off the ice and stitched up in the locker room, even as his bleary eyes struggle to focus on the attending trainer. He returns to the game—blood-stained jersey and all—with just minutes left and the score tied, 1-1. Of course, he scores the series-winning goal on a phenomenal play.

You could label it sappy and implausible. Except that it's true.

41 EIGHT MEN OUT (1988–PG)

SPORT: Baseball ★ **STARS:** David Strathairn, John Cusack, D.B. Sweeney, Charlie Sheen ★ **DIRECTOR:** John Sayles

It's the fall of 1919 and the Chicago White Sox have just clinched the American League pennant. First baseman Chick Gandil (Michael Rooker) is celebrating at a Southside tap room when he is offered a dangerous proposition by gambler "Sport" Sullivan. The out-of-town fixer wants Gandil to round up a half-dozen teammates willing to drop the World Series for $10,000 apiece.

"You can go back to Boston and turn seventy grand at the drop of a hat?" asks Gandil. "I find that hard to believe."

"You say," counters Sullivan, "that you can find seven men on the best club that ever took the field willing to throw the World Series? I find *that* hard to believe."

"Well," says Gandil, "you never played for Charlie Comiskey."

So begins *Eight Men Out*, director John Sayles apologia for the worst scandal in baseball history. Sayles wrote the screenplay off of the highly acclaimed 1963 book by Eliot Asinof.

In this accounting of the infamous 1919 Black Sox, characters come in several easy-to-identify stereotypes. Baseball's owners, led by Charles Comiskey (Clifton James) are greedy and as blind as Mr. Magoo. Reporters (played by Sayles himself and by real Chicago newspaper legend Studs Terkel) are cynical, yet wise. And players are either too angry at ownership to fret about the fix, or too dumb to understand the consequences. Put outfielder Shoeless Joe Jackson (D.B. Sweeney) and third baseman Buck Weaver (John Cusack) in that last category.

The only character with real depth is aging pitcher Eddie Cicotte (David Strathairn), the nominal star of the film. Cicotte agrees to the bribe after Comiskey ostensibly cheats him out of a $10,000 bonus, but his sense of guilt is such that he can't even tell his wife the truth. In one of the best scenes, Cicotte turns over his hotel pillow to find rolls of cash freshly delivered by the fixers. He just stares at the money, too ashamed to even pick it up.

Overall, much of the movie comes across as an excuse for eight men who—regardless of how oppressive and penurious management might have been—conspired to lose sport's biggest event. You almost get the sense that if Sayles were doing a movie on baseball's recent steroid scandal, Roger Clemens would be a victim of all that Congressional bullying and Mark McGwire would become Shoeless (or Witless) Mark.

What *Eight Men Out* fails to address (and what history appears to have forgotten) is how widespread game fixing was in baseball's early days. According to baseball author Bill James, at least 38 players were involved in the scandals of the era. More than 20—not just these eight—were banned from the sport after Judge Kenesaw Mountain Landis became commissioner. Huge names, like Ty Cobb and Tris Speaker, were rumored to be on the take. The shock is not so much that the 1919 World Series was rigged, it's that such a thing hadn't occurred earlier.

There is also historical debate over whether Jackson was involved in the fix or just an innocent bystander. Sayles argues that Shoeless Joe was pushed into the conspiracy by peer pressure, agreed to take money but received none of it, and—by hitting .375 in the Series—did nothing but give his all. Hey, we don't know. We weren't there.

This is not to say that this isn't a worthwhile movie. The settings and old-time baseball scenes are terrific, from the baggy, tobacco-stained flannel uniforms to the National Anthem singer using a megaphone to pump up the volume. Small details are perfect, such as how pitchers' windups were different then than they are today. Notice how the outfielders leave their gloves on the field at the end of each inning. They really did it that way up through the 1940s.

The off-field scenes, as well, take you back to a boozy ragtime era of speakeasies, cigars, and street urchins. Before TV, before radio, drinking men would gather at so-called "gentlemen's clubs" (this is before strip joints co-opted that phrase). As a telegraph wired the game's progress, workers would move cutouts of players around a big wooden board. Hey, it wasn't exactly ESPN.com's "Real Time Scores," but it was the best they could do at the time.

Here's the bottom line: *Eight Men Out* is a worthy watch, full of the color and texture of old-time baseball. The guilty players are misunderstood and management is truly evil. There's a good story here. We just think it could have been told better.

🎬 **PIVOTAL SCENE:** The night before the verdict in the game-fixing trial, Cicotte—without ever actually admitting guilt to his wife—explains to her the roots of his bitterness.

Cicotte: "I always figured it was talent that made a man big. If I was the best at something. I mean, we're the guys they come to see. Without us, there ain't a ball game.

"Yeah, but look at who's holding the money," he continues, "and look at who's facing a jail cell. I mean talent don't mean nothing. And where's Comiskey…and Rothstein? Out in the back room cutting deals. That's the damned conspiracy."

Wife: "You would have won, too. You would have beat those guys easily."

Cicotte: "Well, won't nobody ever know that now."

42 THE FRESHMAN (1925—NR)

SPORT: College Football ★ STAR: Harold Lloyd
DIRECTOR: Fred C. Newmeyer, Sam Taylor

The first truly successful sports movie churned out of Hollywood and the only silent film in this book, *The Freshman* still holds up more than 80 years later for its endearing comedy and well-choreographed football scenes.

Harold "Speedy" Lamb (Lloyd) arrives at Tate University desperate to become the most popular student on campus. Problem is, he's a total geek, from his Harry Potter glasses to the fast-step jig he performs whenever he is introduced. In an effort to make friends, Speedy buys the entire student body an ice cream cone (well, this *was* the Prohibition era), and hosts a party, where he winds up the butt of jokes by the campus troublemaker (Brooks Benedict). About the only person not taking advantage of the well-meaning sap is pretty Peggy (Jobyna Ralston), who finds his naiveté endearing.

In a plotline that combines *Rudy* and *Revenge of the Nerds*, Speedy tries out for the varsity eleven at Tate, a school described as "a large football stadium with a college attached." Hmm, sounds a lot like an early version of USC.

Naturally, he's awful—punting the ball backwards over his head, whiffing on the tackling dummy and steamrolling the head coach. Although the coach (Pat Harmon) appears to have a sadistic streak bordering on Frank Kush ("so tough he shaves with a blowtorch"), he takes a liking to the overeager freshman. Speedy stays on as the waterboy but is allowed to believe he is actually a substitute. He even wears jersey No. 0.

It all leads up to the big game against top-ranked Union State, in front of what appears to be 90,000 fans. Injuries deplete Tate's bench, and in comes our hero. Speedy, of course, scores the winning touchdown (we suspect you saw that plot spoiler coming), gets the girl, and becomes—as he aimed to be—the Big Man on Campus.

If you've never watched a silent movie, *The Freshman* is a good place to start. Think of it as an extended Benny Hill skit. Lloyd was second only to Charlie Chaplin among silent screen comics of the 1920s, the Red Sox to Chaplin's Yankees. This movie grossed $3 million when it came out, which was considered pretty good. Chaplin's *Gold Rush* grossed $4.2 million, which was the record-setting comedy at the time.

66 BEST LINE: The Coach exhorting his squad: "You dubs are dead from the dandruff down." Hey, it isn't much, but it's the most they could fit on a caption card.

★ **SPORTS-ACTION GRADE:** Solid B. Not bad, considering the times. Lloyd, five-foot-ten but pretty scrawny, was a terrific athlete. The football scenes were filmed on the field at the Rose Bowl, and the crowd scenes were shot at California Memorial Stadium during halftime of a game between UC Berkeley and Stanford. Players from USC portrayed members of both Tate and Union, which makes it appear more realistic.

43 MURDERBALL (2005–R)

SPORT: Wheelchair Rugby ★ STARS: Mark Zupan, Joe Soares
DIRECTOR: Henry Alex Rubin, Dana Adam Shapiro

The first clue you get that *Murderball* is not the typical overly sympathetic film about disabled people occurs in its opening minutes. Joe Soares, coach of the Canadian wheelchair rugby team, is shooed from the sidelines of a game between two other countries by a female official.

"---- you," Soares mutters. "Bitch."

Later in the movie, one of the American players is insulted when a guest at a wedding confuses him for a competitor in the Special Olympics—rather than the 2004 Paralympics.

"We're not going for a hug," he snaps. "We're going for a ------- gold medal."

This is not to say that *Murderball* is a documentary about angry young men (well, some of them are). Rather, it's a look at extremely competitive world-class athletes—who can be as intimidating and foul-mouthed as most extremely competitive world-class athletes—but happen to be confined to wheelchairs.

Simply said, *Murderball* is an intimate (sometimes *really* intimate) look at the lives of members of the American and Canadian Paralympic quadriplegic rugby teams over a two-year period leading to the 2004 Paralympic Games in Athens, Greece. It is as entertaining as any documentary you will ever see, more revealing than any A&E biography, and as dramatic as most of the scripted sports movies in this book.

It covers an ultra-violent sport that few know exist. Wheelchair rugby defies the standard belief that people who cannot walk should be handled with care. Rather, the competitors—some paralyzed, others missing limbs—strap themselves into armored chariots, which they propel into each other at high speeds. The game is played on a basketball court, and points are scored when a player crosses the opponent's end line holding the ball. But mostly, it seems, they collide like souped-up bumper cars. As one player says, "It's basically kill the man with the ball."

Indeed, the title of the movie comes from the original name of the sport, which really bears little resemblance to traditional rugby. But, as American star Mark Zupan, one of the subjects of the movie, notes, "We had to change the name. You really can't market something called 'murderball' to corporate sponsors."

Unfortunately, it's equally difficult to market something titled *Murderball* to filmgoers. The movie grossed less than $2 million in theaters after its 2005 release. Perhaps some people equated the name with a *Mad Max*-like futuristic thriller. Others, it seems, were not intrigued about a sports documentary about the disabled.

That's a shame, because there's so much more here. *Murderball* is at its best when it profiles the men who have already won one huge battle in their lives—the mental war of coming to terms with their damaged bodies. That said, this movie is less about people who overcome disability than it is about guys who like banging into other guys, drinking beer, swapping lies about sex, and creating a culture of brotherhood.

There are two central characters here. One is Zupan, a Type-A Texan decked out in tattoos and a menacing red goatee. He's got a compelling personality, an attractive girlfriend (who speaks openly about her attraction to handicapped men), and a brutally honest way of speaking.

Zupan's story begins with the injury that occurred when he was in high school. His best friend's drunk driving caused the accident that threw Zupan into a canal and paralyzed his lower body. Their friendship ended that night—Zupan seething with anger and the friend wracked with guilt. Now, 10 years later, the two attempt a reconciliation. We watch as the nervous friend travels to a tournament to watch Zupan compete, unsure if he can cope with the sight of his old buddy in a wheelchair. You find yourself feeling pain for both men.

If Zupan is the film's hero, the antagonist is Joe Soares, who has been in a wheelchair since his childhood bout with polio. The sport's Babe Ruth from the 1990s, Soare's best playing days are over. After being cut from Team USA, Soares—a Tampa Bay resident—latches on as coach of the Canadian team (taking the American playbook with him). And this provides *Murderball's* spark: The Yanks view him as quad rugby's Benedict Arnold, while Soares is on a vengeful mission to beat the guys whom he believes abandoned him.

Soares, 42 when the movie began filming, resembles a young Robert Duvall, which is perfect since his personality mirrors Bull Meechum from *The Great Santini*. He is manically intense with his men. At home, he seems almost bullying toward his son, a bookish, viola-playing 13-year-old whose aversion to sports seems to embarrass Soares.

Midway through the film, Soares suffers a heart attack. This is amazingly filmed stuff, with the camera peering right into the emergency room. As he recovers, Soares softens, leading to an emotionally touching scene between father and son.

There is no softening, however, on the court. The story arc in *Murderball* leads up to a medal-round match between America and Canada at the 2004 Paralympics. And that's where this terrific movie becomes a terrific sports movie. We won't give away the ending, except to say the old ABC show *Wide World of Sports* could have easily used footage from the game for its "thrill of victory, agony of defeat" montage.

🎬 **CHEERS:** In recent decades, Hollywood has portrayed the disabled—along with Native Americans and Southern blacks—as completely virtuous, entirely sympathetic, and typically victims. This film does not fall into those cliches. As Zupan's friends explain in a scene at his 10-year high school reunion, "Mark was an SOB before the accident, and he's an SOB now." Hey, boys will be boys, even when they happen to be sitting in wheelchairs.

🎬 **PIVOTAL SCENE:** A before-and-after side plot to the movie follows a young man named Keith Cavill, a dead ringer for Jeff Gordon who is paralyzed in a motocross race and undergoes 10 months of painful rehabilitation. You watch over Cavill's shoulder as he moves out of rehab into a modified apartment—sadly absorbing the impact that he will never again be able to use a normal toilet and shower.

Later, Cavill is introduced to Zupan at a quad rugby demonstration. He climbs into one of the souped-up wheelchairs. Suddenly, the former racer is reinvigorated. "It's unreal," he says with a smile. "I could run into anything." As nervous nurses hold him back, Cavill feels his old competitive juices stirring. "This is great," he says. "I want to go hit stuff."

☞ **DON'T FAIL TO NOTICE:** The most honest questions posed to the athletes come not from adults in the film—some of whom seem intimidated by the wheelchair athletes—but from children. During a visit to a grade school, one little boy asks quadruple amputee Bob Lujano, "How do you eat pizza with your elbows?"

"I just pick it up," says Lujano. "No arms, no legs, no problem." And both he and the little boy smile.

★ **SPORTS-ACTION GRADE:** B-plus. While odds are that you know nothing about the rules and strategies of quad rugby, the game sequences can be downright thrilling. The directors rigged cameras onto the players' chairs to give viewers a direct point of view into the speed and violence of the sport.

❝❝ **BEST LINE:** Zupan, summing up his rivalry with Soares: "If Joe was on fire on the side of the road, I wouldn't piss on him to put it out."

44 THE BAD NEWS BEARS (1976–PG)

SPORT: Baseball ★ **STARS:** Walter Matthau, Tatum O'Neal
DIRECTOR: Michael Ritchie

Ahh, the memories of Little League. Alcoholic coaches passed out on the pitcher's mound. Violent parental abuse. Dysfunctional pre-adolescents. Ethnic and racial slurs that would make Howard Stern wince.

What? Your Little League experience wasn't quite like that? No matter. *The Bad News Bears* skewers that most American of youth activities by turning a population of pipsqueak players into a potty-mouthed roster of anarchists. On one level, it's stinging satire. On another, it's downright funny.

To categorize *The Bad News Bears* as a children's movie is not quite right. Most of the actors are kids, of course, and it centers

on a youth activity. But the themes—being an outcast, winning at all costs—are definitely adult in nature. This movie comes by its PG rating honestly.

Indeed, sometimes it pushes that rating to the limit. *Bears* came out in 1976, before the nanny-fication of America, when a movie could have a scene of two pre-teens riding a motorcycle—without helmets!—and the nation wouldn't go into shock. We're not saying that was better; just a sign of the times. So, we get scenes here in which an adult teaches a 10-year-old to mix martinis, a father whacks his son in public, and a cupie-cute little boy refers to his teammates as . . . well, we won't say it, but it's every epithet you ever heard for blacks, Hispanics, Jews, and gays. And no one even blinks.

That's the irony. Much as it seems we live in an increasingly crude society, *The Bad News Bears* is a movie that could not be remade these days. Oh, wait a second. It was, in 2005. But trust us, the second version is much tamer (and less funny) than the original. More about that later.

First, the storyline: Morris Buttermaker (Walter Matthau) is a broken down, alcoholic ex-minor league pitcher who's hired to coach a youth team of non-athletic misfits known as the Bears. Initially, he has no interest beyond collecting a paycheck and watches impassively as his squad loses the first game 26-0. In one inning.

But soon, the familiar Matthau gruff-with-a-heart-of-gold character emerges (see: *The Odd*

"All season long, you've been laughed at, crapped on. Now you've got a chance to spit it back in their faces." —BUTTERMAKER

Couple and *Grumpy Old Men*), and he decides to actually teach the kids some old-school fundamentals. He recruits two ringers—a talented pitcher named Amanda Whurlitzer (Tatum O'Neal, coming off her Oscar in *Paper Moon*), as well as a pint-sized slugger-turned-juvenile-delinquent named Kelly Leak (Jackie Earle Haley).

Of course, the Bears pull off the biggest turnaround this side of the 2004 Red Sox and end up playing for the North Valley League championship. Which is kind of like the Little League World Series, assuming all the coaches

were win-at-all-costs jerks and all the parents were blithering idiots.

Director Michael Ritchie made a career in the 1970s of lampooning American institutions, including politics (the biting *The Candidate*), beauty pageants (the underrated *Smile*), and pro football (the disappointing *Semi-Tough*). He takes a machete to youth baseball here in a way that the establishment still doesn't appreciate all these years later. "That film is something that doesn't portray Little League as it is," the organization's public relations director Lance Van Auken said recently. "It is not something Little League would be proud of."

Perhaps not. But it remains funny. The script was written by Bill Lancaster, son of Burt Lancaster, who said he modeled Coach Buttermaker's personality after his father. Not exactly a tribute to dear old dad.

The Bad News Bears was a success when it came out, grossing $32 million—good enough to inspire two sequels (and almost a third), a spin-off TV show and that 2005 remake. Let's briefly review each of them:

The Bad News Bears in Breaking Training (1977) was a moderately entertaining second effort done without Matthau and O'Neal, which is like redoing the Beatles without Paul and John. Still, it's worth seeing.

The strangest development in *Breaking Training* is that the central character of the fat catcher, Engleberg (played by Gary Lee Cavagnaro in the original), was replaced with another actor (Jeff Starr). Same character name, same persona, no explanation—just a different actor, like revolving Darrins on the TV show *Bewitched*. Cavagnaro later explained that he lost weight after the first film and wasn't willing to put it back on for a movie role. Rather than rewrite a slender catcher into the script, producers axed him.

The Bad News Bears Go to Japan (1978) is a completely unwatchable effort. Don't waste your time seeing it. There were plans for a fourth film, set in Cuba, but Paramount Studios looked at the puny box-office receipts from *Japan* and pulled the plug. Too bad in one sense: The producers had hoped to get Fidel Castro to play a small role as a Cuban coach in Version Four.

In 1979, the movie was made into a TV series starring Jack Warden, and with eight-year-old Corey Feldman as one of the kids. As far as we could determine, not a single person ever watched this show.

Finally, the entire thing was remade, starring Billy Bob Thornton as Morris Buttermaker. It's not bad, and some in the under-15 set might relate better to the lines and characters in this version than the original. But trust us, the politically incorrect snap of the 1976 *Bears* makes that the one worth watching.

15 SEABISCUIT (2003—PG-13)

SPORT: Horse Racing ★ **STARS:** Jeff Bridges, Tobey Maguire, Chris Cooper
DIRECTOR: Gary Ross

It is difficult to imagine the hold that a boxy, bow-legged five-year-old had on America in 1938. At a time when Babe Ruth had left the stage and Joe DiMaggio was entering it, when both Joe Louis and Jesse Owens were destroying the myth of Aryan superiority, the most popular athlete in this country was . . . a horse.

Seabiscuit, an undersized thoroughbred with the humblest of beginnings, was a national sensation. The influential columnist Walter Winchell listed the horse among his top ten newsmakers of the year, right below Roosevelt and Hitler. Seabiscuit had more endorsements than Peyton Manning could ever dream of, hawking everything from fresh fruit to ladies' hats. Thousands swarmed race tracks to cheer him. Millions fiddled with their radio tuners to hear him run.

It boggles the mind, in this era, to recall that horse racing was once so huge that one of its stars—an animal, not a jockey—could be as popular as Michael Jordan or Tom Brady would later become. But that was the case. You can learn the details by reading Laura Hillenbrand's best-selling book, *Seabiscuit.* Or you can get the quicker version—well, 141 minutes worth—in this movie based on her work.

Seabiscuit the movie tells the story not just of one of the history's most remarkable horses, but also of the three men responsible for his success. In fact, if the movie is guilty of anything (beyond running too long) it's that Seabiscuit himself becomes almost a secondary character. Indeed, 45 minutes pass before we even get to meet the title star.

Rather, the film focuses on the three men who create the legend. We meet Seabiscuit's owner, gregarious West Coast automobile magnate Charles Howard (Jeff Bridges). There's an interesting irony—oft pointed out—that Howard's fortune comes from making horses obsolete in American life ("I wouldn't spend more than five dollars for the best horse in America," he tells a potential car buyer early in the film). Howard's salesmanship and flair for publicity—he's Don King with a nice haircut and better ethics—help turn Seabiscuit into America's obsession.

We meet trainer Tom Smith (Chris Cooper), a taciturn, mysterious horse whisperer who sees potential in young Seabiscuit when others just see an unruly loser. Smith doesn't relate much to other humans, but he communicates with animals better than Dr. Dolittle. In one scene, he adopts a broken-down horse about to be euthanized for an injured leg. Asked by Howard why he's wasting his time with an injured animal, Smith says, "You don't throw a whole life away just because he's banged up a little."

And we meet Red Pollard (Tobey Maguire), who has failed both as a jockey and a prize fighter, and regards Seabiscuit as his one last chance for success. Pollard's parents—loving and literary—are ruined at the start of the Depression. As a 16-year-old, he is effectively sold into apprenticeship as a jockey and (as far as we can tell) never sees his family again. He is oversized at five-foot-seven and blind in one eye—which later comes into play.

Each man carries deep wounds. Howard mourns the death of his 11-year-old son (ironically, in a car accident) and the ensuing breakup of his marriage. Pollard grieves the separation from his family. And Smith, well, we're never quite sure of his story, but he sure seems off-kilter about something.

Together these damaged men are brought together by a damaged horse. And that's the story.

Smith is the one who puts everything together, spotting the stumpy Seabiscuit and "seeing greatness in his eyes." He also takes to Pollard, knowing that, despite the jockey's poor history (or maybe because of it), he has the potential to get the most out of a mistreated, troubled horse.

Understand, all of this makes up the first half of the movie. Then we get to the racing. And then *Seabiscuit* shifts into gear.

In rapid time, Seabiscuit goes from being a 70-1 underdog at Santa Anita to the hottest thing on four legs. The undercurrent is that Americans, ravaged by the Depression, regard this unlikely winner as a hero for tough times.

"The movie did a great job of depicting the era and the way people were struggling just to live," said Dick Jauron, the former Buffalo Bills head coach who lists this as his favorite sports movie. "It was a total underdog story—the horse, the owner, the jockey—and the whole country identified with them."

It all leads up to a match race with Triple Crown winner War Admiral, the so-called unbeatable horse. Imagine Rocky vs. Apollo Creed. Hickory High vs. South Bend Central. USA hockey vs. USSR. We've seen this plot line many times before, and yet we still get suckered in. As Howard (Bridges) says, "Sometimes when the little guy doesn't know he's the little guy, he can do great things."

The buildup is terrific and the on-screen race against War Admiral may be the best horse racing scene in the history of film. *Seabiscuit* was nominated for seven Oscars (it won none), and most—like editing and cinematography— must be tied directly to the racing moments.

If the film has a flaw, it's that it tries to be *too* complete. Hillenbrand's book is 480 pages in paperback. Director/screenwriter Gary Ross tries to cram as much as he can into the movie,

so 20 minutes after the first climactic scene, we get another one. And then another. Then a lot of hand wringing. And then a postscript.

Hey, that's a minor quibble. Seabiscuit is a good story well told and brilliantly filmed. The movie, like the horse himself, has a lot of heart.

🎬 **PIVOTAL SCENE:** Smith, the trainer, first lays eyes on Seabiscuit—and falls for his stubborn heart—as he watches the enraged horse wrestle with a half dozen handlers trying to calm him. Seconds later, he turns around and sees an angry Pollard fighting off an entire crowd. The horse and the jockey share a sense of rage, and Smith is wise enough to put them together.

⭐ **SPORTS-ACTION GRADE:** An easy A. The races are breathtaking, with sweat dripping from riders and horses, and clods of dirt flying through the air. You see jockeys whipping the animals and each other, shouting and fighting from their mounts. Cameras are posted everywhere—above, below, between the horses—giving you a real sense of the danger of this sport.

46 THIS SPORTING LIFE (1963–NR)

SPORT: Rugby ★ STARS: Richard Harris, Rachel Roberts
DIRECTOR: Lindsay Anderson

At this point, we're figuring that you're weary of all those uplifting scrappy-longshot-becomes-courageous-hero flicks. How many underdogs can you cheer for, after all? So now for something completely different:

How about a depressing portrait of British working-class angst, centering on a brutish rugby player and his romantic frustration with his emotionally crippled landlady? Now that's a plotline unlike any other on our Top 100.

Trust us, *This Sporting Life*, a 1963 black-and-white British film, is a worthy watch. It may be the most downcast movie in this book (well, perhaps next to *Million Dollar Baby*), but there's a great texture and story to it and the sports action is riveting in its barbarism.

It tells the story of Frank Machin (Richard Harris), a coal miner in England's industrial midlands. In an early scene, Frank picks a bar brawl with the captain of the local rugby

the film. Cinemark Theaters made a similar offer. Still, the film played to mostly empty seats.

It is hard to put a finger on why, but *Cinderella Man* just didn't connect with the public. Maybe *Seabiscuit* stole its thunder, maybe Crowe turned people off, maybe America had its fill of boxing movies. It may have been a combination of all those things. But putting the cash register aside and just looking at *Cinderella Man* as a film, it is quite good. In some ways, it is very good.

"As a boxer," said 10-time world titlist Oscar de la Hoya, "I found it completely inspiring."

It is based on the life of Braddock, a journeyman pugilist who lucks into a few fights and winds up taking the crown from Max Baer. Damon Runyon gave Braddock the name "Cinderella Man" and it was a fit. It is well-documented by Runyon and others that Braddock, like Seabiscuit, became a symbol of hope for millions of Americans suffering through the Depression. He was an uncommon hero, but those were uncommon times.

When we first see Braddock (Crowe), he is handsome and cocky in a three-piece suit riding through Manhattan with his manager Joe Gould (Paul Giamatti) after flattening another opponent at Madison Square Garden. Gould is counting the night's winnings and vowing, "We're going straight to the top, Irishman."

Four years later, it is a totally different picture. The stock market has crashed and Braddock is among the millions who have lost everything. His family lives in a tiny basement apartment and his wife Mae (Zellweger) waters down the milk to make it last. The heat is shut off and as the three children huddle in bed, you can see their breath.

Braddock may have lost his bank account, but he has not lost his honor. When his oldest son steals a salami from the butcher shop, Braddock does not berate him. He takes him by the hand and walks him back to the shop to return it. "No matter what happens," he says, "we don't steal. Not ever." When his daughter says she is hungry, Braddock gives her the one slice of meat from his plate. "It's OK," he says, "I'm full."

His boxing career appears over. He broke his right hand, but continued fighting because he needed the money. Basically fighting with one hand, he loses 20 of 30 bouts and he is so ineffectual the New York boxing commission pulls his license.

Braddock has only one place to go and that's the New Jersey docks where every morning he stands with hundreds of other desperate men outside the gates, hoping to be chosen for the day's work. The film is particularly effective in conveying the feeling of helplessness that grips the men who push and shove and plead as the foreman approaches. He picks a few and slams the gate on the others, saying, "That's it."

Braddock has no celebrity status. He is rejected as often as he is selected. He is forced to go on relief but still doesn't have enough to feed his family, so in one of the film's most painful scenes he goes to Madison Square Garden and asks the promoters gathered there for money. He circles the room, hat literally in hand, as the men dig into their pockets for change. (This really happened. Braddock said it was the lowest point of his life.)

But Braddock's luck changes. A fighter pulls out of a match with the No. 2-ranked heavyweight Corn Griffin. No real contender will take the fight on such short notice so they have to find someone who will. The Garden is desperate enough to give Braddock back his license if he will take the match. They offer him $250, peanuts for a main event, but a windfall for Braddock. He takes the fight even though he has not trained in months and is given no chance at winning.

Braddock shocks everyone by knocking out Griffin. He discovers his left hand, which he rarely used earlier in his career, is a powerful weapon, thanks to working on the docks. The win leads to another bout, which he wins. That leads to another bout, which he also wins. Runyon dubs him "Cinderella Man" and next thing you know, he's signing to fight Baer for the title. At the press conference, Braddock is asked what he's fighting for.

"Milk," he says.

Crowe makes Braddock, the noble working stiff, totally believable. *Variety's* Robert Koehler writes, "As Braddock, Crowe's eyes have never seemed so full of unspoken sadness and ferocity with his body language ranging from spent hopelessness to a single muscle preparing to strike."

Crowe spent a month working with veteran trainer Angelo Dundee, preparing for the role. They studied tapes of Braddock's fights and Crowe sparred with then WBA super middleweight champion Anthony Mundine of Australia. Crowe dislocated his shoulder during one sparring session and needed arthroscopic surgery to repair it. Production was shut down for weeks while he recovered.

Cinderella Man does a fine job of explaining Braddock. Howard leans a little too heavily on the violin bow now and then, but that's OK. The problem comes with the introduction of Baer (Craig Bierko). The film portrays Baer as a sneering savage who revels in the fact he has killed two men in the ring. When he meets the Braddocks, he eyes Mae and says, "You're far too pretty to be a widow." Not content to leave it at that, he says, "Maybe I can comfort you after he's gone." Mae throws a drink in his face.

None of this is true. Yes, Baer did kill one opponent, Frankie Campbell. Another, Ernie Schaaf, died but it was after a bout with Primo Carnera. Schaaf was defeated by Baer before he fought Carnera and it was speculated the injuries he suffered in the bout against Baer had more to do with his death than the fight with Carnera, although that was never proven. However, it was common knowledge that Baer was haunted by the deaths of the two men. In fact, he donated part of his earnings from several bouts to the Campbell family to pay for the children's education.

Baer was never the same fighter after those tragedies. Later in his career, he was more of a clown than a fighter, slapping opponents and playing to the crowd. Newspaper accounts indicate that's how he lost to Braddock. He coasted through the bout, mugging and waving to the fans. Braddock, a 20-1 underdog, landed just enough punches to earn the decision.

There was no need for screenwriters Cliff Hollingsworth and Akiva Goldsman to demonize Baer. Long before the title fight, we were in Braddock's corner; we didn't have to think he was fighting Satan to root for him. Besides, the real villain isn't Max Baer, it's the Great Depression.

🍎 **BET YOU DIDN'T KNOW:** Frankie Campbell, the boxer who died as a result of injuries sustained in a bout with Baer, was born Francisco Camilli. His brother was Dolph Camilli, who played for the Brooklyn Dodgers and won the National League home run crown in 1941.

⭐ **SPORTS-ACTION GRADE:** B. The big fight is well shot, but inaccurate. The film makes it look like an action-packed affair when in reality it was a dull 15 rounds with Baer coasting and Braddock fighting cautiously.

48 ROUNDERS (1998—R)

SPORT: Poker ★ **STARS:** Matt Damon, Edward Norton, John Malkovich
DIRECTOR: John Dahl

Just like the classic *The Hustler* sparked a pool-hall resurgence in America in the early 1960s, *Rounders* is often cited as a key factor in this country's recent mania over poker—particularly Texas hold 'em.

We're not so sure. We would argue that the lipstick camera—which makes it possible to see a player's down cards—has more to do with making Texas hold'em a TV sensation and, subsequently, a mainstream sport. But we recognize that two recent World Series of Poker champions, Chris Moneymaker and Jamie Gold, said that *Rounders* ignited their interest in the game. And we can't argue with this stat: The year the movie came out, there were 350 entrants in the WSOP Main Event; in

2008 there were 6,844.

These days you can't spend five minutes at a table in Vegas or Atlantic City without some wannabe quoting a few lines from the movie.

"It's immoral to let a sucker keep his money," some sunglasses-wearing tourist from Middle America will snicker. And we'll all know where he got his material. If nothing else, that tourist has good taste in movies. Because *Rounders* is an engaging character study, a solid story well told, and the best damn poker film ever made.

It stars Matt Damon as Mike McDermott, a third-year law student trying—not so successfully—to avoid the lure of New York City's mysterious underground poker scene. Damon was on a roll when he shot *Rounders* in 1998, following his work in *The Rainmaker*, *Good Will Hunting*, and *Saving Private Ryan*. That's a run any card player would admire.

Strong performances also come from Edward Norton as Lester "Worm" Murphy, an ex-con more attracted to the hustle than the actual game; from Famke Janssen, as a Russian gambler/vixen named Petra; from John Turturro, as career grinder Joey Knish, and—especially—from John Malkovich, as scenery-chomping Russian mobster Teddy KGB. More about Malkovich later.

The film opens with Mike getting wiped out by Teddy KGB to the tune of $30,000. He agrees to give up the gambling life and please his annoying girlfriend (Gretchen Mol) by sticking to his law studies and working part-time as a delivery driver. That works for nine months—during which time Mike is completely miserable.

He becomes more miserable when old buddy Worm gets out of jail, visits Mike's favorite club, and runs up a massive debt in his name. Making things worse, the debt is owed to the Russian mob. Teddy—and his slovenly enforcer Grama—give Mike and Worm a quick deadline to raise the money. Or else.

And that sets up the second half of the movie, in which the two schemers (they once fixed high school basketball games together) travel from New York's Turkish baths to a preppie cigar bar to a cop's game in Binghamton, N.Y. (bad idea, as it turns out) to win enough money to save their lives. It's not an *Ocean's Eleven*-style caper film, but you'll be rooting for the rascals.

Rounders culminates in a showdown, in which Mike McD must return to the worst site of his card-playing life and try to topple the bully Teddy KGB. It's young Sugar Ray Leonard seeking a rematch against Roberto Duran in 1980, or the Red Sox against the overly smug Yankees in the 2004 ALCS.

There's a lot of nuance here, and if you don't understand the intricacies of poker you may get lost along the way. But the final face-off has as much tension as any Hitchcock film. And it will definitely have you wanting to make that late-night romp to the casino.

Rounders grossed just $23 million at the box office. Director John Dahl suggested that the studio, Miramax, had expected a date film (along the order of *Good Will Hunting II*), and failed to promote the film on its release. Regardless, it is a terrific guy movie and a cult film among millions of card players. It will take you to an intoxicating world.

📣 **CASTING CALL:** If you're doing the seminal movie on poker, who better to step in for a cameo than Johnny Chan—winner of 10 bracelets in the World Series of Poker? Playing himself. Very cool.

❝❝ **BEST LINE:** Mike walks in on The Judges Game involving his law dean (Martin Landau) and five other legal beagles. He glances around the room for about five seconds and then insists his reluctant dean raise the pot. How can he be so sure, asks another player, that the dean has the best hand?

"Well," says Mike, "you were looking for that third three, but you forgot that Professor Green folded it on Fourth Street and now you're representing that you have it. The DA made his two pair, but he knows they're no good. Judge Kaplan was trying to squeeze out a diamond flush but he came up short and Mr. Eisen is futilely hoping that his queens are going to stand up. So like I said, the Dean's bet is $20."

Geez, we don't want to ever play against *that* guy.

SPORT: Soccer ★ STARS: Parminder Nagra, Keira Knightley, Jonathan Rhys-Meyers
DIRECTOR: Gurinder Chadha

Bend It Like Beckham set box office records when it was released in Great Britain. It was the first film by a non-white Briton to reach No. 1 in that country. It became the top-grossing British-financed film ever. It was a triumph for writer-director Gurinder Chadha.

Still, no one expected *Bend It Like Beckham* to find an audience in the United States. It is, after all, a film about soccer and the cast was unknown in this country (although the radiant Keira Knightley has become a major star since then). But the movie was the talk of the Sundance Film Festival and generated enough buzz to make $32 million during its American run.

Simply put, *Bend It Like Beckham* is a charming film that will play anywhere.

"We showed it on our Ebert and Roeper Film Festival at Sea," wrote Roger Ebert in the *Chicago Sun-Times*. "The audience ranged in age from 7 to 81, with a 50ish median, and it was a huge success. The hip Sundance audience, dressed in black and clutching cell phones and cappuccinos, loved it, too. And why not, since its characters and sensibility are so abundantly lovable."

The hero is Jesminder "Jess" Bhamra (Parminder Nagra), a London teenager whose parents fled from Uganda to build a new life in England. They are a strict Sikh Punjabi family which means Jess is expected to attend school, learn how to cook, and settle into a marriage (most likely arranged) with an Indian boy. Her sister Pinky (Archie Panjabi) is doing exactly that.

But Jess has different ambitions. She wants to play soccer—"football," as they call it in England. Her idol is David Beckham and her bedroom is a shrine to the Manchester United superstar. Her parents do not approve. Her father (Anupam Kher) scowls at the Beckham poster over her bed. "That bald man," her father calls him. He does not even care enough to learn his name.

But Jess is more than a fan; she has a gift for playing the game. She plays in the park with the boys and routinely dominates. Juliette "Jules" Paxton (Knightley) sees Jess in action and invites her to try out for the Hounslow Harriers, an all-female club that plays a very serious brand of soccer.

Jess shows up for practice in her running shoes and gym shorts and the coach (Jonathan Rhys-Meyers) is convinced he is wasting his time even looking at her. Jules persuades him to give Jess a chance and, even without proper equipment, she puts on a show, weaving through the defense and making brilliant passes to set up Jules for easy goals. It is clear nothing can stop Jess—except her family.

Her father and mother (Shaheen Khan) do not want Jess playing soccer. "You have played enough, running around showing your bare legs in front of men," her mother says. "You must start behaving like a proper woman."

Beckham benefits by the insight Chadha brings to the story. She grew up in the same Southall neighborhood in a Sikh Punjabi family. She wasn't an athlete, but her interest in filmmaking put her at odds with those around her. That experience allows her to write scenes that feel very natural.

Example: The mother is telling Jess that a woman needs to focus on important things, such as cooking. "What's the good in knowing how to run around kicking a ball," she says, "if you don't know how to make round chapattis?" Meanwhile, Jess is behind her bouncing a head of cabbage from one knee to the other. It is a neatly drawn portrait of a mother and daughter sharing a kitchen yet living in different worlds.

Chadha does a neat balancing act of her own, having fun with Jess's parents, yet never allowing them to become caricatures. She writes them in such a way that you understand they are good people who are doing what they feel is in the best interest of their daughter. The father was a top cricket player in Uganda, but he was discriminated against when he moved to England.

"They would not let me play, they made fun of my turban," he says. "I don't want my daughter to be disappointed as I was."

Jess continues to play but keeps it a secret. Of course, her parents find out and put a stop to it. The Harriers coach tries to convince them to let her play, saying an American coach is coming to scout the final match and there is a good chance he will offer Jess a college scholarship. As it turns out, the match is scheduled for the same day as Pinky's wedding so Jess is forbidden to play.

During the wedding reception, Jess's father softens and allows her slip away to play the second half of the game. When she arrives, she sets up Jules for the tying goal, then scores the game-winner on a penalty kick. On the final shot, she curves the ball around the defense, "bending it" like Beckham. Her father watches, now fully realizing her true potential.

"I saw her play and she is brilliant," he tells his wife. "It would be wrong for us to stop her."

Jess and Jules accept soccer scholarships to Santa Clara. The film ends with their families waving goodbye as the girls leave Heathrow Airport for the United States.

❝❝ BEST LINE: Jules' mother (Juliet Stevenson) worries about her tomboy daughter's passion for soccer. She says, "There is a reason why Sporty Spice is the only one without a boyfriend."

★ SPORTS-ACTION GRADE: C-plus. Most of the action goes like this: Ball is kicked. Cut. Ball is headed. Cut. Ball goes in net. There are not enough wide shots and not enough flow. Jess, however, looks like the real deal handling the ball.

50▷ LONG GONE (1987—NR)

SPORT: Baseball ★ STARS: William Petersen, Virginia Madsen
DIRECTOR: Martin Davidson

Of every movie in this book, *Long Gone* may be the least viewed and the most difficult to obtain. Created and aired by HBO in 1987, the film fell out of release and is nearly impossible to find these days in video stores or online rental clubs.

And that's a shame. Because, 20-plus years after it aired, this forgotten project remains an entertaining, funny, and realistic look at low-level minor league baseball back in the '50s. Take three parts *Bull Durham*, two parts *Slap Shot*, add a dose of *Bingo Long* and a pinch of the *The Longest Yard*, and you've got *Long Gone*.

Those ingredients have made it somewhat of a sports cult movie. While writing this book, we checked Amazon.com for its availability. Not a single copy could be found. Still, 32 people had rated the movie at the site—and

29 of them gave it the maximum five stars.

We would agree. This is an evocative movie, set in the South in 1957, with a smart script and well-staged baseball action. While it did not serve as a star vehicle for either of its main players—William Petersen or Virginia Madsen—both show here why they later went on to become stars. Petersen, of course, is best known as head investigator Gus Grissom on *CSI: Crime Scene Investigation*. And the beautiful Madsen stole the film *Sideways* as Paul Giamatti's out-of-his-league love interest.

Here, Petersen plays Cecil "Stud" Cantrell, pitcher, slugger and manager of the Tampico Stogies, a fictional Class-D franchise on Florida's Gulf Coast. Once, long ago, he was considered a prospect by the St. Louis Cardinals. But first Stan Musial stood in his way, and then World War II. Now he manages a team that must wear dirty uniforms because the owner is too cheap to have them laundered.

Cantrell is part Crash Davis, part Reggie Dunlop (and if those names don't register, you've skipped ahead in this book). At age 38, he lives by three rules:

1. All girls ----.
2. ---- 'em if they can't take a joke.
3. You're the ------- that you have to look at in the mirror.

The third rule, by the way, is often neglected as Cantrell drinks and carouses his way through three Southern states. At a dusty ballpark in Crestview, Ala., he seduces the National Anthem singer, the magnificently named Dixie Lee Boxx (Madsen), who was just elected Miss Strawberry Blossom of 1957. Cantrell thinks he is in for a one-nighter, but Miss Boxx becomes his Annie Savoy—despite his best instincts, he falls hard for her. One year before the release of *Bull Durham*, HBO's *Long Gone* had the Baseball Annie theme down quite well.

The last-place Stogies, meanwhile, get some unexpected help. First, an ambitious teenaged second baseman named Jamie Weeks (Dermot Mulroney) comes knocking. He's a topflight fielder and smart hitter—and also as naïve as he is young. You get a sense he is looking at a naked woman for the first time when he visits Cantrell's hotel room and sees Boxx sprawled across the bed.

Next, the Stogies add Joe Louis Brown (Larry Riley), a power hitter and strong-armed catcher. Problem is, Brown is black. And this is the Jim Crow South of the 1950s. Cantrell decides to pass off Brown as a Spanish-speaking Venezuelan—Jose Luis Brown (a gimmick borrowed from *The Bingo Long Traveling All-Stars & Motor Kings*)—and no one (save a few local Klansmen) is any the wiser.

With the influx of talent, the Stogies go on a winning streak. On the verge of the pennant, however, we learn that Tampico's owner (*Laugh-In* alum Henry Gibson) is as corrupt as he is penurious. Cantrell is given the *The Longest Yard* ultimatum: Lose the championship game, and he will get his chance to hook on as a coach with a major league organization. Win the game, and his career is finished.

There aren't many surprises in the script, and there's nothing original in a story about a scrappy underdog coming out on top. But the screenplay, written by Michael Norell based on a novel by Paul Hemphill (an Atlanta novelist who had a brief minor league career) crafts terrific quirky characters. And director Martin Davidson (*Eddie and the Cruisers*) does a great job of setting time and place, right down to Dixie's daily breakfast of Jax Beer.

We're likely frustrating you by touting a movie that hasn't been on the market since people still used VHS. So we're hoping this little push might convince HBO to re-release *Long Gone*—preferably with some DVD commentary. HBO has a legacy of outstanding movies and mini-series—*Band of Brothers, 61*, Recount,* and many others. Before all of those, there was *Long Gone*. It should not be forgotten to history.

📣 **CASTING CALL:** Jack Nicholson had read the original novel and was interested in playing Cantrell. When this project rolled around, however, he was busy filming a Western comedy called *Goin' South*. We suspect that had Nicholson been available, the movie would have gone to the big screen rather than cable.

⭐ **SPORTS-ACTION GRADE:** B-plus. Mulroney, in his 1950s crew cut and baggy uniform, looks like Bill Mazeroski on second base. And Petersen is a surprisingly good athlete with a smooth pitching delivery (especially when compared with, say, Tim Robbins).

❝❞ **BEST LINE:** Cantrell, explaining the facts of life: "When God made man, He made him out of string. He had a little left, so He left that little thing. When God made woman, He made her out of lace. He didn't have enough, so He left that little space. Thank you, God."

51 THE KARATE KID (1984–PG)

SPORT: Karate ★ **STARS:** Ralph Macchio, Pat Morita
DIRECTOR: John G. Avildsen

The Karate Kid is one of the most predictable and implausible sports movies ever made.

It's also one of the most enjoyable.

It is a movie that requires you to suspend your critical thought process and accept unlikely premises on faith. Like that a five-foot-seven, 105-pound weakling could develop into a martial arts master in a matter of weeks—

mostly by painting fences—and then beat down the linebacker-sized bullies who have been wrecking his life. That plus get the best-looking girl in school who, of course, is on the rebound from dating the most sadistic of the bullies.

If you can just kind of let that all flow by, if you can just say, "Well, sure, it's perfectly reasonable that this old Japanese guy can heal a torn ACL just by touching it," then you'll have two hours of fun with *The Karate Kid*. If not? Well, there are plenty of movies in this book that tell more believable tales. Like, say, the one where the corn stalks tell the Iowa farmer to build a baseball field.

Daniel LaRusso (Ralph Macchio) is a pint-sized New Jersey teenager whose mother takes a job in California, the main attraction of which appears to be that it sinks them to the poverty level. Daniel seems to be a nice enough kid, but everyone west of Newark apparently hates his guts. In fact, he gets roughed up four times within the first 30 minutes of the movie.

His main tormentors are a pack of rich Aryan-Nation-in-training types who study karate under psychotic sensei John Kreese (Martin Kove) at the perfectly named "Cobra Kai" studio. Their mantra is, "Strike first! Strike hard! No mercy—*sir*!" They repeat it often. In unison.

Daniel has just two friends. One is high-school hottie Ali Mills (Elisabeth Shue), who doesn't mind that Daniel is friendless, penniless, frail, socially inept, and goes out on dates with his mother, driving a jalopy that needs to be pushed down the street to get it started. No, Ali loves him for his . . . well, we're not sure, except that it makes her country-clubbing parents uncomfortable. Sometimes that's enough.

His other friend is the handyman at his apartment complex, an aging Okinawan named Mr. Miyagi (Pat Morita), who tries to catch flies with chopsticks and, in one of the movie's best scenes, incapacitates five of the neo-Nazis in about 30 seconds. This persuades Daniel (henceforth known as "Daniel-San") that rather than flee, he must learn to fight.

And learn to fight he does. Kind of. Mr. Miyagi's rigorous training method has Daniel-San sanding his deck, staining his fence, waxing his cars, and painting his house. Somehow, we eventually see, the pipsqueak has learned to defend himself through indentured servitude. Meanwhile, Mr. Miyagi's got some great urban renewal going on.

The Karate Kid leads up to a terrific climactic scene, the standing-room-only All-Valley Karate Tournament, in which Daniel-San must defeat the meaner men from Cobra Kai. First, the bullies taunt him with lines like, "What's the matter, Danny, mommy's not here to dress you?" And, "Put him in a body bag, yeah!" Then one of them takes a deliberate disqualification by snapping Daniel's fibula in half (a violation, we are told, of the all-important Rule 31-point-2).

But our hero isn't done. After Mr. Miyagi lays his healing hands on the kid like the Rev. Benny Hinn, Daniel-San must face the evilest of the Cobra Kai, Johnny Lawrence (magnificently played by William Zabka, who made a career out of portraying jackasses). Let's see, Johnny is 60 pounds heavier, ten times nastier and has studied martial arts for a decade. Plus, he's got his psychotic sensei offering coaching advice like, "Sweep the leg, Johnny. Do you have a problem with that? No mercy."

Now, who's gonna win?

It leads up to one of the great sports movie climaxes, the Crane Kick to Blond Boy's face. A quick flash of a foot and we're done. Cue music, show the girlfriend leaping from the stands, roll credits. Perfect, perfect ending.

In fact, you'll see it all coming a lot quicker than the unfortunate Johnny. This movie telegraphs more than Samuel F.B. Morse. But, hey, don't let that ruin the fun.

❝ BEST LINE: We've got to go with, "Wax on, wax off." It's right up there with, "There's no crying in baseball," and "If you build it, he will come," among the iconic lines in sports movies.

📽 PIVOTAL SCENE: After spending day after day as Mr. Miyagi's personal slave, Daniel-San finally loses his temper.

"I've been busting my ass and haven't learned a ------- thing," he shouts.

"You learn plenty," counters Mr. Miyagi.

"I learn plenty, yeah. I learn how to sand your decks, maybe. I wash your car, paint your house. Paint your fence. I learn plenty."

The angry student storms out, but is called back by his master, who squares off to fight. As Miyagi throws punches and kicks, Daniel-San instinctively blocks them all with the same wax on-wax off and paint up-paint down motions he used on household chores. Miraculously, he has mastered self defense without even knowing it.

We invite you all to try that teaching technique with your teenaged sons.

🔊 CASTING CALL: Chuck Norris was offered the role of evil sensei Kreese, but turned it down because he didn't want karate instructors presented in a bad light.

52 IT HAPPENS EVERY SPRING (1949—NR)

SPORT: Baseball ★ STARS: Ray Milland, Jean Peters, Paul Douglas
DIRECTOR: Lloyd Bacon

In *It Happens Every Spring*, Ray Milland plays a nutty professor (no, not *that* one) who accidentally creates a potion that causes a baseball to be repelled by wood. Armed with his discovery, he becomes a major league pitcher and lifts the St. Louis Cardinals to the World Series.

When the chicanery is discovered, the professor is hauled before a Congressional committee and barred from baseball by Commissioner Bud Selig.

Okay, we lied about the last part. But the rest pretty much outlines the plot. It is a totally silly, utterly preposterous film, but it's also great fun and a delight for audiences of all ages.

While Milland, Jean Peters, and Paul Douglas are the names on the marquee, the real star of the film is Fred Sersen, who created the special effects. Sersen was one of Hollywood's most celebrated artists. He won two Academy Awards and was nominated on six other occasions. He created the catastrophic fire in *Old Chicago* (1938) and the floods in *The Rains Came* (1939).

Sersen was one of the pioneers of cinematic illusion. He was making cities burn, waters rise, and sea creatures attack long before computers made such trickery routine. So when director Lloyd Bacon needed someone to make a baseball leap over a bat, he knew there was only one man for the job. He called Sersen, who had just finished working on *Down to the Sea in Ships* with Lionel Barrymore.

It Happens Every Spring was a different sort of venture for Sersen. In most of his films, Sersen—who worked on more than 200 motion pictures—sought to make the effects realistic. In this film—a fantasy played for laughs—Sersen could have some fun. He made the dancing baseball a comically crude device. It jumps over the bat and into the catcher's mitt with an exaggerated "Whoooosh." You can imagine Sersen chuckling as he put it on film.

Viewed today, the special effects are—well, the word *cheesy* comes to mind. But that only makes the film more amusing. And Ray Milland's pitching motion, which resembles a man falling off a ladder, makes it funnier still. If Milland looked more like a real pitcher and the special effects were slicker, *Spring* wouldn't be nearly as much fun.

Originally a short story written by Shirley Wheeler Smith in 1923, *It Happens Every Spring* was adapted into a screenplay by Valentine Davies, who won an Oscar for writing the Christmas classic *Miracle on 34th Street*.

Spring was such a quirky tale that no one knew how it would be received when it was screened for a test audience in Riverside, Ca. But as Davies wrote to Smith, "There was so much laughter that some of the funniest lines were completely drowned out. The [comment] cards were also most enthusiastic and what pleased us most was that the women and girls seemed to be no less enthusiastic about the picture than the men."

The film's hero is Vernon Simpson (Milland), a nerdy chemistry professor with a crush on Deborah Greenleaf (Peters), daughter of the school president. Vernon is developing a chemical compound designed to keep insects away from wood. One day a baseball crashes through his lab window and lands in a tub filled with the liquid.

While he is cleaning up the mess, Vernon notices something odd. Each time the ball comes near a piece of wood, it veers away. Vernon, a baseball freak, immediately sees the possibilities. If he rubs his secret compound on the ball, the ball will be repelled by the wooden bat and therefore be unhittable.

He tries it out with two members of the varsity baseball team, and when they helplessly flail away at his offerings, he is emboldened to ask the Cardinals for a tryout. When he mows down their best hitters, he is signed to a contract. He takes the name King Kelly, and with a stash of Methyl Ethyl Propin Butyl hidden in his glove, he becomes the top pitcher in baseball.

He wants to keep his identity a secret, so he refuses to be photographed or interviewed, which becomes a real hassle as he piles up win after win. Meanwhile, he is MIA at the university. The professor has disappeared and no one knows why. When he sends Debbie a diamond ring, her parents conclude that he must have turned to a life of crime. It is zany stuff, it doesn't make a whole lot of sense, but it is very entertaining.

It Happens Every Spring is typical of the screwball romantic comedies that were so popular in the 1940s. Cary Grant and Clark Gable made many. But this film was a departure for Milland, who was best known for his performance as an alcoholic writer in *The Lost Weekend*, a role which earned him the Academy Award in 1946.

Milland is good in *Spring*, especially his chemistry with Paul Douglas, who plays his dim-witted catcher and roommate Monk Lanigan. When Monk sees the bottle of Butyl in the bathroom, Vernon tells him it is hair tonic. Monk applies it to his scalp and when he tries to use a wooden brush on his hair, it is time for more of Sersen's special effects.

Over the years, some deep-thinkers have analyzed the film and questioned whether it is a good thing that it celebrates cheating in baseball, as if Gaylord Perry saw *Spring* and that is how he got the idea to smear Vaseline on the ball and become a Hall of Famer.

Lighten up, guys. It's only a movie.

" BEST LINE: The tag line from the movie poster: "The guy who invented the ball nobody could hit. And the girl with the curves nobody could miss."

🍎 BET YOU DIDN'T KNOW: The movie was filmed in Wrigley Field in Los Angeles, home of the minor league Angels. The ballpark was later used for the 1950s TV show *Home Run Derby*.

🔔 CHEERS: The scene in which Monk tries to pick up the ball while wearing a wooden splint on his finger is hilarious.

🤓 "I KNOW THAT GUY": Alan Hale Jr. is the catcher for the on-campus tryout. He is best known as the Skipper in *Gilligan's Island*, so it has a familiar ring when he calls out to Milland, "Hey, Professor...."

53▶ WITHOUT LIMITS (1998–PG-13)

SPORT: Track ★ **STARS:** Billy Crudup, Donald Sutherland, Monica Potter
DIRECTOR: Robert Towne

Robert Towne is a man of considerable influence. He won an Academy Award for his script on *Chinatown*. He received three other Oscar nominations for screenwriting so he had the clout to sell a studio on a film project it might otherwise ignore. *Without Limits* was that kind of project.

Let's count up all the reasons why *Without Limits* was a lousy investment. First of all, it was a movie about track, a sport only a handful of Americans care about. Second, it was the story of a guy who competed in one Olympics and didn't even win a medal. Third, it was tried one year earlier—different studio, different cast—

and flopped at the box office.

But Towne, a track enthusiast, had such a passion for the story of Steve Prefontaine, the charismatic American distance runner, that he was able to convince Warner Brothers to throw its millions behind the film. It helped that Tom Cruise signed on as producer and briefly considered playing the lead before deciding he was too old.

If we go strictly by the bottom line, *Without Limits* was an utter dud. The film cost $25 million to make and it grossed less than $1 million. If Cruise had played Prefontaine—even if he needed an oxygen tank to run the 5,000 meters—it would have sold tickets. But the lead, Billy Crudup, was an unknown, the title was a head-scratcher, and as a result, the film went nowhere.

However, if we evaluate *Without Limits* strictly as a piece of art, it is quite good. Janet Maslin of the *New York Times* called it "the most stirring and unmistakably personal film [Towne] has directed." Owen Gleiberman of *Entertainment Weekly* called it "incisive and enthralling, as deft on its feet as the athletes it's about." Paul Tatara of CNN said it was "one of the best sports movies I've ever seen."

It is fitting in a way because Towne surely viewed this film the way Prefontaine viewed his career on the track; that is, it's not about the bottom line. It's not even about winning; it's about the art. It's about pursuing a higher standard than mere victory or dollars. That was Prefontaine's ethos and Towne was faithful to it in telling his story.

Even without an Olympic medal, Steve Prefontaine will be remembered as a great runner. And even without a big box office return, *Without Limits* will be remembered as a well-crafted piece of film-making, rich with feeling and insight. Towne's gift for writing is one of the film's strengths. With his well-chosen words, Towne makes the character of Prefontaine both understandable and admirable.

Towne tells much of the story in narration written for Donald Sutherland, who plays Prefontaine's coach Bill Bowerman. Towne's writing is lean, yet powerful, and Sutherland's read hits just the right notes. A typical line: "From the beginning, I tried to change Pre. He tried not to change. That was our relationship." Three short sentences, 17 words total, but it sums up everything we need to know about the two men.

"You can call a race any ——————— thing you want, but I wouldn't call it a performance."
—PREFONTAINE

"What would you call it?" —BOWERMAN

"A work of art."
—PREFONTAINE

Prefontaine (Crudup) is a track star of the 1970s. He holds the American record at every distance from 2,000 meters to 10,000 meters. However, "Pre," as he is known, drives his coaches—Bowerman in particular—crazy by refusing to race strategically. He refuses to pace himself. He insists on pushing himself to the limit. That means going to the front of the pack and staying there. Bowerman tells Pre he'll never be able to defeat the world's best runners that way. Pre refuses to bend.

"I don't want to win unless I know I've done my best," Pre says, "and the only way I know how to do that is to run out front, flat out until I have nothing left. Winning any other way is ————————."

Handsome with sun-bleached blonde hair, Pre is like a rock star in the track hotbed of Eugene, Oregon, where he attends college. He is cocky and narcissistic and, frankly, at times he is not all that likable. It's fine to be an iconoclast—it isn't like distance running is a team sport—but Pre takes it to irritating lengths when he says things like: "I can endure more pain than anyone you've ever met. That's why I can beat anyone I've ever met."

But when he is spiked during a race and we see him coming down the final straightaway with the blood soaking through his shoe, we can appreciate the fact that this is a guy who is defined by something more than stopwatches and medals.

Without Limits is at its best when it focuses on the relationship between the headstrong Pre and the coach/psychologist Bowerman. At first, Bowerman thinks he can change Pre simply by ordering him to do so. Then Bowerman realizes Pre is not just another runner, therefore he cannot be coached like one. The coach must find other ways to reach him, and although Towne does not belabor the point, it works as a metaphor for what was going on between the generations in America at the same time.

In the end, the coach and the runner find an understanding and a mutual respect. As Bowerman says in another voiceover: "Pre thought I was a hard case. But he finally got it through my head that the real purpose of running isn't to win a race, it's to test the limits of the human heart. That he did. No one did it more often. No one did it better."

📢 **CASTING CALL:** Towne originally wanted Tommy Lee Jones to play the part of Bill Bowerman, but Jones turned it down.

😎 **"I KNOW THAT GUY":** Jeremy Sisto, who plays American marathoner Frank Shorter, is Detective Cyrus Lupo on TV's *Law and Order*.

✎ **WHAT THEY WROTE AT THE TIME:** "Sutherland, updating the eternal myth of the tough-love coach/drill sergeant/Mr. Miyagi figure, gives a performance of triumphant, winking intelligence. [He] breaks Pre as if he were a beautiful horse."—Owen Gleiberman, *Entertainment Weekly*

📣 **CHEERS:** Donald Sutherland's performance as Bill Bowerman earned him a Golden Globe nomination for Best Supporting Actor. "Sutherland hasn't completely involved himself in all his parts, but he's done so here," wrote Kenneth Turan in the *Los Angeles Times*. "The result is a commanding, almost hypnotic performance that is among the actor's best."

☞ **DON'T FAIL TO NOTICE:** The running shoe that Bowerman develops during the film. It is the prototype of what became the Nike empire.

66 **BEST LINE:** Steve Prefontaine: "I'd like to work it out so that at the end, it's a pure guts race. If it is, I'm the only one who can win it."

🍎 **BET YOU DIDN'T KNOW:** Steve Prefontaine competed in the 1972 Summer Olympics in Munich and finished fourth in the 5,000 meters. He was killed in a car crash in Eugene three years later.

★ **SPORTS-ACTION GRADE:** B. The track scenes are well shot and Crudup has the lean, sinewy stride of a distance runner.

54 THE HARDER THEY FALL (1956–NR)

SPORT: Boxing ★ STARS: Humphrey Bogart, Rod Steiger, Jan Sterling
DIRECTOR: Mark Robson

The Harder They Fall is noteworthy because it was Humphrey Bogart's final screen appearance. Bogart was battling lung cancer, and it was a struggle for him to make it through the long days of filming. But he turned in a moving performance as Eddie Willis, a sportswriter who sells out to the syndicate.

The film is based on a novel by Budd Schulberg, who wrote the screenplay for *On The Waterfront*. The movies are similar in look and feel: gritty black-and-white *films noirs* shot in and around New York with Rod Steiger cast as the mobbed-up heavy. Ads for *The Harder They Fall* read: "If you think *On The Waterfront* hit hard, wait until you see this one."

The Harder They Fall did not reach the heights of *Waterfront*. It earned one Academy Award nomination, and that was for photography (Burnett Guffey). Bogart and Steiger were passed over, although both were deserving. The film earned mixed reviews. Some critics found it heavy-handed. Boxing apologists ripped it as dishonest. But others were blown away by the story, which painted an ugly picture of the fight game.

Sportswriter Reid Cherner of *USA Today* ranks it No. 1 on his all-time list of sports movies. "Bogart and Steiger make this gritty film a success," Cherner wrote.

The film was a thinly disguised biography of

former heavyweight champ Primo Carnera. It was so thinly disguised, in fact, that Carnera sued Columbia pictures, claiming the film makes him look like a fraud. However, most boxing experts agree the film is an accurate account of Carnera's improbable rise to fame.

Carnera was a muscle-bound oaf imported from Italy who was—shall we say—carefully carried up the heavyweight ladder. His nicknames—"Satchel Feet" and "The Ambling Alp"—will give you some idea of his ring artistry. Nat Fleischer of *Ring* magazine described Carnera as "a crude fighter with little skill . . . one of the poorest heavyweight champions." Yet, in 1933-34, he held the title.

In *The Harder They Fall*, the fighter is Toro Moreno (Mike Lane), a 6-10, 270-pound circus strongman, brought in from Argentina to join the boxing stable of Nick Benko (Steiger). There is only one problem: he can't fight. The first time he spars, he's knocked on his butt by a broken-down old pug named George (Jersey Joe Walcott).

"Powder-puff punch and a glass jaw, great combination," says Willis (Bogart), who attends the workout. "That's some discovery you have there, Nick."

Benko has invested a lot of money in Toro, so he refuses to write him off. He asks Willis, an unemployed sportswriter, to sign on as Toro's publicist. His job will be to sell the big galoot, create a phony backstory, and use the press to build Toro into a contender.

"Oh, I get it—Toro Moreno, Wild Man of the Andes," Willis says sarcastically.

"Perfect," Benko says.

"But what happens when you put him in the ring?" Willis asks.

"I'll take care of that," Benko says.

In other words, he'll pay Toro's opponents to take a dive.

Willis has spent years in the newspaper business, working to expose guys like Benko. His instincts are telling him to walk out, but his paper just folded and he needs the money. He stands there agonizing.

"Don't fight it, Eddie," Benko says. "What are you trying to do, hold onto your self-respect? Did your self-respect help you hold onto your job? Did your self-respect give you a new column?"

Reluctantly, Willis takes the job and begins selling a fighter who he knows can't fight a lick. They travel the country on a bus with Toro's likeness on the side. It's like a carnival act with Willis cranking the calliope as Toro piles up one bogus knockout after another.

Willis' conscience continues to nag him, especially as he grows closer to Toro. The big guy is a gentle soul who has no idea what's going on. He thinks he's winning these fights on the level. He is the only one who doesn't know it is a sham. He also is unaware Benko and his crooked associates are stealing his earnings. Willis knows where this is headed and so do we.

Thanks to Willis' publicity campaign and Benko's wheeling and dealing, Toro gets a shot at the heavyweight champion, Buddy Brannen (Max Baer). Here is where art meets life. Baer fought Primo Carnera in 1934, knocking him down 11 times to take his title. In the film, his character does the same thing, pulverizing the helpless Toro who is too proud to follow Willis' advice and take a dive. Toro keeps absorbing punishment until finally he is flat on the canvas, his jaw broken and his face a bloody mess.

After the fight, Willis goes to Benko's office to pick up Toro's purse. "I told him I'd bring it to him in the hospital," he says. Benko's accountant (Nehemiah Persoff) goes through the books and tells Willis that after this deduction and that deduction (all of them phony but notarized) Toro's share of the million dollar gate is $49.07.

That's the last straw. Willis gets Toro out of the hospital, gives him his own share of the purse ($26,000), and puts him on a plane back to Argentina. Then Willis returns to journalism. The last thing we see are the keys of his typewriter pounding out: "Professional boxing should be outlawed in the United States if it takes an act of Congress to do it."

Willis is writing an exposé that presumably will put Benko and the rest of his lowlife ilk out of business. OK, so it didn't work out. Last time we looked, boxing was still around. But, hey, it was a cool way to end the film.

PIVOTAL SCENE: Willis tells Toro the truth: he can't fight, all of his victories have been set-ups, and now he is fighting the heavyweight champion who is eager to destroy him. Willis tries to convince Toro to take a dive and spare himself a beating.

"What will people think of me?" Toro asks.

"What do you care what a bunch of bloodthirsty, screaming people think?" Willis says. "Did you ever get a look at their faces? They pay a few lousy bucks hoping to see a man get killed. To hell with them. Think of yourself. Get your money and get out of this rotten business."

55 SUGAR (2008–R)

SPORT: Baseball ★ STARS: Algenis Perez Soto
DIRECTOR: Anna Boden, Ryan Fleck

Every baseball fan knows of the great Dominicans who have made it in the major leagues: from old-timers like Juan Marichal and the Alou family to current stars who dominate the game—Robinson Cano, Manny Ramirez, Jose Reyes, Hanley Ramirez, and so on. These days, nearly one in seven big leaguers hails from the tiny island nation of fewer than 10 million people.

Less known are the stories of the hopefuls from the Dominican Republic who come over with dreams of playing in Yankee Stadium and driving a Cadillac, but fall short. *Sugar*, a sweet movie filmed mostly in Spanish, focuses on one such young man.

The movie follows the path of Miguel "Sugar" Santos, a 19-year-old pitcher from impoverished San Pedro de Macoris whose noon-to-six knuckle-curve earns him a ticket to Class-A ball with hopes of climbing much further.

Santos is assigned to Bridgetown, Iowa (looking at the abbreviation IA, he wonders, "Where is *Ee-yah*?"), but he might as well be sent to Mars. With no knowledge of the English language or middle-American customs, everything becomes an exotic, often daunting experience—from visiting a diner (he repeatedly orders French toast because that's all he knows to say), to turning on his hotel TV and discovering in-room porn, to dealing with the chaste farm girl (Ellary Porterfield) who flirts with him and then drags him to a Presbyterian youth group meeting.

For a few weeks, Santos is a star pitcher for the Bridgetown Swing. He's signing autographs, pointing at his name in the paper (he can't read the rest of the article), and gaining extra attention from his coaches. But he injures his foot on a fielding play and subsequently loses control of his pitches. He gets knocked out of a few games early. Soon after, his best friend is cut from the team. Santos lacks the maturity and support system to cope with adversity, so he takes a fungo bat to the water cooler and then goes AWOL.

We won't reveal where the movie goes, but suffice it to say that Santos does not realize his goal of starting an All-Star Game and lifting his family from poverty. On the other hand, he does not return home as a defeated man. And he discovers that he can love the sport, even if he may never be pictured on a baseball card.

Sugar is at its best showing Santos as the stranger in a strange land. That, in itself, is a well-traveled movie path, but the story of Latin teens trying to survive in an ultra-competitive sport, while at the same time combating social disorientation, has been vastly underreported over the years. For every one who makes it, there are a dozen who fail because they are homesick, or don't understand what their manager is saying, or spend their entire paycheck (and gain 30 pounds) at the local McDonalds.

True story: Back in the 1980s, I (Glen Macnow) spent time covering baseball in the Dominican Republic. One day, I shared a ride with a young catcher named Francisco Cabrera, who was toiling at the Toronto Blue Jays academy near Santo Domingo. Cabrera had just been handed papers assigning him to play for the Jays' farm team in St. Catherine's, Ontario.

"What is St. Catherine's like?" asked Cabrera, who went on to play parts of five seasons in the majors. "Is it cold? Last year I played in Bradenton [Fla.]. That was freezing. Is this place as cold as Bradenton?"

I didn't have the heart to tell him.

If *Sugar* has a weakness, it's that the movie aims to be inherently neutral about many aspects of Dominican recruitment that warrant a harsher spotlight. The movie refers to the talent brokers known as *buscones* who illegally sign up prospects as young as 12, hide players to keep others from scouting them, and alter birth certificates. But it fails to really delve into the subject. Likewise, *Sugar* makes one quick allusion to drug use but ignores the large issue of commonplace steroid abuse among players seeking to cross the Caribbean.

"Any industry where there's a lot of money to be made and there are poor people involved, there's going to be some exploitation on some level," co-director Ryan Fleck told the *New York Times*. "But we really didn't want to focus on that."

Too bad. A good movie might have become a powerful movie.

Regardless, *Sugar* is a worthy addition to any list of first-rate movies about baseball. It is original, thoughtful, and authentic. If you can handle a film done mostly in subtitles, it is well worth two hours of your time.

CASTING CALL: That's 1990 World Series MVP Jose Rijo playing the head of the fictional Kansas City Knights' Dominican baseball academy.

BET YOU DIDN'T KNOW: Shortly before the film's release, Rijo was fired from his job as special assistant to the Washington Nationals amid a federal investigation into whether he and other baseball officials took kickbacks from signing bonuses promised to Dominican prospects.

BEST LINE: Well, it's a series of lines, actually, when the teens at the Dominican academy recite back baseball phrases like so many Berlitz students.

"I got it!"

"Ground ball!"

"Line drive!"

"Home run!"

And then a pidgin-English chorus of "Take Me Out to the Ball Game."

56 MYSTERY, ALASKA (1999–R)

SPORT: Hockey ★ **STARS:** Russell Crowe, Burt Reynolds, Hank Azaria
DIRECTOR: Jay Roach

Any true hockey head could envision living in Mystery, Alaska, a remote outpost where you skate for miles on frozen rivers and drive to school on a snowmobile. Each weekend, the hamlet's 633 residents gather for the "Saturday game," in which the most talented local men divide into two teams and play pond hockey on natural black ice. The Saturday game has been serving for decades as Mystery's source of identity and pride.

That is, until the town's prodigal son, Charlie Danner (Hank Azaria), writes a *Sports Illustrated* feature extolling the quirky local custom. Despite its hackneyed prose

("legendary players born on skates in a world permanently covered with ice and snow . . ."), the *SI* article prompts the NHL to devise its best marketing gimmick since the glowing puck: The New York Rangers will travel to Alaska to take on the fierce locals in a nationally televised game of shinny.

Yeah, sure they will. The entire premise of *Mystery, Alaska* is far-fetched (similar to, say, a South Philly tomato can being invited to fight for the heavyweight championship), and requires you to suspend critical disbelief. But if you can move beyond the fanciful nature of the script, this is a fun, albeit hokey, hockey movie.

Russell Crowe stars as town sheriff John Biebe, who learns at the film's start that, after 13 years, he is being bumped from the Saturday game for a high school phenom. Crowe must deal with all kinds of conflicted emotions here—how he feels about the kid taking his spot, how he feels about being asked to coach instead, how he feels about his cute wife (Mary McCormack) flirting with former boyfriend Azaria. At one point, he starts babbling like John Nash in *A Beautiful Mind,* while taking out his frustrations on a life-sized metal cutout of former Montreal Canadiens goalie Ken Dryden.

Still, it's a typically solid Crowe performance. Lots of mumbling and pained expressions and a strong physical presence. And, since we don't think we're giving much away here by telling you he ends up lacing them up again to face the Rangers, we'll add that it's fun to see Crowe suffering through the most pain and bleeding he's endured this side of *Gladiator.*

We also learn that he probably didn't play much hockey growing up Down Under. Crowe moves on skates in a lumbering-old-NHL-Moose-Vasko kind of way, at one point using his stick as a rudder. Not as easy as your beloved Australian Rules Football, eh Rusty?

But back to the script. After initially being thrilled by the chance for glory, a few of the locals start to sense that they're being set up for the most lopsided hockey game since Japan played Sweden in the 1960 Winter Olympics (final score: 19-0). Stodgy town judge Walter Burns (Burt Reynolds) implores the Mysterians to back out, saying, "Two things we've always had: Our dignity and our illusions. I suggest we protect them both."

Of course, he's overruled. There wouldn't be a movie otherwise.

That sets up the next crisis. The spoiled pros from New York, naturally, have no interest in filling their few days off during the NHL schedule by flying to the Arctic Circle to get cross-checked by a bunch of yokels. They file a suit to stop the game. In one of the film's best scenes, Mystery's only attorney, Bailey Pruitt (Maury Chaykin), flies to New York, convinces a judge not to pull the plug, and then drops dead of a heart attack.

And so, the disgruntled and freezing Rangers finally arrive.

Like any good sports movie, it all leads up to the big game. And by then, you're fully invested in the quirky underdogs. We won't give away the ending (we probably don't have to), but the actual hockey scenes are well choreographed and offer several highlights that should prompt you to rise from the Barcalounger. Someone among the triumvirate of producer David E. Kelley, head writer Sean O'Byrne, and director Jay Roach knows his hockey.

Actually, the best moments are those in the locker room. Anyone who ever played the game (or any sport, really) will appreciate the bits of clubhouse camaraderie, taping up sticks and swapping lies about sexual conquests. The locker-room language, we suppose, earned this film its R-rating, although there's really nothing here that would be objectionable to anyone over age 12.

In truth, *Mystery, Alaska* comes off like the pilot for a solid television sitcom, and we hope that isn't damning it with faint praise. Producer Kelley is the King of the Emmys, who has created quirky shows with oddball ensemble casts, like *Ally McBeal, Boston Legal,* and *Picket Fences.* Place them in the remotest town on Earth, put them in skates and, well, it's better than 90 percent of the junk on TV.

❝❝ BEST LINE: Skank Marden (Ron Eldard): "I play hockey and I fornicate, because those are the two most fun things to do in cold weather."

57 RUDY (1993–PG)

SPORT: Football ★ STARS: Sean Astin, Jon Favreau, Charles S. Dutton
DIRECTOR: David Anspaugh

It is often said that no one is neutral about Notre Dame. You either love the Irish or you can't stand them. You either tear up at the sound of the Notre Dame victory march or you leave the room. That's the way it is with *Rudy*, the film about a spunky runt named Daniel "Rudy" Ruettiger whose whole life was built on the dream of playing football for the Irish.

Rudy placed fourth in a 2005 ESPN viewers poll of the top sports movies. Leonard Klady of *Variety* lavished it with praise, calling it "a film that hits all the right emotional buttons, an intelligent drama that lifts an audience to its feet cheering."

Other reviewers hated it. Andrew Payne of *Starpulse Entertainment News* called *Rudy* the most overrated sports movie of all-time. "*Rudy* stinks," he wrote. "It is an awful, overly sentimental movie filled with more processed cheese than a supermarket danish."

It is a wide split of opinion based partly on the viewer's feelings about Notre Dame. If you don't buy into the mystique of Rockne and the Gipper, you aren't going to buy into *Rudy*. The only way you can relate to Rudy's story is if you believe there's something special about playing football under the Golden Dome. But if you believe that, even a little bit, *Rudy* will get your Irish up.

The film opens with Rudy as a little kid living in a blue-collar Catholic home in Joliet, IL, where watching Notre Dame football is an autumn ritual. His father, Dan Sr. (Ned Beatty), is parked in his favorite chair every Saturday, and Rudy sits on the floor next to him as they watch the Irish on TV. Rudy's room is decorated with Notre Dame pennants, and he can recite every Knute Rockne pep talk from memory.

Rudy's dream is to attend Notre Dame and become part of that football tradition. The trouble is, he doesn't have the size (he is 5-6 and 170 pounds) or the grades (he is a C student). Even his father tells him: "Notre Dame is for rich kids, smart kids, great athletes. It's not for us. You're a Ruettiger."

Rudy (Sean Astin) joins his father, his brother Frank (Scott Benjaminson), and his best friend Pete (Christopher Reed) working in the steel mill. He isn't happy about it, he still has thoughts about going to Notre Dame, but he is beginning to think, well, like a Ruettiger. He dons his hard hat and heads to the plant every day. "Chasing a dream can only cause you heartache," his father says.

But Pete sees the wistfulness in Rudy's eyes. For his birthday, he gives Rudy a Notre Dame jacket. "You were born to wear that," Pete says. Rudy tells him, "You're the only one who ever took me seriously." Pete replies: "You know what my Dad said: Having dreams is what makes life tolerable."

Pete understands Rudy better than his own family, certainly more than Frank who ridicules his younger brother. "As long as he talks this crazy Notre Dame ---- he deserves anything that comes his way," Frank says.

When Pete is killed on the job, Rudy decides he must pursue his dream for both of them. He hops a bus for South Bend, walks onto the Notre Dame campus, and announces he wants to enroll. He tells his story to a kindly priest (Robert Prosky) who gets him into a junior college. He tells Rudy he could gain admission to Notre Dame if he shows sufficient academic progress.

Of course, just getting into Notre Dame is only part of Rudy's dream. The other part is earning a spot on the football team, which will be next-to-impossible for a defensive end (Rudy's high school position) who is smaller than the ball boys.

That's the second half of the film: Rudy's hard work in the classroom to get the grades (it's a struggle, but he makes it with the help of tutor Jon Favreau), then earning a spot on the practice squad where he joins other walk-ons to serve as human tackling dummies for the Irish regulars. The scenes of a bloodied Rudy being driven into the mud by men twice his size will make you wince, but his tenacity earns the respect of coach Ara Parseghian.

Practice squad players don't dress for games, they sit in the stands with everyone else. Rudy desperately wants to come out of the tunnel at Notre Dame Stadium just once so his family can see that he really is part of the team. (Frank continues to doubt him, saying: "I see the games every week, but I never see you.")

Parseghian promises Rudy that he can dress for a game in his senior year, but then the coach resigns. The new coach, Dan Devine, has no knowledge of the promise and no particular interest in Rudy. To Devine, he is just another faceless body in a blue practice squad jersey. He has no intention of letting Rudy dress for a game until the other players on the team rally behind him.

Rudy is given his chance to dress for the final home game of the 1975 season. As the players gather in the tunnel, co-captain Roland Steele asks Rudy: "Are you ready for this, champ?" Wide-eyed, Rudy replies, "I've been ready for this my whole life."

"Then you lead us onto the field," Steele says.

The slow-motion shot of Rudy coming out of the tunnel, pointing to his family in the stands while the Notre Dame band plays the victory march, should strike a chord even in those who aren't huge fans of the Irish. And when Rudy finally gets a chance to play, as *SI.com* wrote: "Don't be surprised if you find yourself chanting, 'Rudy! Rudy! Rudy!'"

Rudy would seem to be a movie made for the underdog audience, but like *Rocky* it has universal appeal. Chris Long of the St. Louis Rams is certainly no underdog. The son of Hall of Famer Howie Long, an All-America defensive end at Virginia, and second overall pick in the 2008 NFL draft, Chris is no Rudy, yet he relates to his competitive spirit.

"Rudy works for his dream," Long said, "and even though in the end it's just a couple of plays which really isn't that big of a deal, it's huge for him. I loved it."

CHEERS: Charles S. Dutton is very good as Fortune, the stadium groundskeeper who takes Rudy in and provides the tough love to keep him going.

BET YOU DIDN'T KNOW: Charlie Weis opened his first team meeting as head coach at Notre Dame in 2005 by showing the film *Rudy* and asking the real Rudy Ruettiger to address the squad. Rudy's message: "You cannot ever quit on yourself."

BEST LINE: Fortune to Rudy: "You're five foot nothing, a hundred and nothing and you hung in with the best college football team in the land for two years and you're going to walk out of here with a degree from the University of Notre Dame. You don't have to prove nothing to nobody."

REALITY CHECK: The scene where the varsity players all drop their jerseys on Devine's desk and say, "Let Rudy dress in my place," well, it never happened. Devine said he alone made the decision to let Rudy dress for the final home game.

DON'T FAIL TO NOTICE: In the big game, Notre Dame is playing Georgia Tech, but there are fans in the stands wearing Boston College colors. Why? The scenes were shot at halftime of a Notre Dame-Boston College game.

WHAT THEY WROTE AT THE TIME: "Anyone who is not moved by this film has something else in their chest where their heart ought to be."—Robert Roten, *Laramie Movie Scope*

"I KNOW THAT GUY": Chelcie Ross, who plays Dan Devine, is the wanna-be coach who is chased off by Norman Dale (Gene Hackman) in *Hoosiers*. Both films were directed by David Anspaugh and written by Angelo Pizzo.

SPORTS-ACTION GRADE: B-plus. The football scenes were shot by NFL Films, so the coverage is first-rate. Anspaugh hired NFL Films because he knew he had a limited amount of time to shoot his game scenes (it all had to be done in the 15-minute intermission of the Notre Dame-BC game) so the cameramen had to get every shot right the first time. NFL Films crews deal with that pressure every week.

"The football scenes in *Rudy* were the best I've ever seen and I'm a tough critic," said George Martin, who played 14 seasons at defensive end for the New York Giants. "You could almost feel the hits. (Rudy) kept getting knocked down and getting back up. It was sheer brutality, but that's what it's like on the practice field. It's about survival and Rudy is a survivor."

58 ▶ KINGPIN (1996—PG-13)

SPORT: Bowling ★ **STARS:** Woody Harrelson, Bill Murray, Randy Quaid
DIRECTOR: Bobby and Peter Farrelly

Not many folks are ambivalent about *Kingpin*. You either abhor this gross-out comedy (as does one of this book's two authors), or you love it (as does the other). In order to make the second category, you must find humor in the following:

- A sex romp with a skanky yellow-toothed landlady that leaves our protagonist barfing.
- A scene where a city slicker drinks fresh "milk" from a bucket—only to learn that the farm has no cows, only a bull. Think about it.
- A cross-dressing striptease involving a 240-pound man licking a dance pole.

And then there's the *really* distasteful stuff. Kingpin has prosthesis gags, stuttering gags, bong-smoking Amish gags, misuse of urinal gags, and everything else that could be thought up at the frat house over the course of a beer-soaked weekend. That's why we love it—well, at least one of us.

Kingpin represents the first—and perhaps the best—of the recent genre of frat boy sports comedies. *The Waterboy, Talladega Nights, Blades of Glory* and *Dodgeball* all followed, and all grossed over $100 million at the box office (*Kingpin* grossed $25 million). All owe their success to this Farrelly Brothers story of a down-on-his-luck one-handed bowler who finds his second chance in a Pennsylvania Dutch phenom. Kind of like *The Color of Money* meets *Dumb and Dumber*.

The movie opens with the story of young Roy Munson (Woody Harrelson), winner of the 1979 Iowa Odor-Eaters Championship, who heads out to make his name on the professional bowling tour.

He's a natural with a bright future. After his first tournament (which he wins), Roy is tutored in the seamy underbelly of bowling survival by veteran Ernie McCracken (Bill Murray in cinema's greatest-ever combover)—which involves hustling the locals. The hustle goes sour, however, and as McCracken burns rubber escaping, the angry victims feed young Roy's hand to the ball return. Think the wood chipper in *Fargo*.

Fast forward 17 years. We catch up with Roy living in a dilapidated rooming house in Scranton, Pa., drinking rotgut for breakfast and trying to make a living by selling fluorescent condoms ("You don't have a novelty machine in the men's room? And you call yourself a bowling alley?"). He wears a hook where his right hand used to be and also has a screw-on cheap rubber hand that displays his old state championship ring. How the ring survived that reset machine baffles us, but let's not be sticklers.

One day Roy stumbles upon Ishmael Boorg (Randy Quaid), an Amish galoot with great natural talent. The movie slows down a little in the middle as Roy coaches Ish, persuades him to join the tour, teaches him to floss his teeth, and readies him for the $500,000 Reno Open. There's also a secondary plot arc involving a fallen woman with a heart of gold (Vanessa Angel) which doesn't add much to the movie beyond providing the opportunity for a half-dozen good breast jokes.

Act III of the movie heads for Reno, where Roy meets up with his old nemesis. Big Ern is now the sport's top superstar—introduced by Chris Schenkel as "a guy who has done for bowling what Muhammad Ali has done for boxing." You know that the tournament (which has several of the PBA's top stars in cameos) is going to end in a showdown between McCracken and the Amish Kid, and *Kingpin* does not disappoint.

Let's say here that Murray is at his comedic apex in *Kingpin's* final 30 minutes. *Caddyshack*, *Groundhog Day*, and *Stripes* all provided moments of glory for our favorite *Saturday Night Live* alum, but his portrayal of McCracken, the cocky, smarmy narcissist, borders on genius. The Farrellys gave Murray the opportunity to improvise most scenes, and his off-the-head victory speech near the end of *Kingpin* ("Now I can buy my way out of anything! Big Ern is above the law!") should have at least earned him a Best Supporting Actor Oscar nomination in 1996. Hell, who remembers Armin Mueller-Stahl in *Shine* anyway?

Harrelson is also terrific as the once-idealistic, now-cynical schmo. He's such a perpetual victim that he inspires a nationwide verb—to be "munsoned," as the movie says, is to have the world in the palm of your hands and then blow it. And Roy finds a way to blow it every time.

If you're a fan of the Farrelly Brothers, you'll recognize other regular members of their troupe, many of whom could not act their way out of a middle-school musical, but are funny enough to deliver a kick-in-the-pants sight gag. There are also some wooden cameos by jocks, although none approaching Brett Favre's inadvertent fall-down funniness in *There's Something About Mary*. Keep an eye open for pro golfers Brad Faxon and Billy Andrade and a certain steroided-up former major league pitcher with a bad attitude. And make note that the harmonica-playing Amish singer in the final scene is John Popper of Blues Traveler.

The jokes come fast here. Some will make you snicker (like the "Jeffersons on Ice" show at the Reno hotel), some which made you shudder (like the over-exuberant effort to remove a horse's shoes). There's physical humor, stupid jokes, groaners, and throwaways coming at the pace of five a minute for all 113 minutes—which totals 565 jokes. If the Farrellys get you 50 percent of the time, that's a lot of laughs.

😎 **"I KNOW THAT GUY":** The muscled-up white trash trucker named Skidmark who wants to punch out Ishmael at the redneck bar? Why, isn't that Roger Clemens in his funniest appearance this side of a Congressional subcommittee?

❝❝ **BEST LINE:** As Roy prepares to bowl the final frame in his first tournament, you hear what appears to be an announcer in the background. Instead, it's his opponent:

McCracken: "It all comes down to this roll. Roy Munson, a man-child, with a dream to topple bowling giant Ernie McCracken. If he strikes, he's the 1979 Odor-Eaters Champion. He's got one foot in the frying pan and one in the pressure cooker. Believe me, as a bowler, I know that right about now, your bladder feels like an overstuffed vacuum cleaner bag and your butt is kind of like an about-to-explode bratwurst."

Munson: "Hey, do you mind? I wasn't talking when you were bowling."

McCracken: "Was I talking out loud? Was I? Sorry."

59 ▶ ALL THE RIGHT MOVES (1983—R)

SPORT: Football ★ STARS: Tom Cruise, Craig T. Nelson, Lea Thompson
DIRECTOR: Michael Chapman

Tom Cruise was fresh off his success in *Risky Business* when he crossed over to sports films with *All the Right Moves*. Instead of dancing to Bob Seger in his undies, he plays football for Ampipe High School. Cruise describes himself as a "5-foot-10 white cornerback" which seems, oh, about three inches too generous, but that's OK.

Cruise shows real acting chops in his portrayal of Stefen "Steff" Djordjevic, a kid who sees football as a way to earn a college scholarship and escape his depressed Pennsylvania steel town (the movie was shot in Johnstown). His father and brother both work in the steel mill, and unless he gets away, he is going to wind up spending his life there, too.

"I want to study engineering," Steff says. "I want to be the first Djordjevic to have something to say about what happens to the steel after it leaves here."

Steff has the brains to succeed in the classroom and the ability to excel on the football field, but he has an attitude, which irritates Coach Nickerson (Craig T. Nelson). Nickerson is a typical football tyrant who cannot correct a player without first grabbing him by the facemask. When he chews out a player, Nickerson expects the player to say only, "Yes, sir."

Steff is more likely to say, "But I made the play. . ."

Or worse, he'll listen to the instructions and ask, "Why?"

High school coaches, especially the ones in movies, hate players like that. So when Nickerson ends practice by telling the players to take five laps, he grabs Steff and says, "You do ten." It's a scene heavy with symbolism—Steff jogging alone in the twilight with the smoky steel mill looming in the background, the destiny he is trying to escape.

Everyone wants out of Ampipe. Steff's best friend, Brian Riley (Chris Penn), has scored a full scholarship to play football at Southern Cal. Steff's girlfriend, Lisa (Lea Thompson), dreams of going away to college to study music. Even Nickerson is angling for a college coaching job, and he thinks an upset win over Ampipe's unbeaten rival, Walnut Heights, will cinch the deal.

"We're getting out of here" is a phrase we hear repeated over and over again. At times, it seems like a little much. I mean, Ampipe (or Johnstown) doesn't look like Palm Springs, but the way the characters are acting, you'd think it was the Turkish prison in *Midnight Express*. OK, we get it. It's a blue-collar town. The economy isn't exactly booming. But enough already.

Some talented people came together to make *All the Right Moves*. Director Michael Chapman was the cinematographer on *Raging Bull*, so he knows how to shoot sports action. Jan de Bont was director of photography, and he went on to direct *Speed*, which was a big hit (and earned three Academy Award nominations) in 1994. The result is a film that looks good, especially on the field, but sometimes stumbles in the storytelling.

The dialogue is the weakest part of the film. But when Chapman and de Bont let the pictures move the story along, they do so very nicely. The scene in which the Ampipe team rides the bus to Walnut Heights for the big game is a good example. Not a word is spoken as the coaches and players sit on the bus, staring out the window at the rain, each alone with his thoughts. One player fingers his rosary, another squeezes a football, another studies the game plan. Chapman allows the tension to build with just images and music.

When the Ampipe team gets to the locker room, one of the coaches puts a half-deflated football on a table by the door. On the ball is written: Ampipe High, State Champs, 1960. One by one, each player touches the ball for luck as he enters. You never see the players' faces, only their hands, but that makes it more effective. The audience identifies with the team, not the individuals.

The silence in the room as the players go through the ritual of taping and getting dressed knowing what is at stake—for the players, a college scholarship; for the coach, a chance at a better job—resonates more than any pep talk.

📣 **CHEERS:** Steff has the best speech in the film when he tells off Nickerson after the coach kicks him off the team for insubordination.

"You sit in your office and you decide, 'Scholarship here, no scholarship there,'" Steff says. "Who the hell gave you that power? You're not God, Nickerson. You're just a typing teacher."

🍎 **BET YOU DIDN'T KNOW:** Director Michael Chapman asked both Cruise and Lea Thompson to go back to high school "undercover" for a month to get in the right mood for the film. Cruise was recognized the first day. Thompson lasted four days before another student recognized her—and asked her out. (She declined).

🍎 **BET YOU DIDN'T KNOW II:** Walter Briggs, who plays the Ampipe quarterback known as "Rifleman," went on to play college football at Montclair State and played one game for the New York Jets as a replacement player during the 1987 NFL strike. His pro stats: two pass attempts, one incompletion, one interception.

60 FAT CITY (1972—NR)

SPORT: Boxing ★ **STARS:** Stacy Keach, Jeff Bridges, Susan Tyrrell
DIRECTOR: John Huston

Stockton, California is not a fat city; far from it. The opening shots in *Fat City* of down-on-their-luck guys hanging on street corners and sitting on fruit crates set the mood. This is one of those places where dreams come to die. The flop houses and gambling halls are full of life's losers, people who are here because, well, they have nowhere else to go.

Walking amid this human wreckage is Billy Tully, a brooding drifter who was once a decent prizefighter but now, at age 29, spends his days in a boozy haze. When we meet Tully (played by Stacy Keach) he has been fired from his job as a short-order cook and he is forced to work in the fields, picking lettuce for 90 cents an hour.

Tully starts working out in the local gym with the thought of returning to the ring. He meets Ernie Munger, a naive teenager who is competing as an amateur. Tully befriends Munger (played by Jeff Bridges) and talks him into turning professional. Tully introduces the kid to his old manager, Ruben (Nicholas Colasanto, best known as Coach in the TV series *Cheers*), who is happy to take him on.

"I have a good-looking white kid," Ruben tells his wife. He sees Munger as a box office draw on the small-town circuit from Stockton to Reno, where most fighters are black or Hispanic. He gives him the ring name "Irish Ernie Munger" even though he knows the kid isn't Irish. It's all about the marketing.

There was a time when Ruben thought Tully would be his meal ticket, but it didn't quite work out. He lost a few tough fights, then it was a downhill slide to skid row. Tully now lives in a world of "what-might-have-beens."

Most nights, he loses himself in shots and beers and bitterness.

Things don't improve when Tully moves in with Oma, a drunken harpy played by Susan Tyrrell. She does nothing but guzzle sherry and whine all day. Sitting in bed, wrapped in Tully's boxing robe, she scolds and belittles him as he cooks dinner. When he throws up his hands and heads for the door, she bursts into tears and accuses him of leaving her. He can't win and he knows it.

Fat City makes it clear from the beginning there won't be any happy endings. Tully is at a dead end when we meet him and that's where we will leave him when it's all over. Yet for all the heartache, the story is so well told by director John Huston and screenwriter Leonard Gardner (a Stockton native who adapted the script from his novel) you will remember it for a very long time.

The film is beautifully shot by cinematographer Conrad Hall, who won Oscars 30 years apart for *Butch Cassidy and the Sundance Kid* (1969) and *American Beauty* (1999). His photography of the migrant workers and derelicts that wander the street creates an atmosphere that allows us to feel the desperation of people like Tully and Oma.

When Tully talks Ernie into turning pro, he tells him, "Don't waste your green years, kid. Before you know it, your life has slipped away, down the drain." Tully's eyes—sad and hollow—make it clear he is speaking from experience. Tully is only 11 years older than Ernie, but he is weighed down by so much sadness and regret, he seems like an old man.

Ernie needs the money because he marries

his pregnant girlfriend Faye (Candy Clark) and he has some success with Ruben and trainer Babe (former welterweight contender Art Aragon) guiding his career. *Fat City* traces the lives of both fighters, but Tully is the more interesting story because Keach makes it so. He takes a complex—and not very likeable—character and makes him worth rooting for.

Some critics called *Fat City* Huston's most underrated film and compared it to his classics *The Maltese Falcon*, *The Asphalt Jungle,* and *The African Queen*. Huston, who was a successful amateur boxer (and less successful pro), threw himself into the project and was disappointed when *Fat City* flopped at the box office. It is one of the best boxing films ever made, yet it is almost forgotten.

The most memorable scene comes late in the film when a Mexican fighter named Lucero arrives for a bout with Tully. Lucero (played by Sixto Rodriguez, a light heavyweight contender in the '60s) gets off the bus alone carrying a gym bag. Walking to his fleabag hotel, he sees a poster advertising the fight. He stares at it blankly for a few seconds and walks on.

Lucero was once a ranked contender, but now he is on the downside of his career. He travels from one tank town to another, fighting whatever local favorite they put in front of him. In Stockton, it will be Tully. In his hotel room, Lucero goes to the bathroom and as he urinates, the toilet bowl fills with blood. He has kidney damage from a fight in another town just like this one. He shouldn't be going back into the ring, but he will because it is the only way he knows to make money.

In the big fight, whenever Tully pins Lucero against the ropes and pounds him with body punches, you can't help wincing because you know what's happening to the poor guy's insides and you also know what his next trip to the bathroom will be like.

What separates *Fat City* from other boxing films is its lack of glamour. Most films build to a championship fight in a sold-out stadium with millions of dollars at stake. Here, the dimly lit arena is half-empty, the crowd is grungy, and the winner's purse is $100. There is nothing to celebrate, really, no promise of a brighter tomorrow. It's just another night in Stockton—and that's not pretty.

🔔 **CHEERS:** Susan Tyrrell earned an Oscar nomination for Best Supporting Actress for her outstanding performance as Oma. She lost to Eileen Heckart (*Butterflies Are Free*).

🤓 **"I KNOW THAT GUY":** Former welterweight champion Curtis Cokes made his acting debut playing Oma's one-time lover, Earl.

⭐ **SPORTS-ACTION GRADE:** A-minus. The scenes with Keach and Rodriguez are very convincing. Bridges isn't quite as good.

🍎 **BET YOU DIDN'T KNOW:** Huston carefully staged the fights beforehand, but once the cameras started rolling, he said, "All right, boys, now we're going to have two minutes of boxing. Go out there and fight."

Keach, who was in the ring with Sixto Rodriguez, a real fighter, was terrified. He told interviewers that the shot in the film of Rodriguez landing a solid right to the jaw and Keach dropping to the canvas was the real thing. Both Keach and Bridges were bloodied during the filming.

61 > 61* (2001—NR)

SPORT: Baseball ★ STARS: Barry Pepper, Thomas Jane
DIRECTOR: Billy Crystal

For actor/comedian Billy Crystal, directing the film *61** for HBO was a labor of love. Crystal grew up a fan of the New York Yankees and Mickey Mantle in particular. In the film *City Slickers*, Crystal's character, Mitch Robbins, says the best day of his life was when his dad took him to Yankee Stadium for the first time.

"And Mickey hit a homer," he says, as if all his childhood fantasies were fulfilled at once.

Crystal says that scene was borrowed from real life. It happened just that way. To him, Yankee Stadium was a sacred place, and the Mick stood on a pedestal, then and always. So it was no surprise when Crystal did a film about the 1961 baseball season and the home run race between Mantle and teammate Roger Maris.

"I've been in pre-production on this film for 40 years," said Crystal, who was 13 years old

in '61 and hanging on every pitch that season.

The film *61** succeeds largely because Crystal cared so deeply about it. For him, it was like re-opening an old scrapbook, or waving his hand over his baseball card collection and having all of his favorites come alive. It was a chance to relive a wonderful time in his life and it is clear he had a ball.

It is to Crystal's credit that he does not gloss over the failings of his idol. The Mick (Thomas Jane) is shown to be a hard-drinking, foul-mouthed womanizer who is actually less admirable than the straight-arrow family man Maris (Barry Pepper). Crystal does try to put Mantle's self-destructive lifestyle in context, but he does not excuse it.

In the film, Maris confronts Mantle about his drinking and skirt-chasing. Mantle is sharing an apartment in Queens with Maris and teammate Bob Cerv. The arrangement is Maris's idea. He thinks it is better than Mantle living in Manhattan and hanging out at Toots Shor's bar every night. But even in Queens, the Mick can't stay on the wagon and finally Maris loses patience with him.

"Did you ever stop and think if you took better care of yourself, you wouldn't be getting hurt all the time?" Maris says. "You're Mickey Mantle, for Christ's sake."

"What the ---- is that supposed to mean?" Mantle snaps. "You think you know me. You don't know ---- about me."

Mantle tells Maris that his father, uncle, and grandfather all died of Hodgkins Disease before the age of 45. He assumes the same fate awaits him.

"I'm gonna live my life the way I want," Mantle says. "So get off my back. Call your wife and cry to her about it."

The scene provides some understanding and perhaps some sympathy for Mantle, but the way he takes his teammate's concern and throws it back in his face is pretty cold. It's one of many occasions in the film when Mantle acts like a jerk and Maris comes off as a nice, well-intentioned guy.

Of course, the focus of *61** is the duel between Mantle and Maris to break the major league home run record of 60 set by Babe Ruth in 1927. Even the casual sports fan knows how that turned out: Maris hit 61 to break the record, and Mantle finished with 54. Crystal wisely keeps his eye not on the ball, but on the two men and how they deal with the pressure of chasing down the sport's most hallowed record.

Most Yankee fans are rooting for Mantle. He is one of their own, signed and brought to New York in 1951 to succeed Joe DiMaggio in center field. Blonde and movie-star handsome, Mantle is every man's idol and every woman's crush. Maris is seen as an outsider, acquired in a trade from Kansas City in 1960. Shy and uncomfortable in the big city, Maris is not warmly received by the New York fans and media.

The pressure causes Maris to start losing his hair and break out in a rash. He snaps at the sportswriters who crowd around him every day, calling them "bloodsuckers." But even in his worst moments, Maris remains a sympathetic figure: a decent guy who finds himself thrust onto a stage he really did not want.

The asterisk in the title refers to the controversy over the record. Ruth hit his 60 homers in a 154-game season. There was sentiment among some traditionalists that since Maris did not hit No. 61 until the 162nd game that his mark should go into the record books with an asterisk. It did not. There never was an asterisk attached to the Maris record except in some people's minds.

However, the controversy sets up an interesting scene in the film. In the 154th game, Maris needs one more home run for 60. On his last at bat, Baltimore brings in knuckleball pitcher Hoyt Wilhelm to face him. To make it realistic, Crystal hired major league knuckleballer Tom Candiotti to play Wilhelm.

"We shot the scene at the L.A. Coliseum," Candiotti said. "I was there for six days. I had my own trailer just like the stars. They gave me tapes of Wilhelm to study. His nickname was 'Tilt' because he held his head at an angle and he wore his cap a little crooked. I had to get all the details right. Billy was a stickler for that stuff.

"[Crystal] wanted to show how a real knuckleball dances around. The trouble was Barry [Pepper] couldn't hit it and the catcher couldn't catch it. I threw 17 knucklers. Seventeen times Barry swung and missed. Billy came to the mound and said, 'We have a problem. It's called film. We're running out.' He asked if I could lob a slow one so Barry could at least get his bat on it. I blooped one up there and it hit him in the ribs.

"He went down and everyone came running in to see if he was all right. He said, 'How hard was that?' I said, 'It was about fifty miles an hour.' I mean, it was nothing. I'd hate to think what would've happened if he got hit by somebody throwing ninety."

It required several more takes, but finally Pepper got his bat on one of Candiotti's flutterballs. He hit a soft pop foul, Crystal yelled "Cut," and that was it. "He just needed to see Barry make contact," Candiotti said. "He took two or three shots from different angles, put them together and he got what he needed." (Maris was retired on a ground ball.)

"It was an education, seeing how a movie is made," Candiotti said. "Something as simple as me walking in from the bullpen, we shot that scene 30 times. Billy would shoot it from one angle, then another and another. I'd walk to the mound, go back to the bullpen, walk to the mound and back to the bullpen, over and over."

Late in the season, Mantle was hospitalized with a viral infection and an abscessed hip. One of the better scenes in the film is Maris visiting Mantle's bedside. At this point, the race to break Ruth's record is down to one man. Mantle's run is over. Only Maris remains.

What makes the scene effective is that writer Hank Steinberg resists the urge to give either man lines that are eloquent or profound. We know by now that Mantle and Maris are rather inarticulate, so the scene is written with a lot of awkward silences as the two fumble around trying to figure out what to say. But their mutual respect comes through loud and clear.

"I can't play no more," Mantle says. "I'm wore out. Done. I'm out of the race."

"You would've done it, too," Maris says, assuring Mantle that he, too, would have surpassed The Babe.

"Ah, --------," Mantle says. "It was you, Rog. You did it, you --- -- - -----. Nobody can ever take that away from you. That record is yours."

👍 **CHEERS:** Crystal certainly paid attention to details. In addition to Candiotti as Wilhelm, check out the actors who portray Jim Bunning (Detroit) and Dick Hall (Baltimore). They have the same sweeping sidearm delivery as those two pitchers.

62 INVINCIBLE (2006—PG)

SPORT: Football ★ **STARS:** Mark Wahlberg, Greg Kinnear
DIRECTOR: Ericson Core

Take a pinch of *Rudy*, mix in a dash of *The Rookie*, add a '70s rock score (Bachman-Turner Overdrive anyone?), and you have *Invincible*, the story of Vince Papale, the 30-year-old free agent whose raw desire earned him a spot on the Philadelphia Eagles.

The film was produced by Mark Ciardi and Gordon Gray, who previously had success with inspirational sports films *The Rookie* and *Miracle*. They saw an NFL Films feature on Papale and felt his story had all the elements of a feel-good family movie. The only difference was this time the sport was football.

They got the support of the National Football League, who had not partnered on a film since *Jerry Maguire* in 1996. The NFL liked the story—it was wholesome with a positive message—so it allowed them to use the real team names and uniforms, which added authenticity.

The real coup was landing two major stars, Mark Wahlberg and Greg Kinnear, to play Papale and Eagles coach Dick Vermeil. Ciardi and Gray took a shot with a first-time director, Ericson Core (he was head cinematographer on *The Fast and the Furious*) and he delivered a crowd-pleasing film that some compared to *Rocky* in shoulder pads.

That was one of the ironies of Papale's story: he made the Eagles roster as a long shot rookie in 1976, the same year that *Rocky*—Sylvester Stallone's tale of the underdog Philadelphia pug—won the Oscar for Best Motion Picture. Papale was so much like Stallone's character— Philly guy, Italian Stallion, kicked around and written off—that Rocky became his nickname.

Writers Brad Gann and Mike Rich shifted a few locales—they put Papale in South Philly, he actually lived outside the city—and they rewrote a few characters—Vince's love interest Janet (Elizabeth Banks) wasn't related to the bar owner Max. The real Janet was a gymnastics coach who met Vince after he retired from football.

Also, the film makes it appear Vince had no football background other than the rough touch he played with Max and the boys. That's not true either. He played a year with the semi-pro Aston Knights of the Seaboard Football League and two seasons with the Philadelphia Bell of the World Football League before he tried out for the Eagles. So he wasn't quite the rank amateur that *Invincible* suggests.

But those are fairly minor points. The heart of the story is Vince's Rocky-like determination to make it with the Eagles. Like Rocky, he knows he has one shot and one shot only at the brass ring, and while he is fully aware the odds are against him, he's going for it anyway. That spirit comes through.

"There isn't a day that goes by that someone doesn't come up to me and say how much they loved that movie," said Vermeil, who coached the Eagles from 1976 through 1982. "It touched a lot of people. It wasn't 100 percent factual, but so what? It was a great story about someone pursuing their dreams."

Invincible opens with Vince living in a South Philly row house, scuffling to make ends meet after losing his teaching job. His wife walks out, leaving behind a note that tells Vince he's a loser who never will amount to anything. He tends bar at the neighborhood saloon owned by his buddy Max, and he plays rough touch football with his friends on a vacant lot.

The downtrodden Eagles hire a new coach, Dick Vermeil (Kinnear) who holds an open tryout for anyone interested in joining the team, regardless of age or experience. Vince's buddies convince him to go even though he's 30 years old and never played college football. (In real life, Vince attended St. Joseph's University on a track scholarship.)

Several hundred candidates show up, but only Vince catches Vermeil's eye. He runs the 40-yard dash in 4.5 seconds and catches every ball thrown his way. "Where did you play college ball?" Vermeil asks. "I didn't play college ball," Vince replies. Vermeil knows it is a reach, but he invites Vince to join the full squad at training camp.

Most of the film focuses on Vince's daily battle to survive at camp. Vermeil puts the players through two grueling practices a day with lots of hitting and no mercy. The new coach works the players hard because he wants to weed out the weak and uncommitted. Things are particularly rough for Vince because the veterans resent him (some view him as a publicity stunt) and go out of their way to test him with cheap shots and insults.

There are other storylines—Vince's romance

with Janet; his love for his blue-collar father who still talks about watching the Eagles win the NFL title in 1948; a strike that puts his buddies out of work—but, really, it all comes back to Vince and Vermeil.

With Vince, the story is about his willingness to pay any price necessary to make the team, and Wahlberg is very good at conveying his quiet, yet unflinching resolve. With Vermeil, it is about his understanding of what Vince represents. He knows Vince won't be a major contributor on the field—he will be a backup receiver and a special teams player—but Vermeil values the passion he brings to the locker room.

Vermeil tells the Eagles that he believes a team with better character can defeat a team with better talent. In Vince, he sees that character, which is why he decides to keep him while his assistant coaches favor a veteran who has more experience, but less heart.

On the day of the final cut, Vince is called into Vermeil's office. He is convinced he is a goner. "Thanks for sticking with me this long," he says as he slides his playbook across the desk to Vermeil.

Vermeil slides it back to him. "Why don't you hang onto that for a few months," he says. "Welcome to the Philadelphia Eagles."

63 ROLLERBALL (1975—R)

SPORT: Rollerball ★ **STARS:** James Caan, John Houseman
DIRECTOR: Norman Jewison

The future world of Norman Jewison's *Rollerball* is one without crime or disease, poverty or racism. Plus, all the women look like fashion models. That's the good news.

On the other hand, freedom of choice has been replaced by fealty to the corporate states that now rule the world (Hello, Microsoft). And with no wars to stir the masses, the corporations mollify the human bloodlust by televising a brutal soporific sport that's a combination of roller derby, pinball, and Ultimate Fighting.

Welcome to *Rollerball*, a science fiction nightmare set in 2018. The irony here is that Jewison intended for his movie to be a paean to anti-violence. Nice thought. But after it came out (as the horrified director explains on the DVD commentary) some filmgoers so loved the action that there was talk of forming real rollerball leagues.

Even now, what sells this movie is the Roman orgy of plasma and broken bones. Jewison tried to play the brutality so over-the-top that audiences would be repulsed. How could he have anticipated that, 30-plus years later, Americans would become blasé to the mayhem of Grand Theft Auto IV?

The sport of rollerball was invented by writer William Harrison in a short story for *Esquire* in 1973. It was translated to film as an anything-goes game where players in spiked gloves try to throw a softball-sized steel orb into a magnetized goal. Motorcycles race around the elevated oval and any form of defense is allowed—including cracking skulls. At times it's difficult as a viewer to grasp exactly what's taking place, but we'll admit that if Spike TV ever gets its act together, we'd waste a Friday night or two watching this stuff.

There's a melodramatic plot to *Rollerball*, ostensibly about one man's attempt to find freedom in a world devoid of it. Jonathan E (James Caan), the world's top star, is ordered to retire from his Houston-based team by conglomerate bigwig Bartholomew (John Houseman). It seems that Jonathan's success and popularity is undermining the complacency of the world's citizens. As Bartholomew says, Rollerball was designed "to demonstrate the futility of individual effort."

Jonathan refuses to quit, which presents a problem for world leaders, who spend most of their time video conferencing. That must have seemed very futuristic back in 1975.

To be honest, the plot flatlines early in the film. And both Caan and Houseman act with an annoying drugged-out stoicism that, we suppose, is meant to represent the lack of emotion in the future. So, take our advice and don't get hung up on the messagy stuff. Instead, enjoy the ferocious white-knuckle action sequences. *Rollerball*, to be honest, is at its best when Caan, in clunky skates and a 70s bike helmet, is beating the tar out of people.

Another way to enjoy the film is to look at its view of 2018 which, as of now, isn't that many years away. It's always fun to watch antiquated sci-fi to see how accurate its vision was. For example, how come every look at the future has men wearing white jumpsuits?

Anyway, let's run a checklist:

✔ Conniving corporate honchos will destroy the economic markets? Check.
✔ Public address systems will be everywhere you go, blathering in a feminine monotone? Not on our subway system.
✔ TVs will be wall-mounted flat screens, furniture will be ergonomic, and Plexiglas will be the foundation of all interior décor? Check, check, and . . . well, no.
✔ Superstar athletes will get the hottest girls and break all the rules? Check. Some things never change.
✔ Computers will evolve from handheld devices to minivan-sized appliances? No. Although, on the other hand, they'll still crash at the worst times.
✔ Guns will shoot fireballs that set an entire forest to blazes? Please, we hope not.
✔ Librarians will be the sexiest women on Earth? Um, not where we live.

CHEERS: Considering this is a fictitious sport with virtually no rules, the in-game announcing is terrific. Jewison auditioned baseball's Mel Allen and a few NFL announcers, but found they couldn't keep up with the speed of the game. He finally hired hockey broadcaster Bob Miller of the Los Angeles Kings and Dick Enberg, who was 39 years old at the time.

"Jewison allowed me to create things on the fly," Enberg explained. "He said, 'This isn't football or baseball. We're not bound by any rules. If you think of something that sounds good, just say it.' So I did. I created things like the 'Flying Diamond' formation and the 'Flying Triangle.' It wasn't scripted, I just came up with it."

DON'T FAIL TO NOTICE: At one point, Jonathan E's list of all-time records is recited, including most points in a game (18) and greatest number of players put out (13). There is also talk of the legendarily bloody contest between archrivals Rome and Pittsburgh, in which a Rollerball record nine players were killed. Archrivals Rome and Pittsburgh?

BEST LINE: P.A. Announcer: "Your attention please. Rule changes for tonight's World Championship Game: No substitutions, no penalties . . . and no time limit."

CASTING CALL: Jewison says on the DVD commentary that he chose Caan to play the lead here after watching him star as Brian Piccolo in *Brian's Song*.

"I KNOW THAT GUY": Burt Kwouk, the Japanese doctor who tries to pressure Jonathan to turn off another player's life support, played Peter Sellers' karate-smart sidekick, Cato, in the old *Pink Panther* movies.

SPORTS-ACTION GRADE: B-plus. You may not understand the rules, but it is exciting, well-shot stuff.

BET YOU DIDN'T KNOW: Only one Rollerball rink was constructed, in Munich, Germany. It was repainted throughout the movie to suggest rinks from different cities.

64 THE PROGRAM (1993–R)

SPORT: Football ★ **STARS:** James Caan, Omar Epps, Craig Sheffer, Halle Berry
DIRECTOR: David S. Ward

More than 15 years after its release, *The Program* is best remembered for what you *don't* see on the current DVD release.

The original movie had a scene of drunken college football players testing their courage by lying in the street as traffic whizzes by. That scene prompted copycat behavior, and four teenagers from Pennsylvania and Long Island were killed or critically injured trying to similarly prove their bravado. Touchstone Pictures recalled the film and deleted the "dotted-line chicken" moment.

Certainly, that was the necessary move at the time. Unfortunately, those episodes seemed to overshadow the rest of the movie. And that's a shame. Because *The Program* is a film that both takes a serious look at the pressures of big-time college athletics, and offers some excellent football scenes.

The movie follows the 1993 season of Eastern State University (we never are told *which* state), a perennial Top 25 program that has failed to make a major bowl game the two prior years. Coach Sam Winters (James Caan) is warned by his athletic director that ESU's alumni and the state legislature are restless, "and they both wield their checkbooks." In other words, win this season—or else.

Winters has hope, in the form of Heisman Trophy-candidate quarterback Joe Kane (Craig Sheffer) and freshman running back Darnell Jefferson (Omar Epps), who appears to be the second coming of Adrian Peterson—if he can just hold onto the football.

Speaking of Peterson, the Minnesota Vikings star who attended Oklahoma, lists *The Program* as his favorite sports film. "It gives you a taste of what big-time college football is like," he said. "People might think it's not like that, but it is like that."

In some respects, *The Program* parallels the classic sports-movie formula, following a team through its ups and downs, leading up to the must-win game that can salvage its season. That movie template has been copied dozens of times.

But *The Program* also adds human sub-dramas that flash a harsh spotlight on the world surrounding major college athletics. Kane, the star quarterback, is plagued by generations of alcoholism within his family and winds up climbing into a bottle to escape the Heisman hype. His backup, Bobby Collins (Jon Pennell), is an academic dud who beds the coach's daughter and then gets caught having her take his placement test. Enraged, Winters tosses the kid from his squad. But when Kane's foibles push him into alcohol rehab, the coach must swallow his pride and reinstate the backup who humiliated him and his daughter.

Hmmm, perhaps this is starting to sound a little like *The Edge of Night*. Truth be told, there are some soap opera elements here, and many critics initially panned *The Program* as nothing more than a melodrama in shoulder pads. True enough, some of the dialogue seems best suited for weekday afternoon TV (like this romantic line: "I've seen a lot of ------ things in my life, Autumn. That's how I know when I see something good."). Okay, maybe afternoon cable TV.

But we'll disagree with those who quickly dismissed *The Program*. For one, many of the melodramas here pack a punch. The subplot about desperate defensive end Steve Lattimer (Andrew Bryniarski) turning to steroids ("Hey, I just gained 35 pounds in the gym.") is both provocative and tragic—from the coach who closes his eyes ("Let the NCAA do its job," says Winters. "They test, we don't.") to the booster who essentially forgives, even when Lattimer, in a 'roid rage, attempts to rape his daughter.

Plus, there's terrific football action. The producers of *The Program* got the NCAA to sign off on using real college uniforms and stadiums. So when the ESU Timber Wolves are playing the Iowa Hawkeyes, well, you can almost envision old Hayden Fry on the sidelines. The game shots are superbly choreographed and edited, raising the quality to the level seen in *Friday Night Lights* and *Jerry Maguire*.

"For someone like me who was still in high school and just thinking about playing college football, *The Program* gave me a chance to see what to expect," said Brian Westbrook, the former star running back of the Philadelphia Eagles. "It was exaggerated in some ways. I didn't see any [players] getting paid at Villanova. But the action, the hitting and the intensity on the field was pretty authentic."

Except. . . .

Well, except that we're not buying actor Craig Sheffer as a big-time college quarterback. For one, he's got the worst throwing motion since Bernie Kosar. And secondly, you'd think a guy cast in a sports movie might have hit the weight room for a few weeks before filming. Sheffer's got no shoulder/neck muscles at all and his arms are the width of Slim Jims. We're supposed to accept this guy as a *Sports Illustrated* cover boy? Hell, he makes Doug Flutie look like Daunte Culpepper.

On the other hand, Sheffer can act and adroitly pulls off the "young man fighting his demons" role pretty well. Epps, also, is convincing as the cocky freshman who you just know is going to end up as Ricky Watters some day.

Caan doesn't have any heavy lifting here, mostly serving as the base character around whom everything revolves. Mostly, he's asked to react. Fortunately, his face can still move, since he filmed *The Program* before embarking on a decade-long binge of facelifts that left him looking like an escapee from Madame Tussauds.

PIVOTAL SCENE: Coming off a suspension for steroid use, Lattimer tries to play clean. But, in the final seconds against Iowa, he is bulled over at the 1-yard line to lose the game. In the next scene, he is alone in his dorm room, pushing a large needle into his buttocks. "No problem," he reassures himself, as he picks up a barbell and starts compulsively lifting.

DON'T FAIL TO NOTICE: The SEC banner hanging over Eastern State U's stadium at home games. That's because the action footage was filmed at the University of South Carolina's Williams-Brice Stadium, at halftime of several of the Fighting Gamecocks' 1992 contests. In fact, the movie never establishes which conference the Wolves play in. With a schedule that includes Mississippi State, Michigan, Iowa, Boston College, and Georgia Tech, we sure couldn't figure it out.

65 THE BOXER (1997–R)

SPORT: Boxing ★ **STARS:** Daniel Day-Lewis, Emily Watson, Brian Cox
DIRECTOR: Jim Sheridan

The Boxer is Danny Flynn (Daniel Day-Lewis), a man fighting battles that extend far beyond the ring. As a teenager in Belfast, Flynn was active in the IRA and winds up serving 14 years in prison. When he is released, he wants no part of the on-going violence. He wants to return home to resume his boxing career.

Of course, it is not that easy.

Flynn is caught between splintering factions of the IRA, one side led by Joe Hamill (Brian Cox), who is ready to end the sectarian war, the other side led by Harry (Gerald McSorley), a Sinn Fein officer who is determined to continue the bloodshed. Flynn tries to distance himself from the politics, but that is impossible in a city where armed troops patrol the streets

and military helicopters swoop overhead.

Director Jim Sheridan (who teamed with Day-Lewis in *My Left Foot*) establishes Flynn's character in the opening shot. He shows Danny alone in the prison yard, shadow boxing. Dressed in a hooded sweatshirt, dwarfed by the gray walls and guard towers, Danny is a solitary figure. It is the day of his release, yet he is going through the same routine he has followed every day for 14 years. It is the portrait of a hard and stubborn man.

When Danny rebuilds his old gym with the intention of opening it to both Catholics and Protestants, he lights the fuse on an emotional powder keg. It finally explodes on the night of his comeback bout when a car bombing sets off a bloody riot in the streets. The night ends with the gym—and Danny's dreams of a peaceful life—going up in flames.

Adding to the tension is Danny's reunion with his former love Maggie (Emily Watson), who married his best friend while Danny was in prison. The husband, who also was involved in the IRA, is now behind bars and when Danny and Maggie see each other again, it is clear the old feelings are still there. The trouble is, if an IRA wife is unfaithful it is considered an act of treason and there are dire consequences.

The Boxer is three stories in one: a political film, a boxing film, and a love story. It is to Sheridan's credit that he weaves the three together as well as he does, although the politics become muddled at times. Day-Lewis carries the film with his usual beautifully nuanced performance. With his sharp features and piercing eyes, he is well-suited to the role of the distant and embittered Danny, but he is particularly good in several scenes with Maggie where he tries to articulate his feelings.

At one point, he describes the isolation of prison: "At first, you hold onto all the old voices in your head," he says. "Kids you knew in school, friends. When those fade, you talk to yourself. After awhile, it's as if your own voice doesn't belong there. Silence becomes your best friend."

He looks at Maggie and says, "I've lived with your face in silence."

"Those are great lines by any standard," wrote Edward Guthmann in the *San Francisco Chronicle*. "With Day-Lewis speaking them and Watson reacting, they are poetry."

The Boxer has a real life parallel in former featherweight champion Barry McGuigan, who rose to prominence during the Northern Ireland conflict. McGuigan was able to do what Danny Flynn tried to do, that is, use his success in the ring to bring Catholics and Protestants together. He refused to wear the colors of either side. Instead, he entered the ring under a blue flag adorned with a dove, the symbol of peace.

McGuigan served as technical advisor on the film. He worked with Day-Lewis for three years, teaching him how to move and throw a punch. McGuigan said they sparred more than a thousand rounds. The result is some of the most realistic boxing action ever put on film.

In the best scene, Danny goes to London to fight at a swanky dinner club where the spectators wear formal attire and drink champagne. Danny is matched with an African fighter and after several bloody rounds Danny lands a vicious left hook. The African sags against the ropes, out on his feet. Rather than punish him further, Danny drops his hands. The crowd boos and the referee barks, "Fight."

"The fight's over," Danny says, walking away.

Watching on TV, the Sinn Fein officer Harry—who resents Danny for turning his back on the IRA—says, "What do you expect? He's a quitter."

Maggie says nothing, but the look in her eyes says she knows better.

PIVOTAL SCENE: Danny Flynn is released from prison and spends his first night at a homeless shelter where he bumps into his former trainer Ike (Ken Stott). Danny tells Ike he intends to resume his boxing career. Ike scoffs at the idea, reminding Danny he spent the last 14 years behind bars.

"I'm 32 years old," Danny says. "Archie Moore fought for the world championship when he was 42."

Ike smiles.

The next morning, the two men are re-opening the old gym and going back to work.

SPORTS-ACTION GRADE: B-plus. The boxing scenes are first-rate, but there aren't that many of them.

WHITE MEN CAN'T JUMP (1992–R)

SPORT: Basketball ★ STARS: Woody Harrelson, Wesley Snipes, Rosie Perez
DIRECTOR: Ron Shelton

White Men Can't Jump had all the ingredients to be a masterpiece. When it came out in 1992, writer/director Ron Shelton was still basking in the glow from his landmark *Bull Durham*. Actor Wesley Snipes was riding the wave of two powerful films—*Jungle Fever* and *New Jack City*. And Woody Harrelson was making the transition from television after seven years as likable lug Woody Boyd on *Cheers*. With that collection of talent, creating a slick, hip-hop comedy-drama about the hustle of Los Angeles playground basketball—well, how could it not be superb?

Unfortunately, the finished product does not add up to the sum of its parts. Like the Dallas Mavericks of the past decade, it underachieved despite all that talent. While *White Men Can't Jump* is entertaining and occasionally provocative, there is nothing special about it, nothing that pushes it into the pantheon of champions.

Maybe we're being too harsh. Hey, the Mavericks, after all, are a fun team to watch and by most standards have been a success in recent years. Likewise, *White Men* rang up $76 million at the box office, making it the second all-time grossing movie ever about basketball (behind *Space Jam*, but we won't comment on that). Most critics applauded it initially, with Hal Hinson of the *Washington Post* calling it, "a self-celebrating form of verbal jazz."

We're just kind of stuck on what could have been.

White Men takes place on the playground courts of Los Angeles, where crumpled twenties are bet on the outcome of two-on-two basketball games. Sidney Deane (Snipes)

is a world class hoops hustler who successfully destroys opponents by getting into their minds with nonstop bragging and mockery. Billy Hoyle (Harrelson) is the outsider with the backwards cap and Reebok Pumps who uses his geeky looks (and street ball's racial bias) as the lure. Everyone, after all, assumes that a white guy—particularly one appearing this dopey—can't keep up with the blacks.

Billy initially cons Sidney, who cons him back, which leads to a series of back-and-forth revenge games which, quite frankly, grow repetitive. Eventually, the two must join to survive (despite Billy's clever protestation that, "I don't hustle with people who are dishonest."). Sidney needs money to move his wife and son from their oft-robbed apartment at the Vista View ("There's no vista. There's no view," he notes). And Billy, well, he's got to pay off some not-so-goodfellas who threaten to kill him for reneging on a deal to fix a schoolyard game after an opponent taunted him.

Taunting is a huge part of this movie. As important as the game is, destroying your opponent verbally is even more so. Or is supposed to be. Every derivative of every four-letter word is turned into a dagger and any player's warts are there to be exploited, as when Sidney jeers a skinny challenger with, "Shut your anorexic malnutrition tapeworm-having overdose on Dick Gregory Bahamian diet-drinking ass up."

Sometimes the catcalls work, sometimes they just seem contrived. The racial gibes, particularly, grow weary.

What doesn't tire is the performance of Rosie Perez as Sidney's girlfriend, Gloria, a

Puerto Rican one-time disco queen. Gloria believes her destiny is to become a *Jeopardy!* champion, so she crams her mind with endless amounts of useless information—such as foods that begin with the letter Q. With her bombshell body and Muppet voice, Perez steals the movie.

There are other elements to enjoy here. Any soundtrack that includes Duke Ellington, James Brown, Isaac Hayes, Cypress Hill, and Jimi Hendrix gets our attention—although, according to Sidney, "White people can *listen* to Hendrix, but they can't really *hear* him."

In the end, however, *White Men Can't Jump* proves to be not much more than an engaging two-hour diversion. Shelton, who did such a great job of bringing minor league life to the screen in *Bull Durham*, doesn't quite put the ball through the hole in his follow-up.

☞ **DON'T FAIL TO NOTICE:** Early in the film, Billy makes a joking reference to the last words of Lee Harvey Oswald ("It wasn't me, it was the—*bang!*"). In real life, Harrelson's father, Charles Harrelson, was a convicted hitman who boasted several times that he was involved in the conspiracy to kill President John F. Kennedy in 1963.

😎 **"I KNOW THAT GUY":** Sidney's playground pal Junior is played by Kadeem Hardison, who was in the midst of a six-year run on the *Cosby Show* spinoff *A Different World* when *White Men* was shot. Just as Jerry Lewis was once inexplicably a huge idol in France, Hardison's Dwayne Wayne character in *A Different World* was all the rage in Holland. His geeky eyeglasses became that country's top fashion statement in the early 1990s.

67 JUNIOR BONNER (1972–PG)

SPORT: Rodeo ★ STARS: Steve McQueen, Robert Preston, Joe Don Baker
DIRECTOR: Sam Peckinpah

Steve McQueen plays Junior Bonner, a past-his-prime rodeo champion looking to settle down in his hometown of Prescott, AZ. He quickly discovers that his father Ace (Preston) has gambled away his inheritance, and his greedy brother Curly (Baker) is selling off parcels of the family's land to open a mobile home retirement community.

There are all kinds of family dynamics here, and symbolic Old West vs. New West tension, but the real reason to watch the movie is McQueen. At age 42 (although he looks older) he nails the role as aging athlete aiming for one final moment of glory. Think Crash Davis in chaps and spurs.

Bonner signs up for the hometown July 4 rodeo. He's flat busted and doesn't do anything

to enhance his fortunes in the first two riding events. But although he needs the money, he more so needs to regain his pride. So, for the finale, he signs up to ride the prize bull Sunshine—a snorting, slobbering, intimidating menace that no one has ever tamed. Kind of like Mike Tyson circa 1987, if Tyson had a nose ring and was a little better behaved.

We don't want to give away the ending, except to say that in professional bull riding, the magic number appears to be eight seconds. Stretch that out into super-slow-motion, and you can make it last several minutes.

The film moves languidly, in the style of a lot of 1970s westerns. And, despite the opportunity to show cowboys impaled on wild animal horns, it's the least violent film of Sam Peckinpah, the man who directed *The Wild Bunch* and *Bring Me the Head of Alfredo Garcia*. Or was that Damaso Garcia? Anyway, we suspect Peckinpah started feeling guilty about that halfway through the film, because he inserts a gratuitous bar brawl scene that serves no purpose beyond getting a little blood onscreen.

Watch for a brief cameo by genuine cowboy legend Casey Tibbs, who was known as "The Babe Ruth of Rodeo," and whom, we learned from his official bio, died at his home on Jan. 28, 1990, while watching the Super Bowl. It only makes sense that he foolishly bet on the Broncos.

❝❝ BEST LINE: Rodeo promoter Buck Roan (Ben Johnson): "You know something? Old Junior's gonna spoil all my horses if he don't stop blindfolding them with his ass." We don't even know what that means, but we love it.

68 BLUE CHIPS (1994–PG-13)

SPORT: Basketball ★ STARS: Nick Nolte, Shaquille O'Neal, Ed O'Neill
DIRECTOR: William Friedkin

Blue Chips is one of Ron Shelton's earliest writing efforts. He completed the screenplay a decade before *Bull Durham* and *White Men Can't Jump* and it shows. The characters are more one-dimensional and the dialogue doesn't have the same snap. It is a lot more conventional than his later films.

But the subject matter—corruption in big-time athletics—is meaty stuff and the performances by Nick Nolte (expectedly) and Shaquille O'Neal (surprisingly) are very good. The basketball action looks authentic, which makes sense since director William Friedkin has real players—Rodney Rogers, Allan Houston, George Lynch, and Calbert Cheaney, among others—pushing the ball up and down the court.

Friedkin said he basically just rolled the ball out on the floor and let the guys play while the cameras rolled. Sometimes the best directing, he said, is not directing at all. It would not work in every film—it would not have worked in *The French Connection*, for which Friedkin won an Oscar—but it worked well here.

Blue Chips is the story of coach Pete Bell (Nolte), who won two national championships at Western University but now has fallen on hard times. He is coming off his first losing season and the alumni are restless. They want Bell to do what other Division One schools are doing— that is, break the rules and buy the best players.

Reluctantly, Bell succumbs to the pressure. He goes off to recruit an Indiana hotshot named Ricky Roe (Matt Nover). The kid comes right to the point: "I'm a white, blue chip prospect," he says, "and that should be worth $30,000 in one of those athletic bags." He also wants a new tractor for his father's farm.

Bell swallows hard, but says OK.

Butch McRae (Penny Hardaway) is a black inner-city kid from Chicago. He wants a new home with a lawn for his mother. Bell makes it happen. Neon Bodeaux (O'Neal) is a seven-footer with enough game to take any team to the Final Four. He signs with Western and just like that he is driving a new Lexus. All the "gifts" are courtesy of a sleazy alum named Happy Kuykendall (J.T. Walsh).

Kuykendall is a caricature of an unprincipled booster, but he delivers one of the film's better speeches when he tells Bell there is nothing wrong with paying off the players. "Damn it, we owe these kids," he says, pointing out how much money a winning team brings into a university.

Shelton seems to be saying the NCAA rules are a sham. A fair point, but he allows Kuykendall to become such a hateful figure it is impossible to take his side on anything.

With his new recruits, Bell begins piling up the victories and Western is once again a national power. But as the team climbs in the polls and the smiling alumni fat cats pat him on the back, he is overcome with guilt. It doesn't help that sportswriter Ed Axelby (Ed O'Neill) is snooping around looking for the dirt on Bell.

The Western team goes all the way, defeating Indiana (coached by the real Bobby Knight and with Bobby Hurley, the former Duke star, at point guard) and Texas Western (coached by Rick Pitino). But when it is all over, Bell spills his guts at the press conference, admitting to his recruiting violations and lamenting the fact that he traded his principles for a few victories.

If it sounds like a contrived ending, well, it is. As Desson Howe wrote in the *Washington Post*: "I've seen subtler semiology on *Sesame Street*."

But even when *Blue Chips* stumbles, Nolte holds your interest. He makes you feel his pain as he sells his soul to the hustlers and wheeler-dealers he once despised. There is never a doubt that in the end his conscience will win out and he will come clean, but Nolte is so good that he keeps you watching.

Among the players, Shaq brings more depth to his character than you would expect. He is a lot more believable here than he was in, say, *Kazaam*.

🎬 **PIVOTAL SCENE:** Pete (Nolte) has a classic post-game meltdown. He rips the team and storms out of the room, only to return and rip them again. He stalks out, then comes back a third time, screaming, "I can't tell you how sick I am of basketball today." He proceeds to toss chairs and knock over the water cooler as the players cower in their lockers.

SPORT: Football ★ STARS: The Marx Brothers
DIRECTOR: Norman Z. McLeod

The university suffers from a sagging reputation. The incoming president, eager to turn the school around, reasons that beefing up the football team is the easiest way to instant credibility.

Problem is, successful college football doesn't come cheap. Nor do players. So the president siphons off academic funds and hires a shady duo of operators to go out and "buy me some winners."

Sounds like the plot of a 21st century exposé? Could be. But instead, it is the opening of *Horse Feathers*, a Marx Brothers vehicle that is as hilarious over eight decades after it was made as it is prophetic. Hey, who knew that college athletics was a cesspool back then? And never mind those rumors that George Gipp was a paid professional.

Groucho Marx stars as Professor Quincy Wagstaff, newly installed head of troubled Huxley University. Within his first few days in office, he boots a boring biology professor out of the classroom, makes a pass at every pretty young coed, and decides that, well, academics aren't going to keep good old Huxley solvent.

Speaking to his deans, Wagstaff says, "I tell you gentlemen, this college is a failure. The trouble is, we're interrupting football for education."

"Have we got a stadium?" he asks.

Yes, the deans reply.

"Have we got a college?"

Yes.

"Well, we can't support both. Tomorrow we start tearing down the college."

Wagstaff visits the local speakeasy (remember, this movie was made during the Prohibition era)

and enlists bumblers Baravelli (Chico Marx) and Pinky (Harpo Marx) to find "real players" to help win the upcoming game against Darwin University. Turns out, archrival Darwin—which is just as corrupt—has already beaten him to the ringers. So Wagstaff hires the duo to kidnap Darwin's top recruits.

Needless to say, the plan goes awry. The middle part of *Horse Feathers* centers on some double-crossing, a subplot with a so-called "college widow" (which we gather to be 1930s talk for a loose woman), and at least a dozen wonderful sight gags. You can probably figure out one, during a poker game, when someone invites Harpo to "cut the cards."

We wind up at the big game, in front of what appears to be 70,000 fans, and all three Marx Brothers in uniform (Groucho's is under a tuxedo waistcoat). The climactic 10-minute football scene, we would argue, is one of the three funniest in movie history—right there with the original *The Longest Yard* and *M*A*S*H*. Lots of banana peels, a ball on a string, and taunting of priggish referees. Groucho—at quarterback—lights a match on the center's rump and winds up to pass like a baseball pitcher.

And then it really gets silly. Some of the best stuff is Chico's signal calling:

"Humpty Dumpty sat on the wall, Professor Wagstaff gets the ball."

Or, "Hey diddle diddle, the cat and the fiddle, this time I think we go through the middle."

For those too young to have seen a Marx Brothers movie—and we're talking to most of you here—take our advice and rent *Horse Feathers*. Comedy often doesn't hold up through the ages. We can only wonder how the humor in *Kingpin* or *Happy Gilmore* will translate into popular culture around the year 2060.

But this choppy black-and-white film plays well today. With Groucho's searing wit, Chico's wordplay, and Harpo's gift of pantomime, *Horse Feathers* remains a laugh riot even now.

🍎 **BET YOU DIDN'T KNOW:** Thelma Todd, who plays the wanton love interest, died of carbon monoxide poisoning in 1935 at age 30. Although her death was ruled a suicide, many believed she was murdered by mobsters seeking financial control of a Hollywood night club she owned.

70 WE ARE MARSHALL (2006–PG)

SPORT: Football ★ **STARS:** Matthew McConaughey, Matthew Fox, David Strathairn
DIRECTOR: McG

In November 1970, a charter flight carrying the Marshall University football team crashed, killing all 75 people aboard. The tragedy devastated the school and the town of Huntington, WV, where it is located. *We Are Marshall* is the story of how a new coach (Matthew McConaughey) rebuilt the football program and helped the community heal.

Some critics, like Connie Ogle of the *Miami Herald*, loved the film. Ogle wrote: "Equally thrilling and wrenching, [it] is an absolute must for anyone who loves sports and an eloquent explanation for those who don't understand what the fuss is about."

Other critics found it sappy and manipulative. Joseph Williams of the *Boston Globe*, who played football at the University of Richmond, ridiculed the "broadly drawn characters, unlikely scenarios and cornier-than-Iowa dialogue." By the end, Williams said, "I wanted to throw my $4 soda at the screen." Easy there, big fella.

Yes, the film is heavy-handed at times and anyone familiar with the sports movie genre can see certain plot points coming a mile away. Yet there is something to be said for a movie that is so unabashedly heartfelt.

Watching *We Are Marshall*, you get the feeling it was a labor of love. It may have been

just another gig for the cast and crew when they signed on, but in the course of filming on location and meeting the people who still live with the memory of the horrific crash, they clearly became emotionally involved. That caring, which is obvious throughout, is what keeps the movie afloat.

Director McG (real name Joseph McGinty Nichol), whose previous work included two Charlie's Angels films and some music videos, allows us to feel the texture of Huntington. Whether the characters are standing in front of a blast furnace, sitting in a booth at the diner, or walking across campus, it has just the right look and feel. In gentle ways, McG paints the portrait of a small town where everyone knows everyone and a tragedy such as this is particularly painful.

There were 37 players and eight coaches on the doomed flight. There were also many boosters and friends, including a city councilman, a state legislator, and four of the town's six physicians. Seventy children lost one parent in the crash, another 18 were orphaned. It was a trauma that will be felt for generations.

McConaughey plays Jack Lengyel, an outsider brought in to rebuild the football team. Matthew Fox, best known as Dr. Jack

Shephard on *Lost*, plays assistant coach Red Dawson, the lone survivor from the previous staff. Dawson skipped the ill-fated flight to go on a recruiting trip and he has been dealing with guilt issues ever since.

Paul Griffen (Ian McShane), the head of the school board whose son died in the crash, opposes the plan to field a football team. He argues that it is disrespectful to the victims, but in reality he is so caught up in his own grief, he fears the return of football will only add to his suffering. He tries to oust the university president (David Strathairn) who feels football can help the school move forward.

Lengyel knows his team of freshmen and walk-ons will be outmanned and may not win a game, but he also knows the worth of the season cannot be measured on the scoreboard. After the team is crushed in its first game, Lengyel tells Dawson, "Winning is the only thing. I said it all my life and I really believed it. Then I came here and I realized it's not true. Winning doesn't matter. It doesn't even matter how we play the game. What matters is that we suit up and play."

The climax of the film is Marshall winning its first home game, upsetting Xavier on a touchdown pass on the final play. Watching the coaches and players celebrate—and Lengyel is right in the middle of it—you get the feeling that maybe winning does matter, just a little bit.

📖 **CHEERS:** McConaughey isn't normally an actor we associate with subtlety, but he is very good here finding the balance between the tough football coach and sensitive father figure.

🎬 **PIVOTAL SCENE:** The morning of the Xavier game, Lengyel takes the team to the gravesite where the remains of six deceased players are buried. He tells the team, "This is our past, gentlemen. This is where we have been. This is

how we got here. This is who we are. How you play today, from this moment on is how you will be remembered. This is your opportunity to rise from these ashes and grab glory. We are...."

The players respond: "Marshall."

Lengyel repeats: "We are...."

The players answer in one powerful voice: "Marshall."

You get the feeling Xavier might be in trouble.

✔ **REALITY CHECK:** The game-winning touchdown pass was actually a perfectly executed bootleg and the receiver, Terry Gardner, went into the end zone untouched. In the film, it is a desperation pass with the receiver making a spectacular catch. The coach, Jack Lengyel, who was on the set, told the director he should show the play as it really happened. But when Lengyel saw the film, he said, "They were right. The dramatic finish really helps tell the story."

🍎 **BET YOU DIDN'T KNOW:** Kate Mara, who plays Annie Cantrell, the fiancée of one of the crash victims, is the granddaughter of Wellington Mara, the late owner of the New York Giants. A singer and actress, she performed the National Anthem at the first Giants home game following her grandfather's death in 2005.

😎 **"I KNOW THAT GUY":** Ian McShane, who plays the distraught father, is the sadistic sheriff in HBO's Western series *Deadwood*. You may find him difficult to recognize without his mustache and foul mouth.

⭐ **SPORTS-ACTION GRADE:** C. Like many Hollywood directors, McG goes overboard in the football scenes. Too much slow motion, too many bodies cartwheeling through the air, and every hit sounds like a bolt of lightning splitting a telephone pole. Lighten up, man.

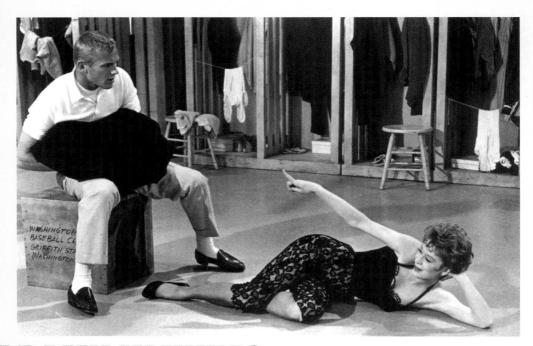

DAMN YANKEES (1958–NR)

SPORT: Baseball ★ **STARS:** Tab Hunter, Gwen Verdon, Ray Walston
DIRECTORS: George Abbott, Stanley Donen

Damn Yankees opened on Broadway in 1955. It was a wonderful play; the critics loved it. There was only one problem: no one was buying tickets. On a good night, the theater was half full. After one month, the play was on the verge of closing.

Desperate, the producers decided to change the billboard. They were using a photo of the show's star, Gwen Verdon, in a Yankees jersey, sweatpants, and sneakers. They replaced it with a shot of Verdon in character as the seductress Lola in a bustier and high heels.

Almost immediately lines began to form, comprised mostly of guys with their tongues hanging out. But they bought tickets. Oh, how they bought tickets. *Damn Yankees* ran for 1,019 performances. When they said the play had legs, they were talking about Gwen Verdon.

In 1958, when Warner Brothers brought *Damn Yankees* to the big screen, they knew what photo to put on the poster. It was Verdon as Lola reprising her vamp routine. It worked once again. The film was a big success.

New York Times critic Bosley Crowther gushed over Verdon's performance, which he called "sizzling." He described Verdon as a "long-legged, swivel-jointed siren [who] manufactures her own strong brand of sex." Wow, you can almost see Bosley mopping his brow between paragraphs.

Verdon's spirited dancing—choreographed by her husband Bob Fosse—spices up a delightful baseball fantasy which still entertains today. Even the premise still applies: the Yankees win all the time and the Washington team stinks. The only difference is now the Washington franchise is in the National League.

The story, written by George Abbott and Douglass Wallop, is a 1950s version of the Faustian deal-with-the-Devil legend. Joe Boyd (Robert Shafer) is a middle-aged guy who lives in D.C. and roots for the Senators. The trouble is the Senators are a last-place team.

One night Joe cries out, "One long ball hitter, that's all we need. I'd sell my soul for one long ball hitter."

Suddenly, a stranger appears in Joe's living room. He introduces himself as Mr. Applegate, but he's really Satan (Ray Walston). He offers to turn couch-potato Joe into that young stud of a slugger who will lead the Senators to glory. All it will cost him is his soul.

Joe accepts the deal, but he puts in an escape clause. If he wants out, he can say so by a certain date and return to his former life. The Devil isn't worried. He will come up with a dirty trick to close the loophole. I mean, he's the Devil, for crying out loud.

What's in this for the Devil? He figures if the Senators win the pennant, it will cause millions of people to freak out. There will be heart attacks and apoplexy galore. "Just like the good old days," he says wistfully.

Joe is transformed into the power-hitting sensation Joe Hardy (Tab Hunter), a blonde Adonis who tears up the American League. Reporters start asking, "Who is this guy? Where did he come from?" Mr. Applegate creates the back story: Joe is a shy farm boy from Hannibal, Missouri. (Cue the song, "Shoeless Joe.")

Joe keeps smacking home runs and the Senators keep winning, but he misses his wife and wants to go home. Fearing Joe will invoke his escape clause, the Devil calls in his version of the closer, the sexy Lola (Verdon). Her job is to shimmy around until Joe's eyes are so glazed over, he forgets all about going home.

What follows is a tender tug of war for Joe's heart and soul with the forces of good—with an unexpected assist from Lola—finally winning out.

The film features 10 songs, including the classics "*You've Gotta Have Heart*," "*Shoeless Joe*," and "*Whatever Lola Wants*." The latter song was written for Verdon to sing while performing a semi-striptease. She did the number more than a thousand times on stage, but Warner Brothers felt it was too racy, so they toned it down. In the film, Verdon does the dance, but without the bumps and grinds.

The film featured virtually the entire Broadway cast. The only change was in the role of Joe Hardy, where Hunter, a '50s matinee idol, replaced Stephen Douglass. Hunter was not a trained singer or dancer, but he was good enough to get by.

"I saw the original on Broadway," Hunter said, "and I was thrilled when Jack Warner bought *Damn Yankees* for me as a make-up gift. I had been on suspension from the studio, so it was sort of a kiss and make-up gift, a damn good one. I was suspended because I had turned down a [previous] role at the studio."

CHEERS: Walston's Mr. Applegate has some great comic moments; for example, the scene in which he drops coin after coin after coin in a pay phone trying to place a call to Hell.

"I KNOW THAT GAL": Sister, the noisy neighbor, is played by Jean Stapleton, who later became famous as TV's Edith Bunker. Stapleton played Sister on Broadway. This was her film debut.

You may find yourself, one late night, clicking around the dial, zapping past the 16th repeat of *SportsCenter* or an infomercial plugging hair replacement, when you stumble on a 30-year-old made-for-TV movie about a prison lifer running sub-four-minute miles.

Put down the remote. *The Jericho Mile*—which occasionally airs during the insomnia hours on ESPN Classic or your local UHF station—is worth pushing back bedtime for a while. It is a gritty, unromantic look at prison life, racism, institutional corruption, and—oh yeah—the joy of running. Invest 97 minutes (plus whatever commercial time) in the entire movie and you'll wonder why you haven't heard about it until now.

The Jericho Mile is the directorial debut of Michael Mann, which is enough reason to watch it. Mann went on to produce TV classics *Miami Vice* and *Crime Story*, and direct movies like *Heat*, *The Insider,* and *Collateral*. Here, you can see his talent in the formative stage, and catch early hints of his trademarks—such as his sympathetic approach to criminals or his use of hard-driving rock music. We've got issues with the music, but we'll get to that later.

There's also a first-rate cast, led by Peter Strauss (coming off his hit-making TV miniseries role as Rudy Jordache in *Rich Man, Poor Man*) and Brian Dennehy (before he appeared in every single movie produced during the 1980s). Dennehy—in slicked-back hair, tight jeans, and a black T-shirt—looks like a refugee from *Grease*, but he plays the role of a cold-blooded prison-yard bully (named Doctor D) with an evil perfection.

Strauss stars as Rain Murphy, a man serving a life sentence at Folsom Prison in California for killing his abusive father. He is a loner, barely talking to other prisoners and balking at any kind of rehabilitative therapy. His own brand of therapy comes from obsessively running the inside of the prison yard perimeter. It allows him—at least for an hour each day—to escape the grim reality that surrounds him.

In another movie, an overzealous tower guard might take a shot at our fast-moving track star. But in *Jericho Mile*, he is instead noticed by a do-gooder psychologist, who decides to time the lifer moving so feverishly in sweat pants and Keds. As it turns out, Murphy is running a prison mile (defined as eight times around the trash cans) in 3:59.

This is brought to the attention of the warden and, well, let's just say that this guy is the flipside of Warden Hazen from *The Longest Yard*. This one is from the school of positive reinforcement. So he hires Murphy a track coach (speaking of *The Longest Yard*, the coach is played by Ed Lauter, whom we fondly remember as Captain Knauer) and tries to arrange for an Olympic tryout.

Not surprisingly, state officials balk at allowing a cold-blooded murderer to go on furlough. So our Mister Rogers of a warden simply buttons up his cardigan sweater and asks for prison volunteers to build an IOC-quality track inside the yard.

"He can be an example to all the other prisoners," suggests Warden Warm-and-Fuzzy, while failing to notice that the Aryan Nation sect within his walls has just shivved Murphy's only buddy, R.C. Stiles (played by Richard

Lawson before he got all gooey in *How Stella Got Her Groove Back*).

The middle part of the movie focuses on the internecine racial battles among prison sects. Give praise to ABC—for a show that aired in 1979, the violence and language are fairly brutal. Eventually, an uneasy coalition of black and Latin prisoners kick Dennehy's butt and decide to work together to build that track.

A qualifying race is arranged, and nervous-looking college runners enter the prison to challenge Murphy. Of course he blows them away, running a 3:52 mile. Los Angeles Olympics, here we come.

It's all leading up to a glorious inspirational ending. We can envision Bob Costas (or, Jim McKay back then), lauding our reformed father-shooter as he ascends the medal stand. Except ... well, give Mann and writer Patrick J. Nolan credit for not going with the obvious and uplifting.

Stick around to the end, even if most networks don't start airing *The Jericho Mile* until 2 a.m. There's a dramatic showdown scene about 90 minutes in, where Murphy must measure his dreams against his principles. And the final coda (which relates to the title of the movie) is so thought provoking, it may just keep you up for the rest of the night.

It's worth the sleep deprivation. Consider yourself forewarned.

🎬 **PIVOTAL SCENE:** Murphy—despite running in a time that qualifies for the Olympic Trials—must convince a civilian board that he is reformed enough to be allowed out of prison. The head of the board insists that Murphy show fealty, as well as remorse for killing his father—even though we learn that the murder occurred after Murphy discovered the old man molesting his 14-year-old sister.

"If you were in the same circumstances, in the same place, would you commit the crime again?" the panel chairman demands. "Would you do it again, Mr. Murphy? I think people would like to know that."

Murphy struggles with the answer, knowing that if he gives the honest response, his Olympic hopes are over. And so. . .

We'll let you wait until you watch it yourself.

🍎 **BET YOU DIDN'T KNOW:** Mann received such positive acclaim in his first directorial effort (as well as a writing Emmy) that he quickly received more than 20 movie offers. He chose to direct *Thief*, a solid 1981 release starring James Caan as a safecracker trying to quit the mob.

73 CHAMPION (1949–NR)

SPORT: Boxing ★ STARS: Kirk Douglas, Arthur Kennedy, Ruth Roman
DIRECTOR: Mark Robson

Kirk Douglas took a chance on the film *Champion*. At 32, Douglas was still trying to crack the big time in Hollywood. He was offered the lead in a big-budget MGM film, *The Great Sinner* but turned it down to do this smaller *film noir* based on a short story by Ring Lardner. He saw *Champion* as a greater risk, but also a greater opportunity.

The role of Midge Kelly, the penniless hobo who achieves fame in boxing, was unusual for its time. Midge is not a hero; quite the opposite. He is a heel, a selfish brute who uses and abuses everyone around him, even his crippled brother Connie (Arthur Kennedy).

Douglas took the part and absolutely nailed it. He received his first Academy Award nomination and the film, which was released with little fanfare, earned six nominations in all. *Variety* called it "a stark, realistic study of the boxing rackets."

The Great Sinner? It was one of the year's biggest disappointments, fizzling at the box office despite a cast that included Gregory Peck and Ava Gardner.

Douglas's gamble paid off. His portrayal of the ruthless boxer made him a major star. It allowed him to break free of the studio system and pick and choose his own films, which he did very well through the '50s and '60s. Midge Kelly may have been a bad guy, but he was very good to Kirk Douglas.

Champion opens with Midge and Connie hitchhiking across the country. Short of cash, Midge agrees to box on a tank town fight card. A total novice, Midge takes a beating, but his toughness impresses veteran trainer Tommy Haley (Paul Stewart). He tells Midge to call if he ever decides to take the sport seriously.

The brothers find work washing dishes at a diner. Midge has a fling with the owner's daughter Emma (Ruth Roman). When the owner finds out, he forces them into a shotgun marriage. Midge has no intention of spending the rest of his life slinging hash, so he takes off and leaves Emma. Connie has fallen for Emma, but tags along with Midge because, well, that's what orphaned brothers do.

With no better options available, Midge decides to call Tommy Haley. Connie tries to talk him out of it. Boxing, he says, "is a rotten business." Midge, who's getting harder and colder by the minute, tells his brother, "Lay off."

"It's like any other business," Midge says. "Only [in boxing] the blood shows."

Tommy agrees to train Midge and soon he is climbing the middleweight ladder. One night in the dressing room, Connie again expresses his misgivings about his brother's involvement in the sport. Midge is now so full of ambition and lust, he refuses to listen.

"For the first time in my life, people are cheering for me," he says. "Were you deaf? Didn't you hear them? We're not hitchhiking any more. We're riding."

Midge is matched against a top contender, Johnny Dunne, but it is with the understanding that Midge will take a dive. When Tommy tells Midge of the terms set down by the mob, he flies into a rage. "I work like a slave," he says, "and then the fat bellies with the big cigars tell me I'm still a bum."

Midge agrees to throw the fight but pulls a double-cross by knocking out Dunne. The mobsters get their revenge by pummeling Midge, Tommy, and even poor Connie in the empty arena afterwards. But the win over Dunne makes Midge a star. He is approached by another manager, Jerry Harris (Luis Van Rooten), who claims he can do more for his career than Tommy. Jerry is richer and slicker, with a fancy office and a hot young wife (Lola Albright), all of which appeals to Midge.

Midge dumps Tommy without a second thought and signs on with Jerry. At this point, he is so consumed with greed that even Connie leaves him. Midge gets his title shot and wins it while beginning an affair with Jerry's wife. Jerry gets Midge to break it off by offering him cash (see "pivotal scene") as he prepares for a fight against a revenge-minded Johnny Dunne.

Midge reaches out and asks Tommy to come back and train him for the Dunne fight. Reluctantly, Tommy agrees. Midge also tries to patch things up with Connie, inviting him to training camp. Connie arrives with Emma and announces they plan to marry. There's a slight complication: Emma never bothered to divorce Midge. He agrees to step aside so Connie and Emma can marry. You start to think that maybe the no-good louse is finally coming around, but the first time they're alone, Midge forces himself on Emma.

The next night in the dressing room, Connie confronts Midge, asking, "Why did you do it? Did you have to prove you were still the champion?" Midge slugs Connie and knocks him to the floor, then throws Connie's cane at him as he puts on his robe and heads for the ring.

In the fight, Midge takes a bloody beating, but just as he is about to be counted out, he looks at the faces at ringside, all the people he stepped on who now are smiling as they watch him brought down to earth. Enraged, he climbs off the canvas and KO's Dunne to retain his title. But in the dressing room, Midge collapses and dies of a brain hemorrhage.

In the final scene, Connie and Emma are surrounded by reporters who ask for a statement on Midge's death. Here is the opportunity to get even. All they have to do is tell the press what a lowlife Midge was. Connie, ever the loyal brother, cannot do it. Instead, he lies through his teeth, saying, "He was a champion. He died like a champion. He was a credit to the fight game right to the very end."

Connie and Emma leave the arena and Midge is allowed to pass into history as a hero.

🎬 **CHEERS:** Arthur Kennedy is superb as Connie. He is nominated for an Academy Award as Best Supporting Actor, the first of five nominations in his distinguished career.

66 BEST LINE: Tommy Haley: "Midge is getting a shot at the title and he's got a new manager. A blonde."

🎬 **PIVOTAL SCENE:** Jerry learns Midge is sleeping with his wife. He tells her Midge is simply using her as he uses everyone else. To prove it, he makes Midge an offer: He'll let Midge keep his manager's commission from the Dunne fight ($65,000) if he agrees to break off the affair.

Midge looks at the wife and says, "You're his wife, I never should have come between you in the first place." Meanwhile, he is stuffing the money in his pockets.

She is speechless. For the first time, she sees how craven Midge really is.

74 LET IT RIDE (1989—PG)

SPORT: Horse Racing ★ STARS: Richard Dreyfuss, Teri Garr, David Johansen DIRECTOR: Joe Pytka

Every $2 bettor who ever spent an afternoon at the track has dreamed of that winning streak, that day when solid tips plus heaven's help allows him to parlay his last few bucks into an instant fortune.

Of course, those days never come. That's how tracks stay in business.

But *Let it Ride* allows us to watch and root as one lifelong loser at the betting windows has that day of days. And we get to enjoy riding along.

Richard Dreyfuss stars as Jay Trotter, a sad-sack Miami cabbie who has just promised his wife (Teri Garr) that he will cut down on his compulsive betting to save their marriage. The opening scene of their reconciliation at a Chinese restaurant ends with him cracking open a fortune cookie that reads, "Sometimes you can be walking around lucky and not know it."

Of course, he immediately receives inside info on a sure thing and heads to the track the next day with his last $50. "This isn't gambling," he rationalizes. "Gambling includes risk. This is just taking advantage of an extraordinary business opportunity."

His horse comes through, returning $710 on his investment. Trotter, who never wins on this kind of long shot, decides that this day is different, and lays down all his earnings on the next race. He wins again. And again, after asking everyone around him which entry to bet on—and going the other way.

His $50 grows to $69,000 and, now, the lifelong schlepper is the hero of Hialeah, attracting buxom, willing women and his own personal bodyguard. Not sure whether he is being guided by divine intervention or just plain luck, he decides to parlay it all on one last pony run. When a horse on its way to the track literally winks his way, he figures he's got his winner.

We won't give away the ending, except to say the final race is a terrific nail-biter. And Dreyfuss makes the whole thing work with his usual likability and nervous energy. He yips with joy or desperation at his fortunes, snaps his fingers, pulls on his tie like Rodney Dangerfield, and makes us root for the rascal who hasn't finished ahead in a decade. Anyone who ever spent time poring over the *Daily Racing Form* can commiserate.

"In order to have a bad feeling, you need to have a good feeling first," he lectures a fellow track degenerate. "So you have no frame of reference."

The movie is held back by an annoying supporting cast. Garr is more of a nag than the horses as the clingy wife, and David Johansen is over-the-top obnoxious as the best friend who's the only guy in America with worse luck than Trotter. Only Robbie Coltrane, as the cynical ticket seller, and a youthful Jennifer Tilly (more about her shortly) give you anyone else worth watching.

Let it Ride is based on a book by Jay Cronley, who covers horse racing for ESPN. The screenplay was written by Nancy Dowd, most famous for penning *Slap Shot,* who is clearly going for a Damon Runyanesque feel here. Often it works, occasionally it falls short.

It also marks the directorial debut of Joe Pytka, who mostly is known for commercials and Michael Jackson videos. In fact, Pytka went on to direct just one more movie, *Space Jam*. We don't know if that stinker was enough to convince him to leave the field.

CHEERS: Well, the real co-stars here are Jennifer Tilly's spectacular breasts, which seem determined to pop out of her skintight red dress through the entire film. We dare you to look away.

At one point, Vicki (Tilly) grabs Trotter from behind and squeezes him tight.

"Guess who?" she asks.

"A football salesman?" he replies.

PIVOTAL SCENE: About to embark on his big day of gambling, and clutching onto his last $50, Trotter pauses in a grungy dive-bar men's room and kneels in front of the toilet to pray for a helping hand.

"I don't belong with these losers, God, you know that," he says, hands clasped in front of him. "I belong in the Jockey Club, around guys with all their teeth. Just let me win this once."

Every day, at every racetrack in the world, a scene like this takes place.

BEST LINE: Trotter (betting his entire winnings of $69,000 on a 40-1 shot): "Let it ride."

Ticket seller: "You've got all kinds of balls."

Bodyguard: "Mostly crystal."

75 THE EXPRESS (2008—PG-13)

SPORT: Football ★ STARS: Rob Brown, Dennis Quaid, Omar Benson Miller
DIRECTOR: Gary Fleder

The Express is the story of Ernie Davis, the Syracuse University halfback who was the first African-American to win the Heisman Trophy. Davis was the top pick in the 1962 NFL draft but he died of leukemia before he ever played a down of professional football.

It is a compelling story and thanks to a fine performance by Rob Brown as Davis, *The Express* will prove enlightening to those who are too young to know the tragic tale. But this is really a film about two men—Davis and his coach Ben Schwartzwalder—and their relationship during that racially divisive time in America.

Dennis Quaid takes on perhaps his most challenging role as Schwartzwalder. He could have played him as the typical hard-nosed, my-way-or-the-highway football coach that we've

seen a million times. But Quaid brings a depth to Schwartzwalder that makes the coach a far more interesting character.

Most of Quaid's other jock roles—Jim Morris (*The Rookie*), Cap Rooney (*Any Given Sunday*), Gavin Grey (*Everybody's All-American*)—are straightforward types, men with a strong sense of self who never waver in their beliefs, even when life kicks them in the teeth. His Schwartzwalder is a far more complex figure and Quaid does a very good job portraying a man who begins to question his values for the first time.

Early in *The Express*, we see Schwartzwalder and Jim Brown, his first All-American, in an uncomfortable conversation. It is clear the two had a difficult relationship: Brown, one of the first black athletes to speak out about race, and Schwartzwalder, the crusty ex-paratrooper who isn't comfortable with players speaking out about, well, anything.

Brown (Darrin Dewitt Henson) has finished his college career and is preparing to leave for Cleveland where he will become a superstar with the Browns. Schwartzwalder asks Brown if he would go with him to Elmira, N.Y. to meet with a high school phenom, Ernie Davis. Schwartzwalder is hoping to sell the kid on attending Syracuse and possibly filling Brown's famed No. 44 jersey.

The coach won't say it straight out, but Brown knows the real story. Davis is black and Schwartzwalder feels Brown would be a more effective salesman. Brown agrees to make the trip and succeeds in convincing Davis to pick Syracuse over Notre Dame and 50 other schools bidding for his services.

Brown tells Davis that Schwartzwalder isn't the easiest man to get along with, but he will make him into the best football player he can possibly be. Davis is swayed by Brown's honesty and accepts the scholarship to Syracuse.

Once Davis reports to campus, *The Express* follows the traditional sports movie playbook. He clashes with hostile upperclassmen who are slow to accept the highly touted newcomer and he bonds with the hulking Jack Buckley (Omar Benson Miller), a fellow African-American who becomes his friend and protector. It is all familiar stuff, competently rendered.

What lifts *The Express* above the pack is the role reversal. In most films, it is the coach who changes the player. Here, the player changes the coach. Davis isn't as outspoken as Jim Brown, but he is equally aware of what's going on around him, such as segregated hotel accommodations on the road. He knows it's wrong and he can't understand why an authority figure like Schwartzwalder just shrugs and goes along with it.

When the Syracuse team—with its handful of black players—travels to West Virginia, Schwartzwalder warns them to expect trouble. "Keep your helmets on," he tells them in the tunnel. As the Orangemen take the field, the redneck crowd pelts them with bottles, garbage, and racial taunts.

When Davis carries the ball to the one-yard line, the coach replaces him with a white halfback who scores the touchdown. Later in the game, the same thing happens. Davis carries the ball to the one and the coach sends the same substitute onto the field. This time Davis waves him off. He stays in the game and scores the touchdown as the West Virginia fans shower him with debris.

Schwartzwalder confronts Davis and the two finally have it out. The coach says he is thinking about the safety of the entire team. He fears the sight of a black player scoring a touchdown could set off a full-scale riot. Davis tells the coach that by bowing to that kind of hate, he is only reinforcing it. Sooner or later, Davis says, a man has to stand against it.

"We're willing to do it," Davis says, "but it doesn't mean a thing unless you're with us."

Quaid allows us to see Schwartzwalder's conscience at work. The way he fiddles with his glasses, the way he addresses the team, not quite as sure of himself, we can see that he is thinking about what Davis said.

Clearly, Schwartzwalder is not a racist. Indeed, he was recruiting black players before most major college coaches. But typical of many white Americans of that time, his idea of tolerance only goes so far. For example, when he sees Davis exchange flirty smiles with a white cheerleader, he calls the player into the office.

"There are certain lines you don't cross," the coach says sternly.

By the time the Orangemen travel to the Cotton Bowl to face Texas—with their race-baiting players and fans—Schwartzwalder knows it is much more than a football game. Even with a national championship and unbeaten season on the line, he sees the larger picture.

When Syracuse wins the game—which features an ugly brawl initiated by the cheap shots and racial taunts of the Texas players—they are told the trophy presentation will take place at a private club in Dallas. No blacks, in other words. When the white players tell the Cotton Bowl officials to shove their trophy and their club, Schwartzwalder nods in approval.

The Express focuses almost entirely on that season, 1959, which was Davis' sophomore year. It flashes forward to his senior season when he becomes the first African-American to win the Heisman Trophy. In his acceptance speech, Davis thanks Schwartzwalder for being "a good coach and a good man."

"I spoke with Jim Brown about his relationship with Ben," Quaid said, "and he told me that Ben was obsessed with football and only football, Xs and Os. He and Jim butted heads over racial issues, but that's because Ben didn't want the civil rights movement spilling over onto his field. He didn't want to hear about it.

"But Ben changed, thanks to people like Jim Brown and Ernie Davis. One of the greatest compliments the film received came from Jim Brown who said he gained new respect for Ben watching the movie. Ben was one of the soldiers who stormed the beach at Normandy. He wore those combat boots at practice; that's who he was. He brought that military experience to his coaching. Jim Brown and later Ernie Davis got Ben to reveal that he actually did have a soft, creamy center."

CHEERS: Quaid steals the film with his portrayal of Schwartzwalder. Michael Granberry of the *Dallas Morning News* wrote: "Quaid has quietly become one of America's finest actors. [He] can say more with a single facial expression than most can with 100 pages of dialog."

WHAT THEY WROTE AT THE TIME: "As Davis, Rob Brown (*Finding Forrester*) is in possession of a kind face and an athletic grace that easily wins over audiences."—Michael Rechtshaffen, *The Hollywood Reporter*

BET YOU DIDN'T KNOW: A statue of Ernie Davis was unveiled on the Syracuse campus in October 2008. The statue had a Nike swoosh on its jersey and Nike cleats on its feet, but Nike wasn't founded until 1972, nine years after Davis' death. The sculptor corrected the errors.

BEST LINE: "I bet about now you're wondering what happened to that nice gentleman who came to your house and asked you to play for Syracuse."—Jack Buckley (Omar Benson Miller) talking to Ernie Davis (Brown) after Schwartzwalder drove them through a tough practice.

"I KNOW THAT GUY": In this case, it is a voice. Clancy Brown, who plays assistant coach Roy Simmons Sr., is the voice for Mr. Krabs on *SpongeBob Squarepants*.

SPORTS-ACTION GRADE: C+. The action is filmed well enough and Rob Brown has the graceful stride of a football player, but the hits are over the top. Only in Hollywood does every hit on the football field sound like a nuclear explosion.

76 ▶ THE CINCINNATI KID (1965—NR)

SPORT: Poker ★ **STARS: Steve McQueen, Edward G. Robinson, Ann-Margret**
DIRECTOR: Norman Jewison

Four years after *The Hustler* shined a spotlight on the seedy world of pool hall gambling, Hollywood decided to turn another pulp novel into a film that would do the same with side-street poker games. Richard Jessup's *The Cincinnati Kid* became that movie.

The two films have striking similarities. Both are stories of a brash young gambler trying to topple his underground sport's reigning king. Both feature charismatic blue-eyed stars in the lead—Paul Newman vs. Steve McQueen—as well as veterans dripping with gravitas as "The Man"—Jackie Gleason vs. Edward G. Robinson.

Both have a meaty "bad girl" role—Piper Laurie vs. Ann-Margret (who even chomps on an apple here, just in case you miss the point)—as well as a corrupting devil figure—George C. Scott vs. Rip Torn. And there's more: marathon games, sleazy side shows, endless cigarette smoke, redemption followed by failure.

The comparisons—well pointed out when *The Cincinnati Kid* was initially released—hurt this movie with critics. It never could reach the classic status that its older brother achieved. Likewise, in recent years, *The Cincinnati Kid* has become the "forgotten" poker movie since the 1998 release of *Rounders*, a grittier story told at a quicker pace.

It's all a bit of a shame. On its own, this is a terrific film. If you can watch it without making the inevitable comparisons, you'll enjoy a hard-boiled story with great action, surprising plot twists, and the honky-tonk atmosphere of New Orleans in the 1930s.

Steve McQueen stars as Eric Stoner—"The Cincinnati Kid"—a crackerjack five-card stud player eager to be seen as the best around. To get there, he must topple Lancey Howard (Robinson)—AKA "The Man"—the incumbent champion. In many ways, this feels like a classic Western, with the young gunslinger stalking Wild Bill Hickok.

The game is arranged through a mutual friend named Shooter (Karl Malden), who agrees to be the dealer. Meanwhile, side deals are made, loyalties tested, and Shooter's wife, Melba (Ann-Margret), aims to sleep with every man in Louisiana.

There are some great side characters here, led by Hollywood veteran Joan Blondell portraying a brassy poker player named Lady Fingers who has some sort of history with Lancey Howard. Also terrific is a young Rip Torn (you won't recognize him as the same guy from *Men in Black*) as the spoiled rich boy who tries to buy 'em when he can't beat 'em.

And, of course, there is McQueen. In the Hall of Fame of Cool, he is a unanimous first-ballot inductee.

The Cincinnati Kid is at its best when it sticks to the poker table (some of the side scenes are downright tedious). Poker is clearly not the most cinematic of sports, but director Norman Jewison makes it work through alternating tension and humor, and showing the reactions of players—all trying so hard not to show their sweat and their tics.

In the DVD commentary, Jewison talks about how tough it was to film a five-hand poker game, and how many of his colleagues in the business told him that no movie about card playing would ever find appeal among audiences. "But poker is about winning," says Jewison. "That's why it's so popular in America, because America is about winning."

🍎 **BET YOU DIDN'T KNOW:** After bombing in the movie *Baby the Rain Must Fall*, McQueen took a year off from filmmaking—riding his motorcycle around the world and having an affair with sexpot Mamie Van Doren. (Not a bad way to kill time.) *The Cincinnati Kid* marked his return. Although it wasn't a smash at the box office, McQueen gained traction and went on to film *The Sand Pebbles*, *The Thomas Crown Affair,* and *Bullitt* in rapid succession.

66 **BEST LINE:** Lancey Howard—The Man— has just beaten The Kid in a huge hand by betting against the odds and all reason.

Lancey: "That gets down to what it's all about, doesn't it? Making the wrong move at the right time."

Cincinnati Kid: "Is that what it's all about?"

Lancey Howard: "Like life, I guess. You're good, Kid, but as long as I'm around you're second best. You might as well learn to live with it."

😎 **"I KNOW THAT GUY":** The fifth player at the big table is the great Cab Calloway, a famed big band leader and jazz pioneer. If you don't recognize him from his old recordings ("Minnie the Moocher," "St. James Infirmary"), perhaps you'll remember him as Curtis from *The Blues Brothers*. Or maybe you grew up marveling at him performing *Hi De Ho Man* with purple muppets on *Sesame Street*. If not, just look up his name on YouTube.com, and you will be rewarded.

SPORT: Golf ★ STARS: Adam Sandler, Carl Weathers, Christopher McDonald
DIRECTOR: Dennis Dugan

In the Hall of Comedy Genius, Adam Sandler does not stand near Charlie Chaplin. He's not up to the standards of W.C. Fields or Richard Pryor or Steve Martin or Eddie Murphy. He's no Jamie Foxx. No Bill Murray. No Robin Williams.

What he is, rather, is a modern-day version of the Three Stooges—an anarchic slapstick performer who trades on class warfare and frathouse humor to create films aiming right into the wheelhouse of the coveted 18-to-34-year-old male demographic.

Sometimes they work (*The Wedding Singer*, *Anger Management*), sometimes they don't (*Little Nicky, You Don't Mess With the Zohan*). Too often, Sandler gets caught up in his impish, man-child persona, sinking into a cutesy-pie voice (*The Waterboy*) and giggling too much at his own cleverness.

In *Happy Gilmore*, Sandler avoids the trap of self-adoration. Instead, he creates a funny, if uneven, story of a working-class slob who invades the pompous professional golf scene and ends up a national hero. All while winning the girl and saving his sweet grandmother's home.

for failure to pay $270,000 in back taxes, he figures this is his best way to earn quick cash—even though, of course, he has never played a round in his life, can't putt, has no short game, and uses every blown shot as an excuse to toss clubs, batter bystanders, and curse like a longshoreman with Tourette's.

Happy's volcanic temper is the one-joke gag of the movie. On the one hand, it keeps threatening to get him tossed from the tour—like when a pin flag he launches like a javelin nearly decapitates a cameraman. On the other hand, it makes him an idol of the crowds. Apparently, golf has been missing a Dennis Rodman-like bad boy, and Happy's just the guy for the job.

"His behavior is completely unacceptable," says tour PR director (and eventual love interest) Virginia (Julie Bowen). "But, you know, golf has been waiting for a player like this—a colorful, emotional, working-class hero."

Happy plays the course in a Boston Bruins jersey and sweat pants. He wears construction boots rather than golf shoes, and uses white

"His behavior is completely unacceptable. But, you know, golf has been waiting for a player like this—a colorful, emotional, working-class hero." —VIRGINIA

The script here could have been written on a cocktail napkin, and probably was. Sandler's title character is a frustrated hockey player (he can fight but he cannot skate), who accidentally discovers that he can drive a golf ball 400 yards.

He has no interest in playing pro golf. But when his grandmother's house is repossessed

tube socks as head covers for the wooden-shaft clubs rescued from his grandfather's attic. His caddy is a homeless beggar who resembles Bigfoot and rinses his tighty-whities in the ball washer.

How could you not root for that guy? Well that, plus Happy's running-start drives that occasionally result in double-eagle holes-in-one. Suddenly, golf fandom is taken over by T-shirt-wearing, pro wrestling devotees, or, as TV announcer Verne Lundquist calls it, "quite a large and economically diverse crowd here today at the Buick Open."

Of course, slobs vs. snobs only succeeds as a theme if you've got an upper-crust worth despising. And actor Christopher McDonald, as tour money leader Shooter McGavin, plays the Judge Smails role to the hilt. He's an arrogant bully who fires his own caddies for laughs and aims to buy Happy's grandmother's house from the taxman just to ensure she stays on the street. When heckled by Happy's gallery, he shouts back, "Damn you people. Go back to your shanties."

One other actor worth noting is Carl Weathers, who devolves here from his role as Apollo Creed to Chubbs, the one-handed golf instructor (the prosthetic hand gag is not to the genius level of *Kingpin*, which also came out in 1996). Chubbs offers less on technique than he does on life lessons. He gets Gilmore to curb his anger by transporting his mind to a tranquilizing "happy place," in this case a world where Grandma gets silver dollars tumbling from the slot machines and the beautiful Virginia reclines on a sofa, wearing a bustier

and hoisting two pitchers of beer.

Ahh, haven't we all been there?

🎬 **PIVOTAL SCENE:** A goofball movie like this has no real pivotal scene, but the funniest moment has to be the pro-am tournament brawl between Happy and *The Price Is Right* host Bob Barker. It begins when Happy's game is crippled by a heckler ("Jackass!"), and Barker—who won the tourney the previous year—dismisses his struggling partner by suggesting, "There's no way that you could have been as bad at hockey as you are at golf."

Happy, still battling anger management issues, sucker punches the 72-year-old game show host. That prompts a three-minute smackdown that has the two men rolling down a hill, head butting each other, and exchanging haymakers in a pond. It's not quite up there with *The Quiet Man* for best movie brawl, but you will laugh at Happy's premature victory cry of, "The price is *wrong*—bitch!"—as well as Barker's physical and verbal retort.

Barker told *Entertainment Weekly* that he initially declined a role in the movie—until he learned he would be duking it out with Sandler. "It seemed a good way for me to get out of character," he said.

78 EVERYBODY'S ALL-AMERICAN (1988—R)

SPORT: Football ★ STARS: Dennis Quaid, Jessica Lange, Timothy Hutton DIRECTOR: Taylor Hackford

Everybody's All-American was originally a fine novel written by Frank Deford of *Sports Illustrated*, but screenwriter Thomas Rickman and director Taylor Hackford fumble in their attempt to bring Deford's story to the screen.

The film is good enough to hold your interest, and there are times when the travails

of Gavin Grey (Dennis Quaid) are achingly real, but too often it is a plodding soap opera that can't decide whether it is about the fall of a football legend or the demise of the Old South. Hackford tries to tie the themes together, but he doesn't quite pull it off.

The story traces 25 years in the lives of

Grey, better known as "The Grey Ghost," All-American halfback at Louisiana University (it's really LSU) and his beauty queen wife Babs (Jessica Lange). What begins as a fairy tale marriage crumbles over time as Grey's football skills decline and, suddenly, it's the bill collectors on the phone and not the fan clubs or the talk shows.

Everybody's All-American is at its best when it is dealing with those issues; that is, what happens to yesterday's hero? What happens to the great athlete when the applause dies and the stage goes dark? What happens when he awakens to discover the public has forgotten his name? This film makes you think about that.

There is a haunting scene late in the film when Grey's old team, the 1956 Sugar Bowl champion, is honored at the homecoming game. As Grey is introduced and walks on the field, the crowd stands and cheers. His smile widens and you can see him thinking "They *do* remember." Then he glances over his shoulder and sees the current varsity emerging from the tunnel. He realizes the cheer wasn't for him but for them. He doesn't say a word but the look of pain and embarrassment on his face is all-too-real.

It is a path most old jocks will travel. Some travel it more gracefully than others. Grey has a particularly difficult time. He goes from college superstar to All-Pro running back with the Washington Redskins to owner of a successful bar and restaurant. However, his old teammate Lawrence (John Goodman), who works at the bar, has a gambling addiction that winds up putting the business in bankruptcy and getting him killed.

Grey, who was able to overcome any challenge on the football field, proves ill-equipped to deal with adversity in the real world. He tries an ill-fated comeback with the Denver Broncos—watching him peel the tape off his battered body will make you wince—

and he picks up a few bucks at a country club, playing golf with the members and entertaining them with his war stories. That will make you wince, too.

"One time we were playing Baltimore and Unitas was kicking our ass," Grey says, beginning a tale he has told too many times. The tone of his voice is flat, bordering on sour.

"Oh, he tells this one real good," the boss says, patting him on the back.

The Grey Ghost has become a pathetic figure and he knows it.

Babs, who was once just a lovely ornament on her husband's arm, must find the strength to keep their family afloat. In college, Babs tells a friend she is majoring in "Gavin and me" and as long as the Grey Ghost is scoring touchdowns, she is content to play the role of Magnolia Queen. But with debts piling up and her husband a disillusioned prisoner of his past, she goes to work and proves to be a capable businesswoman.

This poses even more problems for Grey, who now has to deal with the fact that his wife is the success in the family. There are a lot of issues to resolve and Hackford almost gives himself a sports hernia trying to carry all that baggage to the finish line. The upbeat ending feels forced and the character of Cake (Timothy Hutton), the Ghost's nephew who is meant to serve as the Nick Carraway to Gavin's Gatsby, never really registers.

66 **BEST LINE:** Babs is watching Gavin's first game with the Redskins. She is taken aback at the violence of pro football. One of the other wives says, "Pro's a different game, honey. Not much rah-rah up here. Not like when you were homecoming queen in where was it?"

"Louisiana," Babs says. "I was Magnolia Queen."

"Most of us were the queen of something," the other wife says. "Now? We're just players' wives."

THE GREATEST GAME EVER PLAYED (2005–PG)

SPORT: Golf ★ **STARS:** Shia LaBeouf, Stephen Dillane
DIRECTOR: Bill Paxton

Watching *The Greatest Game Ever Played*, you'll be struck by three things:

1. Once upon a time, Americans were actually considered the scrappy underdogs to their European big brothers.
2. If you think golf is a snooty sport now, you should have seen it a century ago.
3. This story is so outrageous, it can't possibly be true. And if it is, how come you never heard about it before?

Last things first. *The Greatest Game Ever Played* tells the tale of the 1913 U.S. Open at The Country Club in Brookline, MA. The tournament became a showdown between British pro Harry Vardon (Stephen Dillane), who was considered the world's best golfer, and amateur Francis Ouimet (Shia LaBeouf), a 20-year-old former caddy playing in his first major tournament.

Ouimet, as his World Golf Hall of Fame biography notes, "seemed to step from the pages of a Dickens novel." He grew up in a working-class home across the street from The Country Club and learned to play by tapping balls into tomato cans in his backyard. At 11, he began caddying on the famous course and was invited into the Open—as the final participant with the longest odds—after he won the 1913 Massachusetts State Amateur Championship.

Sports history is rife with great upsets—1980 U.S. Olympic hockey team over the Soviets; Buster Douglas over Mike Tyson; the Giants over the Patriots in Super Bowl XLII. Each of those pales in comparison with this one. Imagine Tiger Woods, at the peak of his game, vanquishing all other opponents in a Grand Slam event, only to lose in an extra fifth round to a neighborhood kid making his debut. And just to make it more preposterous, stick the young upstart with a 10-year-old school-skipping caddy, barely bigger than the bags he must lug.

It really happened that way.

The story was largely lost to history until 2002, when author Mark Frost turned it into a best-selling book. Frost's work caught the attention of actor Bill Paxton (*Apollo 13, Twister*), who had himself grown up across the street from a golf course in Texas and whose dad had once played a round with Ouimet's former caddy. Paxton—who had previously directed the horror drama *Frailty*—sold the idea to Disney and hired Frost to turn his book into a screenplay.

The movie plays honest with the actual history—with a story this fantastic, why shouldn't it? In doing so, Paxton takes us back to a time when Brits thoroughly dominated the sport. This is before World War I, before America was a world power, a time when a young United States was just discovering its athletic muscles and still had somewhat of a little-brother complex. Indeed, Ouimet's shocking victory is still regarded as the single largest landmark moment in bringing golf into the mindset of American sports fans, and the 20-year-old became this country's first golf hero.

The Greatest Game Ever Played also succeeds in avoiding a common sports movie trap, which is casting the opponent as a one-dimensional villain—whether he be Ivan Drago, Warden Hazen, or Judge Smails. Vardon, despite being the prohibitive favorite, remains a guy you can root for, in part because he shares Ouimet's underclass roots and must spend his life fighting the sport's upper-crust gatekeepers. One of the movie's best scenes shows a younger Vardon being interviewed at a posh English country club after winning an earlier U.S. Open. Vardon believes he is being asked to join as a member; turns out the snobs just want him to run the pro shop.

The movie follows the two men for a decade (including a brief encounter between them when Ouimet was a young lad), but most of its 120 minutes focuses on the five-round tournament. Golf is not the easiest sport to bring to film; the background may be pastoral, but the action falls closer to chess than it does to football. Paxton tries to speed things up with rapid-fire tee shots, undulating putts, and Fourth of July crowd reaction shots.

"It's kind of like a fireworks show," Paxton said in an About.com interview shortly after the film's release. "We get into that final round, and it's the 'House of Flying Golf Balls' and everything else."

Sometimes that works, sometimes Paxton gets carried away with special effects. In one shot, we get "ball cam"—watching through the golf ball's vantage point as a seeing-eye shot makes its way through the woods. In another, a ladybug comes to alight on a ball, and, in close-up looks like a refugee from Pixar's *Toy Story*. Downright silly. What's next, the *Caddyshack* gopher?

That's a quibble. Generally, this is a story well told and well acted. And, most amazingly, it's true.

🔔 **CHEERS:** With golfers playing in ties and jackets, wooden-shafted mashers, and brassies and balls with pimples rather than dimples,

The Greatest Game Ever Played does a great job re-creating the sport as it was back in 1913.

🎬 **PIVOTAL SCENE:** Ouimet is sailing along midway through the second round of the tournament when a friend informs him that he's tied with Vardon for second place. He begins to feel the pressure and flubs a tee shot when he notices that President Taft is present and watching him. He must collect himself, and with the sage words of his pint-sized caddy (okay, we didn't despise the kid *all* the time), recovers and pars the hole.

☞ **DON'T FAIL TO NOTICE:** In the scene where 12-year-old Francis meets Vardon at a golf demonstration, the older man patiently instructs the boy on how to properly hold the club. The "Vardon Grip," with overlapping fingers, is still the preferred method of gripping the club today.

☺ **GOOFS:** Late in the final round, during a series of rapid-fire shots, Ouimet is shown taking one swing left-handed. We can only assume the film was flopped.

🤓 **"I KNOW THAT GUY":** Graybeard pro shop proprietor Alec Campbell is played by Luke Askew, an older version of the 1960s-70s character actor who had great supporting roles in *Cool Hand Luke* (Boss Paul), *Easy Rider* (the stranger on the highway), and *Pat Garrett & Billy the Kid* (Eno). Look through the white hair and the stubble; you'll spot him.

🍎 **BET YOU DIDN'T KNOW:** As an amateur, Ouimet could not collect the cash prize for winning the U.S. Open, so instead it went to runner-up Vardon. He collected a whopping $300.

80 SEARCHING FOR BOBBY FISCHER (1993–PG)

SPORT: Chess ★ **STARS:** Joe Mantegna, Joan Allen, Ben Kingsley
DIRECTOR: Steven Zaillian

For a brief moment, back in the Cold War year of 1972, chess was the hottest sport in America. A prodigy named Bobby Fischer went to Iceland and wrestled the world championship away from the Soviets in an event that made the gangly New Yorker an unlikely national hero and foreshadowed the 1980 Olympic Miracle on Ice.

Fischer is regarded as the greatest chess player of the 20th century. As a boy, he defeated many of the world's grandmasters and prevailed in public displays, taking on as many as 50 top players simultaneously. He was initially indulged as a quirky personality practicing gamesmanship by complaining about the lighting, the size of the board, even whining that the view from his hotel was "too nice."

Eventually, quirky became paranoid, and Fischer disappeared. He emerged occasionally to emit anti-Semitic outbursts and swear allegiance to religious cults. He renounced his citizenship, won one more competitive match in 1992, and died in exile in Iceland in 2008 at age 64.

There has never been another like him. But in the movie *Searching for Bobby Fischer,* a seven-year-old Manhattan boy suddenly appears to be the Next One.

Josh Waitzkin (Max Pomeranc) is that normal boy who loves baseball and dinosaurs.

One day he stumbles upon the hustlers playing three-minute speed chess at Washington Square Park, and he is hooked. Standing there for five minutes, he absorbs the intricacies of the game in a way that is invisible to even brilliant adults.

His parents are skeptical at first, but quickly convinced when Josh—sitting on two phone books to reach the board—disposes of anyone put against him. Josh's father, Fred (Joe Mantegna), a sportswriter, sees his own competitive juices stirred by his son's success. His mother (Joan Allen) is more concerned about how Josh can enjoy his talent without losing his right to enjoy childhood. The movie does a great job of juggling the blessing and curse that a special gift presents.

Beyond his parents, Josh has two influences: His coach Bruce Pandolfini (Ben Kingsley), who recognizes the youngster's brilliance but wants him to develop a winner's mean streak; and a wizened player from the park named Vinnie (Laurence Fishburne), who wants Josh to enjoy the game for what it is.

This is the kind of story that could be told about any individual sport where young phenoms excel—tennis, gymnastics, figure skating. For a parent, there is a battle between developing the talent to its fullest and letting the child lead a normal life. For the child, there is pressure and the risk of burnout by high school.

"Maybe it's better not to be the best," Josh says to his father early on. "Then you can lose and it's still okay."

Max Pomeranc, who plays Josh, was picked for the role not just for his acting ability (this was his debut), but also because he was ranked among the 100 best child chess players in America. Good choice. It's clear in how he moves the pieces and studies the board that the little boy isn't just faking it.

Presenting chess on film isn't exactly like presenting football. There isn't much action to make it exciting. But director Steve Zaillian, more known as a script writer (*Gangs of New York*, *American Gangster*) than a director, does as best as is possible. Pawns fly off the board, kings are toppled, little hands pound the speed clocks. This is chess as if it were shot by NFL Films. Indeed, it won the 1993 Oscar for best cinematography.

Throughout the movie, grainy black-and-white video footage of Fischer keeps popping up, with young Josh telling the story of Fischer's genius and frailties. It is as if the ghost lurks over the boy, to the point where in one scene, Josh sits under a photo of Fischer's haunting eyes.

Searching for Bobby Fischer is based on a true story and was adapted from a book written by Fred Waitzkin, Josh's true-life father. It is good to know that Josh survived his childhood, went on to win two U.S. Junior Chess championships in his teens and now, in his thirties, has moved on to become a martial arts champion.

"I consider myself a happy and balanced man," he said in a 2003 interview.

The same could never be said of the real Bobby Fischer.

66 **BEST LINE:** In his efforts to make his prodigy more coldblooded, coach Pandolfini explains that he needs to develop "contempt" for his opponents.

Pandolfini: "It means to hate them. You have to hate them, Josh. They hate you."

Josh: "But I don't hate them."

Pandolfini: "Well, you'd better start. Bobby Fischer held the world in contempt."

Josh: "I'm not him."

Pandolfini: "You're telling me."

SPORT: Basketball ★ STARS: Don Cheadle, James Earl Jones, Forest Whitaker, Eriq La Salle ★ DIRECTOR: Eriq La Salle

Rebound, one of HBO's better docudramas, opens with two powerful scenes. First, we visit the actual retirement ceremony of Kareem Abdul-Jabbar. The NBA's all-time scoring leader smiles in an oversized rocking chair presented as a gift, and then answers questions from Los Angeles Lakers broadcasting legend Chick Hearn.

"Who," Hearn wants to know, "was the best player you ever faced?"

"Well," ponders Jabbar, "if I was limited to one—that would have to be The Goat."

Quick cut to a Harlem street, where a drug-addled man stumbles in pain before passing out in the middle of an intersection. A crowd gathers.

"Did you hit him?" a woman asks the closest motorist. "Is he dead? Who is he?"

"Just another junkie," says the dismissive driver, getting back into his car.

The junkie, of course, is the very same "Goat"—Earl Manigault—of whom Jabbar speaks reverentially. And the plotline is quickly established: How did Manigault devolve from a 1960s playground legend to a desperate drug addict in less than a decade?

Consider it a cautionary tale. *Rebound* is a story of poor choices and wasted talent. And, in the end—almost in epilogue form—of redemption. The movie has some plot holes and hoary clichés, but ranks behind only *Hoop Dreams* and *He Got Game* in showing the seduction and exploitation of aspiring inner-city players. It also has some terrific acting turns, most notably by Don Cheadle as The Legend himself.

We first meet Earl Manigault as a 14-year-old who practices hoops all day wearing ankle weights. He quickly shocks the older playground stars with his ability to grab a coin off the backboard rim, as well as perform a double dunk—catching his first jam and stuffing it back in the net before hitting the ground. In a brilliant casting move, the shy teen is played by Colin Cheadle—Don's brother—who looks, well, exactly like a younger version of the star actor.

The older ballplayers lure the kid into their lifestyle. He experiments with drugs and winds up being expelled from Benjamin Franklin High the day before his team is to play for the city championship. That sets up the movie for Rescue No. 1, in which playground director Holcombe Rucker (Forest Whitaker)—the founder of the now-famous pro-am tournament that takes place in Harlem every summer—sees potential in the kid and moves to protect him. Rucker steers The Goat to a South Carolina prep school, where he morphs from Colin to Don Cheadle. He also learns to read, knot a tie, and impregnate Evonne (Monica Calhoun), the prettiest girl in school.

The Goat boasts of scholarship offers from Indiana and North Carolina. But the prep school headmaster (James Earl Jones) pushes him to tiny Johnson C. Smith College because, we later find out, that is where he graduated. So The Goat ends up laboring in a tiny Southern gym for a *Hoosiers*-style "pass-it-four-times" coach (Clarence Williams III), who aims to cure him of his "schoolyard n——r" hot dogging.

Needless to say, this does not work out. And so begins the descent—as The Goat quits the team by heaving the ball deep into the stands—into hell. He returns to Harlem and, within six months, is a full-fledged heroin addict, begging and stealing to support his habit.

"Who was the best player you ever faced?"
—BROADCASTER

"If I was limited to one—that would have to be The Goat."
—KAREEM ABDUL-JABBAR

The back half of the movie is painful to watch. The Goat's pal Diego (Eriq La Salle)—who has returned from Vietnam with no hands—ODs and dies in his arms. In another scene, he hides from wife Evonne and his young son, not wanting them to see him as a junkie. Finally, he is arrested. In a wrenching scene, he goes cold turkey behind prison bars.

And here comes Rescue No. 2. A kindly cop, recognizing The Goat from his playground days, hands him a copy of Pete Axthelm's great book, *The City Game*, which lauds Manigault as Harlem's best-ever street player, but also as "the one star who didn't go on." Reading the book aloud on his cell floor, crying, Manigault becomes inspired to give his failed life another shot.

Ten months later he is out. He confronts neighborhood drug kingpin Legrand (Michael Beach), challenging him to remove the dealers and dice players from Rucker Playground. The Goat, back in his favorite element, starts a youth league and begins his new life.

The rise seems abrupt and a little too easy. *Rebound* is a bit of a roller coaster—straight up, then straight down, then up again. But even if the redemption didn't go quite that smoothly,

it is based on a true story. And it leaves you wondering what might have been if Manigault had made better choices along the way.

The final moments of *Rebound* show the real Earl Manigault in 1996. He is smiling at his playground, directing kids, running the "Goat Walk Away from Drugs Tournament." Sadly, two years later, Manigault died from congestive heart failure. Doctors said his system was weakened by the years of drug abuse.

👍 **CHEERS:** There's a terrific soundtrack running through the movie that reflects the time period, starting with early-60s Motown and ending with Stevie Wonder's "A Place in the Sun."

🎬 **PIVOTAL SCENE:** The Goat, as a college freshman, is chafing under the tough treatment of his coach, who preaches, "Basketball is a team sport, not a personal showcase." At halftime of one game, he is visited by his prep school headmaster, who has come to tell The Goat that his mentor, Mr. Rucker, has died back home in Harlem. The kid is crestfallen.

Moments later, the college coach signals The Goat off the bench, sending him in with the warning, "This is college ball, not one of those Rucker jive-ass tournaments." Of course, the kid goes wild, twisting and dunking and performing every street trick he can. He is immediately pulled, tosses, the ball into the seats and quits.

That was the end of Manigault's brief college career and, ultimately, his hopes of turning pro.

🍎 **BET YOU DIDN'T KNOW:** Actor Daryl Mitchell, who portrays Earl's playground buddy (and future NBA player) Dean Meminger, was paralyzed from the waist down in a 2001 motorcycle accident. He has continued his career, most notably as the wheelchair-bound bowling alley manager in the TV series *Ed*.

THE SANDLOT (1993–PG)

SPORT: **Baseball** ★ STARS: **James Earl Jones, Tom Guiry, Mike Vitar**
DIRECTOR: **David M. Evans**

Perhaps there are still neighborhoods where all the kids gather every day to play baseball from dawn until dusk. Where the joy of the sport is so overriding that no one even bothers to keep score. Where there is no talk of steroids or salary arbitration; not even aluminum bats.

We would love that place.

That utopia does exist in writer/director David M. Evans' *The Sandlot*, a childhood memoir where baseball means everything to a pack of pre-teens. If you grew up—or wish you had—in a world of tree houses and amusement parks, of oiling up a glove and playing a night game under the glare of Fourth of July fireworks, this movie will take you to that balmy, reassuring place.

The story focuses on Scotty Smalls (Tom Guiry), a boy who is uprooted from somewhere in Middle America to a dusty Los Angeles suburb in the summer of 1962. In a neighborhood of baseball-centric boys, he can't fit in. Scotty doesn't know how to play and doesn't know the sport's history (he thinks Babe Ruth was a woman).

Those shortcomings ostracize him from the sandlot gang, until Benjamin Franklin Rodriguez (Mike Vitar), the star and leader of the group of eight, discreetly teaches him how to catch, throw, and hit—as well as the difference between The Bambino and Bambi. The outsider egghead kid becomes one of the crew, headed for the best summer of his life. And eight players conveniently becomes a team of nine.

It takes a huge suspension of critical analysis to believe that Scotty could graduate from nerdy klutz-boy to competent ballplayer within weeks. But *The Sandlot* expects you not to ask too many questions. The movie is told from the perspective of a child (Scotty, as a grownup, serves as narrator), and is full of awe and hyperbole. At its best moments, *The Sandlot* aims to be a warm weather version of *A Christmas Story*. It doesn't quite reach those heights, but some of the better scenes will remind you of that Jean Shepherd memory of childhood or Bill Bryson's great fifties memoir, *The Life and Times of the Thunderbolt Kid*.

The best scene actually has nothing to do with baseball. On a day too hot to play, the boys travel to the municipal pool, where they splash around too much and gawk at chesty lifeguard Wendy Peffercorn (Marley Shelton) in her drop-dead red bathing suit. They all dream of kissing her, but how?

The ballsiest of the boys—marvelously named Squints Palledorous (a dead ringer, you *Nick at Nite* fans, for Ernie from the old *My Three Sons* show), sinks to the bottom of the pool and feigns drowning. Which begins to verge on actual drowning. Wendy dives in, pulls his skinny 12-year-old body out, and begins to administer mouth-to-mouth. Soon enough, the kiss of life turns into the kiss of *his* life, earning Squints a slap and, later, an admiring wink from Miss Peffercorn. (We learn at the end of the movie that, some day, scrawny Squints will marry Wendy and she will bear him nine children. That may be the furthest stretch of reality in the entire film.)

Adolescent fantasies and all-day baseball games make for a pretty good summer. But the fun ends when the boys' last ball is smacked out of the sandlot into a fenced junkyard inhabited by "The Beast"—a fearsome dog that, in some scenes, looks like an escapee from *Jurassic Park* (again, you have to buy the exaggeration). The legend, passed down over the years, is that the snarling, slobbering animal has eaten 173 boys. No one is quite sure, but no one wants to become No. 174.

Scotty, always trying to fit in, runs home to grab a new ball. This time, he becomes the batter who swats it over the fence, right into the waiting jaws of The Beast. Turns out, that baseball, swiped from his stepdad's trophy case, was a prized possession—autographed by the real Babe Ruth. If not recovered, as Scotty recalls in the narration, "My life was over. I had hit a chunk of priceless history into the mouth of a monster."

The back end of the movie focuses on the boys' ingenious ways of trying to rescue the keepsake ball. It concludes with a surprising and poignant meeting with James Earl Jones, who plays the mysterious owner of the junkyard and its larger-than-life canine.

👏 **CHEERS:** Given that the movie is a memoir of childhood, Evans does a nice job of leaving intrusive adults out of the story. The boys play sandlot ball without grownup coaches ruining the fun, as it should be. Only Jones—whose brief role is critical to the plot—has much impact. Karen Allen and Denis Leary, who play Scotty Smalls' parents, are little more than background characters.

🍎 **BET YOU DIDN'T KNOW:** Actor Pablo P. Vitar, who briefly portrays the grown-up Benny Rodriguez as a player for the Dodgers, is the real-life brother of actor Mike Vitar, who portrays Benny as a child.

83 THE LIFE AND TIMES OF HANK GREENBERG (1999—PG)

SPORT: Baseball ★ STARS: Hank Greenberg, Charlie Gehringer, Walter Matthau DIRECTOR: Aviva Kempner

Aviva Kempner spent 13 years researching, editing, and writing *The Life and Times of Hank Greenberg*. That is a long time to spend immersed in one story, but this is quite a story and Kempner tells it extremely well in this meticulously constructed documentary.

Greenberg was a power-hitting first baseman with the Detroit Tigers through the 1930s and '40s. He was also Jewish, which made him a symbol of hope for some and target of hate for others.

When most people think of bigotry in baseball, they think of Jackie Robinson, who broke the color line in 1947, but Greenberg faced many of the same issues more than a decade earlier.

There was rampant anti-Semitism in the 1930s, and it was not uncommon for Jewish people, especially those involved in sports, to change their names. Greenberg, a strapping six-foot-four, 210-pounder from the Bronx, refused to do so. He maintained his name— Henry Benjamin Greenberg—and his pride in his ethnic identity made him a hero in the Jewish community.

Greenberg understood the responsibility that rested on his shoulders. He did not resent it. Rather, he said, it made him work harder. "If

I failed," he said, "I wasn't a bum, I was a Jewish bum." He would not allow that to happen.

Greenberg compiled a .313 batting average in 13 major league seasons. He won four home run titles and was the American League's Most Valuable Player in 1935 and 1940. He made the first serious run at Babe Ruth's single season home run record, banging out 58 homers in 1938. He may have broken the mark of 60 if opposing teams did not stop pitching to him late in the season.

The film suggests the strategy was motivated—at least in part—by anti-Semitism. Greenberg, in an interview recorded before his death in 1986, shrugs off the idea. He says he had a fair shot at The Babe's mark and just fell short. He says he didn't think that much about hitting home runs. He felt the more important statistic was runs batted in and he took pride in the fact that he led the major leagues in that category three times.

"I liked being the guy who comes up in the clutch, changes the ball game, makes all the difference," says Greenberg, who had a career-high 183 RBIs in 1937.

Kempner establishes the context for Greenberg's story through interviews with former teammates and opponents, broadcasters and journalists, and assorted celebrities who bring their own perspective to the film. Some documentaries err in using actors and politicians as talking heads simply for their star power. Kempner uses her celebrities very selectively and effectively. They all have something to say about Greenberg and they say it well.

Actor Walter Matthau says, "Growing up in the Bronx in the '30s, you thought of nothing but baseball and Hank Greenberg." Matthau says after he became a success in Hollywood he joined the Beverly Hills Tennis Club, not because he wanted to play tennis, but because

he wanted to have lunch with Greenberg, who often dined there.

Attorney Alan Dershowitz talks about Greenberg's style of play, his imposing physical stature, and the power he displayed on the field. "He was," Dershowitz says, "what they said Jews could never be." Hammering Hank's success was shared by Jewish people of all ages who still were trying to fit in. "Baseball was our way of showing that we were as American as anyone else," he says.

As Senator Carl Levin says: "Because [Greenberg] was a hero, I was a little bit of a hero, too."

Actor Michael Moriarty, the grandson of a former major league umpire, talks about his grandfather telling stories about the bigotry he saw directed at Greenberg, the taunts of "sheenie" and "kike" coming from the bleachers and the opposing dugout. Greenberg never complained and never responded except to lash out a few more hits and help the Tigers to four World Series.

Kempner's film shows Greenberg to be a gentle and modest man who loved baseball, loved his country (he served in the Army during World War II and missed three seasons right in the prime of his career), and left the game with no regrets.

A highlight of the film is a moment from the 1947 season, Greenberg's final season in baseball. He is playing with Pittsburgh in the National League as Jackie Robinson is in his rookie year with the Brooklyn Dodgers. The scene shows Greenberg, who is on his way out, encouraging Robinson, who is just breaking in, to ignore the bigotry and play his game. It is as if a torch is being passed.

66 **BEST LINE:** Walter Matthau: "Hank Greenberg was part of my dreams, part of my aspirations. I wanted to be Hank Greenberg."

84 TIN CUP (1996—R)

SPORT: Golf ★ **STARS:** Kevin Costner, Rene Russo, Cheech Marin
DIRECTOR: Ron Shelton

Eight years after *Bull Durham*, Kevin Costner and writer-director Ron Shelton teamed up again in *Tin Cup*. But while the original film was a tape-measure home run, the follow-up dies on the warning track.

This time the sport is golf. Shelton again tries for a quirky romantic comedy, but it doesn't work as well. Costner is a dissipated rascal named Roy "Tin Cup" McAvoy, who once was a promising golfer but now runs a broken-down driving range. McAvoy is similar to Crash Davis, the minor league catcher and philosopher Costner played in *Bull Durham*. He just isn't as much fun.

"Greatness courts failure," McAvoy says, explaining why he is content to live in a Winnebago with friend and part-time caddy Romeo (Cheech Marin) rather than pursue a career on the PGA tour. McAvoy has enough game to play with the big boys—and Costner's sweet swing makes you believe it—but he is an undisciplined and none-too-bright screw-up who refuses to play it safe even when it is the right thing to do.

"The word 'normal' and him don't often collide in the same sentence," explains one of McAvoy's cronies. No one says why, exactly. We're just left to wonder.

McAvoy is perfectly content hanging out at the driving range, drinking beer, and counting armadillos until one day when Dr. Molly Griswold (Rene Russo) shows up asking for a golf lesson. She is blonde and willowy and, oh yes, she's dating David Simms (Don Johnson), who just happens to be a star on the PGA tour and McAvoy's former college teammate and nemesis.

Naturally, McAvoy falls for Molly. In an effort to prove he isn't a loser, he cleans himself up and enters the sectional qualifying for the U.S. Open. He makes it into the tournament and quicker than you can say "You Da Man," he is on the leaderboard, shooting a 62—the lowest score in Open history.

That's the problem with *Tin Cup*: Shelton takes ideas that worked in *Bull Durham* and inflates them to the point where you don't buy them anymore. In *Bull Durham*, Crash Davis sets a minor league home run record; in *Tin Cup*, McAvoy turns in the best round ever in a major championship. In *Bull Durham*, Susan Sarandon, the brainy love interest, is a teacher. Here, Sheldon ups the ante by making Molly a psychiatrist. In *Bull Durham*, Crash's romantic rival is a dim-witted teammate. This time his rival is a millionaire stud, exactly the kind of neatly packaged, corporate-sponsored, play-it-by-the-book pro McAvoy despises. So guess who winds up playing together in the final twosome at the Open? Guess who Molly roots for? It's all just a little much.

At the end, McAvoy and Simms are side by side in the 18th fairway, each man knowing a par will give him a tie for the lead. Simms chooses to play it safe. He decides to hit short of the water hazard and play for par. McAvoy sneers as Simms reaches for an iron.

"Fifteen years on the tour and you're still a ------- -----," McAvoy says.

"Thirteen years on the driving range and you still think this game is about your testosterone count," Simms replies.

McAvoy, ever the thick-headed gambler, grabs his three wood. His shot hits the green 240 yards away, but it rolls back into the water. Rather than accept the two-stroke penalty and take a drop, McAvoy hits the three wood again from the same spot with the same result. He tries again. And again. Five shots in all, each one ending up in the water.

Finally on the sixth try, McAvoy not only clears the water but knocks the ball in the hole. The crowd goes wild, McAvoy walks onto the green to a standing ovation. Any sane person is thinking, "Why?" His score for the hole is a 12. He has just thrown away a chance to win the U.S. Open.

How, exactly, is this heroic?

It's not. Frankly, it is dumb.

Tin Cup has some fun moments, but for a Costner-Shelton collaboration it spends a lot of time in the rough.

🍎 **BET YOU DIDN'T KNOW:** The older couple whom Simms (Johnson) berates for asking for an autograph are actually Bill and Sharon Costner, Kevin's parents. The boy who is with them is Joe, Kevin's son.

📢 **CASTING CALL:** John Leguizamo was considered for the part of Romeo, but the jaded and jowly Marin was a better choice.

85 ANY GIVEN SUNDAY (1999—R)

**SPORT: Football ★ STARS: Al Pacino, Cameron Diaz, Jamie Foxx
DIRECTOR: Oliver Stone**

The term "any given Sunday" originated with Bert Bell, the NFL's commissioner from 1946 until his death in 1959. Bell said his goal was to create enough competitive balance that "on any given Sunday" the worst team in the NFL could knock off the best.

If Bell could see the NFL today, he would be pleased, because that's how things are in this era of parity. But he would have a heck of a time figuring out what's going on in the film that hijacks his famous phrase. *Any Given Sunday* isn't for old-school football guys.

One of the team owners is a gorgeous blonde (Cameron Diaz) who strolls through the locker room coolly surveying the prime cuts of beefcake. The team doctor is an evil snake (James Woods) who sends players onto the field with injuries that likely will cripple them for life. The players themselves are hopped-up sociopaths who live by a code that is somewhere between Caligula and Scarface.

And speaking of Scarface, the head coach is Al Pacino minus Tony Montana's white suit and machine-gun, but still in a wild-eyed rage, tossing F-bombs in every direction.

Welcome to the world of professional football as seen by director Oliver Stone.

Anyone familiar with Stone's filmography— *Platoon*, *Wall Street*, *The Doors*, *JFK*, *Natural Born Killers*, etc.—knows his work is provocative, political, and often brutal. That's why the NFL refused to give this film its official blessing, which meant Stone could not use the names or uniforms of real NFL teams. He was forced to create his own league, but since the film is fiction it didn't really matter.

Any Given Sunday focuses on the Miami Sharks, a team owned by Christina Pagniacci (Diaz) and coached by Tony D'Amato (Pacino). She inherited the team from her late father, who was in it for the love of the game. Christina is as different from her father as Jimmy Choos are from Chuck Taylors. She is an ice-cold businesswoman who is into making money.

Christina's father loved Tony, who was his coach and general manager, but she thinks Tony is a fossil whose time has passed. She hires a new offensive coordinator—a whiz kid named Nick Crozier (Aaron Eckhart) who calls plays from his laptop—and it is clear she is grooming him to be the next head coach. Tony sees the writing on the wall.

The clashes between Christina and Tony have the ring of truth. In today's NFL, with new owners taking over and old-timers trying to hang on, such friction is common. Tony tells Christina her father wouldn't approve of the way she is running things. She snaps, "Why the hell do you think my father put me in charge, you bullheaded moron?"

When the film opens, the Sharks are on their way to their third consecutive loss. Veteran quarterback Cap Rooney (Dennis Quaid) is injured and replaced by his untested backup, Willie Beamen (Jamie Foxx). Beamen is so nervous he throws up in the huddle, a scene Stone replays for us several times, including once in slow-motion.

But Beamen recovers to play well. He plays so well, in fact, he becomes an overnight sensation, "Steamin" Willie Beamen. He is on magazine covers and partying in South Beach with groupies tugging at his Armani sleeves. He gets caught up in his fame, which puts him at odds with his coach and teammates.

"You're the ------- quarterback," Tony tells him. "You know what that means? It's the top spot, kid. It's the guy who takes the fall. It's the guy everybody's looking at first, the leader of a team who will support you when they understand you, who'll break their ribs and their necks for you because they believe, because you make them believe."

Cap, of course, is that kind of quarterback. He and Tony have been together for years. They won a title with the Sharks, but that was years ago. Tony is faced with a tough decision: Does he put Cap back in the lineup or does he stay with Beamen, whose hot streak carried the Sharks to the playoffs? What will it be, Tony, loyalty or winning?

Any Given Sunday has the familiar Oliver Stone touches—some call them conceits— such as shots cut so fast and in such bizarre fashion they feel like an LSD trip. (We're still trying to figure out how the Ben-Hur chariot race figures into the Sharks season.)

Opinions on the film break down along generational lines. Most establishment types pan it. Two-time Super Bowl-winning coach Mike Shanahan, who allowed Pacino to shadow him to prepare for his role, said, "I think [Pacino] is great, but I thought the movie was horrible."

Younger audiences, including current NFL players, like it. Detroit Lions running back Reggie Bush calls *Any Given Sunday* "The most realistic and up-to-date football movie I've seen." Veteran running back Rudi Johnson said, "This movie gives you a different look at football. It covers a lot of angles about what it's really like to be a player, on and off the field."

Mike Leach, head coach at Washington State, is somewhere in the middle. "The opening scene did a great job of showing what it's like on the sideline during a game," Leach said. "The emotion, the confusion, the coach trying to keep it together, that's how it is. I didn't care for the movie as a whole—it went off in too many tangents—but those first few minutes my hair was standing up, it felt so real."

👍 **CHEERS:** Two of the best performances are delivered by Pro Football Hall of Famers Jim Brown, who plays a coach, and Lawrence Taylor, who plays an aging All-Pro.

86 BEST IN SHOW (2000–PG-13)

SPORT: Dog Show ★ STARS: Christopher Guest, Catherine O'Hara, Eugene Levy
DIRECTOR: Christopher Guest

Some people will ask why a satire about dog shows is included in a book about sports movies. Well, dog shows are competitions. Most newspapers run the results in the sports section alongside the baseball box scores and the racing form.

Big events, such as the Westminster Kennel Club show at Madison Square Garden, are nationally televised and the highlights are shown on ESPN's *SportsCenter*. Well-known sportscasters like Joe Garagiola call the action. For all those reasons, we feel *Best in Show* belongs in the book.

OK, there's another reason. *Best in Show* made us laugh. So sue us.

It is a breezy 89-minute spoof from the same folks who produced the faux documentaries *This is Spinal Tap, Waiting for Guffman,* and *A Mighty Wind.* This time director Christopher Guest turns his Second City ensemble loose in the dog show world with all of its brushing and fussing, and the results are predictably amusing.

As in his other films, Guest provides the actors with just the sketchiest of scripts. For *Best in Show,* Guest drew up a 15-page outline describing the characters and their assorted quirks, then he put the actors together with the dogs and let them wing it while the cameras rolled. Only the most gifted improv performers could make it work, but like a great coach, Guest knows his players and this is what they do best.

The first part of the film introduces the owners and dogs that will compete at the Mayflower Dog Show in Philadelphia. There is Harlan Pepper (Guest), a fishing shop owner from Pine Nut, N.C., and his bloodhound, Hubert; Stefan Vanderhoof (Michael McKean) and Scott Donlan (John Michael Higgins), a gay couple from New York and their prized Shih Tzu, Agnes; and Gerry and Cookie Fleck (Eugene Levy and Catherine O'Hara) from Fern City, FL, and their Norwich terrier, Winky.

The Flecks are the most fun. Gerry is an uptight clothing salesman who was born (literally) with two left feet. "When I was a kid, they called me 'Loopy' because I walked in little loops," he says. Cookie is a past-her-prime party girl with a trail of old boyfriends who keep popping up with tales of wild escapades ("Remember when we did it on the roller coaster?"). Of course, that makes poor Gerry squirm.

All of this is just building to the big dog show, which Guest stages very well. To make it authentic he insisted the actors take dog handling classes so when they walk the dogs around the ring, they move with the ease and assurance of professionals. The parading and posing for the judges, the primping and perfuming backstage, it all looks like the real thing.

A running gag is the clueless commentary of TV host Buck Laughlin (Fred Willard). The character is loosely based on Garagiola, the former baseball announcer who worked the Westminster show for several years. Laughlin turns the ex-jock into a macho buffoon who often leaves his strait-laced partner Trevor Beckwith (Jim Piddock) appalled and exasperated.

Laughlin blurts out things like: "And to think that in some countries these dogs are eaten."

Speaking of poodles, he wonders aloud, "How do they miniaturize dogs anyway?" When the bloodhound enters the ring, Laughlin says, "Why don't they put a Sherlock Holmes hat on him and stick a pipe in his mouth? That would get the crowd going." Beckwith replies dryly, "I don't think so."

Before the finals, Cookie sprains her ankle so Gerry has to take Winky into the ring. Laughlin says, "Am I nuts? Something's wrong with his feet. He's got two left feet." Beckwith says, "I never thought I'd find myself saying this, but you're right."

Gerry and Winky walk off with the blue ribbon. Improbable, sure, but we were having so much fun at that point, we didn't care.

📖 **CHEERS:** Willard is hilarious doing a riff on his character from the TV series *Fernwood 2Night.* He won the Boston Society of Film Critics Award for Best Supporting Actor.

🍎 **BET YOU DIDN'T KNOW:** *Best in Show* was voted one of the 50 Greatest Comedies of All-Time by *Premiere Magazine.*

87 GLORY ROAD (2006–PG)

SPORT: Basketball ★ **STARS:** Josh Lucas, Jon Voight, Derek Luke
DIRECTOR: James Gartner

The most surprising thing about *Glory Road* is that it took so long to make this momentous event into a movie. It was 1966 when Texas Western became the first school to win the NCAA Tournament with an all-black starting five. The Miners' win—over an all-white Kentucky squad—is often cited as *the* landmark moment in advancing the integration of college sports, particularly in the South.

So why did it take 40 years to put the Earth-changing event onto film? We're not sure. But we've got another problem. With four decades to get the story right, how did producer Jerry Bruckheimer and first-time director James Gartner manage to get so many things so wrong?

Listen, we're not quibblers. We realize that many of our favorite historical movies– *Miracle, Chariots of Fire, Eight Men Out*— take small liberties for dramatic purposes. We can live with Vince Papale being moved from the suburbs to South Philadelphia in *Invincible* because, hey, what's the difference? Maybe Jim Morris wasn't discovered exactly as *The Rookie* lays out, but that's not the important part.

But Disney's *Glory Road* takes a compelling—and significant—story, and performs more renovations than *Extreme Makeover: Home Edition*. In the process, it becomes just another decent sports movie— entertaining, yes, with well-choreographed game scenes and capable acting. Good enough to make this book. But the potential here was for a Top 20 finish. And the bigger sin is that any moviegoer who doesn't know the facts will likely accept what's laid out over 118 minutes in *Glory Road* as a historical document.

Among the more glaring errors:

• The movie portrays Texas Western (now Texas-El Paso) winning the national championship in Coach Don Haskins' first season. In fact, Haskins arrived five years before the 1966 title. Why change that fact?

• *Glory Road* asserts that Haskins (Josh Lucas) broke barriers by recruiting black players. In fact, the school began recruiting blacks in 1956. One player Haskins inherited in 1961 was point guard Nolan Richardson, who went on to become an NCAA Tournament-winning coach himself at Arkansas.

• The movie shows Haskins telling the team of his decision—as a political statement— to start five black players and only use his seven blacks (and not the three whites) in the title game against Adolph Rupp's Kentucky Wildcats. Wrong! Haskins often started an all-black lineup that season and the seven who played that game happened to be his seven best players. (Beyond that, there were five non-black players on the squad—not three as the movie showed.) Haskins has always insisted it did not occur to him he used an all-black rotation until people told him afterward. "I wasn't trying to send a message," he said. "I was trying to win a game."

We've got other misgivings, including the portrayal of Kentucky's players as sore losers (by all accounts they were gracious) and the hip-hop basketball supposedly played by Texas Western when, in fact, Haskins engineered a numbing, slow-down style. It all adds up to one big enchilada of complaints.

On the other hand. . . .

There's enough to like in *Glory Road*. The movie takes a look at 1960s American racism through the prism of sports, much as *Remember the Titans* did six years earlier—upbeat Motown music and all. One of Haskins' players gets beat up in a men's room; others have their motel rooms trashed while the team is on the road. Most viewers of this film will be too young to recall days when episodes like those really happened, and they are certainly worth remembering.

The story, while predictable, builds well. The game scenes work—especially the famous moment in the Midwest regional final when the Miners appear to lose as Kansas star Jo Jo White hits a jumper with seconds left in the first overtime. The referee rules that White's foot was touching the out-of-bounds line. The moment is well shot and dramatic.

The acting here is better than par. Derek Luke, terrific in *Friday Night Lights* and *Antwone Fisher*, once again shows himself as a powerful young actor as player Bobby Joe Hill. And Jon Voight, in a small but critical role as Rupp, portrays a nasty coach without falling into an angry caricature. Our only problem is the prosthetic nose and ears Voight wears, which make him appear more like Ross Perot than "The Baron of the Bluegrass." If you thought Voight looked silly in a Howard Cosell mask in *Ali*, wait until you get a load of this.

☞ **DON'T FAIL TO NOTICE:** The scene where Haskins pulls into the gas station and is asked by the attendant, "Hey, you want me to fill this thing up?" The gas jockey happens to be the real-life Don Haskins.

📢 **CASTING CALL:** Ben Affleck was first hired to play Haskins, but pulled out because of scheduling conflicts.

📀 **CHEERS:** The *Glory Road* DVD has good extras and interviews. Make sure you catch the conversation with former NBA coach Pat Riley, who starred for that losing Kentucky team and proclaims that Haskins and Texas Western "wrote the Emancipation Proclamation of 1966."

★ **SPORTS-ACTION GRADE:** B-minus. As Barra wrote, that's not how they really played back then. But there are rousing moments during the games.

❝❝ **BEST LINE:** Haskins, while trying out recruit Orsten Artis: "Brother, without a little work, I don't think you can get past an old timer like me."

Artis: "Get past you? I will go past you, through you, over you, under you, around you. As a matter of fact I will spin you like a top, twist you in a pretzel, eat your lunch, steal your girl, and kick your dog at the same time."

🍎 **BET YOU DIDN'T KNOW:** Many of the basketball scenes were filmed at a high school in suburban New Orleans that had been devastated by Hurricane Katrina. The gym at Chalmette High was rebuilt prior to shooting the movie.

🥸 **"I KNOW THAT GUY":** Red West, who plays trainer Ross Moore, is a veteran character actor once best known for being Elvis Presley's close friend. You've probably seen him as the grieving dad in *The Rainmaker*, or the Southern wag in *Road House* who gets to utter that movie's best line: "I got married to an ugly woman. Don't ever do that. It just takes the energy right out of you. She left me, though. Found somebody even uglier than she was."

SPORT: Baseball ★ STARS: James Stewart, June Allyson
DIRECTOR: Sam Wood

When MGM announced it was making a movie about the life of Monty Stratton, some of Hollywood's biggest stars expressed an interest. Gregory Peck and Van Johnson read for the part. Ronald Reagan wanted the role, but he was under contract to Warner Brothers and they would not let him work for another studio.

It is easy to see why so many actors wanted the part. It is the true story of a shy country boy who becomes a star pitcher with the Chicago White Sox, only to have his career cut short by a hunting accident which results in the loss of his leg. Outfitted with a prosthesis, Stratton comes back to pitch winning baseball again, albeit in the minor leagues.

It is an inspiring tale, it had a solid script (writer Douglas Morrow won the Oscar) and it was directed by Sam Wood, who had success with similar material in *The Pride of the Yankees*, the Lou Gehrig story which was nominated for five Academy Awards in 1942.

The real Monty Stratton was technical advisor on the film and he wanted Jimmy Stewart to play him. He had nothing against the other actors; he just thought the lanky, soft-spoken Stewart was a better fit. Once Stewart tested for the role, the producers agreed. It was a wise decision.

Johnson and Reagan weren't the right body type to play the spindly 6-5, 180-pound Stratton. Peck was lean enough, but he was too urbane. Stewart, with his scarecrow physique and aw-shucks manner, was perfect. He didn't throw the ball like a big league all-star, which Stratton was, but moviegoers weren't sticklers for that sort of thing back then.

The result is a satisfying little film that launched the on-screen partnership of Stewart and leading lady June Allyson. They paired so effectively as Monty and Ethel Stratton, they later were cast as husband and wife in *The Glenn Miller Story* (1953) and *Strategic Air Command* (1955). In the latter film, Stewart plays a major leaguer recalled to active duty with the Air Force. He wears his baseball cap while piloting his plane.

When we first see Monty Stratton (Stewart), he is pitching semi-pro baseball in Texas for three dollars a game. A broken-down old scout who happens to be riding a freight train through town spots Stratton and takes him to the White Sox spring training camp for a tryout. Manager Jimmy Dykes (playing himself) works him out and signs him to a contract.

In reality, Stratton bounced back and forth between the White Sox and the minors for the better part of three seasons, but in the film, he has a much faster rise to stardom. He is shown getting rocked in his major league debut and being sent down to Omaha, but when he rejoins the White Sox, he immediately starts mowing down the best hitters in the American League.

Stratton did not really hit his stride until 1937, his fourth season, but when he did, he was very good. That year, he was 15-5 with a 2.40 ERA, second in the American League behind only Lefty Gomez (2.33) of the Yankees. Stratton had five shutouts and 14 complete games in his 21 starts. He was named to the American League All-Star team and at age 25, he had a bright future.

The film makes that point. In one scene, Dykes is watching from the dugout and says, "I wouldn't trade him for any pitcher in baseball."

Stratton returns to his Texas farm in the off-season. One day while hunting rabbits, he stumbles and accidentally shoots himself in the leg. The wound becomes infected and the doctors are forced to amputate. Ethel (Allyson) is shown agonizing as she is asked to sign the paper giving the permission to operate.

"We have to remove the leg to save his life," the doctor says.

"His legs are his life," she replies.

She signs the paper, knowing she has ended her husband's major league career. When the doctor leaves the room, she puts her head on the desk and cries.

When Stratton returns to the farm, he is bitter and self-pitying. Ethel brings him a stack of get-well messages and he pushes them away, saying, "If you like 'em so much, you read 'em." When their baby boy takes his first step, Monty snaps, "What's so wonderful about that? He's got two legs, doesn't he?"

With Ethel's steadfast support, Monty slowly comes around. He agrees to put his crutches away and wear the prosthesis, which he has resisted. One day she coaxes him into having a catch. Pretty soon, he is throwing strikes into a bucket he nailed to the barn. Without telling anyone, he arranges to pitch in a minor league all-star game just to see if he can still do it. He wins the game even though the opposing team tries to take advantage of his bad leg by bunting on him.

The film ends with Stratton throwing out the final hitter and walking off as the crowd cheers and the narrator says, "Monty Stratton has won an even greater victory. He has shown what a man can do when he refuses to admit defeat."

It was true in Monty Stratton's case. He actually returned to pitching in the minor leagues and in 1946, eight years after losing his leg, he won 18 games for Sherman in the Class C East Texas League.

CHEERS: Jimmy Stewart is well cast as Monty Stratton and he is convincing in his struggling attempts to walk after his injury.

89 ▶ VARSITY BLUES (1999–R)

SPORT: Football ★ **STARS:** James Van Der Beek, Jon Voight, Amy Smart
DIRECTOR: Brian Robbins

Varsity Blues is *Friday Night Lights* made for the MTV audience. It takes the same slice of American life (Texas high school football) and the same point of view (the values are corrupt, the kids are exploited), but it adds more booze, sex, and rock and roll. It also rolls back the intelligence factor by half.

Friday Night Lights had the advantage of being a true story with real characters who were first introduced in Buzz Bissinger's best-selling book. The coaches and players from the dusty back roads of Odessa, TX., had the feel of authenticity. The characters in *Varsity Blues* are fictional, but that isn't the problem. Fictional characters can feel real. Rocky Balboa, Norman Dale, Happy Gilmore. Well, maybe not Happy Gilmore, but you get the idea.

In *Varsity Blues*, the characters are more like a checklist. Stud quarterback, check. Vixen cheerleader, check. Nasty coach, check. Fat, dumb lineman, check. Let's see, did we miss anything? Oh yeah, we need some clueless parents who are still living out their own high school fantasies through their children. Got 'em. Check and double check.

In *Varsity Blues*, the only interesting character is Jonathan "Mox" Moxon (James Van Der Beek) but he is enough to keep you involved. In football-crazed West Canaan, Mox is unusual—heretical, almost—in that he is ambivalent about the game. He likes to play, but he is smart enough to realize coach Bud Kilmer (Jon Voight) has taken his win-at-all-costs philosophy to unhealthy extremes.

Kilmer is a living legend in West Canaan, thanks to his 23 district championships. The high school stadium bears his name and there is a statue of him behind the end zone. He has coached several generations of West Canaan Coyotes, which means he coached most of the fathers of the kids who are playing for him now. Mox's father (Thomas F. Duffy) is one of them, and he remains a loyal disciple who believes the coach can do no wrong.

Mox is the second-string quarterback buried behind all-state superstar Lance Harbor (Paul Walker), which means he has virtually no chance of ever taking a snap. He is so detached, he sits on the bench reading novels that he hides inside a playbook. You see, Mox is a brain who hopes to attend Brown University on an academic scholarship.

two. For one thing, Mox finds that he likes the perks of being a star. He is a nice guy—in fact, his teammates tease him about being a goody-goody—and with his Ivy League brainpower, he should be above it all, but when girls start covering his car windshield with rollouts that say "Mox is a fox," well, he kind of digs it. He even takes a few teammates out for a night of shots and beers at the local strip joint. For awhile, you actually wonder if he is going over to the other side.

Van Der Beek, who became a star on the TV series *Dawson's Creek*, is quite effective portraying a kid who suddenly finds himself pulled in a lot of unexpected directions. His girlfriend Jules (Amy Smart) serves as his conscience, although she is more than a little judgmental. One night after leading the team to a

"Never show weakness. The only pain that matters is the pain you inflict." —COACH KILMER

But all that changes when Lance goes down with a knee injury and suddenly Mox has to put down *Slaughterhouse-Five* (yes, he was reading Kurt Vonnegut while the game was going on), pick up his helmet, and save the season. Kilmer doesn't like Mox. He has seen the flicker of skepticism in his eyes and, besides, he gets all A's in class. Shoot, who can trust a kid like that? Not Bud Kilmer, that's for dang sure.

Mox confirms Kilmer's worst fears. He pulls out the victory, but he does it with a razzle-dazzle play he improvises in the huddle. While the team and fans mob Mox, Kilmer gives him the evil eye. You can see where this is headed. Mox is good enough to keep the team winning, but he and Kilmer are headed for a showdown. And given the nature of the film, you know Mox will win out (which he does).

But director Brian Robbins tosses in a few wrinkles which offer a welcome surprise or

66-3 victory, Mox is stopped by a radio reporter. Jules rolls her eyes as Mox does the interview.

"You're enjoying this, aren't you?" she says.

"Well, yeah," he says.

She offers a chilly "Look, I gotta go" and takes off, leaving Mox standing there, feeling like he should apologize but not quite sure what he is apologizing for. We get her point. Their quiet, down-to-earth, boy-girl dynamic is being intruded upon, but, geez, the guy is entitled to enjoy his success a little bit, isn't he?

One scene tells you all you need to know about Mox. The night Lance suffers his knee injury, Mox drives Lance's girlfriend, bombshell cheerleader Darcy Sears (Ali Larter), home. She starts flirting with Mox in the car and later comes on to him wearing a bikini made of whipped cream. With Lance's football career shot, Darcy is looking for another ticket out of town.

Mox pulls away, saying: "I can't do this."

Darcy is crushed. "Great, now I can look forward to a life working at the Wal-Mart," she says.

Such are the career options in West Canaan.

❝❞ BEST LINE: Coach Kilmer talking to Mox: "You got to be the dumbest smart kid I know."

👍 CHEERS: Van Der Beek is a likeable actor who brings credibility to a film where most of the characters are made out of cardboard.

☞ DON'T FAIL TO NOTICE: Mox wears uniform No. 4. That's because James Van Der Beek's favorite NFL player was Brett Favre.

90 DIGGSTOWN (1992–R)

SPORT: Boxing ★ STARS: James Woods, Louis Gossett, Jr., Bruce Dern
DIRECTOR: Michael Ritchie

Director Michael Ritchie has a hit-and-miss history with sports movies. He started with two good ones (*Downhill Racer* and *The Bad News Bears*) but stumbled badly with the next two (*Semi-Tough* and *Wildcats*). His fifth effort, *Diggstown*, falls somewhere in between.

A combination boxing and caper film, *Diggstown* has an interesting premise and at times it is on the verge of being terrific, but there are too many shifts in tone and too many layers to the con.

Other sports films balance comedy and drama effectively—*The Longest Yard* is a good example—and that's what Ritchie tries to do here, but he's only partly successful. *Diggstown* veers from slapstick humor to ugly violence and back again, and it does so in a way that is hurried and often jarring.

Still, James Woods is very funny as con man Gabriel Caine and Louis Gossett Jr. brings just the right note of "here-we-go-again" cynicism to his role as "Honey" Roy Palmer, who supplies the muscle for Caine's latest scam.

The film opens in prison where Caine is about to be released. He has pocketed a cool $50 grand during his time behind bars, arranging (and betting on) fights among the prisoners. Caine is one of those 24/7 hustlers who talks fast and knows how to work an angle.

Caine's cellmate, Wolf Forrester (boxer Randall "Tex" Cobb), tells him about Diggstown, a little hamlet in Georgia where they hold what are known as "cash fights." There is no TV or newspaper coverage, no records kept. These fights exist for one reason: betting. They are the equivalent of cockfights, except with people. It is a nasty business, but it is made to order for a sharpie like Caine.

Diggstown is controlled by John Gillon (Bruce Dern), a rich guy who loves the action and does whatever is necessary to win. Gillon got control of the town by fixing the last fight of Charles Diggs—the boxer for whom the town was named—and betting against him. To insure his bet, Gillon doped Diggs, which resulted in a savage beating that left Diggs in a vegetative state. He now sits in a wheelchair staring into space while Gillon runs the town.

Flush with his prison money, Caine heads for Diggstown with a card shark named Fitz (Oliver Platt). They offer Gillon a bet: they will find a boxer who will meet and defeat 10 Diggstown men in a 24-hour period. Caine already has a boxer in mind—Palmer (Gossett), a heavyweight who fought for Caine in the past. The trouble is it was the distant past. He is now 48 years old, and he hasn't fought in more than a decade.

Palmer is appropriately skeptical when Caine tries to enlist him in the scheme. But talking is what Caine does best, and he's able to convince Palmer that the town is full of hicks with no real boxing ability, which means they'll be easy pickings for a professional, even a rusty one such as Palmer. Of course, Caine knows Gillon will have a trick or two up his sleeve, so the real question is: which hustler gets hustled?

The early bouts are played mostly for laughs. Palmer knocks off the first few opponents with little difficulty. Gossett has just the right look: He isn't in his prime, but he's still an imposing figure in a pair of trunks and he's clearly too much for the ham-and-eggers who normally fight in Diggstown.

As Palmer piles up the wins, the maneuvering starts between Caine and Gillon. Caine wants to raise the ante, Gillon wants to change the rules. Soon they are discussing what, exactly, is a Diggstown man? Did he have to be born there? Live there? What if he just happens to show up that day? Yes, we're talking ringers. (You get the feeling Gillon has Lennox Lewis on speed dial.)

"Do I look stupid to you?" Fitz asks as Gillon tries to expand the boundaries of Diggstown to include roughly two-thirds of North America. "Or have you people been breeding too close to the gene pool again?"

The battle of wits between Caine and Gillon becomes the real heavyweight struggle, with each man trying to outfox the other. Gillon does succeed in back-loading his lineup card so Palmer's opponents get progressively bigger and tougher. Bleeding and sucking wind, Palmer starts looking his age and Caine has to do some fast thinking to save the day.

In a movie full of twists, Ritchie saves the best for last, so we don't want to ruin it. Stick with *Diggstown* to the end and you'll be rewarded. It is one of the better double-crosses you are likely to see.

👍 **CHEERS:** Ritchie, a director known for his keen eye, really delivers the look and feel of a small town fight club. The visual elements make up for some of the weak spots in the narrative.

★ **SPORTS-ACTION GRADE:** C. Gossett is fine as the ex-heavyweight, but some of his opponents take the clumsy country boy thing to extremes.

91 THE CLUB (1980—NR)

SPORT: Australian Rules Football ★ **STARS:** Jack Thompson, Graham Kennedy, Frank Wilson ★ **DIRECTOR:** Brian Robbins

The Club, a film about Australian Rules Football, is next-to-impossible to find in the United States, but if you can track down a VHS copy, you will be rewarded with an entertaining trip to the Land Down Under.

Think *North Dallas Forty* with an Aussie accent. *The Club* is a look inside the lives of the coaches and players, the bickering and backstabbing of the front office, the jealousies that destroy teams regardless of sport or hemisphere. It is well-acted—especially by Jack Thompson, who plays the idealistic coach Laurie Holden—and it is often very funny.

> "Do you remember your first game?"
> —TED
> "Not all that well."
> —LAURIE
> "I do. I was there and we all knew we were seeing our next great star."
> —TED

Americans will be amused to see star players holding out for more money, owners meddling with coaches, and coaches telling players, "The sportswriters say we're a lot of hacks and has-beens. Let's make them eat their words." Gee, it's all so familiar.

There are semantic differences. When the star player tells off the coach, he says, "Get stuffed." That's Aussie for, well, you know. The film was made for Australian audiences, so they don't bother to explain the rules of the game, but that's OK. An American viewer won't have trouble following the action. It's a cross between soccer and rugby.

The Club spends a fictional season with Collingwood, a real club in the Victorian Football League. The owner, Ted Parker, (Graham Kennedy), signs Geoff Hayward (John Howard), a highly touted player from Tasmania without ever consulting Holden, as the head coach.

Holden is irritated by Parker's actions and the other players are resentful of Hayward's fat contract. In the first game, Hayward whiffs on his first few scoring opportunities and Holden benches him. Enraged, Parker leaves his box to confront the coach, who tells him to take a hike.

"I won't be told how to run my team by the owner of a meat pie factory who never played a game of football in his life," Holden tells Gerry Cooper (Alan Cassell), a member of the executive committee.

Collingwood loses its first five games, and the committee—a nasty basket of vipers—begins plotting to dump Holden. Jock Riley (Frank Wilson) is Holden's most vocal critic. He has resented Holden ever since Holden succeeded him as coach. Riley believes Holden went behind his back to get the job, which isn't true, but the old timer has held a grudge ever since.

It doesn't help that Riley and Holden are the two most celebrated players in Collingwood history. Riley holds the club record for most games played (282). Holden was on his way to breaking the mark when a knee injury ended his career. "I wasn't sorry to see it," Riley says with a wicked smile.

Unlike Riley, Holden is a decent fellow who loves and honors the game. He is infuriated by Hayward, who comes late to practices and shows up stoned for a game. When Holden confronts him, Hayward blows him off, saying he doesn't care about football. "It's a lot of macho --------," he says.

Holden has seen glimpses of the kid's talent and he is determined to unlock his potential. When Hayward says, "Football bores me," the coach tells him he's full of it.

"You know what you are? You're scared," Holden says. "You were a star in Tasmania. Big deal. Anyone who's not cross-eyed or bandy-kneed could be a star in Tasmania. You're in the big leagues now, Geoff, and you're scared. You're scared that you're not half the player you thought you were."

The message gets through and Hayward commits to Holden and the team. The coach demonstrates his confidence by putting Hayward at center and moving the aging team captain Danny Rowe (Harold Hopkins) to the backline. Hayward begins to play like an all-star and Collingwood becomes unstoppable, advancing to the championship game.

The Club was originally a successful play that debuted on the Melbourne stage in 1977. When it was made into a film, it reunited Thompson and director Bruce Beresford, who were coming off the hit *Breaker Morant*.

That courtroom drama set in the Boer War was Australia's entry in the 1980 Cannes Film Festival and earned honors for both Thompson (Best Supporting Actor) and Beresford (nominated for Best Director).

It seemed like a can't-miss: a popular story, a hot actor, a talented director. Yet *The Club* flopped on the big screen. It brought in less than $900,000 during a brief run in Australia, but it attracted a cult following when it aired on American TV. It isn't easy to follow all the dialogue—the Aussie accents are thick—but the theme of why athletes devote their lives to a sport, and each other, comes through very clearly.

🏴 **JEERS:** Sorry, but the scene where Geoff—stoned out of his mind—stops in the middle of a game to watch a seagull fly over the stadium takes his hippie jock too far.

🎬 **PIVOTAL SCENE:** Laurie Holden tells Jock Riley that he saw Riley's last game as a player. He was standing a few feet away when Riley deliberately slammed his knee into the face of a defenseless opponent.

"They said you retired out of remorse," Holden says, "but I know the truth. All the other teams said if you ever stepped on the field again, you'd be lucky to get off alive."

"It didn't scare me," Riley says.

"Not much it didn't," Holden replies.

"No one's ever called me a coward," Riley snaps.

"I'm calling you one now," Holden says.

⭐ **SPORTS-ACTION GRADE:** D. Beresford tries inter-cutting real game action with staged action involving the actors and it doesn't work. The difference between real players and actors trying to execute the same moves is obvious even to the untrained American eye.

COACH CARTER (2005–PG-13)

SPORT: Basketball ★ **STARS:** Samuel L. Jackson, Rob Brown, Robert Ri'chard
DIRECTOR: Thomas Carter

It is hardly an original story: the idealistic teacher/coach takes on a classroom/team full of problem kids and with a firm hand and an open mind turns their lives around. We've seen it in *Blackboard Jungle* (1955), *To Sir, with Love* (1967), *Stand and Deliver* (1988), *Lean on Me* (1989), and *Dangerous Minds* (1995)—just to name a few.

The film plays out the same way every time: Teacher/coach is appalled by the apathy of the school administration, which is prepared to write the kids off as a hopeless cause. The teacher/coach vows to whip the students/players into shape. His/her tough-love methods are met with hostility. There is a physical confrontation in which the teacher/coach prevails and the kids have that "Uh oh, Teach is a former Green Beret" epiphany.

Next thing you know, it is graduation day and all the kids who once appeared headed for the slammer are in their caps and gowns, thanking the teacher/coach for putting them on the right path. It doesn't matter if the film is based on a true story or if it is fiction, the dramatic arc and moral are always the same.

So what distinguishes the good ones? The lead, mostly.

Michelle Pfeiffer (*Dangerous Minds*) in a karate pose doesn't work. Sidney Poitier (*To Sir, with Love*) in boxing gloves? That doesn't work either. But Samuel L. Jackson slamming a Latino gang-banger against the gym wall? Yeah, that works. And for that reason, so does *Coach Carter*.

The film is not without its flaws. For one thing, it is much too long at two hours and 16 minutes. Also, there are too many subplots and secondary storylines. Each time the film builds up a little momentum, director Thomas Carter veers off in another direction and it seems to take forever getting underway again. But when Jackson is on the screen, *Coach Carter* comes to life.

"Jackson is the ideal star to deliver a lecture with fire, brimstone and panache, which he's demonstrated in everything from his DJ in *Do The Right Thing* and reformed hit man in *Pulp Fiction* to the besieged teacher in *187*," wrote Wesley Morris in the *Boston Globe*. "In *Coach Carter*, he delivers a castigating speech on the problem with kids who use the 'n' word on each other. Any owner of *Jackie Brown* will recognize the powerful irony in that."

Coach Carter is based on the true story of Ken Carter, a former hoops star who returns to his old high school in Oakland to rebuild the basketball program. The real Carter earned national publicity with his discipline, which included forcing players to sign a "contract" calling for them to maintain a 2.3 GPA (higher than the required 2.0), sit in the front row of class, and wear a jacket and tie to every game.

At first, the players—all inner city kids with major attitudes—sneer at his rules. They aren't impressed by the fact he was the school's last All-American and still holds the record for points, assists, and steals. But when Carter puts the mouthy Timo Cruz (Rick Gonzalez) up against the wall and announces, "I'm not a teacher, I'm the new basketball coach," they realize he means business.

Conducting tougher practices—with pushups and sprints as punishment for showing up late or talking back—Carter gets astonishing results. A team that was 4-22 the previous season becomes unbeatable. But with the newfound success comes arrogance. Players begin cutting class and skipping tests. When Carter finds out, he literally locks the gym and shuts down the basketball program, resulting in the forfeiture of several games.

Parents accuse Carter of ruining their sons' chances for a college scholarship. Even school administrators feel Carter has gone too far. The team was undefeated and bringing the mostly black and Hispanic community a welcome measure of pride. The school board votes to end the lockout, sending Carter into a rage.

"You need to consider the message you're sending the boys by ending the lockout," Carter says in the film. "It's the same message that we as a culture send to our professional athletes, and that is they are above the law. If these boys cannot honor the simple rules of a basketball contract, how long do you think it will be before they're out there breaking the law?

"I played ball here at Richmond High 30 years ago. It was the same thing then. Some of my teammates went to prison, some of them even ended up dead. If you vote to end the lockout, you won't have to terminate me. I'll quit."

When Carter returns to the school to clean out his office, he finds that although the administration has unlocked the gym, the players will not practice. Instead, they have turned the gym into a study hall. The same kids who were cutting class are now sitting with tutors doing algebra and reading *Beowulf*.

"Sir, they can cut the chains off the door, but they can't make us play," says Jason Lyle (Channing Tatum).

"We've decided we're going to finish what you started, sir," says Damien Carter (Robert Ri'chard), who transferred from a private school to Richmond High to play for his father.

Heartened by the team's support, Carter stays on as coach and leads the team to the state championship game.

"I came to coach basketball players and you became students," he says. "I came to teach boys and you became men."

It's the kind of cornball line that could have fallen flat, but the cool and commanding Jackson sells it big-time.

CHEERS: The closing credits include an epilogue saying six of the Richmond basketball players went on to college, five on basketball scholarships. As Roger Ebert of the *Chicago Sun-Times* writes: "Lives, not games, were won."

BET YOU DIDN'T KNOW: *Coach Carter* has the highest opening weekend of any MTV film release, $24.1 million. Its total box office was a solid $67.1 million.

PIVOTAL SCENE: Carter kicks point guard Timo Cruz off the team. When Cruz asks for a second chance, Carter says he will consider it if Cruz does 2,500 pushups and 1,000 sprints. Cruz falls short by 500 pushups and 100 sprints. As he is leaving the gym, another player says, "I'll do some." He drops to the floor and starts doing pushups. Several other players join in. Soon the whole team is doing pushups and sprints to help pay off Cruz's debt.

Carter's message of "team" is finally getting through.

DON'T FAIL TO NOTICE: Pop singer Ashanti makes her acting debut as Kyra, the pregnant girlfriend of the star player.

SPORTS-ACTION GRADE: C-plus. It gets a little repetitive after the 300th lob pass followed by a slam dunk.

93 LAGAAN: ONCE UPON A TIME IN INDIA (2001—PG)

SPORT: Cricket ★ STARS: Aamir Khan, Gracy Singh, Paul Blackthorne, Rachel Shelley ★ DIRECTOR: Ashutosh Gowariker

A 224-minute movie about cricket in British colonial India is, pardon the phrase, not everyone's cup of tea. You must be willing to accept subtitles (unless you speak the three distinct Hindi dialects in the film) and a different acting style (melodramatic to say the least), and you must appreciate that most Indian movies come with lengthy singing and dancing scenes appearing to have little to do with the script.

Here's hoping that doesn't chase you away. Because if you can embrace (or at least cope with) the cultural differences in a movie out of Bollywood, Mumbai's version of Hollywood, you'll be rewarded with a rich, textured, epic—part drama, part comedy—that meshes sports, politics and romance. It's a period piece (uh-oh, we're risking losing audience again) that's as beautifully shot as anything out of Hollywood in years.

The story centers on an Indian village in 1893. The residents are impoverished by a drought but still forced to pay a steep land tax (or "lagaan") to their British overlords. When brash young Bhuvan (Aamir Khan) implores for a break from the levy, he is goaded by the bullying English Capt. Andrew Russell (Paul Blackthorne) into a bet—if the Indians can beat the Brits in a game of cricket, they will be excused from the tax for three years. If not, they must pay triple.

Problem is, the natives have never played the game. So Bhuvan must assemble a ragtag squad of villagers that includes a mentally disturbed semi-feral fortune teller, a Sikh mercenary with a nasty grudge against the British Army, and an Untouchable with a gimpy arm that allows him to throw a breaking ball better than Daisuke Matsuzaka. Rudyard Kipling could have written these characters.

Hopeless as it seems, there are a few athletes among the group. And Capt. Russell's sister (Rachel Shelley), appalled by his sadistic treatment of the Indians, sneaks away to teach them the sport. The underdogs start to believe. It's the same basic plot we've seen before—in *Rocky, Hoosiers,* and dozens of other movies in this book—but it works here as it works in those.

Along the way, there are subplots involving a romantic triangle and a double agent. Plus singing. Lots of singing. And dancing. We've come to learn that moviegoers in India can't get enough of the Busby Berkeley stuff, and that they also like to get their money's worth at the theater. Hence the nearly four-hour length.

Indeed, it takes more than two hours before we get to the dramatic cricket match, and that lasts nearly 90 minutes itself (in the plot it covers a full three days!). We'll be honest and tell you that we don't understand the intricacies of the sport—when the score stood "295 for 6" we thought it was looking hopeless for the locals. We were wrong.

But you know what? Even if cricket is foreign to you, it's fun for a sports fan to watch. There's a batter and a pitcher (one of the meaner Brits will remind you of vintage Goose Gossage) and fielders. They might just be called by other names.

Lagaan was a huge box office success worldwide, but did poorly in the United States, grossing just over $1 million. We invite you to add to that total.

👍 **CHEERS:** To the training scene where the villagers run up stairs and chase down chickens. Wait a second, haven't we seen that somewhere before?

❝❝ **BEST LINE:** When the angry villager Guran comes to bat he shouts at his opponents, "You tea drinkers! Flea bags! Boot wearers! Tormentors of the weak—beware, you will pay!"

🍎 **BET YOU DIDN'T KNOW:** The small village where the movie was shot, named Bhuj, was destroyed by an earthquake six months after filming.

94 GRAND PRIX (1966–PG)

SPORT: Auto Racing ★ **STARS:** James Garner, Eva Marie Saint, Yves Montand, Jessica Walter ★ **DIRECTOR:** John Frankenheimer

Grand Prix was an expensive and not particularly successful venture for MGM. The film was shot on the Formula One circuit during the 1965 season and cost a fortune, with director John Frankenheimer and his crew (which included every existing Panavision 65mm camera) traveling across Europe and North America.

It was marketed as a Cinerama film, which meant it could only play in theaters with oversized screens and surround sound. It also was three hours long (including an intermission) which meant many locations could show it only twice a day. The result was higher ticket prices and that, along with mixed reviews, discouraged business.

From a technical standpoint, *Grand Prix* is a triumph. Frankenheimer mounted cameras on the cars, allowing the audience to see the race from the driver's seat. This perspective—the tight turns on the streets of Monte Carlo, the open road flashing by in Belgium, the terrifying vibration on the banked track in Italy—had theatergoers clutching their seats and, in some cases, dashing for the bathroom.

When the movie leaves the race track, however, it slows to a crawl. The soap opera plot follows four drivers—one American (James Garner), one Corsican (Yves Montand), one Englishman (Brian Bedford), and one Sicilian (Antonio Sabato)—as they compete for the world championship.

Naturally, they all have romantic issues that threaten to distract them at the worst possible time—for example, when Pete Aron (Garner) and Scott Stoddard (Bedford), are coming down the final straightaway. They look at each other and realize not only are they dueling for the Grand Prix title but they also slept with the same woman in the last reel. And, well, that's a lot to sort out at 180 miles per hour.

Stoddard has a decent back-story—he is trying to escape the shadow of his older brother who was world champion before he was killed—and the suave Jean-Pierre Sarti (Montand) is likable enough, but every time the actors start talking, we're looking at our watches wondering when they are going to get back in their cars and stomp on the gas.

For Frankenheimer, who also directed classics like *Birdman of Alcatraz* and *The Manchurian Candidate*, this was his first venture into color. He uses multiple images, taking a mundane shot (say, a wrench tightening a bolt), then dividing the screen into rows of smaller boxes until there are dozens of wrenches tightening dozens of bolts and, suddenly, what was mundane isn't mundane anymore.

Frankenheimer also uses a split screen to show a driver talking while at the same time steering his car around the track. One of the better moments is seeing Nino Barlini (Sabato) drive his Ferrari through a blinding rain while he tells an interviewer, "These cars, you sit in a box, a coffin, gasoline all around you. It is like being inside a bomb."

🎬 **PIVOTAL SCENE:** Sarti begins to question his involvement in the sport. He tells his lover, Louise (Eva Marie Saint), "I've begun to see the absurdity of it. All of us, proving what? That we can go faster and perhaps remain alive?"

At that point, you know Jean-Pierre is a goner.

95 WIND (1992–PG-13)

SPORT: Sailing ★ STARS: Matthew Modine, Jennifer Grey
DIRECTOR: Carroll Ballard

We're not America's Cup guys. Truth be told, we wouldn't know a spinnaker from a winch. Sailboats are for catching sunrays and studying bikinis—not so much for racing.

Nor are we chick-flick guys. Our idea of romance is Crash Davis breaking down Annie Savoy's defenses with The Speech—you know, "I believe in the sweet spot, soft-core pornography. . . . "

But somehow, despite our best instincts, we were drawn in by this little movie about the men who race 12-meter yachts and the women who love them (or, in some cases, race better than them). In fact, it wasn't the plot line, the characters, or the romance that got us. Definitely not the romance, which seems like something lifted right off the Lifetime Network. Rather, it was the spectacular racing scenes. Who knew that a bunch of snooty, zinc-oxide-wearing bluebloods on boats could create some of the most intense sports action we've seen?

Wind centers on the story of young Will Parker (Matthew Modine), a talented sailor intent on winning the America's Cup. In case your sports interests lie elsewhere, the America's Cup is a 150-year-old competition among the world's top yachtsmen, most of whom appear attracted to ascots and silly hats. Up through the 1990s the United States has dominated the event.

Anyway, Will tries out for the crew of the legendary Morgan Weld (Cliff Robertson), who's kind of a combination of Howard Hughes and Robert Shaw's Quint from *Jaws*. He gets the nod to skipper Weld's yacht but is told he must send home his sassy girlfriend, Kate Bass (Jennifer Grey), even though she's clearly the best sailor in the movie. So Will dumps the girl. Then he blows the race and loses the America's Cup to Australia. Bad month.

Flash forward six months. Will tracks down Kate on the salt flats of Utah (not exactly sailing country), where she's living with her new boyfriend, an aircraft designer named Joe (played by the always annoying Stellan Skarsgard). In a bit of a leap of faith, Will persuades them both to help him design a revolutionary yacht that might win back the Cup. To make things more unlikely, their chief financial backer is Weld's nutty daughter, Abigail (Rebecca Miller), who likes to dance to *Madame Butterfly* while seducing a bag of Fritos.

Yeesh, this is starting to sound like the fall plotline of *The Young and the Restless*. Unfortunately, *Wind* loses its energy during all of the on-land sequences. But once we get into the water—wow, suddenly we have a heart-pounding thriller.

The racing scenes are absolutely invigorating. We witness head-on crashes, spectacularly torn sails, and one moment when a crewman ends up strung helplessly upside down from the mast during a critical competition. This is not serene ocean sailing; these are real athletes—sweating, grunting, fighting fatigue to be the best in the world at their sport.

Give credit to director Carroll Ballard, who has shown a great visual sense in *The Black Stallion* and *Fly Away Home*. And give credit to cinematographer John Toll, whose daredevil camera shots—above, below, beside the yachts—put you right in the crashing waves.

The payoff here is the final race in Freemantle, Australia—the "Game Seven," if you will, between Will and Jack Neville (smartly played by Aussie Jack Thompson). Ballard devotes a full 21 minutes to the contest and you won't be bored for a second. Pay attention to the sound effects—creaking masts, crashing hulls, splashing water—which add to the tension. And notice how you keep inching forward in your chair as the tight race moves along. This is good stuff, comparable to *Raging Bull*, *Friday Night Lights*, or any other brilliantly shot sports action.

PIVOTAL SCENE: In the penultimate America's Cup race, the American and Australian boats crash. Both skippers blame each other. But at a post-race news conference, Will—who earlier insisted he would do *anything* to win—suddenly gets a pang of conscience. He admits the truth: it was his poor sailing that caused the accident. He even offers to lend the Aussies an extra mast to replace the one his yacht busted.

"We're almost there and you betray us with your principles," Abigail screams at him. But to Will, winning the right way is the only way.

SPORT: High School Wrestling ★ **STARS:** Matthew Modine, Linda Fiorentino
DIRECTOR: Harold Becker

Vision Quest is the all-too-familiar tale of a teenager from a working class town trying to prove his mettle (and win the girl) by taking on a near-impossible athletic challenge. We saw the same story on bikes in *Breaking Away* and on the mat in *The Karate Kid*.

Here, the venue shifts to high school wrestling. And while *Vision Quest* breaks no new ground in the coming-of-age sports genre, it succeeds for two reasons: First, the sport translates well onto film, with all of its rituals and practice methods and the easy-to-shoot action.

Second, and more importantly, *Vision Quest* works because you get to a watch a talented young cast of actors before they became famous. Matthew Modine stars as 17-year-old Louden Swain, two years before his breakout role in *Full Metal Jacket*. Linda Fiorentino, making her first movie appearance, displays the same icy sexiness (and no, that's not an oxymoron in this case) that she would later perfect in *The Last Seduction*. And that nightclub singer with the teased up hair and fingerless black gloves? That's Madonna, also in her screen debut (well, not counting any soft-core porn indie movies she may or may not have made).

Daphne Zuniga, a future star of *Melrose Place*, and Forest Whitaker, the future Oscar winner for his role as Idi Amin, also have small roles here. Whitaker doesn't get to say much, but his character's name—Balldozer—is worth a few laughs.

The story centers on Louden, a smart and quirky high school senior who can pin most anyone as a 190-pound wrestler. But that doesn't interest him. He wants a tougher challenge, so he decides to cut all the way down to 168 pounds to battle the nastiest monster in Spokane, WA, the undefeated Brian Shute. (It may be unrealistic to believe that any 168 pounder would be tougher than the wrestlers in Louden's weight class, but, hey, that's the movies.)

Shute (Frank Jasper) is kind of like Ivan Drago in a singlet. When we first meet him, he is running up rows of stadium bleachers carrying a log on his back the size of a telephone pole. When Louden goes to scout him, Shute humiliates a highly ranked opponent in 12 seconds.

Clearly, this is craziness. Everyone tries to talk Louden out of his folly—his coach (Charles Hallahan), who worries that he risks losing a college scholarship; his dad (Ronny Cox), who grows alarmed when Louden's quest to drop 22 pounds leads to fainting spells and constant nosebleeds; his teammates, who are all panic-stricken by the Mighty Shute. "I hear when he pins a guy, his coach has to stop him from biting the guy's throat open," says one.

The only support comes from two people. Louden's best friend, Kuch (Michael Schoeffling), a self-proclaimed half-Native American, touts Louden's goal as a "vision quest," an Inuit rite where a young man goes on a personal, spiritual endeavor. Kuch explains that Louden will be guided by the "Everywhere Spirit," although the first few times he says it in the film, it sure sounds like the "Underwear Spirit."

Louden's other ally is Carla (Fiorentino), a 20-year-old drifter who—at first—doesn't take him too seriously, but winds up falling for his boyish optimism. Suffice it to say, Louden earns his victory in bed before he ever gets to the match against Shute.

It leads up to an all-too-predictable finish, which would leave you flat if *Vision Quest* didn't have such full characters and a fine cast. Instead, this is a fun, well-written film that's worth the 105 minutes it asks you to invest.

PIVOTAL SCENE: When he discovers that Carla has skipped town right before his big match, Louden considers ditching his whole dream. He seeks solace at the home of his friend, short-order cook Elmo (J.C. Quinn). Louden discovers, to his amazement, that Elmo plans to skip work that night and attend the meet.

"Why bother?" asks a dejected Louden. "It's only six minutes."

Elmo, struggling to knot his only tie, launches into a pep talk about how he once stumbled upon a soccer game on TV and saw Pelé score an amazing goal that left him in tears.

"That's right, I started crying," says Elmo. "Because another human being, a species that I happen to belong to, could kick a ball, and lift himself—and the rest of us sad-assed human beings—up to a better place, if only for a minute. Let me tell you kid, it was pretty --------- glorious. It ain't the six minutes—it's what happens in those six minutes. Anyway, that's why I'm getting dressed up and giving up a night's pay for this."

Quick cut to the next scene: Louden sprinting to get to his pre-match weigh-in on time.

BET YOU DIDN'T KNOW: Modine said in later interviews that he believes the movie's poor box office showing ($13 million) stemmed from an inapt title. "It sounds more like a science fiction film than anything else," he told *Entertainment Weekly*.

97 COOL RUNNINGS (1993–PG)

SPORT: Bobsledding ★ STARS: John Candy, Leon
DIRECTOR: Jon Turteltaub

Silly as it sounds, there really is a Jamaican bobsled team. In 1988, the tropical nation best known for reggae and ganja amused the world by sending a four-man crew to the Winter Olympics. The team crashed and finished last, but it delighted the crowds in Calgary, Canada, and more than earned its way home by selling souvenir T-shirts.

Since then, the country that has never seen a flake of snow has tried out for each Olympics. In 1994, in Lillehammer, Norway, Jamaica's four-man sled finished in 14th place—ahead of both American crews. These days, the team maintains a training base in Wyoming, which must create a good deal of culture shock for both locals and visitors.

All of which makes you ask: How does a tiny speck in the Atlantic—where the average daily temperature is 82 degrees—get involved in this most winterish of winter sports?

There's a good tale here, told—although not especially accurately—in the 1993 comedy *Cool Runnings*. In the movie, disgraced American bobsled coach Irv Blitzer (John Candy) recruits three Jamaican sprinters and their pushcart racing buddy (more about that in a second) to form a sledding team after they fail to qualify to run in the Summer Olympics.

With no funds, no support, and no training facilities, they practice by lugging a rickety go-kart up and down hills laden with palm trees.

In real life, two Americans with business ties to Jamaica witnessed the annual Pushcart Derby in Kingston, in which street vendors race their carts down steep roads at speeds reaching 60 miles per hour. They signed up the champion racer, enlisted three soldiers from the Jamaican Army, and moved the operation north to Canada, where they received top-flight training. Their coach, Howard Siler, was one of the best in the world and certainly not a pariah to the sport, as Candy's character portrayal would have us believe.

In the movie, the Jamaicans are met with ridicule and hostility at Calgary. This leads to the inevitable bar brawl between our naive dreadlocked Rastas and the evil East German bobsled team, led by a Josef Grool, the nastiest movie Teuton this side of Hans Gruber. (On a side note, we've lost much of our interest in the Olympics since the Communist empire crumbled. It was more fun when you could root against the baddies from the GDR and USSR. Who's the enemy now? Until Iran forms a hockey team, we're boycotting the Winter Games.)

In real life, the Jamaicans became somewhat of a pet to the other competitors, who offered them everything from coaching tips to extra clothing to survive the minus-10 chill of Alberta. No one tried at the last minute to have them barred from the Games, as the movie suggests. No one considered them an embarrassment to the sport.

But, hey, that's Hollywood. "The movie accurately depicted the spirit of the team," original team member Devon Harris told *Entertainment Weekly*. "For purposes of entertainment, they portrayed us with far more color, let us say."

Okay, so it's not exactly *Hoop Dreams* on ice. But *Cool Runnings* is an upbeat, entertaining movie that's very appealing to kids and at least palatable for adults. While it uses every underdog movie cliché, you will find yourself attracted to the characters and, ultimately, cheering their *Rocky*-like finish in Calgary. We won't give that ending away, except to say it, too, wasn't exactly true to real life.

CHEERS: The quartet of Jamaican sledders—played by actors Doug E. Doug, Leon, Rawle D. Lewis, and Usain Bolt look-alike Malik Yoba—are an engaging, likable crew. They fit every TV stereotype so well (the goofy one, the angry young man, the earnest guy, the rich kid struggling for his own identity) that we're surprised they weren't later cast together in a recurring sitcom.

DON'T FAIL TO NOTICE: In an apparent money-grabbing nod to product placement, there are many Red Stripe beer bottles shown throughout the movie. But given that this is a Disney production, no one is ever seen actually drinking the beer.

BET YOU DIDN'T KNOW: At the 1998 Olympics in Nagano, Japan, the Jamaican team crashed again. The driver suffered a concussion and for several days could not remember the last 11 years of his life. His brother, who was also on the team, had to inform him that he was married and had two children.

BEST LINE: Coach Blitzer giving his pep talk: "Always remember, your bones will not break in a bobsled. No, they will shatter. So, who wants in?"

98 YOUNGBLOOD (1986–R)

SPORT: Hockey ★ STARS: Rob Lowe, Patrick Swayze
DIRECTOR: Peter Markle

Some movies set out with great ambitions. They have a social conscience and their goal is to make people think. Those are the movies that are celebrated at Sundance and maybe win an Oscar.

Then there are movies like *Youngblood*.

The producers knew they weren't going to impress the critics with this film, and they didn't really care. They weren't worried about satisfying hardcore hockey fans who would take note of certain details, such as Rob Lowe's wobbly skating. *Youngblood* wasn't made for purists, either artistic or athletic.

This is the kind of movie a bunch of guys watch together with a few beers. It's the kind of movie women will enjoy because it's full of hunky guys who spend a lot of time in the locker room getting undressed. There is no great message here. It's like a fast food meal: it goes down easy, and while it doesn't offer much in the way of nutrition, it tastes pretty good.

Parts of the film are downright hilarious, some intentionally, some not. For example, there is the moment the Hamilton Mustangs goalie appears and you realize, "Hey, that's Keanu Reeves." Then he begins talking in a mangled French-Canadian accent and you are on the floor laughing.

Also, there is the scene where Dean Youngblood (Lowe) shows up at his boarding house and is greeted by a landlady who looks like she just slid off a pole at the local gentlemen's club. When she knocks on Dean's door and says she has his afternoon tea, we know the room service won't stop there. The bewildered "Is-this-included-in-my-room-and-board?" look on Lowe's face is priceless.

Youngblood is the story of a hotshot teenage hockey player (Lowe) from upstate New York who is invited to try out for a junior team in Hamilton, Ontario, where he hopes to impress the pro scouts. His older brother Kelly (Jim Youngs) had the same dream, but his career ended when an opponent's high stick left him blind in one eye.

As they sit in the car outside the rink in Hamilton, Kelly tells Dean, "Don't take any ---- from them Canucks. To them, you're just another wetback crossing the border to play their game."

"They'll never catch me," Dean says.

"Oh, they'll catch you," Kelly says, speaking from experience.

The Mustangs are looking to fill one roster spot, and coach Murray Chadwick (Ed Lauter) narrows the candidates down to Dean and a bearded tough guy named Carl Racki (George J. Finn). Chadwick picks Dean, even though Racki punches him out during a scrimmage.

Dean goes through the ritual of proving himself to his teammates, surviving a hazing and a drunken night on the town, but eventually the other players decide he's okay, even though he is an American and he looks, well, like Rob Lowe. Dean bonds with team captain Derek Sutton (Patrick Swayze), who also has his eyes on a pro career.

"I want to go No. 1 in the draft and sign the biggest contract I can," Sutton says. "I've been busting my ass in this league for four years and I'm gonna get what's coming to me."

That's a tip-off to what's coming. So is the sight of Racki skating in warm-ups with the Thunder Bay Bombers, Hamilton's opponent in the Memorial Cup Playoffs. He caught on with the Bombers after Chadwick decided to let him go.

When Racki casts an evil stare in the direction of the Hamilton bench, you know it's only a matter of time before he lowers the boom on one of the Mustangs. It's no surprise when Racki drops Sutton on his mullet and sends him off the ice on a stretcher.

So much for Derek's hockey career. Guess he'll have to settle for being a dance instructor in the Catskills. Oh wait, that's another movie.

At this point, *Youngblood* lapses into melodrama and suffers for it. Dean is so sickened by Derek's injury that he quits the team and returns to the family farm. His brother can't believe it.

"With one eye, I begged them to let me play," he says. "Would you rather spread manure or play hockey in Madison Square Garden?"

"Spread manure," Dean says.

"You candy-ass," his brother growls.

But soon Dean is in the barn smacking a punching bag and he is on the pond practicing his slap shot. It is clear he is planning to rejoin his team. One night Dean is on the ice and his father (former NHL star Eric Nesterenko) comes out to join him.

"You can learn to punch in the barn," the father says, "but you gotta learn to survive on the ice."

Dad shows Dean how to win a hockey fight (just wondering, but why didn't he give him this lesson *before* he went to Canada?) and just like that, he's headed back to Hamilton. In real life, an old-school coach like Chadwick would have slammed the locker room door in Dean's face. He wouldn't have let the kid fool around with his precious daughter (Cynthia Gibb) either, but that's a different story.

But Chadwick allows Dean to dress for the deciding game. The movie tries for the big finish, but it turns out to be the weakest part of the film because so much of it doesn't make sense.

CHEERS: *Youngblood* was filmed in Hamilton, so you get a feel for life on the junior hockey circuit.

BET YOU DIDN'T KNOW: Keanu Reeves played goalie on his high school hockey team, De La Salle in Toronto. We're not kidding.

99 ▶ VICTORY (1981–PG)

SPORT: Soccer ★ **STARS:** Michael Caine, Sylvester Stallone
DIRECTOR: John Huston

John Huston won an Academy Award as best director for *The Treasure of the Sierra Madre*. He was nominated for an Oscar for his work in other classic films, such as *The Asphalt Jungle*, *The African Queen,* and *The Man Who Would Be King*.

So why did he sign on to direct *Victory*, a rather far-fetched tale of soccer and World War II?

"I only make pictures if I like the story, or they offer me a lot of money," Huston said, "and in this case it happened to be both."

Producer Freddie Fields wanted to hire Brian G. Hutton, who directed two other hit films about World War II: *Where Eagles Dare* and *Kelly's Heroes*. But Sylvester Stallone, one of *Victory*'s co-stars, persuaded Fields to hire the 74-year-old Huston.

Vincent Canby of the *New York Times* saw *Victory* as a lark for Huston. Canby said the film "represents [Huston] in an expansive and almost carefree mood. It's as if Mr. Huston—having just directed two of the most imaginative and difficult films of his long career, *The Man Who Would Be King* and *Wise Blood*—had decided to work on something more conventional and less demanding."

Victory was shot entirely on location in Hungary. It cost $15 million to produce and it earned just $10 million at the box office. Most critics dismissed it—Canby said the film "is not meant to be taken too seriously"—and soccer purists laughed at Stallone as a goalkeeper even if he did drop 40 pounds of his *Rocky* weight to look the part.

But *Victory* developed a cult following

among the DVD crowd because it does have some enjoyable elements. The soccer action—we know, that seems like a contradiction in terms—is actually entertaining. It helps that eight goals are scored in the big game. That's seven more than are scored in most soccer games. And while Michael Caine doesn't look like David Beckham, his acting helps the script overcome its rough spots.

The film begins in an Allied POW camp where the prisoners, most of them British, pass the time by playing soccer. John Colby (Caine) is a former World Cup player now held captive by the Nazis. Major Karl Von Steiner (Max von Sydow) is a former German soccer star who recognizes Colby. Von Steiner proposes an exhibition match between the POWs and the German National team.

The Germans intend to use the game as a propaganda vehicle. They will play the game in Paris and show the world how well they are treating the POWs—"Look, we let them play soccer and everything"—then, of course, the Germans will win the game and wave the banner of Aryan superiority.

The Allies have a different agenda. Their plan is to use the game as a means to escape. Robert Hatch (Stallone), an American POW, breaks out of the prison camp, makes his way to Paris to meet with the French Resistance and plan the big escape. Hatch allows himself to be re-captured by the Nazis—hey, we said this is far-fetched—so that he can be returned to the POW camp where he tells Colby and the others what the French have in mind.

This all works out a little too conveniently. As Canby notes, Hatch escapes the camp and gets to Paris "all with less difficulty than your average journey to and from East Hampton via railroad." And how does an American POW find the Resistance in Paris? Is it listed in the Yellow Pages? Also when Hatch allows himself to be re-captured, how can he be sure the Nazis will return him to the camp? Suppose they just shoot him on the spot?

The same thoughts probably occurred to John Huston and then he decided: "Aw, the hell with it. Let's just get to the soccer game." Good idea.

Think *The Longest Yard* meets *Stalag 17*—that's *Victory*. The ending doesn't make a lot of sense—one critic called it "egregiously silly"—but it is fun in a shamelessly rah-rah kind of way. And the slow-motion shot of the legendary Brazilian soccer star Pelé (he is one of the Allied POWs) executing his famous bicycle kick is one that will amaze even non-soccer fans.

👍 **CHEERS:** The photography is often stunning, in particular the first shot of the POW camp. It is shown from a distance and the camp looks like a barbed wire cage set in the middle of nowhere. It really conveys the feeling of isolation and loneliness.

☞ **DON'T FAIL TO NOTICE:** The music score sounds very *Rocky*-like. That's because it was written by Bill Conti, who did the original theme for *Rocky*.

⭐ **SPORTS-ACTION GRADE:** C. There are a number of world class soccer players on the Allied team, including Bobby Moore, the captain of England's 1966 World Cup team, and the great Pelé, but Stallone and Caine are the leads and they are woefully out of place on a soccer field.

☺ **GOOFS:** In the crowd scenes at the big game, many of the fans are wearing clothes and hair styles that are clearly from the 1980s, not the 1940s.

THE FISH THAT SAVED PITTSBURGH (1979—PG)

SPORT: Basketball ★ STARS: Julius Erving, Jonathan Winters, Meadowlark Lemon ★ DIRECTOR: Gilbert Moses

The Fish That Saved Pittsburgh is by no means a great movie. It is not even considered by critics to be a good movie. The script is inane, the acting barely passable, the directing spotty. But if you give it a chance, *Fish* proves to be a campy period piece that prompts you to smile nostalgically at times and guffaw out loud at others. Just set aside any rational disbelief and let the movie wash over you. A beer or two wouldn't hurt.

We'll call it a guilty pleasure, a quirky film with enough redeeming value that we recommend it, warts and all. (See the next chapter for more guilty pleasures.)

Take the plot. (Please.) The Pittsburgh Pythons, led by selfish star Moses Guthrie (Julius Erving) are the worst team in pro basketball. When Guthrie's teammates all walk out in protest, the 12-year-old towel boy (James Bond III) convinces the addle-brained owner (Jonathan Winters) to hold open tryouts. The one requirement is that only players sharing Guthrie's astrological sign are selected.

And then it gets weird.

The team, which now resembles the Village People and is renamed the "Pisces," is guided by an astrologer, Miss Mona Mondieu (a slumming Stockard Channing), whose game plans have more to do with Jupiter rising than Dr. J dunking. And, of course, the team starts winning every game.

We are not making this up.

It all culminates with a kidnapping by the owner's evil twin, a showdown against Kareem Abdul-Jabbar, and a lot of disco dancing in platform shoes. What's not to like?

If you're too young to remember the 1970s, *Fish* will take you to a time of huge Afros and pastel clothes, of breakdancing to a disco beat. Somehow, between the turbulence of the 60s and the greed of the 80s, there was this decade of totally fun nonsense.

It was also a decade of fun basketball. Much of the joy here is seeing Erving—an iconic figure of the era—lope his way through the game scenes and also try to look serious reacting to dialogue like, "The man is only interested in playing for the bucks—the spending kind."

Leroy Burrell, the 1992 Olympic Gold-Medal sprinter who once held the title of world's fastest human, said, "I grew up in the Philadelphia area, Doctor J was my favorite athlete, I love basketball, and I'm a Pisces. How could I not love that movie?"

An enjoyable sidelight here is spotting all the NBA players who show up for anonymous cameos as opposing players. Some are easy—Connie Hawkins, Bob Lanier, Spencer Haywood—but others will challenge your hoops memory. Isn't that Chris Ford in a Detroit uniform? Hey, did Cornbread Maxwell just pass the ball to Kevin Stacom? We were able to count 20 of Doc's old colleagues who dropped in for a scene or two. Perhaps you'll spot more.

PIVOTAL SCENE: Well, with a plot this dopey, it's hard to say anything is pivotal. But the funniest scene has to be the Pisces open tryout scene, which resembles the intergalactic bar scene in *Star Wars*.

★ GUILTY PLEASURES ★

***Idol of the Crowds* (1937).** John Wayne made 147 films during his 50-year acting career. Next to *The Conqueror*—a radioactively bad cult movie in which he portrays Genghis Khan—this one may be the least credible of them all. The Duke plays Johnny Hanson—the original "Hanson Brother"—a retired hockey player living out his dream as a Maine chicken farmer. He needs some scratch to keep the farm going, so he agrees to play one more season for the New York Panthers. The plot gets tough to follow from there, but there's a brush with cliché-spouting gangsters, an attempt on our hero's life and, of course, a B-movie finale that ends with Wayne getting the girl.

If you're a hockey fan, you'll enjoy *Idol* because, well, there just aren't many hockey movies. If you're a John Wayne fan, you'll enjoy watching him skate around on the insides of his ankles and shoot the puck like a guy sweeping out his garage. Let's just say it's a good thing for his career that Wayne learned how to ride a horse.

***Kid Galahad* (1962).** Speaking of which, it's good that Elvis Presley learned to play the guitar, because he wouldn't have made it as a prize fighter. He plays one here, however, or at least attempts to. In a plot that seems eerily similar to *Idol of the Crowds*, Elvis plays an ex-GI aiming for the dream of becoming a car mechanic. Instead, he's lured into becoming a boxer, and ends up as a top contender and crowd favorite. And, once again, here comes to mob trying to muscle in on his success.

You'll enjoy Elvis in satin trunks and gloves, enjoy the great Charles Bronson as his loyal trainer and—of course—enjoy the soundtrack, which includes six tunes by Elvis (none, however, were one of his classics). How he strums that guitar while wearing boxing gloves is beyond belief.

Best line? Elvis as Walter Gulick to sleazy manager Willy Grogan (Gig Young): "Don't push me, Willy. I'm a grease monkey that won't slide so easily."

***Kansas City Bomber* (1972).** No one under age 40 is likely to remember Raquel Welch as anything more than a violent diva from the famous 1997 "The Summer of George" episode of *Seinfeld*. Trust us, once upon a time, she was Pamela Anderson—without the silicone. Likewise, no one under 40 is likely to remember when Roller Derby was a staple of local UHF programming on weekend afternoons.

Damn shame, because both Raquel and Roller Derby were fun eye candy. They combine in this sometimes laughable flick about a star skater trying to balance her shaky personal life (watch for Jodie Foster as her 10-year-old daughter) and her barbaric professional world. There are occasional ham-handed early women's liberation references, but mostly there's a lot of cleavage heaving and cat fights. Extra points to anyone who recognizes Judy Arnold, former Philadelphia Warrior, as the one doing Ms. Welch's dirty work as a stunt double.

***Death Race 2000* (1975).** The ultimate in cult flicks, starring a pre-*Rocky* Sylvester Stallone, Louisa Moritz's crowd-pleasing breasts, and David Carradine as a cyborg-like character named Frankenstein in a leather suit resembling The Gimp from *Pulp Fiction*. All with cool cars and a 1970s porn-style soundtrack in the background. What's not to enjoy?

The setup here is a futuristic national sport/television series in the so-called United Provinces of America, in which cross-country drivers gain points for running over pedestrians ("The big score: anyone, any sex, over 75 years old, has been upped to 100 points!"). There are rivalries, soap operas, and sexcapades among the drivers, but, mostly, innocent fools getting flattened on the highway.

It all sounds ridiculous, until you turn on most contemporary reality shows, or take a peek at video games like *Grand Theft Auto*. But, hey, we don't want to get preachy. Where else do you get dialog like, "Well America, there you have it, Frankenstein has just been attacked by the French Air Force and he's whipped their derrieres." Occasionally, well after midnight, Turner Classic Movies will run the uncut version of *Death Race 2000*. Be there.

And don't bother with the version remade in 2008, starring Jason Statham and Ian McShane. It's not as sexy or as campy.

Brewster's Millions (1985). Any movie starring both Richard Pryor and John Candy has to be a hoot and this one does not disappoint. Pryor plays minor-league pitcher Monty Brewster, who comes into a huge inheritance—with a catch. He must spend (waste, actually) $30 million in 30 days. If successful, he'll inherit $300 million; if not, he gets nothing. And, of course, Brewster can't tell anyone about the deal. That includes Candy, his catcher and goofy sidekick, who helps Brewster by spending too much money on a gold-plated catcher's mask once owned by Johnny Bench.

The best sports scene comes when Brewster pays to stage an exhibition game between his low-level club and the New York Yankees. The best waste of money comes when he spends $1.25 million on a valuable stamp and then uses it to send a letter. But, really, the best part of the movie is the chemistry between Pryor and Candy—two comic masters who left the stage too early.

Bloodsport (1988). Jean-Claude Van Damme's pantheon movie, narrowly edging *Sudden Death* (in which the Belgian badass somehow ends up as the Pittsburgh Penguins

goalie). In this classic, J.C.V.D. plays U.S. Army soldier Frank Dux, who goes AWOL to compete in a mysterious martial arts tournament, where, we keep being reminded, "Death is a single punch away." Or kick.

We don't want to say the movie plays to old-time stereotypes, but it seems that every Asian character was cast by the same Hollywood folks who produced those World War II propaganda films. Many of the cast members speak no English, which leaves the bulk of the dialog to the Shakespearean Van Damme. Like, "I ain't your pal, dickface."

It's all about as dumb as a 1980s martial arts shlockfest can be, complete with a lame love interest (Leah Ayres), and some of the greatest fighting scenes ever choreographed. Overall, it's wonderful mindless fun.

Men With Brooms (2002). The best curling movie ever made. Actually, the only curling movie ever made. An all-Canadian cast—including a slumming Leslie Neilsen (in what aims to be a serious role) and the popular Ontario rock band The Tragically Hip—tell the story of a small-town team that reunites in the wake of its coach's death and tries to win the Golden Broom, curling's version of the Stanley Cup. All while making gags about beavers and frozen groins and drinking lots of Molson Ale.

Montreal Mirror critic Matthew Hays dismissed *Men With Brooms* as "a movie that's clearly meant to have everything in it: romance, comedy, pathos, suspense [who will win that final curling match?—stay tuned!], etc. You'll laugh, you'll cry, you know the drill. Sadly, everything about the film feels clumsy in that ultra-Canadian way."

We don't know about that. We didn't even know that Canadians were ultra-clumsy. (Take that, Sidney Crosby!) We'll just argue that this movie is so bad that it's good. Just make sure to bring the Molson Ale.

Dodgeball: A True Underdog Story **(2004).** A great cast, led by Vince Vaughn and Ben Stiller, would seem to promise a great film about that junior high school gym staple that allowed bullies to terrorize glasses-wearing eggheads for decades. But *Dodgeball* doesn't quite get there. It's a funny concept—the owner of a gym perfectly named "Average Joe's" tries to save the shabby place by entering his clients into a $50,000 dodgeball tournament. But a funny concept doesn't always make for 92 minutes of on-screen entertainment.

Still, it has its moments. Stiller is hysterical as the mullet-wearing owner of the rival macho-man training spa. He's the guy who fights off his own little-man complex by surrounding himself with steroid-addled bodybuilders and wearing an inflatable crotch under his gym shorts. Vaughn aims to play the same wiseguy role he perfected in *Old School*—but doesn't quite get there. The movie is stolen by Rip Torn as Patches O'Houlihan, a legendary former dodgeballer (we didn't know dodgeball had legends) who reinvents himself as a sadistic, wrench-throwing coach.

Nacho Libre **(2006).** Jack Black's ode to Mexican professional wrestling is kind of a cross between *Wrestlemania IV* and *Napoleon Dynamite*. We're not saying that the movie was made under the influence of drugs but, well, since Black has referred to himself as a modern-day Jeff Spicoli, we can't really rule it out, either.

Black plays Brother Ignacio, a monk living in the outposts of Mexico, who tends after impoverished orphans. He wants to give them better food than the slop he's providing, so he begins a second career as a *luchador*, or masked wrestler, in the sport of *Lucha Libre*, a freestyle form of wrestling that makes what you see on American cable resemble *Wall Street Week* in comparison. Brother Ignacio also spends a lot of time inappropriately lusting after Sister Encarnacion, played by the incredibly beautiful Ana de la Reguera. You'll share his lust. You'll also share a few laughs. You may also share a bong, but we advise against it.

Talladega Nights: The Ballad of Ricky Bobby **(2006).** We happen to think that Will Ferrell is a talented comic—first for his work on *Saturday Night Live*, and later for the aforementioned *Old School* and *Anchorman: The Legend of Ron Burgundy*. His attempts at sports comedies, however, have largely fallen flat. *Blades of Glory*? Weird, but not amusing. *Semi-Pro*? A great script idea (focusing on the old American Basketball Association) that proved to be a snoozer. *Kicking & Screaming*? That's how we'll react if ever forced to watch that youth soccer nonsense again.

The best of the bunch, and the only Ferrell-inspired sports movie worth seeing, is *Talladega Nights*. Ferrell lets 'er rip as Ricky Bobby, a proudly dumb race driver from North Carolina. John C. Reilly, who can do lowbrow with the best of them, is his best bud, Cal. There's a white trash wife, some bratty kids, and a tasteless (but funny) homage to the Baby Jesus. And then, unfortunately, director Adam McKay tries to get fluffy and sentimental with the ending. Take it from us—watch the first 90 minutes and then turn off your DVD player. You'll have a few laughs and won't miss a thing.

★ ★ ✦ ABOUT THE AUTHORS ✦

RAY DIDINGER won six Emmy® Awards as a senior producer for NFL Films, authored nine books, and was a sports columnist for the *Philadelphia Bulletin* and *Philadelphia Daily News.* He is a host on 610-WIP all-sports radio in Philadelphia and a football analyst on Comcast Sports-Net. He lives in Philadelphia.

GLEN MACNOW is a popular sports-radio host on 610-WIP, and had an accomplished career as a sportswriter for the *Philadelphia Inquirer* and *Detroit Free* Press. He is the author of three best-selling sports books, and is a frequent guest on local and national television sports shows. He lives outside Philadelphia.

If you have enjoyed this book
or it has touched your life in some way,
we would love to hear from you.

Please send your comments to:
Hallmark Book Feedback
P.O. Box 419034
Mail Drop 100
Kansas City, MO 64141

Or e-mail us at:
booknotes@hallmark.com